T0234174

INTRODUCTION TO SURFACE AND SUPERLATTICE EXCITATIONS
Second Edition

Graduate Student Series in Physics
Other books in the series

GRADUATE STUDENT SERIES IN PHYSICS

Series Editor:
Professor Douglas F Brewer, MA, DPhil
Emeritus Professor of Experimental Physics, University of Sussex

INTRODUCTION TO SURFACE AND SUPERLATTICE EXCITATIONS

Second Edition

MICHAEL G COTTAM

Department of Physics and Astronomy
University of Western Ontario, London, Ontario, Canada

DAVID R TILLEY

School of Physics
Universiti Sains Malaysia, Penang, Malaysia

Routledge
Taylor & Francis Group

LONDON AND NEW YORK

First published 2005 by IOP Publishing Ltd
First edition published 1989 by Cambridge University Press

Published 2018 by Routledge
2 Park Square, Milton Park, Abingdon, Oxon OX14 4RN
52 Vanderbilt Avenue, New York, NY 10017

Routledge is an imprint of the Taylor & Francis Group, an informa business

British Library Cataloguing-in-Publication Data

A catalogue record for this book is available from the British Library.

Library of Congress Cataloging-in-Publication Data are available

Cover Design: Victoria Le Billon

Typeset by Academic + Technical, Bristol

ISBN 13: 978-0-7503-0588-4 (pbk)

Contents

Preface to the first edition

The past twenty years have seen a great expansion in the study of surface properties. One part of this activity has been concerned with the various acoustic, magnetic and optic modes that propagate at the surface of a solid or liquid. These modes have a great deal in common; for example, they are often characterized by an amplitude that decays as an exponential (or sometimes a sum of exponentials) with distance from the surface. The generality of the concepts is well known to research groups working on surface modes, and most of them have made contributions across the board. However, although a number of excellent advanced monographs and review articles have appeared, there is no introduction to the field. The present work is designed to fill this gap.

Our intention is to provide an introductory text for someone starting research on surface modes or extending their range from one type of surface mode to another. It is hoped in addition that much of the material will be useful for advanced undergraduate teaching. In keeping with this pedagogical character, we have provided problems at the end of each chapter, and the lists of references are extensive, although we do not claim that they are comprehensive.

The experimental techniques employed for the study of surface modes are described, some in chapter 1, and the more specialized techniques at the appropriate points in later chapters. The experimental data that are shown have been selected to clarify the discussion and not with any intention, for example, of always showing the latest available results.

The theoretical description of surface modes can be given at two levels. First, homogeneous equations of motion can be solved for a dispersion equation (frequency versus wavevector) and related properties such as variation of amplitude with distance from the surface; this level is adequate for understanding most of this book. Second, inhomogeneous equations can be solved to find Green functions by means of linear response theory. This method is harder, but it gives more complete information, including, for example, thermal fluctuation spectra and scattering cross sections. For those who are interested, we have given the basic formalism in the Appendix, and details of a Green function calculation are given just once, in chapter 2.

Somewhere in the middle of the gestation period of this book the authors got involved in the worldwide effort on superlattices. A system with a large number of interfaces is an obvious generalization of a system with one or two interfaces, so naturally many ideas about surface modes carry over for superlattices. We have therefore included a chapter on this topic.

The book is primarily concerned with acoustic, magnetic and optic properties, mainly because they are closely related, but also because that is where the bulk of our own experience lies. This means that some topics, although substantial, are dealt with only briefly. In particular, electronic properties do not feature largely, since to do more than we have would have required a lengthy digression into electron-band theory. Likewise the discussion of liquid surfaces is restricted in scope.

Our indebtedness to a large number of friends and collaborators will be clear from the book. It has been our privilege for the past ten years or more to be part of the theoretical surfaces group at Essex, and to develop many ideas with Professor Rodney Loudon, Dr Mohamed Babiker and Dr Stephen Smith. Our graduate students, now established in their own careers, include John Nkoma, Enaldo Sarmento, Karsono bin Dasuki, Latiff bin Awang, Eudenilson Albuquerque, Fernando Oliveira, Bob Moul, Demosthenes Kontos, Marcilio Oliveros, Arnobio dos Santos, Aurino Ribeiro Filho, Nilesh Raj, Nic Constantinou, Roger Philp, Monkami Masale, Hossein Heidarpour and Heidar Khosravi. We are grateful to all of them. People elsewhere who have contributed to our understanding include the surface groups in Irvine, California; Natal, Brazil; Exeter, England; and Royal Holloway and Bedford New College, England. Their work features in many parts of the book.

Finally, we should like to express our thanks to two people. Simon Capelin, the commissioning editor has given us unfailing encouragement and has allowed us a free hand over content. Carol Snape has typed the book with an accuracy which is no less amazing because we have become accustomed to it.

Mike Cottam
David Tilley
June 1988

Preface to the second edition

Our aim is, as previously, to provide a comprehensive introduction to dynamic properties of surfaces, superlattices and related structures. We have tried to reflect the advances and changes that have occurred since the first edition was written. Some modes that were then discussed theoretically, like magnetic polaritons and exchange magnons in superlattices, have now been observed in experiments. Also, perhaps causing more excitement, some quite unforeseen effects have been found, like the oscillatory exchange underlying giant magnetoresistance in metallic multilayers. In general, the burgeoning field of nanotechnology has led to led to many exciting advances, including the fabrication and study of high-frequency devices with length scales in the sub-micrometre range.

When we wrote the first edition, the study of superlattices was still fairly new, so we were able to discuss them in a single chapter. The subject is now much more complete, and parts of the book have been expanded accordingly. We give a unified treatment based largely on the transfer-matrix formalism and we believe this clarifies the theory. Structures that are periodic in more than one dimension appear in a section on photonic band gaps, and an introductory account is given of quasiperiodic structures.

The first edition contained a relatively brief account of nonlinearities. It is now possible to give an extended account of nonlinear properties of surfaces, interfaces and superlattices. This account, which includes recent theoretical and experimental developments, forms a major part of the new material.

While the motivation and target readership for this book remains the same, we felt that the best way to proceed was to change the organization of the book into what we hope is a more logical structure. Part One (consisting of two chapters) sets the scene by dealing mainly with bulk properties. Part Two (five chapters) develops the discussion for surface and interface modes, and Part Three (two chapters) takes matters forward for superlattices and other multilayers. These first three parts are mainly concerned with linear excitations and Part Four (three chapters) contains the account of nonlinear properties, emphasizing the dynamical aspects. Following the pattern of the rest of the book, this starts with a chapter on relevant non-linear properties of bulk media. Then the subsequent chapters cover the extension and application to surfaces, films and superlattices. The problems at the end of the chapters have been removed, but instead we have indicated in the text where the reader might work through problems and derivations.

We have continued to benefit from working with friends and colleagues, and special thanks are due for the unfailing support we have received at our home institutions, namely the University of Western Ontario and Universiti Sains Malaysia. As in the case of the first edition, we gratefully acknowledge the input of our colleagues, research associates and graduate students. Above all we deeply appreciate the encouragement and tolerance of our wives and families during the lengthy time it took to prepare this second edition.

Mike Cottam
David Tilley
May 2004

PART ONE

BASIC CONCEPTS AND BULK PROPERTIES

Chapter 1

Introduction

In this book we are concerned with the ways in which surfaces and interfaces modify the properties of solids and liquids. Various different effects may be identified. First, there may be a modification to the equilibrium configuration in a medium close to a surface; this is known as *surface reconstruction*. For example, the atoms near to a surface may have a different crystallographic arrangement compared with those in the bulk, or they may be disordered. Another example is a ferromagnetic solid, in which the interactions between the magnetic moments at the surface may differ from those in the bulk, leading to a different value of the local magnetization. Clearly this type of effect may be temperature dependent, and it is particularly relevant when the system exhibits a phase transition (e.g. close to the Curie temperature in a ferromagnet). Second, the *excitations* within the system (such as the phonons in the lattice dynamics of a crystal or the magnons of a ferromagnet) are modified by a surface. In an infinite medium the bulk (or volume) excitations are characterized by an amplitude that varies in a wave-like fashion in three dimensions. When surfaces are present the bulk excitations are required to satisfy appropriate boundary conditions. Consequently there will, in general, be changes to the density of states and, in some cases, to the frequencies of the bulk excitations. However, a more interesting effect of a surface on the excitation spectrum is that it can give rise to localized surface excitations (e.g. surface phonons or surface magnons). By contrast with bulk excitations, the surface excitations are wave-like for propagation parallel to the surface (or interface) and have an amplitude that decays with distance away from the surface.

If there are nonlinear interactions in a material, their effect at a surface may be different from that in the bulk region. For example, some materials exhibit a *nonlinear* component to the dielectric function (i.e. a part that depends on the electric field strength). In turn, that may lead to a type of non-linear electromagnetic surface wave that has no counterpart in a linear material. In other cases the nonlinearity may result in a modification of the properties of the excitations produced by the linear terms. For example, in ferromagnetic systems the nonlinear effects due to pumping with a microwave field can give rise to magnon instabilities and a transition to chaotic behaviour. Nonlinear effects and the excitations associated with them are discussed in chapters 10 to 12.

As indicated by the title of this book, we are primarily concerned with the *dynamic* properties of finite media, i.e. with the excitations in these structures. For the most part, we include discussion of the static properties of surfaces (the surface reconstruction and disorder effects) only to the extent that these may influence the phase behaviour and the nature of the excitation spectrum. The general characteristics of bulk and surface excitations, together with their symmetry aspects, are described in section 1.1 of this chapter, together with the generalization to multilayer systems such as super-lattices. Section 1.2 deals briefly with the topics of sample growth, surface reconstruction and surface characterization. This is followed in section 1.3 by an outline of some of the theoretical methods applied to bulk and surface excitations in linear materials. In section 1.4 a preliminary survey is given of the experimental methods for detecting surface excitations. Specific types of excitations and/or specific surface structures are dealt with in the later chapters.

1.1   Excitations in crystals

We now introduce some of the general properties of surface excitations, stressing the symmetry aspects. It is instructive to do this by comparing and contrasting them with bulk excitations in infinite media, for which the concepts are more familiar. The results for bulk excitations will be summarized first; details are to be found in any of the standard textbooks on solid-state physics (e.g. Ashcroft and Mermin 1976, Madelung 1981, Kittel 1995, Grosso and Parravicini 2000).

1.1.1   Bulk excitations

An ideal crystal can be described as a *basis* of one or more atoms or ions located at each point of a *lattice*. A lattice is an infinitely-extended regular periodic array of points in space. All lattice points are equivalent, and the system possesses translational *symmetry*. The lattice may be defined in terms of three fundamental translation vectors, $\mathbf{a}_1$, $\mathbf{a}_2$ and $\mathbf{a}_3$, which are non-coplanar. If a translation is made through any vector that is a combination of integral multiples of these basic vectors the crystal appears unchanged. The end points of such vectors $\mathbf{R}$, defined by

$$\mathbf{R} = n_1\mathbf{a}_1 + n_2\mathbf{a}_2 + n_3\mathbf{a}_3 \tag{1.1}$$

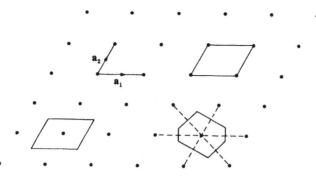

Figure 1.1. General oblique lattice in 2D showing the basic vectors a_1 and a_2. Three possible forms of the unit cell are indicated. Two are parallelograms with different centres and the other is the Wigner–Seitz cell.

with n_1, n_2 and n_3 integers, form the *space lattice*. In addition to the translation operations, there may also in general be other symmetry operations, such as certain rotations and reflections, which leave the crystal apparently unchanged. All such symmetry operations must leave both the space lattice and basis unchanged.

An important part is played by the *unit cell*, defined as the smallest volume based on one lattice point such that the whole of space is filled by repetitions of the unit cell at each lattice point. The specification of the unit cell is not unique: one possible choice would be the parallelepiped subtended by the basic vectors a_1, a_2 and a_3, with the cell arbitrarily centred on one of the atomic positions. Some choices of unit cell are illustrated in figure 1.1, which depicts for ease of representation a two-dimensional (2D) lattice with the basic vectors a_1 and a_2. Two examples of parallelograms as the unit cell are shown, both with sides defined by a_1 and a_2 but centred at different positions on the space lattice. A hexagonal unit cell also is indicated, obtained by drawing the perpendicular bisectors of the lattice vectors from a central point to the nearby equivalent sites (a construction known generally as the *Wigner–Seitz cell*).

The translational symmetry of the crystal structure implies that various position-dependent physical quantities, such as the electron density or the electrostatic potential, are the same within each unit cell. These quantities must be multiply-periodic functions satisfying

$$f(\mathbf{r} + \mathbf{R}) = f(\mathbf{r}) \tag{1.2}$$

for all points $\mathbf{r}$ in space and for all vectors $\mathbf{R}$ defined by (1.1). In 1D it is a well-known mathematical result that a periodic function can be expanded as a Fourier series of complex exponentials. The analogous result in 3D enables us to write

$$f(\mathbf{r}) = \sum_{\mathbf{Q}} F(\mathbf{Q}) \exp(i\mathbf{Q} \cdot \mathbf{r}) \tag{1.3}$$

where the vectors $\mathbf{Q}$ must satisfy

$$\exp(i\mathbf{Q} \cdot \mathbf{R}) = 1 \tag{1.4}$$

for all lattice vectors $\mathbf{R}$. The end points of all vectors $\mathbf{Q}$ satisfying (1.4) form a lattice known as the *reciprocal lattice* (since $\mathbf{Q}$ has the dimension of a reciprocal length). The reciprocal lattice can be generated from the basic vectors $\mathbf{b}_1$, $\mathbf{b}_2$ and $\mathbf{b}_3$ satisfying

$$\mathbf{a}_i \cdot \mathbf{b}_j = 2\pi\delta_{ij} \qquad (i,j = 1,2,3) \tag{1.5}$$

where δ_{ij} is the Kronecker delta (defined by $\delta_{ij} = 1$ if $i = j$, $\delta_{ij} = 0$ if $i \neq j$). Explicitly, the definitions are

$$\mathbf{b}_1 = (2\pi/V_c)\mathbf{a}_2 \times \mathbf{a}_3, \qquad \mathbf{b}_2 = (2\pi/V_c)\mathbf{a}_3 \times \mathbf{a}_1, \qquad \mathbf{b}_3 = (2\pi/V_c)\mathbf{a}_1 \times \mathbf{a}_1, \tag{1.6}$$

denoting $V_c = \mathbf{a}_1 \cdot (\mathbf{a}_2 \times \mathbf{a}_3)$. A general reciprocal lattice vector $\mathbf{Q}$ then takes the form

$$\mathbf{Q} = \nu_1\mathbf{b}_1 + \nu_2\mathbf{b}_2 + \nu_3\mathbf{b}_3 \tag{1.7}$$

where ν_1, ν_2 and ν_3 are integers. The reader may easily prove that (1.6) and (1.7) satisfy the property (1.4).

The unit cell of the reciprocal lattice is conventionally obtained using the Wigner–Seitz construction described earlier; it is known as the *Brillouin zone*. Its volume is the same as the parallelepiped formed by the basic vectors $\mathbf{b}_1$, $\mathbf{b}_2$ and $\mathbf{b}_3$, which can be shown from (1.6) to be $8\pi^3/V_0$ where $V_0 \equiv |V_c|$ is the volume of the unit cell in real space. The easiest example is a simple cubic space lattice, for which the vectors $\mathbf{a}_1$, $\mathbf{a}_2$ and $\mathbf{a}_3$ may be taken as

$$\mathbf{a}_1 = a(1,0,0), \qquad \mathbf{a}_2 = a(0,1,0), \qquad \mathbf{a}_3 = a(0,0,1) \tag{1.8}$$

where a is the lattice parameter (the nearest-neighbour separation). The basic vectors of the reciprocal lattice are found to be

$$\mathbf{b}_1 = \frac{2\pi}{a}(1,0,0), \qquad \mathbf{b}_2 = \frac{2\pi}{a}(0,1,0), \qquad \mathbf{b}_3 = \frac{2\pi}{a}(0,0,1). \tag{1.9}$$

Hence the reciprocal lattice is also simple cubic in this case. It is fairly straightforward to show that the reciprocal lattice of a body-centred cubic (bcc) space lattice is face-centred cubic (fcc) and the reciprocal lattice of a fcc space lattice is bcc. The details are given in solid-state textbooks.

We are now in a position to discuss how the elementary excitations of a crystal are influenced by the symmetry. The important property is embodied in a result that is generally known to solid-state physicists as *Bloch's theorem* (Bloch 1928); it is also related to Floquet's theorem in mathematics (e.g. see Whittaker and Watson 1963). A statement of Bloch's theorem appears as (1.13) below: here we outline a simple proof applicable to any type of excitation in a crystal. Examples of excitations that we shall consider in later chapters include lattice vibrations (phonons), spin waves (magnons), electronic modes (such as plasmons), and so forth. All of them are excitations of the whole region, rather than localized excitations of particular atoms, and this leads to common symmetry features.

A consequence of the periodicity of the crystal lattice is that a position-dependent quantity, such as the Hamiltonian operator $H(\mathbf{r})$, is unaltered by a translation through the vector $\mathbf{R}$ in (1.1). For example, the Hamiltonian might take the form $-(\hbar^2/2m)\nabla^2 + V(\mathbf{r})$ appropriate to an electron (mass m) in a periodic potential $V(\mathbf{r})$. The translational invariance property of the Hamiltonian can be stated as $H(\mathbf{r} + \mathbf{R}) = H(\mathbf{r})$. It follows that the corresponding Schrödinger's equation

$$H(\mathbf{r})\psi(\mathbf{r}) = E\psi(\mathbf{r}) \tag{1.10}$$

must be invariant under the translation $\mathbf{r} \rightarrow \mathbf{r} + \mathbf{R}$. Hence, if $\psi(\mathbf{r})$ denotes the wave function of a stationary state (with energy eigenvalue E), then $\psi(\mathbf{r} + \mathbf{R})$ is also a solution describing the same state of the system. This implies that the two functions must be related by a multiplicative factor, and we write

$$\psi(\mathbf{r} + \mathbf{R}) = c\psi(\mathbf{r}). \tag{1.11}$$

It is evident that c must have unit modulus, otherwise the wave function would tend to infinity if the translation through $\mathbf{R}$ (or $-\mathbf{R}$) were repeated indefinitely. Hence c must be expressible as

$$c = \exp(i\mathbf{q} \cdot \mathbf{R}) \tag{1.12}$$

where $\mathbf{q}$ is an *arbitrary* (real) constant vector having the dimensions of reciprocal length. The general form of the wave function having the above property is

$$\psi(\mathbf{r}) = \exp(i\mathbf{q} \cdot \mathbf{r})U_{\mathbf{q}}(\mathbf{r}) \tag{1.13}$$

where $U_{\mathbf{q}}(\mathbf{r})$ is a periodic function: $U_{\mathbf{q}}(\mathbf{r} + \mathbf{R}) = U_{\mathbf{q}}(\mathbf{r})$. The overall phase factor $\exp(i\mathbf{q} \cdot \mathbf{r})$ gives a plane-wave variation to the wave function in accordance with the properties (1.11) and (1.12) deduced from the translational symmetry of the lattice.

An alternative (and more rigorous) proof of Bloch's theorem (1.13) can be formulated using group theory (e.g. see Wherrett 1986). For convenience we have expressed the result in terms of the wave function $\psi(\mathbf{r})$, as would be the case for electronic states. More generally, Bloch's theorem would be written in terms of whatever variable is used to describe the amplitude of the excitation (e.g. an atomic displacement in the case of phonons).

The wave vector $\mathbf{q}$ is clearly not unique: values differing by any reciprocal lattice vector $\mathbf{Q}$ as defined in (1.7) would equally satisfy (1.12) because $\exp[i(\mathbf{q} + \mathbf{Q}) \cdot \mathbf{R}] = \exp(i\mathbf{q} \cdot \mathbf{R})$. In general, the energy eigenvalue in (1.10) will depend on $\mathbf{q}$, and it is periodic in the reciprocal lattice. Denoting $E = \hbar\omega(\mathbf{q})$, where $\omega(\mathbf{q})$ is the excitation frequency, we must therefore have

$$\omega(\mathbf{q} + \mathbf{Q}) = \omega(\mathbf{q}). \qquad (1.14)$$

For an infinite crystal $\mathbf{q}$ can take any real value but, because of the above property, it is unnecessary to define it outside the first Brillouin zone, defined to be the Brillouin zone centred at $\mathbf{q} = 0$. For example, earlier in this section we showed that the Brillouin zone of a simple cubic lattice is a cube with sides of dimension $2\pi/a$. Hence it is sufficient in this case to take $\mathbf{q}$ values corresponding to

$$-\pi/a < q_x \leq \pi/a, \qquad -\pi/a < q_y \leq \pi/a, \qquad -\pi/a < q_z \leq \pi/a. \quad (1.15)$$

In a finite but macroscopically large crystal, the number of excitations (and hence the number of allowed values of $\mathbf{q}$ in the Brillouin zone) is finite. For ease of discussion, suppose the crystal has a simple cubic structure and that the sample is macroscopically large forming a cube of side Na. Strictly in such problems it would become necessary to introduce boundary conditions describing the surfaces (as discussed later). However, if N is very large and only bulk properties of the crystal are being examined, this difficulty is often circumvented by the use of cyclic (or Born–von Karman) boundary conditions. To illustrate this we take a 1D case of a 'crystal' of length Na, with the ends joined to form a ring. The condition that

$$\psi(z + Na) = \psi(z) \qquad (1.16)$$

where z is a position coordinate, follows from the requirement that the wave function is single-valued. However, the 1D case of (1.11) is

$$\psi(z + a) = c\psi(z) \qquad (1.17)$$

for a symmetry translation through a. Thus it follows that $c^N = 1$, which has the roots $c = \exp(iq_z a)$ where

$$q_z = \frac{2\pi m_z}{Na} \qquad (1.18)$$

and m_z is an integer. Within the first Brillouin zone, m_z will take the N consecutive integer values in the range $-\frac{1}{2}N < m_z \leq \frac{1}{2}N$ (if N is even). The generalization of the above argument to 3D is obvious: each component of the vector $\mathbf{q}$ will be given by an expression analogous to (1.18). Thus we obtain for the allowed wave vectors a set of uniformly distributed points in $\mathbf{q}$-space. The 'volume' per allowed wave vector is a constant, namely $(2\pi/Na)^3$. Since N is supposed to be very large this distribution is effectively continuous, and it is permissible to convert summations over discrete $\mathbf{q}$

values to integrations over continuous variables by the prescription

$$\sum_{\mathbf{q}} \rightarrow \frac{V}{(2\pi)^3} \iiint dq_x \, dq_y \, dq_z \qquad (1.19)$$

where $V = (Na)^3$ is the volume of the crystal.

1.1.2 Surface excitations

In the presence of one or more surfaces or interfaces, many of the symmetry arguments presented in the preceding section need to be modified. To illustrate this we restrict attention initially to the effect of a single plane surface. Specifically we consider the situation represented in figure 1.2, where the surface is taken to lie in the plane $z = 0$ of a Cartesian coordinate system. The thickness of the crystal and its dimensions in the x and y directions will be taken as sufficiently large (effectively infinite) that the crystal may be regarded as filling the half-space $z < 0$. For the present we ignore any rearrangement or distortion of the surface atoms (i.e. surface reconstruction) and deal with an 'idealized' surface.

If the orientation of the surface relative to the crystal axes is such that two of the basic translation vectors ($\mathbf{a}_1$ and $\mathbf{a}_2$, say) are parallel to the surface, while $\mathbf{a}_3$ is not, then the set of translations $\mathbf{R}_\parallel$ defined by the vectors

$$\mathbf{R}_\parallel = n_1 \mathbf{a}_1 + n_2 \mathbf{a}_2 \qquad (1.20)$$

with n_1 and n_2 integers, are symmetry operations of the crystal. However, translation operations involving $\mathbf{a}_3$, such as those given by (1.1) with $n_3 \neq 0$, are not symmetry operations: they do not leave the crystal apparently unchanged, because they connect points that have different positions relative to the surface plane. The end points of the set of vectors $\mathbf{R}_\parallel$ form the 2D space lattice. There are in fact only five space lattices (called *Bravais lattices*) in 2D, compared with fourteen in 3D. These are depicted in figure 1.3.

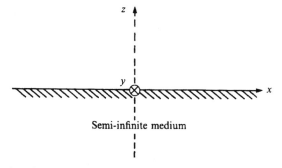

Figure 1.2. Choice of coordinate axes for a semi-infinite medium.

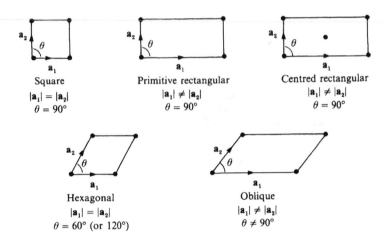

Figure 1.3. The five Bravais lattices in 2D, defined in terms of the basic vectors $\mathbf{a}_1$ and $\mathbf{a}_2$.

The corresponding 2D reciprocal lattice has just two basic vectors $\mathbf{b}_1$ and $\mathbf{b}_2$. They are defined as the vectors in the xy plane that satisfy

$$\mathbf{a}_i \cdot \mathbf{b}_j = 2\pi\delta_{ij} \qquad (i,j = 1,2). \tag{1.21}$$

A general 2D reciprocal lattice vector $\mathbf{Q}_\parallel$ can then be expressed as

$$\mathbf{Q}_\parallel = \nu_1 \mathbf{b}_1 + \nu_2 \mathbf{b}_2 \tag{1.22}$$

with ν_1 and ν_2 integers. The Wigner–Seitz unit cell of the 2D reciprocal lattice defines the 2D Brillouin zone: its area is equal to $4\pi^2/A_0$ where A_0 is the area of the unit cell for the lattice planes parallel to the surface. It is a useful exercise to deduce the basic vectors and hence the Brillouin zone for each of the 2D Bravais lattices in figure 1.3.

The periodicity property (1.2), that is satisfied by any position-dependent physical quantity $f(\mathbf{r})$ in an infinite 3D system, no longer holds in the present case because there is translational symmetry only in the xy plane. The more limited property applicable to the semi-infinite crystal is that

$$f(\mathbf{r}_\parallel + \mathbf{R}_\parallel, z) = f(\mathbf{r}_\parallel, z) \tag{1.23}$$

where we denote $\mathbf{r} = (\mathbf{r}_\parallel, z)$ for the 3D position vector and $\mathbf{r}_\parallel = (x, y)$. From the properties of the 2D reciprocal lattice vectors and from the lattice periodicity in the xy plane we may define a 2D Fourier series expansion by

$$f(\mathbf{r}_\parallel, z) = \sum_{\mathbf{Q}_\parallel} F(\mathbf{Q}_\parallel, z) \exp(i\mathbf{Q}_\parallel \cdot \mathbf{r}_\parallel). \tag{1.24}$$

This is the analogue of (1.3) for the infinite 3D case.

The proof of Bloch's theorem also requires modification in the presence of a surface. We may follow the same type of argument as in the preceding section: equation (1.11) must be replaced by $\psi(\mathbf{r}_{||} + \mathbf{R}_{||}, z) = c\psi(\mathbf{r}_{||}, z)$ because $\mathbf{R}_{||}$ is now the general operation of translation symmetry. This leads to the conclusion that $c = \exp(i\mathbf{q}_{||} \cdot \mathbf{R}_{||})$ where $\mathbf{q}_{||}$ is a *real* 2D vector, and the 2D analogue of Bloch's theorem becomes

$$\psi(\mathbf{r}_{||}, z) = \exp(i\mathbf{q}_{||} \cdot \mathbf{r}_{||}) U_{\mathbf{q}_{||}}(\mathbf{r}_{||}, z). \tag{1.25}$$

Here $U_{\mathbf{q}_{||}}(\mathbf{r}_{||}, z)$ has the periodicity property in 2D that $U_{\mathbf{q}_{||}}(\mathbf{r}_{||} + \mathbf{R}_{||}, z) = U_{\mathbf{q}_{||}}(\mathbf{r}_{||}, z)$.

It is not possible from symmetry arguments to make a rigorous statement about the dependence of $U_{\mathbf{q}_{||}}(\mathbf{r}_{||}, z)$ on the coordinate z perpendicular to the surface. The procedure for deducing the z dependence in any specific problem is frequently to obtain a differential equation (or, in some cases, a finite-difference equation) for $U_{\mathbf{q}_{||}}(\mathbf{r}_{||}, z)$; this would be Schrödinger's equation in the case of electronic states or the appropriate equation of motion in the case of other excitations. The equation must be solved subject to the relevant boundary conditions; in the present example, these apply at the surface $z = 0$ and at $z \to -\infty$. It is usually found that there are more admissible solutions than is the case in the corresponding infinite crystal problem. Suppose, for simplicity, that the equation for the semi-infinite medium is a linear equation and it admits solutions of the form $\exp(iq_z^{(j)}z)$ where $q_z^{(j)}$ may be *complex* and take a series of values (labelled by j) determined from the differential equation. The solution for $U_{\mathbf{q}_{||}}(\mathbf{r}_{||}, z)$ would be formed from a linear combination of such terms:

$$U_{\mathbf{q}_{||}}(\mathbf{r}_{||}, z) = \sum_j B_{\mathbf{q}_{||}}^{(j)}(\mathbf{r}_{||}) \exp(iq_z^{(j)}z) \tag{1.26}$$

where $B_{\mathbf{q}_{||}}^{(j)}(\mathbf{r}_{||})$ is an amplitude factor. Because the right-hand side of (1.26) must remain finite at large distances from the surface ($z \to -\infty$), we have *either* $q_z^{(j)}$ is real *or* $q_z^{(j)}$ is complex with $\mathrm{Im}(q_z^{(j)}) < 0$.

The former possibility corresponds to a *bulk excitation*, since $\exp(iq_z^{(j)}z)$ has a constant modulus, equal to unity, for all z. The quantity $q_z^{(j)}$ is just the third component of the 3D wave vector $\mathbf{q} = (\mathbf{q}_{||}, q_z^{(j)})$ describing the propagation of the excitation. The bulk excitation is affected by the surface in that it must satisfy a boundary condition at $z = 0$, and this will enter into the calculation of the corresponding B coefficients in (1.26).

The second of the above possibilities, i.e. $\mathrm{Im}(q_z^{(j)}) < 0$, corresponds to a *surface excitation* localized near the $z = 0$ surface, because $\exp(iq_z^{(j)}z) \to 0$ as $z \to -\infty$ within the crystal. If $q_z^{(j)}$ is pure imaginary we may denote $q_z^{(j)} = -i\kappa$ (with κ real and positive), and the surface excitation has a simple exponential decay proportional to $\exp(\kappa z)$ as $z \to -\infty$. The attenuation length (or decay length) is $1/\kappa$. More generally, $q_z^{(j)}$ might be complex and this would correspond to an excitation that oscillates within an exponentially decaying envelope function as $z \to -\infty$. Hence, in such cases, the surface excitation can be characterized by a 2D wave vector $\mathbf{q}_{||}$ describing its propagation parallel to the surface and by a decaying amplitude in the direction perpendicular to the surface.

The concepts of bulk and surface excitations may be illustrated physically by considering some aspects of bulk and surface waves in a semi-infinite isotropic elastic medium. This problem was first examined by Rayleigh (1887) and we return to it in chapter 3. The general equation of motion is the wave equation

$$\frac{\partial^2 u}{\partial t^2} - v^2 \nabla^2 u = 0 \tag{1.27}$$

where u denotes any component of the vectors $\mathbf{u}_L$ and $\mathbf{u}_T$ for longitudinal and transverse displacements in the elastic medium, and $v = v_L$ or v_T is the corresponding velocity. Assuming coordinate axes as in figure 1.2, we seek a plane-wave solution to (1.27) propagating parallel to the surface with wave vector $\mathbf{q}_{\parallel}$ and frequency ω:

$$u(\mathbf{r}, t) = \exp(\mathrm{i}\mathbf{q}_{\parallel} \cdot \mathbf{r}_{\parallel}) f(z) \exp(-\mathrm{i}\omega t). \tag{1.28}$$

On substituting (1.28) into (1.27) the equation for $f(z)$ is

$$\frac{\mathrm{d}^2 f}{\mathrm{d}z^2} - \left(q_{\parallel}^2 - \frac{\omega^2}{v^2} \right) f = 0. \tag{1.29}$$

It can now be seen that there are two types of solution for $f(z)$. If $q_{\parallel}^2 < \omega^2/v^2$ we have

$$f(z) = B_1 \exp(\mathrm{i}q_z z) + B_2 \exp(-\mathrm{i}q_z z) \tag{1.30}$$

where

$$q_z = \left(\frac{\omega^2}{v^2} - q_{\parallel}^2 \right)^{1/2}. \tag{1.31}$$

This describes a bulk wave. The two terms in (1.30) describe a wave propagating towards the surface and a reflected wave. The other type of solution of (1.29) occurs when $q_{\parallel}^2 > \omega^2/v^2$ and it corresponds to

$$f(z) = B_3 \exp(\kappa z) \tag{1.32}$$

where

$$\kappa = \left(q_{\parallel}^2 - \frac{\omega^2}{v^2} \right)^{1/2}. \tag{1.33}$$

This describes a surface wave decaying with distance from the surface (we recall that $z < 0$ within the solid). To proceed further one would need to form the total displacement $(\mathbf{u}_L + \mathbf{u}_T)$ and then use a boundary conditions at $z = 0$. The details are given in section 3.2 of chapter 3.

In the discussion so far, we have restricted attention to the simplest case of a semi-infinite medium, so that only one surface is involved. The extension to a parallel-sided slab (or film) of finite thickness involves boundary conditions at two surfaces. However, the symmetry considerations in terms of the 2D Brillouin zones and the modified Bloch's theorem (1.25) are still applicable. Numerous applications to thin films and slabs are given in the following chapters. A general point, which we may usefully emphasize here, is that (1.18) for the discrete q_z values of a bulk excitation is an artefact of employing cyclic boundary conditions. When the proper boundary conditions appropriate to the surface problem in a film are applied, (1.18) is modified and the discrete q_z values may no longer be evenly separated in reciprocal space. A simple example of this occurs in section 4.2 of chapter 4, where we show how to derive the discrete q_z values for a ferromagnetic film.

1.1.3 Superlattice excitations

In some of the later chapters we shall also be interested in multilayer systems, where there are many surfaces or interfaces between different media. Of particular interest are *superlattices*: these are structures in which the composition and thickness of the layers are arranged to form a particular mathematical sequence. The usual case is that of a *periodic* superlattice. For example, figure 1.4 shows a two-component superlattice consisting of

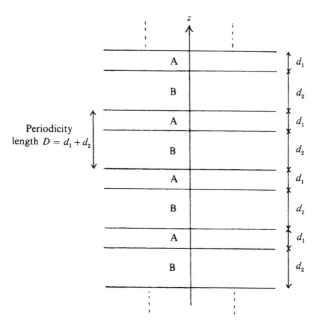

Figure 1.4. Schematic illustration of a periodic superlattice with alternate layers (labelled A and B) of thickness d_1 and d_2.

alternate layers of two media, labelled A and B, with thicknesses d_1 and d_2 respectively. They are stacked to form a sequence ABABAB..., so the periodicity length (or repeat distance) of the superlattice is

$$D = d_1 + d_2. \tag{1.34}$$

If the superlattice has very many layers (so that it may be assumed to extend indefinitely) then translations through distance D, or multiples thereof, in the z direction are well-defined symmetry operations of the structure. This property, together with the ability to control the magnitude of D by varying the growth conditions, gives rise to useful and distinctive properties. Periodic superlattices were originally envisaged as structures consisting of alternating ultrathin semiconductor layers with period less than the electron mean free path (Esaki and Tsu 1970). The term is now applied to many different kinds of materials, and periodic superlattices are currently of widespread interest for device applications.

In recent years there have also been experimental and theoretical studies of so-called *quasiperiodic* superlattices (see e.g. MacDonald 1987). They can be formed in various ways, as will be discussed later, but basically their structure along the growth direction is quasiperiodic in the sense that it is characterized by two different fundamental periods whose ratio is irrational. For example, this is often implemented in terms associating the layer structure with the Fibonacci mathematical sequence. The Fibonacci numbers are defined by the recursion relation

$$F_{n+2} = F_{n+1} + F_n \tag{1.35}$$

with the initial condition $F_0 = 1$ and $F_1 = 1$. The ratio F_{n+1}/F_n of successive Fibonacci numbers converges rapidly (for large n) to a certain irrational number known as the golden mean with a value $(1 + \sqrt{5})/2 \approx 1.618$. Quasiperiodic superlattices are of particular interest with regard to studies of the localization and multifractal aspects of the excitations (see Albuquerque and Cottam 2003).

The excitations in superlattices, including both periodic and quasiperiodic structures, form the topic of chapters 8 and 9.

1.2 Sample growth and surface analysis

We begin with brief reviews of some of the techniques used for sample growth and preparation. Then surface reconstruction is discussed, followed by an account of analytical methods for the surface crystallographic structure and composition. Each of these is a major topic in its own right, and so we do not attempt to present a comprehensive account. Instead, we give a summary of those aspects that are relevant for the material in this book, and we provide other references where appropriate.

1.2.1 Growth and preparation

Several different methods for surface and/or film preparation are available, including evaporation, sputtering, metallo-organic chemical vapour deposition (MOCVD), molecular-beam epitaxy (MBE), liquid-phase epitaxy (LPE), and Langmuir–Blodgett (LB) methods, amongst others. The suitability of any of these, or of the various hybrid methods that are being developed, depends on the type of material and film thickness being grown and on the need (if any) for *epitaxy*. Epitaxial growth of layers occurs when the deposited atoms are aligned with (i.e. are commensurate with) the atoms in the underlying single-crystal substrate. Successful heteroepitaxy occurs between materials with nearly the same lattice constant, as is the case in the III–V semiconductors GaAs and $Al_xGa_{1-x}As$ for certain ranges of the Al concentration x. Some good general references for film growth are the books by George (1992) and Zangwill *et al* (1996); other references for specific techniques are given in the descriptions below.

Vacuum evaporation, although one of the oldest techniques, is still commonly used because of its simplicity and convenience. The basic steps involved in the deposition are generation of vapour from the condensed phase (solid or liquid), transfer of the vapour from the source to the substrate, and condensation of the vapour on the substrate surface to form the solid film. The pressure used for normal evaporation is about 10^{-5} torr. The conventional method for evaporating the material to be deposited is evaporation from a 'boat' or wire of a refractory material, but other more recent techniques include flash evaporation, electron beam bombardment and laser evaporation (George 1992).

Sputtering is the process in which atoms are ejected from a surface when it is bombarded by high-velocity positive ions. When the ejected atoms are made to condense on a substrate, thin film deposition takes place. Again there are several variants of the technique, but we mention in particular ion-beam sputtering. Here the ion beam generated at an ion source is extracted into a high-vacuum chamber and directed at the target material, which is sputtered and deposited on a nearby substrate. The directionality of the beam facilitates varying the angles of incidence (on the target) and deposition (on the substrate). Other advantages over conventional sputtering include the lower background pressure and better isolation of the substrate from the ion-production stage.

MOCVD is a technique for growing thin layers of compound semiconductors for use in optoelectronic and microwave devices. It is a development of earlier chemical vapour deposition (CVD) methods in which various combinations of organo-metallic compounds and hydrides are used for the growth of epitaxial layers (see George 1992, Tu *et al* 1992). The technique of depositing, say, $Al_xGa_{1-x}As$ layers for different x values is based on a chemical reaction in which $(CH)_3Ga$, $(CH)_3Al$ and AsH_3 decompose to form $Al_xGa_{1-x}As$ and CH_4. The layer quality and composition, as well as growth rate, can be computer controlled via flowmeters.

MBE is a sophisticated method of epitaxial film growth that offers sensitive and flexible control for the production of microstructures (see e.g. Foxon 1987, Jaros 1989, Tu *et al* 1992, George 1992). It requires an ultrahigh vacuum (UHV) chamber, an atomically clean surface on which the growth is to take place, and one or more impinging beams of atoms supplied by evaporation cells. A heavy-ion gun mounted in the chamber can be used to bombard the surface and remove unwanted features such as structureless conglomerates on the surface. An Auger analyser can be used to monitor the quality of the surface. MBE can be used to produce ultrathin layers and multilayers with precisely controlled characteristics and abrupt interfaces.

LPE is a thermally controlled technique to grow high purity epitaxial films of semiconductor compounds and alloys. In principle it is growth from solution, so a solvent for the material is needed. The material from such a saturated solution, when allowed to cool at a suitable rate, is grown on to an underlying substrate (see George 1992). In general, the quality of the surface is inferior to that obtainable by MBE.

Finally we briefly mention the LB technique, which makes use of molecules that have carefully balanced hydrophilic (affinity for water) and hydrophobic (no such affinity) parts, e.g. as is the case for the opposite ends of certain fatty acid chain molecules. A monomer film may be formed on the surface of water, and transferred to a solid substrate to produce high quality monomer films or, with repeated application, bilayers and multilayers (see George 1992).

1.2.2 Surface reconstruction

We have already mentioned at the beginning of this chapter that under certain circumstances the atoms at a surface or interface may have a different equilibrium configuration from those in the bulk, a phenomenon known as surface reconstruction. As a result the 2D unit cell of the surface layer (and possibly adjacent layers) may be different from that of a parallel layer well inside the crystal. The surface layer may even be disordered. If the reconstructed surface region is sufficiently thin (e.g. corresponding to a single layer of adsorbed atoms) it could be argued that the effect on the excitations will be through a modification to the boundary conditions at the surface. In this case a necessary condition on the thickness of the reconstructed region is that it should be small compared with the attenuation length $1/\kappa$ for a surface excitation or compared with the wavelength $2\pi/q_z$ for a bulk excitation. Such requirements can often be satisfied for long-wavelength excitations. In other cases the reconstructed surface region would have to be treated explicitly as a finite-thickness layer (sometimes called the *selvedge*).

Some circumstances where surface reconstruction occurs include the adsorption of overlayer foreign atoms on to the surface of a crystal, or an impure material where the concentration of impurities may be different at the surface. However, surface reconstruction can also take place in a chemically clean material, as we shall discuss. We mention next a few situations in which surface reconstruction is known to occur. For general reviews see Somorjai (1975) and Zangwill (1988). The evidence for the occurrence of surface reconstruction comes from the experimental techniques summarized later in section 1.2.3.

The clean crystal surfaces of metals have been thoroughly studied using low-energy electron diffraction (LEED). In Al, Cu and Ni the surfaces corresponding to low Miller indices (as defined e.g. in Kittel 1995) each have a 2D unit cell that appears to be very similar to a layer in the bulk. However, in some cases there is evidence for an expansion or contraction of the outermost atomic layer in the direction *normal* to the surface. For example, the outer layer of the (110) face of Al is apparently moved inwards by 10–15% relative to the bulk spacing, while for the (100) and (111) faces the effect is of order 5% or less (Martin and Somorjai 1973). Similar changes in the interlayer spacing at the surface occur in some of the alkali halides, for example in LiF (Laramore and Switendick 1973).

It is appropriate here to mention some notational conventions for classifying a surface layer. Cases such as those just described above, where the surface layer is identical to the 2D symmetry of a parallel plane in the bulk of the sample, are referred to as (1×1). This indicates that the basic vectors $\mathbf{a}_1$ and $\mathbf{a}_2$ in the surface layer are identical to those in the underlying bulk region (or substrate). In other cases the surface structure may have unit cells that are integral multiples of the substrate unit cell. For example, the notation $W(211) - (2 \times 2)$ applied to the (211) face of tungsten is used to describe the situation of a surface layer having basic vectors $\mathbf{a}_1$ and $\mathbf{a}_2$ twice that of the substrate. This would, in fact, be the structure appropriate to an adsorbate layer of hydrogen atoms on a W(211) surface. A similar notation can also be employed for cases where the surface unit cell is rotated with respect to the substrate. This is illustrated in figure 1.5, where the $c(2 \times 2)$ structure on a square lattice substrate (with the letter c denoting a centred unit cell) can equivalently be designated as $(\sqrt{2} \times \sqrt{2}) - R45°$. The latter notation means that the primitive unit cell of the surface structure has sides that are a $\sqrt{2}$ multiple of those for the substrate and that it is rotated through 45° with respect to the substrate.

Other examples of surface reconstruction are provided by the (111), (100) and (110) faces of Si and Ge. In particular, detailed studies have been made of the Si(111) surface (e.g. see Zangwill 1988). For a freshly cleaved sample prepared at room temperature the structure is $Si(111) - (2 \times 1)$. On heating to about 700–1000 K the structure eventually converts to $Si(111) - (7 \times 7)$. The temperature at which this transformation

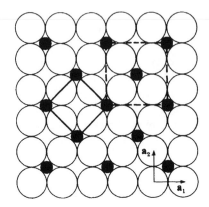

Figure 1.5. Surface layer of adsorbed atoms (full circles) on a square lattice substrate (open circles). The basic vectors $\mathbf{a}_1$ and $\mathbf{a}_2$ of the substrate are shown. The unit cell of the adsorbed layer can alternatively be chosen as the square marked in broken lines, designated as $c(2 \times 2)$, or the square marked in full lines, designated as $(\sqrt{2} \times \sqrt{2}) - R45°$. After Somerjai (1975).

takes place seems to depend sensitively on the presence of trace impurities (such as Fe or Ni) and there may be other intermediate structures. The cause of surface reconstruction in Si(111) is the covalent bonding between atoms. A Si atom in the bulk is bonded to its four nearest neighbours in a diamond cubic structure, and consequently an 'unreconstructed' (111) surface will have a layer of atoms with one bond 'dangling' into the vacuum in the direction normal to the surface (see figure 1.6). Such an arrangement of dangling bonds is unfavourable energetically, and surface reconstruction may take place in such a way as to reduce the overall energy of the bonds.

A different type of surface reconstruction may occur in magnetically ordered solids. For example, even if there is no distortion or rearrangement of the atomic positions at the surface of a ferromagnet or antiferromagnet, there may be a different orientation of the spins (or magnetic moments) compared with the bulk material. This could occur due to thermal effects at the surface being different from those in the bulk or due to modified exchange interactions and anisotropy energies at the surface. Examples are given later in chapter 4.

1.2.3   Analysis of surface structure

There is now a wide range of techniques available for characterizing the crystallographic structure and other static properties of surfaces. Full accounts are to be found in the books by Prutton (1983), Feldman and Mayer (1986), Woodruff and Delchar (1986), and Tu *et al* (1992). The book edited by Bland and Heinrich (1994) emphasizes applications to

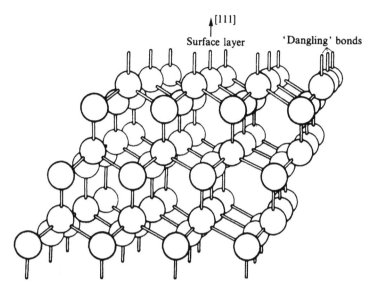

Figure 1.6. Dangling bonds from the 'unreconstructed' (111) surface of a covalently bonded diamond cubic structure. After Prutton (1983).

magnetic materials. Here we will focus on the techniques of low-energy electron diffraction (LEED) and Auger electron spectroscopy (AES). The former method provides an accurate determination of the surface crystallography, while the latter is sensitive to the chemical composition. Other techniques will be briefly mentioned.

In LEED a beam of monoenergetic electrons (typically with energies 20–500 eV) is directed at the surface of a sample under ultra-high vacuum (UHV) conditions, and a diffraction pattern is produced. The high atomic scattering cross sections for electrons in this energy range makes LEED extremely sensitive to surface atomic arrangements (the penetration depth is typically less than 1 nm). A variation of the technique is reflection high-energy electron diffraction (RHEED) in which electrons with energies between 1 and 10 keV are used. In order to keep the penetration depth small in this case, thereby retaining surface sensitivity, it is necessary to use grazing incidence and emergence. The diffraction patterns from both LEED and RHEED can be interpreted by standard methods, e.g. using the Ewald sphere construction, but a proper theory of the intensities is extremely complicated.

The basic principle of AES is rather different and can be understood by reference to figure 1.7. When an atom is ionized by the production of a core hole in level A, typically by incident electrons of sufficient energy (~1.5–5 keV), the ion may lose some of its potential energy by filling this core with an electron from a shallower level (B) together with the emission of

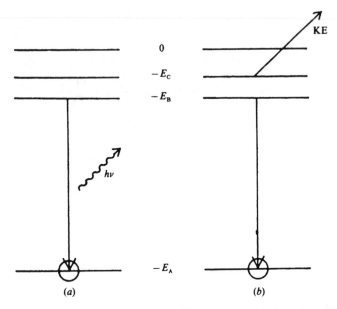

Figure 1.7. Energy level diagram showing the filling of a core hole in level A, giving rise to (a) x-ray photon emission or (b) Auger electron emission. The levels are labelled with their one-electron binding energies. After Woodruff and Delchar (1986).

energy. This energy may appear either as a photon, as in figure 1.7(a), or alternatively as kinetic energy given to another shallowly bound electron (C), as in figure 1.7(b). The latter process is known as the *Auger effect* and is allowed energetically provided $E_A > E_B + E_C$. The kinetic energy of the emitted electron (which is roughly equal to $E_A - E_B - E_C$) is characteristic of the atom from which it originates. The Auger electrons have a short mean-free-path, and so their detection outside the sample provides a surface-sensitive probe of chemical composition. The Auger effect also plays a role in the technique of surface extended x-ray absorption fine structure (SEXAFS), in which surface-specific features of the x-ray absorption coefficient in thin films are measured.

Another spectroscopic technique is photoelectron spectroscopy, in which the surface is bombarded with photons of sufficient energy to ionize an electronic shell so that an electron is ejected into the vacuum. The emission energies are characteristic of the surface atoms, and the angular dependence gives information about the surface structure. Usually either ultraviolet photons are employed (ultraviolet photoelectron spectroscopy, or UPS) or x-ray photons (x-ray photoelectron spectroscopy, or XPS).

Spin-polarized versions of many of these electron spectroscopies, along with polarized neutron reflectivity, have been developed for studying surfaces of magnetic materials (see e.g. Bland and Heinrich 1994).

The above techniques, together with other methods such as transmission electron microscopy (TEM) and secondary ion mass spectroscopy (SIMS), sometimes used in conjunction with chemical etching, are used mainly to study the *static* properties of surfaces. They are extensively reviewed in the books mentioned earlier in this section.

1.3 Theoretical methods for excitations

Here we summarize some of the theoretical considerations relevant for studying bulk and surface excitations. The details of the calculations will, of course, depend on the type of excitation being considered (e.g. phonons, magnons etc.), and several different cases are dealt with individually in the subsequent chapters. Nevertheless, it is helpful to mention some features that they have in common. The following will refer to *linear* excitations; the techniques employed to calculate properties of nonlinear excitations will be covered later (in chapters 10 to 12).

1.3.1 Dispersion relations

In many cases one is interested only in calculating the dispersion relations (i.e. frequency versus wave vector relations) for the bulk and surface excitations. As discussed in section 1.1.2, the frequency of a bulk excitation depends on a 3D wave vector $\mathbf{q} = (\mathbf{q}_{\|}, q_z)$, while the frequency of a surface excitation depends on a 2D wave vector $\mathbf{q}_{\|}$. The surface mode is also characterized by one or more *attenuation factors* (usually dependent on $\mathbf{q}_{\|}$). The calculation of the excitation frequencies normally involves obtaining a set of *homogeneous* equations satisfied by the amplitude variable, which we denote here by $u(\mathbf{r}, t)$, describing the excitation. For example, for a phonon $u(\mathbf{r}, t)$ would simply be an atomic displacement and for a magnon (or spin wave) $u(\mathbf{r}, t)$ would be a component of a spin vector. The next step is to utilize symmetry considerations in accordance with Bloch's theorem: if all the surfaces and interfaces are planar and perpendicular to the z axis, then $u(\mathbf{r}, t)$ can be written as $\exp[i(\mathbf{q}_{\|} \cdot \mathbf{r}_{\|} - \omega t)] f(z)$ as in (1.28). This leads to a set of homogeneous equations to be solved for the spatial dependence of $f(z)$.

To reach this stage we may have followed either a microscopic approach or a continuum (macroscopic) approach. In the former case, the discrete lattice structure is taken into account when forming the equations of motion for $u(\mathbf{r}, t)$, usually from a Hamiltonian. This leads to a set of *finite-difference equations* satisfied by $f(z)$ in the different lattice planes parallel to the surface. In the latter case, however, the medium is treated as a continuum and the description of the excitation is in terms of macroscopic variables. The approach is appropriate only for small wave vectors $\mathbf{q}$ or $\mathbf{q}_{\|}$, such that the excitation wavelength (and the attenuation length in the case of a surface excitation) are large compared with the interatomic distances. This is usually a simplification that leads to a *differential equation* for $f(z)$, rather than finite-difference equations. An example of a continuum calculation was already given in section 1.1.2 for vibrational waves in a semi-infinite elastic medium. We shall use both of the above methods in the following chapters. The connection between the two methods is explored in particular detail for the case of magnons in semi-infinite ferromagnets and ferromagnetic films (see chapters 4 and 5).

Solutions of the differential equation, or the set of finite-difference equations, for $f(z)$ may be found by various standard mathematical techniques. In many examples (such as the semi-infinite isotropic elastic medium mentioned earlier) it is convenient to study the bulk excitations by seeking wave-like solutions of the form (1.30). This introduces the real wave vector q_z in the z direction. For a surface excitation in a semi-infinite medium (occupying the half-space $z < 0$), one would seek localized solutions for $f(z)$ having the form of (1.32). In general, κ is not necessarily real, but it must satisfy $\mathrm{Re}(\kappa) > 0$ to ensure the decay of $|\exp(\kappa z)|$ as $z \to -\infty$ in the medium. More generally, for a medium of finite thickness (e.g. a slab of thickness L confined between the surface planes $z = 0$ and $z = -L$) it would be necessary to generalize (1.32) for the surface excitation to

$$f(z) = B_3 \exp(\kappa z) + B_4 \exp(-\kappa z). \tag{1.36}$$

The boundary conditions would need to be applied at the surfaces $z = 0$ and $z = -L$.

1.3.2   Linear-response theory

For many applications one may need to know more about the excitations than just their dispersion relations: it may be important to determine the statistical weightings of the bulk and surface excitations and the dependences on the wave vector $q_\parallel$. For example, the statistical weighting would enter into the calculation of the thermodynamic properties of the excitations or the intensities measured by inelastic scattering (e.g. of light or a beam of particles) from the excitations.

Linear-response theory is a convenient method for investigating the spectrum of excitations. Basically it involves calculating the response of a system to a small applied stimulus. The result may be expressed in terms of a *response function*. This provides information about the dispersion relation of any excitation and, in addition, the power spectrum of the thermally excited fluctuations in the excitation amplitude may be deduced. The response functions are directly related to quantum-mechanical Green functions, which can be calculated by various other methods. However, linear-response theory has a more direct physical appeal than many of the alternative methods.

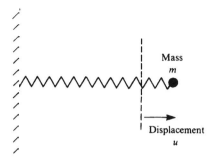

Figure 1.8. The mechanical damped harmonic oscillator used to illustrate linear-response theory.

It is helpful to give a simple mathematical example to illustrate some of the basic concepts of linear-response theory. The more formal aspects, including the relation to Green functions, are treated in the Appendix. General accounts of linear-response theory have been given by Landau and Lifshitz (1980) and Lovesey (1986). Applications to surface problems have been comprehensively treated by Cottam and Maradudin (1984). Other references are quoted in the Appendix.

Consider a 1D damped harmonic oscillator, where u specifies the instantaneous displacement of a mass m subject to an elastic restoring force and a damping force (see figure 1.8). We let $f(t)$ be a fictitious force that acts on the mass in such a way that the interaction Hamiltonian has the following simple form, linear in both $f(t)$ and the displacement:

$$H = -uf(t). \tag{1.37}$$

The driven equation of motion for the oscillator can then be written as

$$m\left(\frac{\mathrm{d}^2 u}{\mathrm{d}t^2} + \Gamma\frac{\mathrm{d}u}{\mathrm{d}t} + \omega_0^2 u\right) = f(t) \tag{1.38}$$

where ω_0 denotes the natural resonance frequency and the term proportional to Γ describes the effect of damping ($\Gamma > 0$). Fourier transforms of the time-dependent quantities $u(t)$ and $f(t)$ can be defined by

$$U(\omega) = \frac{1}{2\pi}\int_{\infty}^{\infty} u(t)\exp(\mathrm{i}\omega t)\,\mathrm{d}t \tag{1.39}$$

$$F(\omega) = \frac{1}{2\pi}\int_{-\infty}^{\infty} f(t)\exp(\mathrm{i}\omega t)\,\mathrm{d}t \tag{1.40}$$

where ω is a label for angular frequency.

Suppose the Fourier-transformed displacement $U(\omega)$ is averaged over an ensemble of oscillators (with randomly assigned phases) to give a value denoted by $\bar{U}(\omega)$. In the absence of a driving force we would have simply $\bar{U}(\omega) = 0$, but when $F(\omega)$ is non-zero $\bar{U}(\omega)$ acquires a nonzero value that

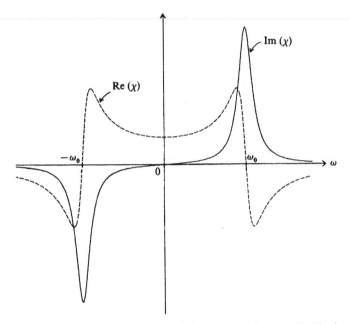

Figure 1.9. The real and imaginary parts of the response function (1.42) plotted as a function of ω, taking $\Gamma/\omega_0 = 0.2$.

is linear in $F(\omega)$. We may use this to define a linear-response function (or generalized susceptibility function) by

$$\bar{U}(\omega) = \chi(\omega)F(\omega). \tag{1.41}$$

From (1.38)–(1.41) it is easy to show that

$$\chi(\omega) = \frac{1}{m(\omega_0^2 - \omega^2 - i\Gamma\omega)}. \tag{1.42}$$

This complex function of ω displays a resonant behaviour at $\omega = \pm\omega_0$, as can be seen from figure 1.9 where the real and imaginary parts of $\chi(\omega)$ are plotted. Note that the symmetry property

$$\chi(-\omega) = \chi^*(\omega) \tag{1.43}$$

is satisfied, implying that Re $[\chi(\omega)]$ and Im $[\chi(\omega)]$ are respectively symmetric and antisymmetric functions of ω. The asterisk in (1.43) denotes complex conjugation. Equation (1.41) may be written in an alternative (but less convenient) form as

$$\bar{u}(t) = \int_0^\infty X(t')f(t - t')\,dt' \tag{1.44}$$

where $X(t')$ is the time-dependent Fourier transform of $\chi(\omega)$. The limits of integration in (1.44) are in accordance with causality: the value of $\bar{u}$ at time t can depend only on the force at *preceding* times, i.e. $t' > 0$.

A close connection can be shown to exist between the response function $\chi(\omega)$ and the absorption (or dissipation) of energy from the fictitious force f. A clear and concise general description of this property is given by Landau and Lifshitz (1980). To illustrate this point, we continue with the example of the damped harmonic oscillator and take the case where the driving force has the simple form

$$f(t) = f_0 \cos(\omega t). \tag{1.45}$$

Using (1.39), (1.41) and (1.43), we may then prove

$$\bar{u}(t) = \tfrac{1}{2} f_0 [\chi(\omega) \exp(-i\omega t) + \chi^*(\omega) \exp(i\omega t)]. \tag{1.46}$$

With the interaction Hamiltonian taking the form given in (1.37), the average energy dissipation Q per unit time due to the force is

$$Q = -\bar{u}(t) \frac{\mathrm{d}}{\mathrm{d}t} f(t) = -\tfrac{1}{4} i f_0^2 \{ \chi(\omega)[1 - \exp(-2i\omega t)] + \chi^*(\omega)[1 - \exp(2i\omega t)] \} \tag{1.47}$$

using (1.45) and (1.46). The terms proportional to $\exp(-2i\omega t)$ and $\exp(-2i\omega t)$ give zero when averaged over the period $2\pi/\omega$ of the driving force, and the energy dissipation becomes

$$Q = \tfrac{1}{2} f_0^2 \omega \, \mathrm{Im}[\chi(\omega)]. \tag{1.48}$$

It can be checked directly from (1.42), or seen from figure 1.9, that $\omega \, \mathrm{Im}[\chi(\omega)] > 0$ so that $Q > 0$ as required on physical grounds.

Another connection with the dissipative properties comes from considering the mean square displacement $\langle u^2(t) \rangle$, where the angular brackets indicate a statistical mechanical average. The frequency Fourier transform $\langle u^2 \rangle_\omega$ of this quantity is often referred to as the *power spectrum* of the thermal fluctuations. It can be shown (e.g. see Landau and Lifshitz 1980) that $\langle u^2 \rangle_\omega$ is directly related to the energy dissipation Q per unit time, and so it follows from (1.48) that there is a relationship between $\langle u^2 \rangle_\omega$ and $\mathrm{Im}[\chi(\omega)]$. In the 'classical' (or high-temperature) regime of $k_\mathrm{B} T \gg \hbar \omega$ the result is

$$\langle u^2 \rangle_\omega = (k_\mathrm{B} T / \pi \omega) \, \mathrm{Im}[\chi(\omega)]. \tag{1.49}$$

This is an example of the *fluctuation–dissipation theorem*; its general quantum-mechanical form is proved in section A.3 of the Appendix, and in the present application it becomes

$$\langle u^2 \rangle_\omega = (\hbar / \pi)[n(\omega) + 1] \, \mathrm{Im}[\chi(\omega)] \tag{1.50}$$

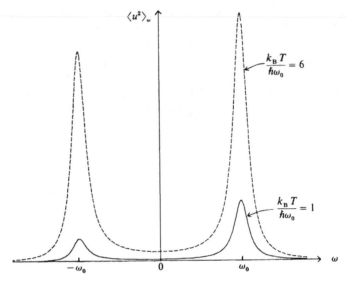

Figure 1.10. The power spectrum of $\langle u^2 \rangle_\omega$ versus ω for two different values of the ratio $k_B T/\hbar\omega_0$, taking $\Gamma/\omega_0 = 0.2$.

where $n(\omega)$ is the Bose–Einstein thermal factor:

$$n(\omega) = [\exp(\hbar\omega/k_B T) - 1]^{-1}. \qquad (1.51)$$

Equation (1.50) reduces to give (1.49) in the limit of $k_B T \gg \hbar\omega$. We may note that, in the classical limit, $\langle u^2 \rangle_\omega$ is a symmetric function of ω, but this is not the case generally for (1.50) because

$$n(-\omega) = -n(\omega) - 1. \qquad (1.52)$$

This behaviour is illustrated for two different temperatures in figure 1.10 using the response function $\chi(\omega)$ of (1.42).

In the context of excitations in solids the preceding mathematical example can be generalized as described in the Appendix. Briefly, the stimulus (or 'force') could be applied at one position and the response calculated at another position. This would yield a *position-dependent* response function (or Green function) by analogy with the relationship (1.41). The calculation will generally involve solving *inhomogeneous* equations (finite-difference equations or differential equations), whereas evaluating the dispersion relations as discussed in section 1.3.1 corresponded to the simpler task of solving homogeneous equations. The response functions exhibit a resonant behaviour corresponding to any bulk or surface excitations, and this property enables the dispersion relations to be deduced from them. Also, from the response functions together with the fluctuation–dissipation theorem, the power spectra of the various excitations can be calculated. An example of this type of calculation is given in section 3.2.3 where response functions (Green functions) are derived for acoustic waves in a semi-infinite elastic medium.

1.4   Experimental methods for excitations

It is appropriate now to describe some of the experimental methods used for studying the properties of the excitations that may occur in the vicinity of surfaces and interfaces. In doing so we are concerned with the *dynamics* of surfaces and interfaces, contrasting with the experimental methods discussed in section 1.2.3 that are used to characterize the static properties of surfaces. Certain techniques are specific to particular types of excitation and are best introduced in the context of the relevant later chapter, so at this stage we discuss only some of the methods of more general applicability. In particular, we emphasize the techniques of inelastic light scattering, which has proved to be an extremely sensitive probe of long-wavelength surface excitations, and various types of inelastic particle scattering.

1.4.1   Inelastic light scattering

Raman and Brillouin scattering of light by dense media were first demonstrated in the 1920s and 1930s, but it was not until the advent of the laser, together with other technical developments, that these methods were widely applied to bulk excitations in solids and liquids. In recent years they have been used to study *surface* excitations, particularly in relatively opaque materials where the light penetrates only to a limited surface region. General references for light scattering are the books by Hayes and Loudon (1978) and Cottam and Lockwood (1986), and there have been numerous review articles emphasizing the applications to surface excitations (e.g. Sandercock 1982, Cardona and Güntherodt 1989, Dutcher 1994).

The essential distinction between the techniques of Raman and Brillouin scattering lies in the frequency analysis employed for the scattered light. In Raman scattering this is achieved by use of a grating spectrometer, typically with shifts in the wave number of the light in the range 5–4000 cm^{-1}. Conversions of cm^{-1} to other frequency- and energy-related units are given in table 1.1. In Brillouin scattering a Fabry–Pérot interferometer is used and the wave number shifts are typically in a range up to about 5 cm^{-1}. The instrumental resolutions obtainable are generally of order 1 cm^{-1} in the case of conventional Raman scattering and several orders of magnitude smaller for Brillouin scattering. As an example, we show in figure 1.11 a schematic arrangement for Brillouin scattering off the surface of an opaque sample. It was the development of the high-contrast multipass Fabry–Pérot interferometer (Sandercock 1970) that made such experiments

Table 1.1. Conversion factors between frequency- and energy-related units

	Frequency (GHz)	Wave number (cm^{-1})	Electron volts (meV)	Temperature (K)
1 GHz	1	0.033 356	0.0041 357	0.047 99
1 cm^{-1}	29.979	1	0.123 99	1.438 8
1 meV	241.80	8.065 5	1	11.605
1 K	20.836	0.695 03	0.086 173	1

possible, and shortly after this the first measurements on Si and Ge surfaces were reported (see section 3.3.1). More recently it has become possible to achieve higher-resolution Raman scattering and to extend the range of wave-number shifts measurable (see e.g. Zhang *et al* 1992).

We consider first the kinematics of light scattering in a bulk (effectively infinite) transparent medium. The simplest processes involve the incident light of frequency ω_I and wave vector $\mathbf{k}_I$ creating or absorbing a single excitation of frequency ω and wave vector $\mathbf{q}$, thereby scattering into light of frequency ω_S and wave vector $\mathbf{k}_S$. These are represented in figures 1.12(a) and (b), and are known as the *Stokes* and *anti-Stokes processes*

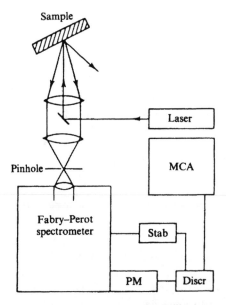

Figure 1.11. Schematic experimental arrangement for Brillouin scattering off the surface of a sample. The detection system includes a photomultiplier (PM), discriminator (Discr), stabilizer (Stab) and multichannel analyser (MCA). After Sandercock (1982).

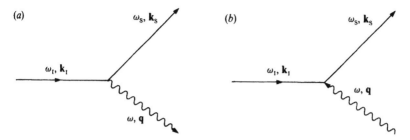

Figure 1.12. The (a) Stokes and (b) anti-Stokes scattering processes corresponding to creation and absorption, respectively, of an excitation (wavy line).

respectively. The conservation of energy and momentum imply that

$$\hbar\omega_I = \hbar\omega_S \pm \hbar\omega, \qquad \hbar k_I = \hbar k_S \pm \hbar q \qquad (1.53)$$

where the upper and lower signs refer to Stokes ($\omega_S < \omega_I$) and anti-Stokes ($\omega_S > \omega_I$) scattering respectively.

In many simple cases involving bulk media, the ratio of intensities for anti-Stokes and Stokes scattering, I_{AS}/I_S, is given by the thermal factor

$$I_{AS}/I_S = n(\omega)/[n(\omega) + 1] = \exp(-\hbar\omega/k_B T) \qquad (1.54)$$

where $n(\omega)$ is defined in (1.51). Hence anti-Stokes scattering is less intense, in general, than Stokes scattering from the same excitation. However, as emphasized by Loudon (1978), (1.54) holds only provided the light scattering mechanism has certain symmetry properties under time reversal, and exceptions to (1.54) have been observed (e.g. in certain bulk and surface magnetic systems).

The conservation conditions in (1.53) impose limitations on the wave vector q, so that only excitations near the centre of the Brillouin zone are detectable by light scattering. This follows from noting that the optical wave vectors are related to their frequencies by

$$|k_I| = \eta_I \omega_I/c, \qquad |k_S| = \eta_S \omega_S/c \qquad (1.55)$$

where η_I and η_S are the refractive indices corresponding to frequencies ω_I and ω_S. Typically $\omega \ll \omega_I$ and $\eta_S \simeq \eta_I$, from which it is a simple exercise to prove that $|q| \simeq 2\eta_I \omega_I \sin(\theta/2)/c$ where θ is the angle between k_I and k_S.

We now turn to the backscattering of light from the surface of a semi-infinite medium, taken as before to be in the half-space $z < 0$. A scattering geometry with the incident and scattered light beams in the same vertical plane (the xz plane) is assumed, as shown in figure 1.13. Medium 1 has real constant relative permittivity ε_1 (typically $\varepsilon_1 = 1$ corresponding to vacuum or air) and medium 2 (the scattering medium) has relative permittivity $\varepsilon_2(\omega)$ which may be frequency dependent. The incident light beam of wave vector k_{1I} in medium 1 is partially transmitted through the surface to medium

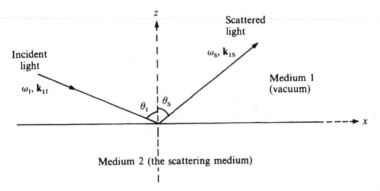

Figure 1.13. Assumed geometry for light scattering from the surface of a semi-infinite medium (medium 2). The incident and scattered light beams in medium 1 are indicated.

2 where the wave vector becomes $\mathbf{k}_{2I}$. For the wave-vector components perpendicular to the surface we have

$$k_{1I}^{z} = (\omega_I/c)\varepsilon_1^{1/2}\cos\theta_I \tag{1.56}$$

$$k_{2I}^{z} = (\omega_I/c)[\varepsilon_2(\omega) - \varepsilon_1\sin^2\theta_I]^{1/2} \tag{1.57}$$

while the property of translational invariance parallel to the surface implies

$$\mathbf{k}_{1I.\|} = \mathbf{k}_{2I.\|} = ((\omega_I/c)\varepsilon_1^{1/2}\sin\theta_I, 0) \equiv \mathbf{k}_{I.\|}. \tag{1.58}$$

There are similar expressions for wave vectors $\mathbf{k}_{1S}$ and $\mathbf{k}_{2S}$ of the scattered light beam, but with ω_I and θ_I replaced by ω_S, and θ_S respectively.

The light-scattering process is subject to the same energy conservation condition as in (1.53) for a bulk medium, but the second part of (1.53) is modified to

$$\hbar\mathbf{k}_{I,\|} = \hbar\mathbf{k}_{S,\|} \pm \hbar\mathbf{q}_{\|}. \tag{1.59}$$

This determines $\mathbf{q}_{\|}$, while the other wave-vector component q_z of the excitation is not fixed because the semi-infinite system does not possess translational invariance symmetry in the z direction. Consequently there is a spread of values of the q_z wave vector component, and this will lead to a broadened peak being observed in the light-scattering spectrum if the excitation frequency depends explicitly on q_z. This effect is called *opacity broadening* since it depends on the optical absorption in the medium, and examples are given in chapters 3 and 5.

The connection between theory and experiment is provided by the *scattering cross section* σ, which is defined as the rate at which energy is removed from the incident light beam by the scattering, divided by the power flow in the incident beam. A proper calculation of σ for a semi-infinite scattering medium must take account of the possibility of scattering from either bulk or surface excitations and of the transmission of the incident and scattered light beams through the surface at $z = 0$. The formalism is described in detail by Mills *et al* (1970) and Bennett *et al* (1972); here we outline a derivation following Nkoma and Loudon (1975), Cottam (1976) and Tilley (1980). If $\bar{A}$ is the area of the sample surface through which the scattered beam emerges, the beam cross-sectional area is $\bar{A} \cos \theta_S$ and from the definition we have

$$\sigma = \left(\frac{\omega_I}{\omega_S}\right) \frac{\bar{A} \cos \theta_S \langle |\mathbf{E}_{S1}|^2 \rangle}{|\mathbf{E}_{I1}|^2}. \tag{1.60}$$

Here the factor (ω_I/ω_S) allows for the change in quantum energy of the scattered photons compared with the incident photons, and $\mathbf{E}_{I1}$ and $\mathbf{E}_{S1}$ are the incident and scattered electric fields, respectively, in medium 1 ($z > 0$) where the measurements are made. If we write for the scattered electric field

$$\mathbf{E}_{S1} = \sum_{\mathbf{Q}} \mathbf{E}_{S1}(\mathbf{Q}) \exp[\mathrm{i}(\mathbf{k}_{S,\|} \cdot \mathbf{r}_\| - k_{S1}^z z)] \tag{1.61}$$

where $\mathbf{r}_\| = (x, y)$ and the summation is over all wave vectors $\mathbf{Q}$ of the thermal excitations, the cross section becomes

$$\sigma = \frac{\omega_I \bar{A}}{|\mathbf{E}_{I1}|^2} \sum_{\mathbf{Q},\mathbf{Q}'} \frac{\cos \theta_S \langle \mathbf{E}_{S1}(\mathbf{Q}) \cdot \mathbf{E}_{S1}^*(\mathbf{Q}') \rangle}{\omega_S}. \tag{1.62}$$

The wave vectors $\mathbf{Q}$ and $\mathbf{Q}'$ can be split into their components parallel and perpendicular to the surface as $\mathbf{Q} = (\mathbf{Q}_\|, Q_z)$ and $\mathbf{Q}' = (\mathbf{Q}'_\|, Q'_z)$. The summations over $\mathbf{Q}_\|$ and $\mathbf{Q}'_\|$ may be carried out (by using the property of translational symmetry parallel to the surface) and the final result may eventually be expressed as a *differential cross section* $\mathrm{d}^2\sigma/\mathrm{d}\Omega \, \mathrm{d}\omega_S$. This describes the scattering of light into an elementary solid angle $\mathrm{d}\Omega$ with scattered frequency between ω_S and $\omega_S + \mathrm{d}\omega_S$:

$$\frac{\mathrm{d}^2\sigma}{\mathrm{d}\Omega \, \mathrm{d}\omega_S} = \frac{\varepsilon_I \omega_I \omega_S A \bar{A} \cos^2 \theta_S}{4\pi^2 c^2 |\mathbf{E}_{I1}|^2} \sum_{\mathbf{Q},\mathbf{Q}'} \langle \mathbf{E}_{S1}(\mathbf{q}_\|, Q_z) \cdot \mathbf{E}_{S1}^*(\mathbf{q}_\|, Q'_z) \rangle_{\omega_S} \tag{1.63}$$

where A is the surface area of the scattering medium, the 2D wave vector $\mathbf{q}_\|$ is given by the conservation condition (1.59), and the correlation function is evaluated at the frequency ω_S of the scattered light.

The next stage in evaluating $\mathrm{d}^2\sigma/\mathrm{d}\Omega \, \mathrm{d}\omega_S$ would be to relate the electric field variables in medium 1 to the corresponding quantities inside the scattering medium 2. The incident light transmitted to medium 2 can be regarded as interacting with a crystal excitation to produce a polarization. It is the radiation of light by the polarization that produces the scattered beam transmitted to medium 1. An example of such a calculation is given in section 3.3.1 using linear-response theory. In some cases there may be contributions to the light scattering due to a diffraction grating being ruled on the surface or due to a periodic corrugation ('surface ripple') of the surface caused by the propagating excitation.

Equation (1.63) for the differential cross section can be generalized to apply to multiple-interface geometries (see the references cited earlier) for comparison with experiments on thin films and multilayers.

1.4.2   Inelastic particle scattering

Although inelastic light scattering offers very high sensitivity and resolution for studying surface excitations, it is limited to those excitations with wave vectors very close to the Brillouin zone centre. Under appropriate conditions the inelastic scattering of particles by crystal excitations can involve larger momentum changes, and hence this type of scattering may provide a probe of surface excitations at wave vectors extending throughout the Brillouin zone (although generally with less favourable resolution than in light scattering). Here we shall be concerned mainly with the scattering of electrons and neutral atoms.

Electron energy loss spectroscopy (EELS) has proved to be successful for studying the dynamics of surfaces, i.e. the elementary excitations. For the early work in the 1960s, which was applied principally to plasmons and electronic transitions, the energy losses in the inelastic scattering process were typically many eV and the resolution was of order 250 meV. The development of high-resolution EELS (or HREELS) in the 1980s was motivated by studies of surface phonons and led to much improved sensitivity, corresponding to an energy resolution of ~ 7 meV (or ~ 55–60 cm^{-1} in wave-number units). Reviews are given by Ibach and Mills (1982), Feldman and Mayer (1986) and Woodruff and Delchar (1986). The technique can be used to measure dispersion relations of excitations for wave vectors throughout the entire 2D Brillouin zone associated with the layers parallel to the surface. This is exemplified by the work of Szeftel *et al* (1984) for surface vibrational modes (phonons) on Ni. In their experiments an incoming electron of kinetic energy E_I and wave vector $\mathbf{k}_I$ impinges on the surface, emits or absorbs a surface phonon of frequency ω and 2D wave vector $\mathbf{q}_\parallel$, and emerges with energy E_S and wave vector $\mathbf{k}_S$ (see figure 1.14). The conditions for conservation of energy and of momentum parallel to the surface give

$$E_I - E_S \pm \hbar\omega = 0 \tag{1.64}$$

$$\hbar k_I(\sin\theta_I - \sin\theta_S) = \hbar q_\parallel \tag{1.65}$$

where the upper and lower signs in (1.64) refer to annihilation and creation of a phonon, respectively, which in turn corresponds in the EELS spectrum to a gain and a loss. Equation (1.65) has been approximated by using $E_I \gg \hbar\omega$

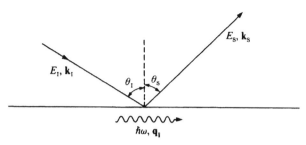

Figure 1.14. Geometry for inelastic particle scattering, used in the discussion of EELS and the scattering of atoms.

(typically $E_{\mathrm{I}} \approx 100\,\mathrm{eV}$ while $\hbar\omega$ might be of the order of tens of meV), implying $|\mathbf{k}_{\mathrm{I}}| \cong |\mathbf{k}_{\mathrm{S}}| \equiv k_{\mathrm{I}}$. Also by omitting a reciprocal lattice vector term from (1.65) we have ignored any Umklapp processes. By varying the angles θ_{I} or θ_{S} the 2D wave vector $\mathbf{q}_{\|}$ can be scanned over its whole Brillouin zone and the dispersion relation of ω versus $\mathbf{q}_{\|}$ can be deduced. Some experimental results are given in chapter 3.

Next we discuss the inelastic scattering of heavier particles, which would make the study of excitations at large wave vectors more easily accessible than is the case with electron scattering. Inelastic neutron scattering has been widely applied to investigate bulk excitations, but in the context of surface studies it has the disadvantage that neutrons interact only weakly with the atoms in the sample. Hence the penetration depth of the neutrons is large (several metres) and the technique is relatively insensitive to surface effects. There are two obvious ways in which the neutron intensity scattered from surfaces and interfaces can be enhanced: the first is to increase the amount of viewed surface, and the second is to select a favourable scattering geometry. Multilayer systems such as superlattices are an ideal way of achieving a large amount of surface or interface area. As regards the geometry, a method of enhancing the surface signal is to send the neutron beam at grazing incidence to the surface, and to measure the intensity of the Fresnel reflected beam. Recent progress along these lines, particularly for magnetic materials, has been reviewed by Schreyer *et al* (2000).

However, a technique that has so far proved more useful for studying surface excitations is inelastic scattering of light neutral atoms, such as He atoms. At low energies ($\sim$20 meV) the He atoms are scattered only by the first monolayer at the surface, and also their de Broglie wavelength is comparable with the lattice dimensions so that the entire Brillouin zone can be probed. It was the development of a He nozzle beam with good velocity resolution ($\Delta v/v < 0.01$) by Brusdeylins *et al* (1980, 1981) that first enabled time-of-flight (TOF) spectra to be obtained for atomic beam scattering from surface phonons. The apparatus consists of a beam source, a target chamber and a mass spectrometer detector at a fixed angle of 90°

to the incident beam (i.e. $\theta_I + \theta_S = 90°$ in figure 1.14). The beam is chopped into pulses before impinging on the target and producing a scattered signal. For the kinematics we may modify the preceding discussion of EELS. On putting $E_I = \hbar^2 k_I^2 / 2M$ and $E_S = \hbar^2 k_S^2 / 2M$ for the incident and scattered kinetic energies (M = mass of He atom), the energy conservation condition for a scattering process in which an excitation is created becomes

$$\frac{\hbar^2}{2M}(k_I^2 - k_S^2) + \hbar\omega = 0. \tag{1.66}$$

Conservation of in-plane momentum (for the 90° scattering geometry) gives

$$k_I \sin\theta_I - k_S \cos\theta_I = q_{||} \tag{1.67}$$

where we have again, for simplicity, ignored Umklapp processes. Note that, since $\hbar\omega$ is not necessarily small compared with E_I in atomic beam scattering, we cannot put $k_S \approx k_I$ as was done in (1.65) for EELS. Eliminating k_S from (1.66) and (1.67) yields the following relationship between ω and $q_{||}$:

$$\omega = \frac{\hbar^2 k_I^2}{2M}\left[\left(\frac{q_{||}}{k_I \cos\theta_I} + \tan\theta_I\right)^2 - 1\right]. \tag{1.68}$$

Hence a TOF spectrum provides a scan along a parabolic path in $(\omega, q_{||})$ space, from which the dispersion relation of the surface excitation may be deduced. Examples are discussed in chapter 3.

In spin-polarized EELS (or SPEELS) a polarized primary electron beam is employed, together with a polarization analyser in the scattered beam. It has been applied to study excitations at ferromagnetic surfaces (Hopster 1994).

1.4.3   Other techniques

Other more specific experimental methods, to be described later in this book, include measurements of thermodynamic properties, such as the vibrational or magnetic contributions to the surface specific heat. When such quantities are calculated (e.g. see chapter 4) they often have a temperature dependence that is different from that of the corresponding bulk effect. This is due to the localized surface excitations and/or surface perturbations in the density of states of the bulk excitations. Measurements of these temperature-dependent effects can, in appropriate cases, yield information about the excitations.

Surface excitations that generate electromagnetic fields can be investigated with optical techniques utilizing evanescent wave coupling between the juxtaposed surfaces of two media of different refractive index. This is the principle that underlies coupling by attenuated total reflection (ATR) and by gratings. These methods have been successfully applied to surface plasmons and surface polaritons, and we reserve discussion of the method until sections 6.3 and 6.4 of chapter 6. An important experimental development has been to use the coupling by means of ATR or a grating to produce greatly enhanced Raman scattering from surface excitations (see section 6.5).

Magnetic resonance techniques can be used to study the excitations (magnons) in ordered magnetic materials. In particular, spin wave resonance (SWR) in thin films can provide information about surface magnetic properties (see Dutcher 1994 for a review). The technique depends on applying an oscillating magnetic field to excite selectively magnons (or spin waves) with small but nonzero wave vector. This will be discussed in chapters 4 and 5.

References

Albuquerque E L and Cottam M G 2003 *Phys. Rep.* **376** 225

Ashcroft N W and Mermin N D 1976 *Solid State Phys.* (New York: Holt, Rinehart and Winston)

Bennett B I, Maradudin A A and Swanson L R 1972 *Ann. Phys. NY* **71** 357

Bland J A C and Heinrich B (eds) 1994 *Ultrathin Magnetic Structures I* (Berlin: Springer)

Bloch F 1928 *Z. Physik* **52** 555

Brusdeylins G, Doak R B and Toennies J P 1980 *Phys. Rev. Lett.* **44** 1417

Brusdeylins G, Doak R B and Toennies J P 1981 *Phys. Rev. Lett.* **46** 437

Cardona M and Güntherodt G (eds) 1989 *Light Scattering in Solids V* (Berlin: Springer)

Cottam M G 1976 *J. Phys. C* **9** 2137

Cottam M G and Lockwood D J 1986 *Light Scattering in Magnetic Solids* (New York: Wiley)

Cottam M G and Maradudin, A A 1984 in *Surface Excitations* ed V M Agranovich and R Loudon (Amsterdam: North-Holland) p 1

Dutcher J R 1994 in *Linear and Nonlinear Spin Waves in Magnetic Films and Superlattices* (Singapore: World Scientific) p 287

Esaki L and Tsu R 1970 *IBM J. Dev.* **14** 61

Feldman L C and Mayer J W 1986 *Fundamentals of Surface and Thin Film Analysis* (Amsterdam: North-Holland)

Foxon C T 1987 in *Interfaces, Quantum Wells, and Superlattices* ed C R Leavens and R Taylor (New York: Plenum) p 11

George J 1992 *Preparation of Thin Films* (New York: Marcel Dekker)

Grosso G and Parravicini G P 2000 *Solid State Physics* (San Diego: Academic)

Hayes W and Loudon R 1978 *Scattering of Light by Crystals* (New York: Wiley)

Hopster H 1994 in *Ultrathin Magnetic Structures I* ed J A C Bland and B Heinrich (Berlin: Springer) p 123

Ibach H and Mills D L 1982 *Electron Energy Loss Spectroscopy and Surface Vibrations* (New York: Academic)

Jaros M 1989 *Physics and Applications of Semiconductor Microstructures* (Oxford: Oxford University Press)

Kittel C 1995 *Introduction to Solid State Physics* 7th edition (New York: Wiley)

Landau L D and Lifshitz E M 1980 *Statistical Physics* (Oxford: Pergamon)

Laramore G E and Switendick A C 1973 *Phys. Rev. B* **7** 3615

Loudon R 1978 *J. Raman Spectrosc.* **7** 10

Lovesey S W 1986 *Condensed Matter Physics: Dynamic Correlations* 2nd edition (Menlo Park, CA: Benjamin)

MacDonald A H 1987 in *Interfaces, Quantum Wells, and Superlattices* ed C R Leavens and R Taylor (Plenum: New York) p 347

Madelung O 1981 *Introduction to Solid-State Theory* (Berlin: Springer)

Martin M R and Somorjai G A 1973 *Phys. Rev. B* **7** 3607

Mills D L, Maradudin A A and Burstein E 1970 *Ann. Phys. NY* **56** 504

Nkoma J S and Loudon R 1975 *J. Phys. C* **8** 1950

Prutton M 1983 *Surface Physics* 2nd edition (Oxford: Oxford University Press)

Rayleigh, Lord 1887 *Proc. Lond. Math. Soc.* **17** 4

Sandercock J R 1970 *Optics Commun.* **2** 73

Sandercock J R 1982 in *Light Scattering in Solids III* ed M Cardona and G Güntherodt (Berlin: Springer) p 173

Schreyer A, Schmitte T, Siebrecht R, Bödeker P, Zabel H, Lee S H, Erwin R W, Majkrzak C F, Kwo J and Hong M 2000 *J. Appl. Phys.* **87** 5443

Somorjai G A 1975 in *Surface Science* vol I (Vienna: IAEA) p 173

Szeftel J, Lehwald S and Ibach H 1984 *J. de Physique* **45** C5–109

Tilley D R 1980 *J. Phys. C* **13** 781

Tu K-N, Mayer J W and Feldman L C 1992 *Electronic Thin Film Science for Electrical Engineers and Materials Scientists* (New York: Macmillan)

Wherrett B S 1986 *Group Theory for Atoms, Molecules and Solids* (London: Prentice-Hall)

Whittaker E T and Watson G N 1963 *A Course of Modern Analysis* 4th edition (Cambridge: Cambridge University Press)

Woodruff D P and Delchar T A 1986 *Modern Techniques of Surface Science* (Cambridge: Cambridge University Press)

Zangwill A 1988 *Physics at Surfaces* (Cambridge: Cambridge University Press)

Zangwill A, Jesson D, Chambliss D and Clarke R (eds) 1996 *Evolution of Epitaxial Structure and Morphology* (Pittsburgh: MRS)

Zhang P X, Lockwood D J, Labbé H J and Baribeau J-M 1992 *Phys. Rev. B* **46** 9881

Chapter 2

Bulk models and properties

Most of the models used in the description of surfaces, films and super-lattices are based on corresponding models for bulk materials, so in this preliminary chapter we summarize the latter and review their properties. We deal in turn with lattice dynamics and elasticity theory, with magnetic Hamiltonians and with the frequency-dependent dielectric function (and magnetic susceptibility). Also, in sections 2.5 and 2.6, we discuss some of the implications for interaction with an electromagnetic field; in particular, we introduce the mixed mode called a *polariton* and see how its properties are manifested in the reflection of light from the bulk. This material then forms the basis for making the generalizations to the surface-related geometries in the subsequent chapters.

2.1 One-dimensional lattice dynamics

In a full discussion of lattice dynamics it is necessary to include the 3D character of the crystal structure in question. For our purposes, however, it is sufficient to review the properties of the simplest 1D models, described in terms of 'mass-and-spring' arrangements (see figure 2.1).

2.1.1 Monatomic lattice

The simplest case is the monatomic lattice shown in figure 2.1(a), where an infinite number of identical masses m are joined by identical springs of spring constant C. If u_n is the longitudinal displacement of mass n from its equilibrium position, the equation of motion is

$$m\frac{\partial^2 u_n}{\partial t^2} = C(u_{n+1} - u_n) + C(u_{n-1} - u_n) = C(u_{n+1} + u_{n-1} - 2u_n). \quad (2.1)$$

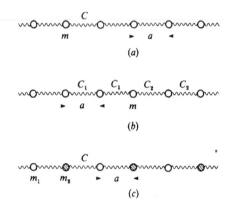

Figure 2.1. The 1D models discussed in section 2.1: (a) monatomic lattice with masses m coupled by springs C; (b) interface between two monatomic lattices; (c) diatomic lattice with alternating masses m_1 and m_2 coupled by springs C.

For normal mode solutions, all the masses vibrate at the same angular frequency: $u_n \propto \exp(-i\omega t)$. Then (2.1) becomes

$$-m\omega^2 u_n = C(u_{n+1} + u_{n-1} - 2u_n). \tag{2.2}$$

This equation is the typical member of an infinite set of coupled difference equations. The set is solved by use of Bloch's theorem (see section 1.1), according to which

$$u_n = u_0 \exp(inqa) \tag{2.3}$$

where a is the equilibrium distance between the masses. The solution corresponds to a wave travelling along the 1D array. Substitution of (2.3) into (2.2) gives

$$\omega^2 = (4C/m) \sin^2(qa/2). \tag{2.4}$$

This result is derived and discussed in all introductory books on solid-state physics. We pause only to note the limiting expressions

$$\omega^2 = (Ca^2/m)q^2 \quad \text{for } qa \ll 1$$
$$\omega^2 = 4C/m \qquad \text{when } qa = \pi. \tag{2.5}$$

The first of these shows that the mode is a sound wave, $\omega = vq$ with $v^2 = Ca^2/m$, in the long-wavelength limit, and the second gives the frequency at the Brillouin-zone boundary $q = \pi/a$ in 1D.

Two further consequences of (2.1) may be noted. First, one can pass directly from (2.1) to a continuum limit which is applicable at long wavelengths, $qa \ll 1$. Assuming slow spatial variation, we write

$$u_{n \pm 1} = u_n \pm a \frac{\partial u_n}{\partial z} + \frac{1}{2} a^2 \frac{\partial^2 u_n}{\partial z^2} \tag{2.6}$$

and (2.1) becomes

$$\frac{m}{a} \frac{\partial^2 u}{\partial t^2} = Ca \frac{\partial^2 u}{\partial z^2}. \tag{2.7}$$

This has the form of the 1D wave equation with density $\rho = m/a$ and elastic modulus $K = Ca$. It will be recalled that the velocity of sound is $v = (K/\rho)^{1/2}$, so (2.7) is consistent with the first part of (2.5).

For a second elaboration of (2.1) we consider an interface, as depicted in figure 2.1(b). We take all masses as equal, but the spring constant changes abruptly from C_1 to C_2 at one of them, say u_0. The equation of motion for the mass at the interface is

$$m \frac{\partial^2 u_0}{\partial t^2} = C_1(u_{-1} - u_0) + C_2(u_1 - u_0). \tag{2.8}$$

To lowest order in the continuum approximation, (2.6), this becomes

$$m \frac{\partial^2 u}{\partial t^2} = -C_1 a \frac{\partial u_-}{\partial z} + C_2 a \frac{\partial u_+}{\partial z} \tag{2.9}$$

where u_- and u_+ are the continuum displacements to left and right of the interface mass. In terms of continuum parameters, (2.9) is

$$\rho a \frac{\partial^2 u}{\partial t^2} = -K_1 \frac{\partial u_-}{\partial z} + K_2 \frac{\partial u_+}{\partial z}. \tag{2.10}$$

In frequency ω and wave vector q, the left side of (2.10) is of order ω^2 while the right side is of order q. Since ω and q are proportional in the acoustic mode under discussion, the left side is of second order in the small quantity qa and can therefore be neglected. Thus the equation of motion of the interface mass yields the continuum boundary condition

$$K_1 \frac{\partial u_-}{\partial z} = K_2 \frac{\partial u_+}{\partial z}. \tag{2.11}$$

It will be seen in section 2.2 that (2.11) ensures that the net force on the interface is zero. The connection between the equation of motion of an interface mass in a discrete formulation and the boundary condition in a continuum formulation will occur again in the discussion of superlattices in chapter 8.

2.1.2 Diatomic lattice

We now discuss the 1D diatomic lattice depicted in figure 2.1(c), first because it is the simplest model in which a surface mode arises and second to set up the formalism for the later discussion of superlattices. We take the equilibrium positions as $(2n+1)a$ for masses m_1 and $2na$ for masses m_2 ($n = $ integer). The equations of motion of adjacent masses are

$$m_1 \frac{d^2 u_{2n+1}}{dt^2} = C(u_{2n+2} + u_{2n} - 2u_{2n+1}) \tag{2.12}$$

$$m_2 \frac{d^2 u_{2n}}{dt^2} = C(u_{2n+1} + u_{2n-1} - 2u_{2n}). \tag{2.13}$$

Following Bloch's theorem, we write solutions

$$u_{2n+1} = U_1 \exp\{i[(2n+1)qa - \omega t]\} \tag{2.14}$$

$$u_{2n} = U_2 \exp[i(2qa - \omega t)]. \tag{2.15}$$

Substitution into (2.12) and (2.13) gives two alternative expressions for U_1/U_2 and equating these we find a quadratic equation for ω^2. The solution leads to the dispersion relation as it is usually quoted:

$$\omega^2 = C\left(\frac{1}{m_1} + \frac{1}{m_2}\right) \pm C\left[\left(\frac{1}{m_1} + \frac{1}{m_2}\right)^2 - \frac{4\sin^2(qa)}{m_1 m_2}\right]^{1/2}. \tag{2.16}$$

A typical graph relating ω to q is shown in figure 2.2.

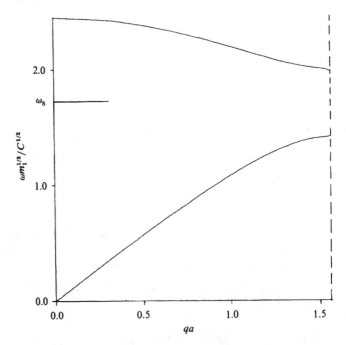

Figure 2.2. Dispersion graph, ω versus q, for the diatomic lattice of figure 2.1(c). Mass values $m_2 = 0.5m_1$. The surface-mode frequency $\omega_s = C^{1/2}(m_1^{-1} + m_2^{-1})^{1/2}$, to be derived in chapter 3, is shown for reference.

Like (2.4) for the monatomic lattice, (2.16) and figure 2.2 are discussed in solid-state texts. The *acoustic branch* occupies the frequency interval $\omega \le (2C/m_1)^{1/2}$ and the *optic branch* occupies $(2C/m_2)^{1/2} \le \omega \le [2C(m_1 + m_2)/m_1 m_2]^{1/2}$. The interval $(2C/m_1)^{1/2} < \omega < (2C/m_2)^{1/2}$ is a *stop band* for bulk modes.

In order to prepare for later chapters we need to go further and probe the nature of the stop band; as well as the high frequency region $\omega > [2C(m_1 + m_2)/m_1 m_2]^{1/2}$. Equation (2.16) can be reorganized as an equation for qa:

$$4C^2 \sin^2(qa) = 2C(m_1 + m_2)\omega^2 - m_1 m_2 \omega^4. \qquad (2.17)$$

The solutions for real q are those shown in figure 2.2. However, in the stop band and in the high-frequency region (2.17) admits solutions for *complex* q. In the former range (2.17) gives $\sin^2(qa) > 1$ while in the latter $\sin^2(qa) < 0$. The corresponding expressions for q are $\pi/2a + \mathrm{i}y$ and $\mathrm{i}y$, where y is real. It will be seen later that these complex values of q play an important part in surface and superlattice modes.

2.2 Bulk elasticity theory

In the previous section we showed for the 1D monatomic crystal that in the long-wavelength limit the difference equations for the atomic displacements go over into the elastic wave equation. Furthermore, the equation of motion of an interface atom generates the boundary condition of elasticity theory. In a similar way, for long wavelengths the difference equations of motion of masses in a 3D crystal become the elastic equations of motion, and the equations of motion of interface masses yield the elasticity-theory boundary conditions. The interested reader is referred to Maradudin *et al* (1971) for details. Here we follow an easier path starting from the standard equations of an elastic medium. It should be noted that, like (2.7), the equations of elasticity apply only to the small q part of the dispersion curve in the acoustic-mode branch. The notation we use is that of Landau and Lifshitz (1970).

2.2.1 General formalism

The state of an elastic medium is characterized by two symmetric second-rank tensors, namely the strain $\bar{u}_{ij}$ and the stress σ_{ij}. With $\mathbf{u}(\mathbf{r})$ denoting the displacement from equilibrium at position $\mathbf{r}$, the strain is defined by

$$\bar{u}_{ij} = \frac{1}{2}\left(\frac{\partial u_i}{\partial x_j} + \frac{\partial u_j}{\partial x_i}\right). \qquad (2.18)$$

The definition of the stress is that on a surface of area df whose normal is the unit vector $\hat{\mathbf{n}}$, component i of the force is

$$F_i = \sum_k \sigma_{ik} \hat{n}_k \, df = \sigma_{ik} \hat{n}_k \, df \qquad (2.19)$$

where the tensor σ_{ij} is symmetric ($\sigma_{ij} = \sigma_{ji}$). The basic assumption of linear elasticity theory is the generalized Hooke's Law that stress and strain are linearly related, that is,

$$\sigma_{ij} = \sum_{k,l} \lambda_{ijkl} \bar{u}_{kl} = \lambda_{ijkl} \bar{u}_{kl}. \qquad (2.20)$$

This introduces the elasticity tensor λ_{ijkl}. In the second part of (2.19) and (2.20) we have introduced the *summation convention*, which will be used subsequently when convenient. The convention is that where an index is repeated, summation over that index is implied. Thus in $\sigma_{ik} \hat{n}_k \, df$ summation is over k, and in $\lambda_{ijkl} \bar{u}_{kl}$ summation is over k and l.

Equation (2.20) shows that λ_{ijkl} is of fourth rank, and since σ_{ij} and $\bar{u}_{ij}$ are symmetric, λ_{ijkl} is symmetric on interchange of i and j and of k and l. It also satisfies $\lambda_{ijkl} = \lambda_{klij}$. Given these symmetries, it can be seen that λ_{ijkl} may have up to 21 independent components. However, in a crystal the point-group symmetry generally reduces this number; the forms of λ_{ijkl} for all the different crystal classes are given by Lovett (1999), for example. We shall consider only isotropic media, for which there are just two independent components. These are conventionally taken as Young's modulus E and Poisson's ratio σ, or sometimes as the Lamé parameters λ and μ. The stress–strain relations are

$$\sigma_{xx} = \frac{E}{(1+\sigma)(1-2\sigma)} [(1-\sigma)\bar{u}_{xx} + \sigma(\bar{u}_{yy} + \bar{u}_{zz})]$$

$$= (\lambda + 2\mu)\bar{u}_{xx} + \lambda(\bar{u}_{yy} + \bar{u}_{zz}) \qquad (2.21)$$

$$\sigma_{xy} = \frac{E}{1+\sigma} \bar{u}_{xy} = 2\mu\bar{u}_{xy} \qquad (2.22)$$

with relations for σ_{yy}, σ_{yz} etc. being found by appropriate permutation of suffixes. The alternative sets of parameters are related by

$$\lambda = \frac{E\sigma}{(1-2\sigma)(1+\sigma)}, \qquad \mu = \frac{E}{2(1+\sigma)}. \qquad (2.23)$$

Poisson's ratio satisfies the inequality $-1 \leq \sigma \leq \frac{1}{2}$ for reasons of hydrostatic stability, and in practice the stronger inequality $0 \leq \sigma \leq \frac{1}{2}$ is satisfied.

The form of propagating waves in a bulk elastic medium is found from the equation of motion for a mass element, namely

$$\rho \frac{\partial^2 \mathbf{u}}{\partial t^2} = \sum_k \frac{\partial \sigma_{ik}}{\partial x_k}. \qquad (2.24)$$

With the use of (2.21) and (2.22) this becomes

$$\rho \frac{\partial^2 \mathbf{u}}{\partial t^2} = \frac{E}{2(1 + \sigma)} \nabla^2 \mathbf{u} + \frac{E}{2(1 + \sigma)(1 - 2\sigma)} \nabla(\nabla \cdot \mathbf{u}). \tag{2.25}$$

It can be seen from this that longitudinal and transverse waves propagate with different velocities. A longitudinal wave satisfies $\nabla \times \mathbf{u}_L = 0$, since for a plane wave this implies that $\mathbf{q} \times \mathbf{u}_L = 0$, i.e. the displacement $\mathbf{u}_L$ is parallel to the propagation vector $\mathbf{q}$. Hence $\nabla \times (\nabla \times \mathbf{u}_L) = 0$, so that $\nabla(\nabla \cdot \mathbf{u}_L) = \nabla^2 \mathbf{u}_L$ and (2.25) reduces to the ordinary wave equation

$$\frac{\partial^2 \mathbf{u}_L}{\partial t^2} = v_L^2 \nabla^2 \mathbf{u}_L \tag{2.26}$$

where the longitudinal velocity v_L is given by

$$v_L^2 = \frac{E(1 - \sigma)}{\rho(1 + \sigma)(1 - 2\sigma)}. \tag{2.27}$$

For a transverse wave $\nabla \cdot \mathbf{u}_T = 0$, since for a plane wave this gives $\mathbf{q} \cdot \mathbf{u}_T = 0$. The last term in (2.25) then vanishes, so that the transverse velocity is given by

$$v_T^2 = \frac{E}{2\rho(1 + \sigma)}. \tag{2.28}$$

Equations (2.27) and (2.28), together with the inequality satisfied by σ, show that

$$v_L \geq \sqrt{2} v_T. \tag{2.29}$$

2.2.2 Reflection of acoustic waves at a free surface

As a preliminary to the later discussions of surface and superlattice modes, we now deal with the reflection of bulk acoustic waves at a free surface. We take the elastic medium to be in the half-space $z < 0$, with the surface as the plane $z = 0$. The surface plane is assumed to be stress-free, which in view of the definition of the stress tensor means $\sigma_{xz} = \sigma_{yz} = \sigma_{zz} = 0$. With the use of the relations (2.21) and (2.22) these may be written in terms of the strain tensor as

$$\bar{u}_{xz} = 0 \tag{2.30}$$

$$\bar{u}_{yz} = 0 \tag{2.31}$$

$$\sigma(\bar{u}_{xx} + \bar{u}_{yy}) + (1 - \sigma)\bar{u}_{zz} = 0. \tag{2.32}$$

Consider now a wave incident on the surface at some angle θ to the normal. The propagation vector $\mathbf{q}$ and the normal to the plane define the plane of incidence, which as shown in figure 2.3 we take as the xz plane. To begin

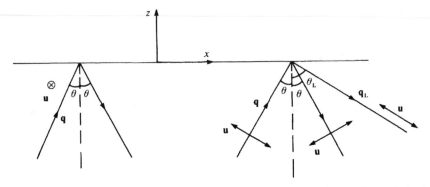

Figure 2.3. To illustrate reflection of acoustic waves at a free surface. Left: s-polarized transverse wave incident. Right: p-polarized transverse wave incident.

with, assume that the wave is transverse, with displacement $\mathbf{u}$ in the y direction; this is sometimes referred to as *s-polarization*. The incident wave is

$$\mathbf{u} = \mathbf{u}_0 \exp[i(q_x x + q_z z - \omega t)] \tag{2.33}$$

where

$$q_x/q_z = \tan\theta, \qquad q_x^2 + q_z^2 = \omega^2/v_T^2 \tag{2.34}$$

$$\mathbf{u}_0 = (0, a, 0). \tag{2.35}$$

The reflected wave is

$$\mathbf{u} = \mathbf{u}_1 \exp[i(q_x x - q_z z - \omega t)] \tag{2.36}$$

with $\mathbf{u}_1 = (0, b, 0)$. The same q_x appears in (2.33) and (2.36) because the boundary condition (2.31) must be satisfied for all x. The fact that the q_z values are the same follows from (2.35), implying that the angles of incidence and reflection are the same, as indicated in figure 2.3. Finally, application of the boundary condition (2.31) shows that $b = a$ for the amplitudes.

Matters become more complicated if we turn to an incident transverse wave polarized in the xz plane, known as *p-polarization*. Equations (2.33) and (2.34) continue to describe the incident wave, but now

$$\mathbf{u}_0 = (a, 0, -a\tan\theta). \tag{2.37}$$

We might attempt to describe the reflection by means of a reflected transverse wave given by (2.36) with $\mathbf{u}_1 = (b, 0, b\tan\theta)$. However, this introduces only one amplitude b, whereas two boundary conditions, (2.30) and (2.32), have to be satisfied. The problem is resolved by observing that, as sketched on the right of figure 2.3, a reflected longitudinal wave is generated in addition to the reflected transverse wave. In order that the boundary conditions may be satisfied for all x, this wave must have the same q_x value as the incident wave. It is therefore described by

$$\mathbf{u} = \mathbf{u}_2 \exp[i(q_x x - q_L z - \omega t)] \tag{2.38}$$

with

$$q_x/q_L = \tan \theta_L, \qquad q_x^2 + q_L^2 = \omega^2/v_L^2 \tag{2.39}$$

$$\mathbf{u}_2 = (c, 0, -c \cot \theta_L). \tag{2.40}$$

Since $v_L > v_T$, the above equations imply that $\theta_L > \theta$, as drawn in figure 2.3. In fact

$$\frac{\sin^2 \theta_L}{\sin^2 \theta} = \frac{v_L^2}{v_T^2}. \tag{2.41}$$

With the introduction of the longitudinal wave with independent amplitude c, it becomes possible to find the reflection coefficients b/a and c/a. This is left as an exercise.

One consequence of (2.41) should be noted. If the angle of incidence θ is larger than θ_c, where

$$\sin^2 \theta_c = v_T^2/v_L^2 \tag{2.42}$$

then $\sin^2 \theta_L > 1$. The meaning of this can be seen from (2.39). For $\theta > \theta_c$, $q_x^2 > \omega^2/v_L^2$, so (2.39) can be satisfied only if q_L^2 is negative. We put $q_L = i\kappa_L$ and the longitudinal wave of (2.38) becomes

$$\mathbf{u} = \mathbf{u}_2 \exp[i(q_x x - \omega t)] \exp(\kappa_L z). \tag{2.43}$$

This is a wave travelling along the surface, with amplitude decaying exponentially with distance into the medium. It is accompanied by the reflected transverse wave, and the combination is called a *pseudo-surface wave*.

We have discussed incident transverse waves of either polarization. An incident longitudinal wave generates reflected longitudinal and p-transverse waves whose amplitudes can be found in a similar way to those for an incident p-transverse wave.

2.2.3 Acoustic Green function in one dimension

So far we have been dealing with homogeneous equations of motion, for example (2.1) and (2.25). These led to dispersion equations, ω versus $\mathbf{q}$, like (2.4) and (2.16). Considerably more information is available from the calculation of a Green function, which is obtained as the solution of an *inhomogeneous* wave equation. The formal theory of Green functions, as far as it is needed, is given in the Appendix. In particular it is shown there that an appropriately defined linear-response function is the same as a

Green function, and that it may be used to calculate the power spectrum of thermally excited fluctuations. It was seen in section 1.4.1 that inelastic light scattering is governed by the fluctuation spectrum, so that an evaluation of the relevant Green function is central to the theory of any given light-scattering experiment. We now present the simple example of the calculation of the Green function for the 1D acoustic wave equation. A much more comprehensive account of acoustic Green functions is given by Cottam and Maradudin (1984).

We start with the equation for a 1D longitudinal wave travelling in the z direction (i.e. (2.26) with $\nabla^2 \to \partial^2/\partial z^2$) and we write the displacement simply as u. We augment the equation with a periodic point force of frequency ω at position z':

$$F = (f/A)\exp(-\mathrm{i}\omega t)\delta(z - z') \tag{2.44}$$

where the surface area in the xy plane is taken as A. The energy of interaction of this force with the elastic medium is

$$H_1 = -\int F(z)u(z)\,\mathrm{d}x\,\mathrm{d}y\,\mathrm{d}z = -u(z')f\exp(-\mathrm{i}\omega t) \tag{2.45}$$

since the integral over x and y gives A. The linear-response formalism of section A.2 then shows that the displacement at position z is related to f by the retarded Green function:

$$u(z) = \langle\langle u(z); u(z')^*\rangle\rangle_\omega f. \tag{2.46}$$

This is a specific illustration of the relation (A.19).

With the introduction of the driving force, the wave equation becomes

$$\rho\frac{\partial^2 u}{\partial t^2} + \frac{\rho}{\tau}\frac{\partial u}{\partial t} - \rho v_{\mathrm{L}}^2\frac{\partial^2 u}{\partial z^2} = \frac{f}{A}\exp(-\mathrm{i}\omega t)\delta(z - z'). \tag{2.47}$$

Here a damping term has been added. The displacement u must be proportional to $\exp(-\mathrm{i}\omega t)$ so (2.47) reduces to

$$\frac{\mathrm{d}^2 u}{\mathrm{d}z^2} + q^2 u = -\frac{f}{\rho v_{\mathrm{L}}^2 A}\delta(z - z') \tag{2.48}$$

where

$$q^2 = (\omega^2 + \mathrm{i}\omega/\tau)/v_{\mathrm{L}}^2. \tag{2.49}$$

A consequence of (2.49) is that the root with $\mathrm{Re}(q) > 0$ also has $\mathrm{Im}(q) > 0$. Thus, for example, the solution $\exp(\mathrm{i}qz)$ of the homogeneous wave equation travels in the $+z$ direction with attenuating amplitude.

The solution of (2.48) includes the complementary functions $\exp(\pm\mathrm{i}qz)$. However, with $\mathrm{Im}(q) > 0$ these diverge as $z \to \mp\infty$ and are therefore discarded if we require that u should be finite at infinity. We are left with the particular integral

$$u = \frac{if}{\rho v_L^2 q A} \exp(iq|z - z'|). \tag{2.50}$$

This is easily verified as follows. Away from $z = z'$, the form is $\exp[\pm iq(z - z')]$, which satisfies the homogeneous equation. As $z \to \pm\infty$, $u \to 0$ because $\mathrm{Im}(q) > 0$. Finally, the first derivative du/dz has a step discontinuity at $z = z'$, so the second derivative is proportional to $\delta(z - z')$. Its coefficient, as shown in (2.50), may easily be found, so (2.46) and (2.50) now give for the Green function

$$\langle\langle u(z); u(z')^* \rangle\rangle_\omega = (iq/2\rho\omega^2 A) \exp(iq|z - z'|) \tag{2.51}$$

where $\omega^2 + i\omega/\tau$ has been replaced by ω^2 in the denominator. This assumes small damping, $\omega\tau \gg 1$, and is well satisfied in the applications to light scattering in chapter 3.

It may be helpful to illustrate some general properties of Green functions with the aid of this simple example. First, the power spectrum of thermally excited acoustic waves is

$$\langle u(z)u(z')^* \rangle_\omega = (k_B T/\pi\omega) \, \mathrm{Im}\langle\langle u(z); u(z')^* \rangle\rangle_\omega$$

$$= (k_B T/2\pi\rho\omega^3 A)[q_1 \cos(q_1|z - z'|) - q_2 \sin(q_1|z - z'|)]$$

$$\times \exp(-q_2|z - z'|) \tag{2.52}$$

where we have written $q = q_1 + iq_2$ and used (A.31). Thus the correlations between the thermally excited field amplitudes at z and z' depend only on $z - z'$ (as follows from translational invariance) and have the form of damped oscillations.

The wavevector Fourier transform of (2.51) is

$$G(Q, \omega) = \int_{-\infty}^{\infty} \exp[-iQ(z - z')]\langle\langle u(z); u(z')^* \rangle\rangle_\omega \, d(z - z'). \tag{2.53}$$

It gives the response of the medium to a driving force proportional to $\exp(iQz - i\omega t)$. The integral is readily evaluated when the range is split into $(-\infty, 0)$ and $(0, \infty)$; the contributions at infinity vanish because $\mathrm{Im}(q) > 0$. The result is

$$G(Q, \omega) = \frac{1}{\rho\omega^2 A} \frac{q^2}{(Q^2 - q^2)}. \tag{2.54}$$

In the absence of damping ($\tau \to \infty$) this may be written

$$G(Q, \omega) = \frac{1}{\rho v_L^2 A} \frac{1}{(Q^2 - \omega^2/v_L^2)} \tag{2.55}$$

from which it is seen that the pole of $G(Q, \omega)$ is at $\omega = v_L Q$, the excitation energy of the system.

2.3   Bulk magnetism

Magnons (or spin waves) are the low-lying excitations that occur in ordered magnetic materials. The concept of spin waves was introduced by Bloch (1930), who envisaged some of the spins as deviating slightly from their ground-state alignment and this disturbance propagating with a wave-like behaviour through the solid. Since the spins are properly described by quantum-mechanical operators, the spin waves are quantized with the basic quantum being referred to as the *magnon* (by analogy with the phonon for quantized lattice vibrations). Magnons can be studied through their contribution to thermodynamic properties like specific heat or directly by techniques such as light scattering, neutron scattering and magnetic resonance.

In this section we give an introductory account of the theoretical models required to describe magnons in bulk ferromagnets and antiferromagnets. We then discuss the magnon spectra for three of the most important models, namely Heisenberg (or exchange) model, the dipole-exchange model and the Ising model in a transverse field. Apart from its status in magnetism, the last case is relevant to ferroelectricity, since it is a model for order–disorder ferroelectrics like potassium dihydrogen phosphate (KDP).

2.3.1   Magnetic Hamiltonians

Ferromagnets are characterized by an interaction between neighbouring electronic spins that gives rise to a parallel alignment at low enough temperatures. In simple antiferromagnets the sign of the interaction is such that antiparallel ordering of spins is favoured. As the temperature is raised the long-range magnetic order in either material decreases, until eventually there is a phase transition to a paramagnetic state. The critical temperature at which this occurs is known as the Curie temperature T_C in ferromagnets and the Néel temperature T_N in antiferromagnets.

The nature of the interaction that is responsible for the magnetic ordering was first explained by Heisenberg (1928), who was able to show that it is electrostatic in origin and due to quantum-mechanical exchange. Details are given in most textbooks on quantum mechanics or magnetism (e.g. Bransden and Joachain 2000, White 1983), but the essential steps of the argument may be summarized as follows. We consider two neighbouring ions, each of which has one electron. The electrons have spin-$\frac{1}{2}$ and obey Fermi–Dirac statistics. Therefore their total wavefunction, which may be expressed as the product of a spatial part ψ and a spin part χ_S, must be antisymmetric with respect to exchange of the electrons. The two possible spin states, which are the symmetric $\chi_{S=1}$ and the antisymmetric $\chi_{S=0}$ with S denoting the total spin quantum number, have to be combined respectively with the anti-symmetric (ψ_-) and symmetric (ψ_+) spatial wavefunctions. Expressions for ψ_- and ψ_+ can be written down in the Heitler–London theory, and are different provided there is appreciable overlap of individual electronic wave-functions. Consequently when the total energy is calculated from first-order perturbation theory, there is a difference of energy between the $S = 1$ and $S = 0$ states. With an appropriate choice of energy origin this can be represented by the operator $-J\mathbf{S}_1 \cdot \mathbf{S}_2$ where $\mathbf{S}_1$ and $\mathbf{S}_2$ are spin operators for the two electrons. The quantity J is known as the *exchange interaction*, and its magnitude is a measure of the degree of overlap of the electronic wavefunctions. If $J > 0$ the $S = 1$ state will be favoured, corresponding to parallel spins, as in ferromagnetism. Similarly $J < 0$ is associated with anti-ferromagnetism. When the above simplified argument is generalized to the whole system of spins in a magnetic medium we arrive at the Heisenberg Hamiltonian

$$H = -\sum_{\langle i, j \rangle} J_{ij}\mathbf{S}_i \cdot \mathbf{S}_j \qquad (2.56)$$

where the summation is over all distinct pairs i and j. The exchange inter-action J_{ij} is short range, because it depends on the wavefunction overlap, and it is often sufficient to consider only nearest-neighbour pairs i and j. A proper treatment of exchange is much more complicated than indicated here: the subject has been reviewed by Anderson (1963), Herring (1966) and Mattis (1981).

More generally, in some magnetic materials the Heisenberg Hamilto-nian may take the anisotropic form

$$H = -\sum_{\langle i, j \rangle} J_{ij}(\varepsilon_x S_i^x S_j^x + \varepsilon_y S_i^y S_j^y + \varepsilon_z S_i^z S_j^z) \qquad (2.57)$$

with dimensionless parameters ε_x, ε_y and ε_z. Cases of particular interest are $\varepsilon_x = \varepsilon_y = \varepsilon_z = 1$ (the isotropic Heisenberg Hamiltonian) and $\varepsilon_x = \varepsilon_y = 0$, $\varepsilon_z = 1$ (the Ising Hamiltonian). As mentioned, the latter is a basic model in ferroelectricity.

In addition to the exchange interaction there are magnetic dipole–dipole interactions. Each spin $\mathbf{S}_i$ has magnetic moment $g\mu_B\mathbf{S}_i$, and the classical dipole–dipole interaction between these moments gives a contribution to the energy of (see e.g. Jackson 1998)

$$\frac{\mu_0}{4\pi}g^2\mu_B^2 \sum_{\langle i, j \rangle} \left(\frac{\mathbf{S}_i \cdot \mathbf{S}_j}{r_{ij}^3} - \frac{3(\mathbf{S}_i \cdot \mathbf{r}_{ij})(\mathbf{S}_j \cdot \mathbf{r}_{ij})}{r_{ij}^5} \right) \qquad (2.58)$$

where $\mathbf{r}_{ij}$ is the vector joining sites i and j. Here μ_0 is the permeability of free space, g is the Landé g-factor and μ_B is the Bohr magneton. Typically the dipole–dipole interactions are much weaker than the exchange interactions (since usually $\mu_0 g^2 \mu_B^2 / r_{ij}^3 \ll J_{ij}$ for neighbouring ions) and have a small effect on the *static* magnetic properties such as the magnetization. Similarly they produce only a small perturbation in the magnon frequency for most wavevectors $\mathbf{q}$ in the Brillouin zone. However, in the limit of small $\mathbf{q}$ (i.e. long wavelengths) the influence of dipole–dipole interactions on the magnon frequency becomes important and eventually dominates. This is essentially because the exchange is short range whilst the dipole–dipole interactions are long range. Typically exchange interactions are dominant for magnons with $|\mathbf{q}| > 10^8 \, \mathrm{m}^{-1}$, compared with about $10^{10} \, \mathrm{m}^{-1}$ for a Brillouin zone boundary wavevector. For $|\mathbf{q}| < 10^7 \, \mathrm{m}^{-1}$ the dipole–dipole interactions are dominant, and in the intermediate régime, typically $10^7 \, \mathrm{m}^{-1} < |\mathbf{q}| < 10^8 \, \mathrm{m}^{-1}$, the two types of interaction may be comparable. The above numbers are approximate and may vary for different magnetic materials. For bulk effects exchange interactions are dominant in many cases and therefore in the present chapter we concentrate on the exchange-dominant case and treat the dipole–dipole interactions more briefly (see sections 2.3.4 and 2.6). The latter, however, are of great importance in surface and superlattice modes, as will be seen in chapters 5 and 9.

2.3.2 Magnons in bulk Heisenberg ferromagnets

As the simplest illustration we consider the magnons in an isotropic Heisenberg ferromagnet which is assumed to be infinite in extent. The Hamiltonian can be written

$$H = -\tfrac{1}{2} \sum_{i,j} J_{ij} \mathbf{S}_i \cdot \mathbf{S}_j - g\mu_B B_0 \sum_i S_i^z \qquad (2.59)$$

where the factor of $\tfrac{1}{2}$ in the first term has been included to allow for the fact that each pair of sites is counted twice in the summation over i and j. The second term describes the effect (the Zeeman energy) of an applied magnetic field B_0 in the z direction, which is also the direction of average spin alignment.

The magnon excitations can be calculated from (2.59) by a variety of different techniques (see e.g. White 1983), many of which involve transforming from spin operators to other operators such as boson creation and annihilation operators. However, in the present case it is convenient to work directly in terms of the spin operators since this will allow us to introduce temperature-dependent effects in a straightforward manner. We define operators S_j^+ and S_j^- at site j by

$$S_j^{\pm} = S_j^x \pm iS_j^y \qquad (2.60)$$

and the commutation relations satisfied by the spin operators can then be written as

$$[S_i^+, S_j^-] = 2S_i^z \delta_{ij}, \qquad [S_i^z, S_j^\pm] = \pm S_i^\pm \delta_{ij}. \tag{2.61}$$

Here and in later equations concerning magnetism we use units such that $\hbar = 1$. This convention, which has the effect of equating angular frequency and energy, simplifies notation significantly. Factors of $\hbar$ can be restored in final equations by dimensional analysis.

The notation in (2.61) is that $[A, B]$ is the commutator $(AB - BA)$ between any two operators A and B. We also require the equation of motion for any operator A, which from elementary quantum mechanics (e.g. Bransden and Joachain 2000) is

$$i \frac{dA}{dt} = [A, H], \qquad (\hbar = 1) \tag{2.62}$$

where H is the Hamiltonian. Choosing $A = S_j^+$, and using (2.59) and (2.61), we obtain

$$i \frac{d}{dt} S_j^+ = g\mu_B B_0 S_j^+ + \sum_i J_{ij}(S_i^z S_j^+ - S_j^z S_i^+). \tag{2.63}$$

The product of spin operators in (2.63) may be simplified by means of the *random-phase approximation* (RPA):

$$(S_i^z S_j^+ - S_j^z S_i^+) \rightarrow \langle S_i^z \rangle S_j^+ - \langle S_j^z \rangle S_i^+ \tag{2.64}$$

where the angular brackets $\langle \ldots \rangle$ denote a thermal average. Equation (2.64) represents a 'decoupling' of the product of operators, whereby each S^z is replaced by its nonzero thermal average. It should be a satisfactory approximation provided the spins are fairly well aligned in the z direction.

If we now define a wavevector Fourier transform by

$$S_j^+ = N^{-1/2} \sum_q S_q^+ \exp(i\mathbf{q} \cdot \mathbf{r}_j) \tag{2.65}$$

where N is the number of sites, and we use the property that $\langle S_i^z \rangle = \langle S_j^z \rangle = \langle S^z \rangle$ due to the equivalence of lattice sites, the approximated (or linearized) equation of motion becomes

$$i \frac{d}{dt} S_q^+ = \omega_B(\mathbf{q}) S_q^+. \tag{2.66}$$

This describes a simple harmonic oscillator with angular frequency $\omega_B(\mathbf{q})$ where

$$\omega_B(\mathbf{q}) = g\mu_B B_0 + \langle S^z \rangle [J(0) - J(\mathbf{q})]. \tag{2.67}$$

The quantity $J(\mathbf{q})$ is defined by

$$J_{ij} = N^{-1} \sum_{\mathbf{q}} J(\mathbf{q}) \exp[i\mathbf{q} \cdot (\mathbf{r}_i - \mathbf{r}_j)]. \qquad (2.68)$$

Typically it is found (as in the example below) that $J(\mathbf{q})$ decreases with increasing $|\mathbf{q}|$. Consequently $\omega_B(\mathbf{q})$, representing the bulk-magnon frequency, has its minimum value $g\mu_B B_0$ at $\mathbf{q} = 0$ and increases with $|\mathbf{q}|$ (for a fixed direction of $\mathbf{q}$). The precise form of $J(\mathbf{q})$ depends on the range of exchange interactions and on the crystal symmetry. In the case of a simple cubic lattice (with lattice parameter a) and exchange coupling J to the six nearest neighbours only, it is easily verified that

$$J(\mathbf{q}) = 2J[\cos(q_x a) + \cos(q_y a) + \cos(q_z a)]. \qquad (2.69)$$

For small values of $q = |\mathbf{q}|$ (such that $qa \ll 1$) the magnon dispersion relation for cubic systems takes the form

$$\omega_B(\mathbf{q}) = g\mu_B B_0 + Dq^2 + O(q^4) \qquad (2.70)$$

where expansion of the cosine functions shows that $D = \langle S^z \rangle J a^2$ in the case of (2.69).

It is seen from either (2.67) or (2.70) that the long-wavelength ($\mathbf{q} \to 0$) limit of the magnon frequency is the *ferromagnetic resonance frequency* $g\mu_B B_0$, which corresponds to the classical picture of all the spins precessing with uniform phase around the applied field B_0. For B_0 values of a few Tesla, the ferromagnetic resonance frequency lies in the microwave region so that ferromagnetic resonance and related effects are generally studied by the microwave technique of field sweep at fixed frequency.

As the temperature T is increased, the thermal average $\langle S^z \rangle$ (which is proportional to the magnetization) decreases monotonically. The temperature dependence of $\langle S^z \rangle$ can be estimated by mean-field theory (also known as molecular-field theory), which was first proposed by classical arguments (Weiss 1907). If spin fluctuations are ignored in (2.59), the mean-field Hamiltonian is obtained as

$$H_{MF} = -g\mu_B \sum_i [B_0 + B_E(i)]S_i^z \qquad (2.71)$$

where $B_E(i)$ is the average effective field acting on spin $\mathbf{S}_i$ due to the exchange interactions:

$$g\mu_B B_E(i) = \sum_j \langle S_j^z \rangle J_{ij}. \qquad (2.72)$$

There are no mean-field terms in (2.71) proportional to S_i^x or S_i^y because $\langle S_i^x \rangle$ and $\langle S_i^y \rangle$ vanish for the Hamiltonian (2.59). In the present case $B_E(i) = \langle S^z \rangle J(0)/g\mu_B$ independent of site i. For a system with spin $S = \frac{1}{2}$ the calculation of $\langle S^z \rangle$ is now straightforward. Corresponding to the

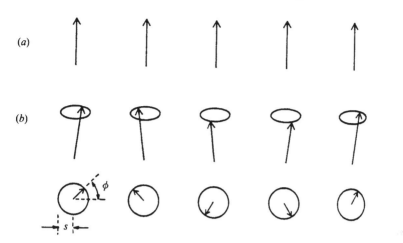

Figure 2.4. Semi-classical representation of magnons in a ferromagnet: (a) the ground state; (b) a bulk magnon, showing the precessing spin vectors viewed in perspective and from above.

eigenvalues $\pm\frac{1}{2}$ of the S^z operator, the mean-field energies from (2.71) are $\mp g\mu_B(B_0 + B_E)$, and it is easy to show that

$$\langle S^z \rangle = \tfrac{1}{2}\tanh[g\mu_B(B_0 + B_E)/2k_B T]. \tag{2.73}$$

Since B_E is proportional to $\langle S^z \rangle$ the above equation can in principle be solved self-consistently (e.g. numerically) to obtain $\langle S^z \rangle$ for any values of B_0 and T. For general values of S, the right-hand side of (2.73) would be written in terms of the Brillouin function for spin S (see e.g. Mattis 1981). It is predicted that $\langle S^z \rangle \to S$ as $T \to 0$, and $\langle S^z \rangle$ decreases as T is increased, falling to zero at $T_C = S(S + 1)J(0)/3k_B$ if the applied field B_0 is zero.

Although we have followed a quantum-mechanical approach here, magnons can also be interpreted semi-classically in terms of precessing spins, as represented schematically in figure 2.4. For bulk magnons the spins precess with constant amplitude s but varying phase angle ϕ, leading to a spin wave propagating through the crystal.

The preceding microscopic treatment is relatively straightforward for the Heisenberg model in a cubic lattice, but this is not necessarily the case for more complicated materials. It is therefore helpful to outline an alternative approach based on using a *continuum approximation* for the ferromagnetic medium. The method is similar to the transition from lattice dynamics to elasticity theory discussed in section 2.1.1, and it is applicable provided the wavelength of any excitation is large compared with the lattice spacing a, i.e. $qa \ll 1$. In contrast to the earlier approach, we choose to follow a classical derivation with the rate of change of the spin angular momentum equated to the torque acting upon it:

$$\frac{\mathrm{d}\mathbf{S}_i}{\mathrm{d}t} = \gamma(\mathbf{S}_i \times \mathbf{B}_i) \qquad (2.74)$$

where $\gamma(= g\mu_B/\hbar)$ is the gyromagnetic ratio, while $\mathbf{B}_i$ (an *instantaneous* effective field acting on $\mathbf{S}_i$) has components derived from the Hamiltonian (2.59) by

$$B_i^\alpha = -\gamma^{-1}\partial H/\partial S_i^\alpha \qquad (\alpha = x, y, z). \qquad (2.75)$$

Equation (2.74) can be regarded as the classical equivalent of the quantum-mechanical equation (2.62). Either form of the equation of motion may be employed and they give the same linear magnon dispersion relations. This will be seen subsequently. The classical equation is sometimes preferred in a macroscopic treatment because, if required, it can readily be augmented by terms that describe a phenomenological damping (e.g. see Kittel 1986). At low temperatures $T \ll T_C$ we write

$$\mathbf{S}_i = S\hat{\mathbf{z}} + \boldsymbol{\mu}_i \exp(-\mathrm{i}\omega t) \qquad (2.76)$$

where $\hat{\mathbf{z}}$ is a unit vector in the z direction and $\boldsymbol{\mu}_i$ denotes the fluctuating component of $\mathbf{S}_i$ at frequency ω. A similar separation of $\mathbf{B}_i$ into a static part and a fluctuating part can be made:

$$\mathbf{B}_i = \left(B_0 + \gamma^{-1}S\sum_j J_{ij} \right)\hat{\mathbf{z}} + \mathbf{b}_i \exp(-\mathrm{i}\omega t) \qquad (2.77)$$

and the connection between $\mathbf{b}_i$ and $\boldsymbol{\mu}_i$ is provided by

$$\mathbf{b}_i = \gamma^{-1}\sum_j J_{ij}\boldsymbol{\mu}_j. \qquad (2.78)$$

From (2.74), (2.76) and (2.77) and making the linear-magnon approximation of neglecting terms that are of second order in $\boldsymbol{\mu}_i$ (because $|\boldsymbol{\mu}_i| \ll S$) we obtain

$$-\mathrm{i}\omega\boldsymbol{\mu}_i = \hat{\mathbf{z}} \times \left[\gamma S\mathbf{b}_i - \left(g\mu_B B_0 + S\sum_j J_{ij} \right)\boldsymbol{\mu}_i \right]. \qquad (2.79)$$

Next we may go over to a continuum representation in which $\boldsymbol{\mu}_i$ and $\mathbf{b}_i$ at magnetic site i are replaced by position-dependent functions $\boldsymbol{\mu}(\mathbf{r})$ and $\mathbf{b}(\mathbf{r})$. Formally this can be achieved by writing

$$\boldsymbol{\mu}(\mathbf{r}) = \sum_i \boldsymbol{\mu}_i \delta(\mathbf{r} - \mathbf{r}_i) \qquad (2.80)$$

and averaging on the microscopic scale (over distances of order a few times a). We assume as before a simple cubic ferromagnet with exchange coupling J to the six nearest neighbours. On using (2.78) and making a Taylor-series expansion appropriate to the continuum representation we obtain

$$\mathbf{b}(\mathbf{r}) = \gamma^{-1}(6J + a^2 J^2 \nabla^2 + \cdots)\boldsymbol{\mu}(\mathbf{r}). \tag{2.81}$$

The above result is proved in a way that is analogous to the treatment of vibrational modes in the 1D monatomic lattice in the continuum limit (cf. equation (2.6)). With (2.79) it leads to

$$i\omega\boldsymbol{\mu}(\mathbf{r}) = \hat{\mathbf{z}} \times (\gamma B_0 - SJa^2\nabla^2)\boldsymbol{\mu}(\mathbf{r}). \tag{2.82}$$

With the use of translational invariance, implying $\boldsymbol{\mu}(\mathbf{r}) = \boldsymbol{\mu}_0 \exp(i\mathbf{q} \cdot \mathbf{r})$, we derive from (2.82) the bulk-magnon dispersion equation in the continuum limit:

$$\omega_B(\mathbf{q}) = \gamma B_0 + SJa^2q^2. \tag{2.83}$$

This is, as might be expected, just the long-wavelength (small $|\mathbf{q}|$) limit (2.70) of the previous general result given by (2.67).

The continuum approach can be a useful simplification when we are concerned only with long-wavelength excitations. For example, contributions of the magnons to the bulk magnetization at low-temperature are dominated by the long-wavelength region, where the excitations have lower frequencies; similarly it is these excitations that participate in one-magnon light-scattering processes or in magnetic resonance.

2.3.3 Magnons in bulk Heisenberg antiferromagnets

In a Heisenberg antiferromagnet the sign of the exchange interaction is such that an antiparallel ordering of neighbouring spins is favoured at temperatures below the Néel temperature T_N. The material can be regarded as two interpenetrating sublattices corresponding to each spin type (loosely referred to as 'spin-up' and 'spin-down'). There is no overall magnetization in zero applied magnetic field because the contributions from each sublattice cancel one another, but there is long-range magnetic ordering below T_N.

The Heisenberg Hamiltonian may be expressed as

$$H = \sum_{i,j} J_{ij}\mathbf{S}_i \cdot \mathbf{S}_j - g\mu_B B_0 \left[\sum_i S_i^z + \sum_j S_j^z \right]$$
$$- g\mu_B B_A \sum_i S_i^z + g\mu_B B_A \sum_j S_j^z \tag{2.84}$$

where i and j refer to sites on sublattices 1 (spin-up) and 2 (spin-down) respectively, and $J_{ij} > 0$ is the corresponding exchange interaction. The second term in (2.84) is the interaction with a field B_0 in the z direction and the last two terms describe the effect of an anisotropy field B_A, taken to have the same magnitude at each spin site but to be oppositely directed on each sublattice. The role of the anisotropy is important in antiferromagnets (particularly those of non-cubic structure) and its origin is usually due to crystal-field effects (see e.g. Bland and Heinrich 1994). In (2.84) we ignore, for simplicity, the effect of exchange between sites on the same sublattice.

We now derive the bulk-magnon dispersion relation in an infinite medium. The equation of motion for an S^+ operator can be constructed as in section 2.3.2, except that the Hamiltonian (2.84) must now be employed and there is an equation of motion for each sublattice. For a single-frequency wave $\exp(-i\omega t)$ and after using RPA decoupling, as in (2.64), we have

$$\omega S_{1\mathbf{q}}^+ = [\omega_0 + \omega_A + \omega_E]S_{1\mathbf{q}}^+ - \langle S^z \rangle J(\mathbf{q})S_{2\mathbf{q}}^+ \qquad (2.85)$$

$$\omega S_{2\mathbf{q}}^+ = [\omega_0 - \omega_A - \omega_E]S_{2\mathbf{q}}^+ + \langle S^z \rangle J(\mathbf{q})S_{1\mathbf{q}}^+ \qquad (2.86)$$

where wavevector Fourier transforms have been defined by analogy with (2.65) and an additional subscript (1 or 2) has been introduced as a sublattice label. We have also assumed the property that $\langle S^z \rangle_1 = -\langle S^z \rangle_2 \equiv \langle S^z \rangle$, which holds exactly by symmetry if $B_0 = 0$ and holds approximately at low temperatures even when B_0 is nonzero. We employ the convenient notations that $\omega_0 = g\mu_B B_0$, $\omega_A = g\mu_B B_A$ and $\omega_E = \langle S^z \rangle J(0)$ for the external-field, anisotropy and exchange frequencies respectively. The excitation frequencies may now be found from solutions of the coupled equations (2.85) and (2.86). The result is $\omega = \omega_0 \pm \omega_B(\mathbf{q})$ where

$$\omega_B(\mathbf{q}) = [(\omega_A + \omega_E)^2 - \langle S_z \rangle^2 J^2(\mathbf{q})]^{1/2}. \qquad (2.87)$$

For wavevector $\mathbf{q} = 0$ this frequency becomes

$$\omega_B(0) = [\omega_A(2\omega_E + \omega_A)]^{1/2}. \qquad (2.88)$$

This gives the well-known *antiferromagnetic resonance (AFMR) frequency*; for materials with $\omega_A \ll \omega_E$ it approximates to $(2\omega_A\omega_E)^{1/2}$. Exchange fields in most materials are much higher than practical applied fields, so because of the appearance of ω_E in (2.88) AFMR frequencies are usually much higher than ferromagnetic resonance frequencies. The much-studied uniaxial antiferromagnets MnF_2 and FeF_2 have resonance frequencies (in wave-number units – see table 1.1 for conversions) equal to $8.4\,\mathrm{cm}^{-1}$ and $52\,\mathrm{cm}^{-1}$ respectively. Both these frequencies are accessible to far-infrared spectroscopic techniques (the former only just) so it is possible to investigate AFMR and related properties by means of frequency-scan spectra.

2.3.4 Dipole–dipole interactions

In section 2.3.1 it was pointed out that the relative importance of the exchange interaction (2.56) and the dipolar interaction (2.58) for the *spin–wave dynamics* depends on the magnitude of the wavevector $\mathbf{q}$. The discussion there may be summarized (and extended) by reference to figure 2.5. It will be

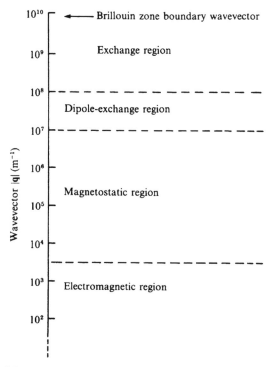

Figure 2.5. The different regions of magnetic behaviour in terms of the magnitude $|\mathbf{q}|$ of the excitation wavevector. The numbers are approximate and may vary for different materials.

seen later that the magnetostatic region (where the dipolar effects are dominant) and the dipole-exchange region (where the dipolar and exchange effects are comparable) are of special importance in films and superlattices. Here we discuss the relevant results for bulk materials by combining the Hamiltonian contributions from (2.58) and (2.59). An alternative (continuum) method for the dipolar region will be given in section 2.6. The electromagnetic region indicated in figure 2.5 is related to the discussion of magnetic polaritons (see section 2.6.2).

The magnon dispersion relation for ferromagnets can be investigated by following the operator equation-of-motion method as in section 2.3.2. Taking $A = S_j^+$ in (2.62) and using the RPA as before, we obtain

$$i\frac{d}{dt}S_{\mathbf{q}}^+ = U(\mathbf{q})S_{\mathbf{q}}^+ + V(\mathbf{q})S_{\mathbf{q}}^- \qquad (2.89)$$

which generalizes (2.66). Here we have defined

$$U(\mathbf{q}) = g\mu_B B_0 + \langle S^z \rangle [J(0) - J(\mathbf{q})] + \langle S^z \rangle [D^{zz}(0) + \tfrac{1}{2}D^{zz}(\mathbf{q})] \qquad (2.90)$$

$$V(\mathbf{q}) = \tfrac{1}{2}\langle S^z \rangle [D^{xx}(\mathbf{q}) - D^{yy}(\mathbf{q}) - 2iD^{xy}(\mathbf{q})]. \qquad (2.91)$$

The dependence on exchange and dipole–dipole interactions comes from $J(\mathbf{q})$ and $D^{\mu\nu}(\mathbf{q})$, respectively, where the latter quantities are defined by

$$D^{\mu\nu}(\mathbf{q}) = g^2 \mu_B^2 \left(\frac{\mu_0}{4\pi}\right) \sum_{\mathbf{r}} \frac{3r^\mu r^\nu - |\mathbf{r}|^2 \delta_{\mu\nu}}{|\mathbf{r}|^5} \exp(-i\mathbf{q} \cdot \mathbf{r}). \qquad (2.92)$$

The summation is over all lattice translation vectors $\mathbf{r}$, excluding $\mathbf{r} = 0$, and μ and ν are Cartesian components. We see that the equation of motion (2.89) involves a coupling to $S_{\mathbf{q}}^-$, by contrast with (2.66) in the absence of dipole–dipole interactions. The equation of motion for $S_{\mathbf{q}}^-$ is easily shown to be

$$i\frac{d}{dt}S_{\mathbf{q}}^- = -U(\mathbf{q})S_{\mathbf{q}}^- - V^*(\mathbf{q})S_{\mathbf{q}}^+ \qquad (2.93)$$

and the normal mode solutions of (2.89) and (2.93) with time dependence $\exp(-i\omega t)$ correspond to $\omega = \pm\omega_B(\mathbf{q})$ with

$$\omega_B(\mathbf{q}) = [U^2(\mathbf{q}) - |V(\mathbf{q})|^2]^{1/2}. \qquad (2.94)$$

This is the modified dispersion for bulk magnons in the presence of dipole–dipole interactions. At low temperatures where $\langle S^z \rangle \to S$ we recover the result that can be derived using boson operators instead of the spin operators (e.g. see Keffer 1966).

The summations in (2.92) are complicated to evaluate because of the long-range nature of the dipolar interactions, a problem that was fully discussed by Cohen and Keffer (1955). For a cubic lattice and small wavevectors ($L^{-1} \ll |\mathbf{q}| \ll a^{-1}$, where L is of the order of the sample dimensions and a is the lattice constant) the dispersion relation takes the form

$$\omega_B(\mathbf{q}) = [g\mu_B B_0 + Dq^2 - g\mu_B\mu_0 M_0(N_z - \sin^2\theta)]^{1/2}$$
$$\times [g\mu_B B_0 + Dq^2 - g\mu_B\mu_0 M_0 N_z]^{1/2} \qquad (2.95)$$

where D is an exchange factor (the spin wave 'stiffness') defined as in (2.70), M_0 is the magnetization, and N_z is the classical demagnetizing factor of the sample. The magnon frequency depends not just on the *magnitude* of wavevector $\mathbf{q}$ but also on its *direction* through angle θ, which is the angle between $\mathbf{q}$ and the z axis.

The magnetostatic limit, in which exchange effects become negligible, corresponds to $Dq^2 \ll g\mu_B\mu_0 M_0$ in (2.95). For example, if $N_z = 0$ then

$$\omega_B(\mathbf{q}) = [g\mu_B B_0(g\mu_B B_0 + g\mu_B\mu_0 M_0 \sin^2\theta]^{1/2}. \qquad (2.96)$$

Dipole–dipole interactions can play a significant part in antiferromagnets as well as in ferromagnets, and the above arguments may readily be generalized to that case. We return to this later when discussing the dynamic susceptibility function for both ferromagnets and antiferromagnets in section 2.6, as well as in chapter 5.

2.3.5 Magnons in Ising ferromagnets

In section 2.3.1 we introduced the Ising model as the special case of the anisotropic exchange Hamiltonian (2.57) in which $\varepsilon_x = \varepsilon_y = 0$ and $\varepsilon_z = 1$. If there is also an applied magnetic field in the x direction the Hamiltonian has the form

$$H = -\tfrac{1}{2}\sum_{i,j} J_{ij} S_i^z S_j^z - \sum_i \Omega_i S_i^x. \tag{2.97}$$

Here J_{ij} is the exchange interaction as before, and Ω_i is related to the *transverse* field ($\Omega_i = g\mu_B B_0$ where B_0 is the magnitude of the field). Although this Hamiltonian provides a good description of some real anisotropic magnetic materials in a transverse field, it is of much wider applicability as a simple pseudo-spin model for *hydrogen-bonded ferroelectrics* as discussed below.

Ferroelectrics are materials that exhibit a phase transition to a low-temperature state with a spontaneous electric dipole moment. Two good books dealing with the properties of such systems are by Blinc and Zeks (1974) and Lines and Glass (1979). The hydrogen-bonded ferroelectrics are materials such as KDP (potassium di-hydrogen phosphate) in which the occurrence of the electric dipole moment is associated with the ordering of protons in the hydrogen bonds in the solid. Essentially each proton is considered as being in a double potential well, as represented schematically in figure 2.6, where there are two positions A and B of stable equilibrium. For temperature T above a critical temperature T_C the protons in the solid are distributed at random. However, for $T < T_C$ the occupation of the sites becomes ordered in the sense that the occupation probability of an A position differs from that of a B position. It is convenient to describe the two positions available to the proton at any site i in terms of the eigenstates $S^z = \pm\tfrac{1}{2}$ of an effective spin $S = \tfrac{1}{2}$. This leads to the Hamiltonian (2.97) in which the exchange term accounts for the cooperative interaction between adjacent hydrogen bonds that leads to the phase transition and the transverse

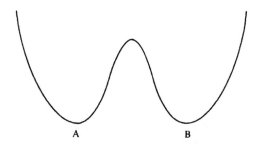

Figure 2.6. Schematic representation of the double well potential in the model for hydrogen-bonded ferroelectrics.

field term represents the ability of a proton to tunnel from position A to B or vice versa. As is clear from this description, the transverse field is disordering and it will be seen shortly that T_C decreases as Ω_i increases. A fuller discussion of the applicability of the model to KDP-type ferroelectrics is given by Levstik *et al* (1984).

For an infinite system we can take $\Omega_i \equiv \Omega$ independent of i. The phase transition can be analysed in an analogous fashion to the calculation in section 2.3.2 for a Heisenberg ferromagnet. It is found that the critical temperature T_C in the mean-field approximation satisfies

$$\tanh(\Omega/2k_B T_C) = 2\Omega/J(0) \tag{2.98}$$

where the exchange term $J(0)$ is defined as in (2.68). For $T < T_C$ the average spin orientation has components in the x and z directions, but for $T > T_C$ it lies along the x direction. In the pseudo-spin model of a ferroelectric the electric dipole moment is proportional to $\langle S^z \rangle$, so T_C represents the critical temperature between para-electric ($T > T_C$) and ferro-electric ($T < T_C$) phases. Because $\tanh(\Omega/2k_B T_C) \leq 1$ (2.98) has a solution only if $2\Omega \leq J(0)$, otherwise the para-electric phase exists at all temperatures. It is easily seen from (2.98) that T_C is a decreasing function of Ω, taking the value of $J(0)$ as fixed. As stated, the transverse field is disordering. A striking example of this is given by the comparison between KDP and deuterated KDP (or KD*P), in which the protons are replaced by deuterons. The chemical behaviour of a deuteron is the same as that of a proton so the crystal structure and pseudo-exchange $J(0)$ are identical in the two cases. However, because of its larger mass the tunnelling probability of the deuteron between the two minima in the potential of figure 2.6 is lower, and therefore the transverse field Ω is lower, in KD*P. This is reflected in the higher critical temperature (213 K) of KD*P, as against 123 K for KDP.

The magnons of the Ising model can be calculated by various techniques, but it is convenient to employ the torque equation of motion (2.74). The instantaneous field $\mathbf{B}_i$ is found using (2.75) and (2.97). The thermal average $\langle \mathbf{S}_i \rangle$ is independent of site i and will be denoted by $\mathbf{R} = (R^x, 0, R^z)$. Mean-field theory for $S = \frac{1}{2}$ gives

$$\mathbf{R} = (\mathbf{B}^0/2|\mathbf{B}^0|) \tanh(|\mathbf{B}^0|/2k_B T) \tag{2.99}$$

where $\mathbf{B}^0 = (\Omega, 0, R^z J(0))$ is the static part of the field $\mathbf{B}_i$. From this it is seen that R^x is constant for $T < T_C$ and decreases with increasing temperature for $T > T_C$, while R^z has the property of being zero for $T \geq T_C$. These properties are illustrated in figure 2.7. The equations of motion, derived from (2.74), (2.75) and (2.97), can be linearized by first writing

$$\mathbf{S}_i = \mathbf{R} + \boldsymbol{\mu}(\mathbf{q}) \exp(i\mathbf{q} \cdot \mathbf{r}_i - i\omega t) \tag{2.100}$$

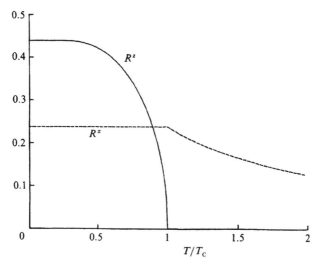

Figure 2.7. The temperature dependence of the spin averages R^x and R^z, calculated using mean-field theory, for an infinite bulk sample described by the Ising model in a transverse field. We have assumed the ratio $\Omega/J(0) = 0.24$. The zero-temperature values of R^x and R^z are 0.44 and 0.24, corresponding to $\{\frac{1}{4} - [\Omega/J(0)]^2\}^{1/2}$ and $\Omega/J(0)$ respectively.

where $\boldsymbol{\mu}$ is the fluctuating component at wavevector $\mathbf{q}$ and frequency ω. Then expanding to first order in $\boldsymbol{\mu}$ we obtain in component form

$$-\mathrm{i}\omega\mu^x = R^z J(0)\mu^y \tag{2.101}$$

$$-\mathrm{i}\omega\mu^y = [\Omega - R^x J(\mathbf{q})]\mu^z - R^z J(0)\mu^x \tag{2.102}$$

$$-\mathrm{i}\omega\mu^z = -\Omega\mu^y. \tag{2.103}$$

The solution for ω of the above equations is easily shown to be $\omega = \pm\omega_B(\mathbf{q})$ where

$$\omega_B(\mathbf{q}) = \{\Omega^2 - \Omega R^x J(\mathbf{q}) + [R^z J(0)]^2\}^{1/2}. \tag{2.104}$$

This is the standard expression for the bulk excitations of the Ising model in a transverse field.

2.4　Dielectric functions

Later in this book, we shall be concerned with electromagnetic surface waves, particularly the mode known as the *surface phonon–polariton* which occurs at the surface of an ionic solid in a frequency range just above the long-wavelength optic-phonon frequency ω_T. A related mode is the *surface plasmon–polariton* which is found at the surface of a conducting solid, either a metal or a doped semiconductor. These modes may be seen as mixed modes of the electromagnetic field and the excitations of the solid. The response of a nonmagnetic solid to an electromagnetic field is governed by the dielectric function, which is a generalization of the ordinary dielectric constant to include frequency dependence and if necessary wavevector dependence. Thus an understanding of the dielectric function is basic to a discussion of surface–polariton and related modes. In this section we first give a general account of the dielectric function and then review the properties of *bulk* polaritons in section 2.5. The case of magnetic systems is covered at the end of the chapter.

2.4.1 Dielectric function: definitions and basic properties

We use the standard equations of macroscopic electrodynamics (Landau and Lifshitz 1971, Jackson 1998) so that there are two electric-field vectors **E** and **D**. The latter is determined by the external charge density ρ_{ext} through

$$\nabla \cdot \mathbf{D} = \rho_{\text{ext}} \tag{2.105}$$

while **E** gives the force $q\mathbf{E}$ on a test charge q. The two vectors are related by

$$\mathbf{D} = \varepsilon_0 \mathbf{E} + \mathbf{P} \tag{2.106}$$

where **P** is the induced electric polarization, or dipole moment per unit volume. The properties of the medium enter via a symmetric second-rank tensor γ_{ij}, which is the electric susceptibility in space and time arguments:

$$P_i(\mathbf{r}, t) = \varepsilon_0 \int d^3 r' \int_{-\infty}^{t} dt' \, \gamma_{ij}(\mathbf{r} - \mathbf{r}', t - t') E_j(\mathbf{r}', t'). \tag{2.107}$$

Here (and throughout this section) we use the summation convention of summing over the repeated index $j = x, y, z$. A similar relation to (2.107) holds for **D**:

$$D_i(\mathbf{r}, t) = \varepsilon_0 \int d^3 r' \int_{-\infty}^{t} dt' \, f_{ij}(\mathbf{r} - \mathbf{r}', t - t') E_j(\mathbf{r}', t') \tag{2.108}$$

where f_{ij}, which like γ_{ij} is symmetric, is the dielectric tensor in space and time arguments. It is seen from (2.107) and (2.108) that

$$f_{ij}(\mathbf{r} - \mathbf{r}', t - t') = \delta_{ij} \delta(\mathbf{r} - \mathbf{r}') \delta(t - t') + \gamma_{ij}(\mathbf{r} - \mathbf{r}', t - t'). \tag{2.109}$$

In a medium with conductivity σ_{ij}, the response to **E** includes a current density **J** given by

$$J_i = \sigma_{ij} E_j \tag{2.110}$$

where integrals over $\mathbf{r}'$ and t', as in (2.107) and (2.108), are implied. However, **J** and **P** are interchangeable via

$$\mathbf{J} \equiv \partial \mathbf{P} / \partial t \tag{2.111}$$

and we therefore assume that any current-density term is included in (2.107).

In (2.107) and (2.108) we are making the assumption that the field $\mathbf{E}$ is small enough in magnitude that $\mathbf{P}$ and $\mathbf{D}$ are linear in $\mathbf{E}$. In practice, for most media this assumption is warranted except for the $\mathbf{E}$ fields of very intense laser beams. We are therefore dealing with *linear optics*, for which as will be seen many important and useful results can be obtained. The generalization to nonlinear optics will be discussed in chapter 10.

Several points about γ_{ij} and f_{ij} may be noted. First, as stated they are second-rank tensors, although for an isotropic or cubic medium they reduce to scalars (see e.g. Lovett 1999). The translational invariance of a bulk medium means that they are functions of $\mathbf{r} - \mathbf{r}'$, not of $\mathbf{r}$ and $\mathbf{r}'$ separately. A surface breaks translational invariance, so in principle γ_{ij} and f_{ij} may change from the bulk form when $\mathbf{r}$ and $\mathbf{r}'$ are within a few atomic spacings of a surface. However, most of the modes with which we deal in subsequent chapters have decay lengths within the medium that are typically hundreds of interatomic spacings. Their properties are therefore determined by the bulk value of f_{ij}, and there is usually no appreciable error in assuming that this bulk value describes the medium right up to the surface. The limits on the t' integrals in (2.107) and (2.108) imply that $\mathbf{P}(\mathbf{r}, t)$ and $\mathbf{D}(\mathbf{r}, t)$ depend on the values of $\mathbf{E}(\mathbf{r}', t')$ at earlier times $t' < t$. The response of the medium itself is given by $\mathbf{P}$ rather than $\mathbf{D}$ since the latter contains the 'vacuum' term $\varepsilon_0 \mathbf{E}$. This means that $\gamma_{ij}(\mathbf{r} - \mathbf{r}', t - t')$ is a response function like the part proportional to $\langle [A(t), B(t')] \rangle$ in (A.17).

Since f_{ij} depends on $\mathbf{r} - \mathbf{r}'$ and $t - t'$, it is convenient to Fourier transform (2.108) as

$$D_i(\mathbf{q}, \omega) = \varepsilon_0 \varepsilon_{ij}(\mathbf{q}, \omega) E_j(\mathbf{q}, \omega) \qquad (2.112)$$

where $\varepsilon_{ij}(\mathbf{q}, \omega)$ is the Fourier transform of $f_{ij}(\mathbf{r}, t)$; a similar equation involving a susceptibility tensor $\chi_{ij}(\mathbf{q}, \omega)$ relates $\mathbf{P}(\mathbf{q}, \omega)$ to $\mathbf{E}(\mathbf{q}, \omega)$. The dependence of ε_{ij} on $\mathbf{q}$ is referred to as *spatial dispersion*, since its origin is seen in the dependence on $\mathbf{r} - \mathbf{r}'$ in (2.108). Most of the electromagnetic modes with which we are concerned are of sufficiently long wavelength (small $\mathbf{q}$) that spatial dispersion may be neglected, although an exception arises for exciton–polaritons (see section 2.5).

Thus, when spatial dispersion is neglected, the medium is characterized by a frequency-dependent tensor function $\varepsilon_{ij}(\omega)$. A further simplification is that in an isotropic or cubic medium the dielectric function reduces to a scalar function of frequency, with $\varepsilon(\omega) = 1 + \chi(\omega)$ where $\chi(\omega)$ is the susceptibility. An elementary consequence of the frequency dependence of ε is that the refractive index $\eta = \varepsilon^{1/2}$ is also frequency dependent; hence the name *optical dispersion* applied to ε.

A number of general properties may be deduced from (2.107) and (2.108). For simplicity we restrict attention to scalar functions $\varepsilon(\omega)$

and $\chi(\omega)$; the general results for tensors can be found elsewhere (Landau and Lifshitz 1971). First, since the physical fields **P** and **E** are real the function γ is real. From the Fourier transform defining $\varepsilon(\omega)$ it follows that

$$\varepsilon(-\omega) = \varepsilon^*(\omega) \tag{2.113}$$

or in terms of the real and imaginary parts $\varepsilon'(\omega)$ and $\varepsilon''(\omega)$,

$$\varepsilon'(-\omega) = \varepsilon'(\omega), \qquad \varepsilon''(-\omega) = -\varepsilon''(\omega). \tag{2.114}$$

A second result follows from regarding ω formally as complex, $\omega = \omega' + i\omega''$. The Fourier transform defining $\varepsilon(\omega)$ can be written

$$\varepsilon(\omega' + i\omega'') = 1 + \int_0^\infty \gamma(t) \exp(i\omega't) \exp(-\omega''t)\, dt. \tag{2.115}$$

For any physical response $\gamma(t) \to 0$ as $t \to \infty$, so because of the factor $\exp(-\omega''t)$ the integral in (2.115) is convergent for any positive value of ω''. Thus $\varepsilon(\omega)$ has a finite value and is analytic for any ω in the upper half plane; therefore the singularities of $\varepsilon(\omega)$ lie in the lower half plane. This analyticity in the upper half plane is all that is required in the proof of the Kramers–Kronig relations, (A.35) and (A.36), which express $\varepsilon'(\omega) - 1$ in terms of an integral involving $\varepsilon''(\omega)$ and vice versa. Finally we state that the high-frequency limit is

$$\varepsilon(\omega) \to 1 \qquad \text{as } \omega \to \infty. \tag{2.116}$$

For this we use the intuitive argument that at sufficiently high frequency **P** cannot follow the variations of **E** so that $\chi \to 0$ as $\omega \to \infty$.

We may use these general results to find two basic forms of $\varepsilon(\omega)$ by looking for simple distributions of singularities in the lower half plane that are consistent with (2.114) and (2.116). There is a difference between materials with a nonzero d.c. conductivity σ and insulators with $\sigma = 0$. For the former, at low frequency,

$$\mathbf{J} = -i\omega\mathbf{P} = \sigma\mathbf{E} \tag{2.117}$$

where (2.111) has been applied for a single frequency ω. Since $\mathbf{P} = \varepsilon_0\chi\mathbf{E}$, it follows that $\chi = i\sigma/\varepsilon_0\omega$ and, given $\varepsilon = 1 + \chi$, it follows that the low frequency limit of ε is

$$\varepsilon \approx i\sigma/\varepsilon_0\omega. \tag{2.118}$$

This cannot hold for all frequencies, since it is inconsistent with (2.116) and with the Kramers–Kronig relations. The simplest extension that can be made to meet all the requirements is to put in the high-frequency limit by hand and add a simple pole at a point $\omega = -i/\tau$ on the negative imaginary axis, to yield

$$\varepsilon(\omega) = 1 - \frac{\omega_p^2}{\omega^2 + i\omega/\tau} \tag{2.119}$$

where the conventional parameters ω_p (known as the plasma frequency) and τ (the relaxation time) are related to σ by $\sigma = \varepsilon_0 \omega_p^2 \tau$. We shall see in section 2.4.2 how (2.119) may be derived from a classical model of electron dynamics but the present discussion shows that it has a wider validity than the result of one particular model.

In an insulator, $\varepsilon(\omega)$ does not have the low-frequency divergence of (2.118). The simplest distribution of singularities that leads to a form satisfying the various requirements is a single pole at $\omega = \Omega - i\Gamma$ and another at $\omega = -\Omega - i\Gamma$ with equal and opposite residues $\pm A$. It can then be shown that the expression for $\varepsilon(\omega)$ is

$$\varepsilon(\omega) = 1 + \frac{S\omega_T^2}{\omega_T^2 - \omega^2 - i\omega\Gamma} \qquad (2.120)$$

where $\omega_T^2 = \Omega^2 + \Gamma^2$ and $S\omega_T^2 = 2A\Omega$. Like (2.119), the above can be derived from a specific model, in fact the extension of the diatomic-lattice dynamics of section 2.1.2 to include coupling to the electromagnetic field, but the discussion here shows that it is a rather general form. The lattice-dynamic derivation will be given in section 2.4.3.

2.4.2 Electron gas: hydrodynamic treatment

The response of the carriers in a metal or a doped semiconductor to a time-varying electric field is a problem in quantum-mechanical many-body theory. A treatment by perturbation theory leads to an expression for the response in terms of the Lindhard function (Lindhard 1954); details can be found in standard texts (e.g. Ashcroft and Mermin 1976) and in monographs (e.g. Keldysh *et al* 1989); a brief summary of the results is given in the next section. First we outline a simpler, hydrodynamic approach that enables us to derive the required results in a rather transparent way. This method has the further advantage that the effects of an applied magnetic field are easily included. Fuller accounts can be found elsewhere (e.g. Boardman 1982). The electron gas is treated as a continuous fluid having density $n(\mathbf{r}, t)$ and velocity $\mathbf{v}(\mathbf{r}, t)$ with associated electric current density $\mathbf{j}(\mathbf{r}, t)$:

$$\mathbf{j} = -ne\mathbf{v} \qquad (2.121)$$

where dependence on $\mathbf{r}$ and t is understood and we assume that the carriers are electrons (charge $-e$). The particle current density is $n\mathbf{v}$, so n and $\mathbf{v}$ are related by the continuity equation

$$\partial n/\partial t + \nabla \cdot (n\mathbf{v}) = 0. \qquad (2.122)$$

The velocity is assumed to obey the acceleration equation

$$m(\partial/\partial t + \mathbf{v} \cdot \nabla)\mathbf{v} = -e(\mathbf{E} + \mathbf{v} \times \mathbf{B}) - m\mathbf{v}/\tau - (m\beta^2/n)\nabla n. \qquad (2.123)$$

As in all hydrodynamics, the total derivative $D/Dt = \partial/\partial t + \mathbf{v} \cdot \nabla$ is used; however, the term $\mathbf{v} \cdot \nabla \mathbf{v}$ is nonlinear, and is usually negligible. The term $-m\mathbf{v}/\tau$ introduces the collision, or damping, time τ. The final term results from the compressibility of the electron gas. The value of the parameter β is discussed by Boardman (1982) and the value $\beta^2 = 3v_F^2/5$, where v_F is the Fermi velocity, is often assumed.

The dielectric function describes the linear response of the electron gas to an external electric field, and we therefore proceed by linearizing (2.121) to (2.123). We treat the electric field as the driving force, so that $\mathbf{E}$ and $\mathbf{v}$ are first-order quantities, whereas n is written as a sum of zero- and first-order terms: $n = n_0 + n_1$. A single-frequency field $\mathbf{E}\exp(-i\omega t)$ is considered, so $\partial/\partial t$ is replaced by $-i\omega$. The linearized forms are then

$$\mathbf{j} = -n_0 e \mathbf{v} \tag{2.124}$$

$$-i\omega n_1 + n_0 \nabla \cdot \mathbf{v} = 0 \tag{2.125}$$

$$i\omega m\mathbf{v} = m\mathbf{v}/\tau + e\mathbf{E} + e\mathbf{v} \times \mathbf{B}_0 + (m\beta^2/n_0)\nabla n_1 \tag{2.126}$$

where $\mathbf{B}_0$ is a possible static magnetic field. Substitution of n_1 from (2.125) into (2.126) gives the basic equation for $\mathbf{v}$:

$$i\omega m\mathbf{v} = m\mathbf{v}/\tau + e\mathbf{E} + e\mathbf{v} \times \mathbf{B}_0 - (im\beta^2/\omega)\nabla\nabla \cdot \mathbf{v}. \tag{2.127}$$

For now, we drop magnetic field effects by putting $\mathbf{B}_0 = 0$ and recognize that since the electron gas is isotropic the dielectric function is a scalar $\varepsilon(\mathbf{q}, \omega)$. To derive an explicit expression we take $\mathbf{E}$ in the plane-wave form $\mathbf{E}\exp(i\mathbf{q} \cdot \mathbf{r} - i\omega t)$. The operator ∇ is then replaced by $i\mathbf{q}$, (2.127) is solved for $\mathbf{v}$ and application of (2.124) gives

$$\mathbf{j} = -\frac{e^2 n_0}{m} \frac{1}{i\omega - 1/\tau - i\beta^2 q^2/\omega} \mathbf{E}. \tag{2.128}$$

To determine $\varepsilon(\mathbf{q}, \omega)$, we write the relevant Maxwell equation in two equivalent forms:

$$i\mathbf{q} \times \mathbf{H} = \mathbf{j} - i\omega\varepsilon_0\varepsilon_\infty\mathbf{E} = -i\omega\varepsilon_0\varepsilon(\mathbf{q}, \omega)\mathbf{E}. \tag{2.129}$$

Here we have included the constant ε_∞ to allow for the effects of excitations of higher frequency than the plasma mode under consideration. This is necessary for doped semiconductors, in which the plasma frequency is typically in the far infrared region, but it is not needed for metals since the plasma frequency is in the ultraviolet, and is therefore the highest frequency involved in typical optical effects. Comparison of the two forms in (2.129) with use of (2.128) gives the required expression

$$\varepsilon(\mathbf{q}, \omega) = \varepsilon_\infty\left(1 - \frac{\omega_p^2}{\omega^2 + i\omega/\tau - \beta^2 q^2}\right) \tag{2.130}$$

where the *plasma frequency* ω_p is given by

$$\omega_p^2 = e^2 n_0 / m \varepsilon_0 \varepsilon_\infty. \tag{2.131}$$

A few comments about this basic result are in order. Since ω_p depends on e^2, the same expression is found for holes (charge $+e$) as for electrons $(-e)$. Substitution of typical values of the effective mass m and carrier density n_0, say $10^{18}\,\text{cm}^{-3}$ for semiconductors and $10^{23}\,\text{cm}^{-3}$ for metals, shows that as mentioned already ω_p typically lies in the infrared or far infrared for semiconductors and in the ultraviolet for metals. In order to understand (2.130), it is helpful to quote the simplified form without spatial dispersion, i.e. $\beta = 0$:

$$\varepsilon(\omega) = \varepsilon_\infty \left(1 - \frac{\omega_p^2}{\omega^2 + i\omega/\tau} \right). \tag{2.132}$$

Apart from the multiplicative constant ε_∞ this has the same form as (2.119), which was derived on the basis of a simple distribution of singularities consistent with causality. Within that method, the factor ε_∞ corresponds to additional singularities. If damping is omitted ($\tau \to \infty$) (2.132) expresses $\varepsilon(\omega)/\varepsilon_\infty$ as a function of ω/ω_p, with the form shown in figure 2.8.

Inclusion of the effect of a static magnetic field $\mathbf{B}_0$ in the analysis requires simply the retention of the first-order term $e\mathbf{v} \times \mathbf{B}_0$ in (2.127). This is left as an exercise for the reader; it is found that ε_{ij} takes the *gyrotropic* form

$$\bar{\varepsilon} = \begin{pmatrix} \varepsilon_1 & i\varepsilon_2 & 0 \\ -i\varepsilon_2 & \varepsilon_1 & 0 \\ 0 & 0 & \varepsilon_3 \end{pmatrix} \tag{2.133}$$

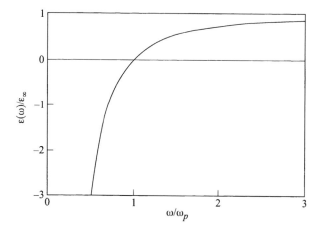

Figure 2.8. Plasma dielectric function of (2.132).

where

$$\varepsilon_1 = \varepsilon_\infty \left\{ 1 - \frac{\omega_p^2(\omega + i/\tau)}{\omega[(\omega + i/\tau)^2 - \omega_c^2]} \right\} \tag{2.134}$$

$$\varepsilon_2 = \frac{\varepsilon_\infty \omega_p^2 \omega_c}{\omega[(\omega + i/\tau)^2 - \omega_c^2]} \tag{2.135}$$

and ε_3 is given by (2.130). The z axis was taken in the direction of $\mathbf{B}_0$ and the (positive) cyclotron frequency $\omega_c = -eB_0/m$ has been introduced. Damping has been included but spatial dispersion neglected.

2.4.3   Electron gas: Lindhard function

As has been said, a more fundamental discussion of the dielectric function is based on statistical perturbation theory. Accounts are given in the books cited at the beginning of the previous section and also in books on many-body theory, such as Abrikosov *et al* (1963), Fetter and Walecka (1971) and Rickayzen (1980). Here we give a summary of the main results. The theory of a large number N of electrons interacting via the Coulomb force cannot be solved exactly. The analysis starts with the equation of motion of the Green function for the electron density which is then simplified by a random-phase approximation similar to that applied to the spin operators in (2.63).

The results involve two Green functions $D(\mathbf{q}, \omega)$ and $D^0(\mathbf{q}, \omega)$ related by

$$D(\mathbf{q}, \omega) = \frac{D^0(\mathbf{q}, \omega)}{1 - 2\pi v(\mathbf{q})D^0(\mathbf{q}, \omega)} \tag{2.136}$$

where $v(\mathbf{q})$ is the Fourier transform of the Coulomb potential in 3D,

$$v(\mathbf{q}) = e^2/\varepsilon_0 V q^2 \tag{2.137}$$

in which V is the volume of the system and

$$D^0(\mathbf{q}, \omega) = \frac{2V}{(2\pi)^4} \int d^3k \, \frac{f_\mathbf{k} - f_{\mathbf{k}+\mathbf{q}}}{\hbar\omega - \varepsilon_{\mathbf{k}+\mathbf{q}} + \varepsilon_\mathbf{k}}. \tag{2.138}$$

Here $\varepsilon_\mathbf{k} = \hbar^2 k^2/2m$ is the single-particle energy and $f_\mathbf{k}$ is the Fermi function. At zero temperature (where $f_\mathbf{k}$ becomes a step function) the above integral can be evaluated explicitly. The dielectric function is given in terms of these functions by

$$\varepsilon(\mathbf{q}, \omega) = [1 + 2\pi v(\mathbf{q})D(\mathbf{q}, \omega)]^{-1} = 1 - 2\pi D^0(\mathbf{q}, \omega). \tag{2.139}$$

The two forms are equivalent because of (2.136).

Equation (2.139) is the general RPA expression for the dielectric function of the 3D electron gas. Its properties are quite involved, and in particular the low-frequency, long-wavelength limit depends on the order in which the limits are taken:

$$\lim_{\mathbf{q}\to 0}\lim_{\omega\to 0}\varepsilon(\mathbf{q},\omega)\neq\lim_{\omega\to 0}\lim_{\mathbf{q}\to 0}\varepsilon(\mathbf{q},\omega). \qquad (2.140)$$

The former is the static dielectric function, which describes the screening of the Coulomb interaction, and the latter is the dynamic long-wavelength function. We are concerned with the second, and for the free-electron gas the form of $\varepsilon(\mathbf{q},\omega)$ is found by using the definition in (2.139) together with the appropriate limiting value of $D^0(\mathbf{q},\omega)$ found at zero temperature. The result (see references quoted earlier for details) is the undamped form of the plasma function of (2.119). It may be seen as a justification of the hydrodynamic method used in section 2.4.2 that it leads to the same form of $\varepsilon(\mathbf{q},\omega)$ as the RPA analysis.

2.4.4 Dielectric function of ionic crystals

In section 2.4.1 it was seen that in an insulator, with zero conductivity, the simplest distribution of singularities consistent with the general properties of the dielectric function led to an expression for $\varepsilon(\omega)$ of the form of (2.120). We now rederive this expression from a specific lattice-dynamical model. Our account is quite brief, and is based on the method introduced by Born and Huang (1985). Most solid-state textbooks deal with the topic, and Ashcroft and Mermin (1976) give a particularly full account. For simplicity, we restrict attention to an isotropic or cubic medium, in which vector quantities, such as **P** and **E**, are parallel. We may then use the lattice dynamics of a 1D diatomic lattice, given in section 2.1.2, as our starting point. Extension to anisotropic media does not involve any new points of principle.

As previously, we consider an infinite diatomic lattice in which masses m_1 and m_2 alternate. We now take the crystal to be ionic, so that the two sublattices are associated with opposite electric charges. The mode that couples to electromagnetic radiation is the long-wavelength ($q = 0$) optic phonon at frequency ω_T, say. This follows from the expression for the ratio U_1/U_2 that can be derived from (2.12) to (2.15). This ratio is seen to be negative for $q = 0$ so that the two sublattices move in antiphase. Since the sublattices carry opposite charges, the optic phonon therefore carries an oscillating dipole moment. The relevant wave number is $q \approx 2\pi/\lambda$, where λ is the wavelength of the electromagnetic radiation; on the scale of the Brillouin-zone edge $\pi/2a$ this is effectively $q = 0$, as stated.

We denote by **u** the relative displacement of the two sublattices. The polarization **P** contains a term proportional to **u**, as well as a term due to the electronic susceptibility rather than the ionic displacements:

$$\mathbf{P} = \varepsilon_0(\alpha\mathbf{u} + \chi\mathbf{E}). \qquad (2.141)$$

As mentioned, for an isotropic medium **P**, **u** and **E** are parallel. In (2.141) **E** is the *macroscopic* electric field, the value found by averaging the *local* field $\mathbf{E}_{loc}$ over many unit cells. Evaluation of the constants of proportionality α and χ depends upon details of the lattice dynamics and electronic polarizability, and need not concern us.

With the electric field **E**, and therefore the other field variables, proportional to $\exp(-i\omega t)$, the equation of motion for **u** is

$$(-\omega^2 - i\omega\Gamma)\mathbf{u} = -\omega_T^2\mathbf{u} + \beta\mathbf{E}_{loc}. \tag{2.142}$$

On the left-hand side a damping term in Γ is included; it arises from an assumed damping force $-\Gamma\, d\mathbf{u}/dt$. The first term on the right-hand side is a restoring force that ensures that in the absence of coupling to the electric field the mode frequency is ω_T, and the second is the driving force due to the electric field. In order to derive an expression for the dielectric function from (2.141) and (2.142) it is necessary to find the relation of $\mathbf{E}_{loc}$ to **E**. This is discussed by Ashcroft and Mermin (1976), for example. What Born and Huang pointed out, however, is that this relationship must be linear. Equation (2.142) can therefore be replaced by an equation involving **E** rather than $\mathbf{E}_{loc}$:

$$(-\omega^2 - i\omega\Gamma)\mathbf{u} = -\omega_T^2\mathbf{u} + \gamma\mathbf{E}. \tag{2.143}$$

Equations (2.141) and (2.143) are readily solved for **P**:

$$\mathbf{P} = \varepsilon_0\left(\frac{\alpha\gamma\mathbf{E}}{\omega_T^2 - \omega^2 - i\omega\Gamma} + \chi\mathbf{E}\right). \tag{2.144}$$

The defining relations for the dielectric function (see section 2.4.1) give

$$\mathbf{D} = \varepsilon_0\mathbf{E} + \mathbf{P} = \varepsilon_0\varepsilon(\omega)\mathbf{E}. \tag{2.145}$$

Combining (2.144) and (2.145) leads to

$$\varepsilon(\omega) = \varepsilon_\infty\left(1 + \frac{\omega_L^2 - \omega_T^2}{\omega_T^2 - \omega^2 - i\omega\Gamma}\right) \tag{2.146}$$

where

$$\varepsilon_\infty = 1 + \chi, \qquad \omega_L^2 = \omega_T^2 + \alpha\gamma/(1 + \chi). \tag{2.147}$$

For $\varepsilon_\infty = 1$, apart from notational changes, (2.146) is identical to (2.120), which was derived earlier on rather general grounds. As seen from (2.147) the difference between ε_∞ and unity is the electronic polarizability, which was inserted here as a simple, frequency-independent, constant χ. A more complete treatment of this contribution, along the lines of section 2.4.1, would introduce additional singularities in $\varepsilon(\omega)$ in the lower half frequency plane.

Equation (2.146) uses the conventional notation; it is written in terms of ε_∞, the high-frequency (electronic) dielectric constant, and ω_L, the LO

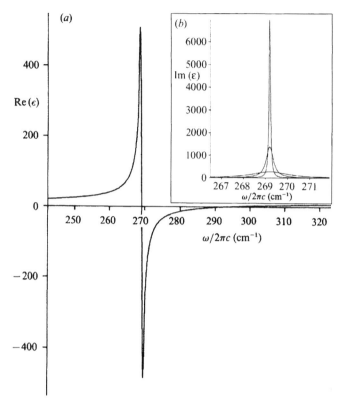

Figure 2.9. (a) real and (b) imaginary parts of dielectric function for GaAs, which is accurately described by (2.146). Numerical values are $\omega_T/2\pi c = 269.2\,\text{cm}^{-1}$, $\omega_L/2\pi c = 292.8\,\text{cm}^{-1}$, $\Gamma/2\pi c = 2.5\,\text{cm}^{-1}$ and $\varepsilon_\infty = 10.9$ (Kim and Spitzer 1979). In (b) the sharpest curve is drawn for the quoted value of Γ; the two broader curves, drawn for comparison, correspond to values 5 and 25 times larger.

(longitudinal-optic) phonon frequency. The zero-frequency value of $\varepsilon(\omega)$ is

$$\varepsilon(0) = \varepsilon_\infty \omega_L^2/\omega_T^2 \tag{2.148}$$

which is known as the *Lyddane–Sachs–Teller* (LST) *relation*. For typical materials, both dielectrics and partially ionic semiconductors, ω_T lies in the far infrared, generally with $\omega_T/2\pi c$ in the region 100 to $1000\,\text{cm}^{-1}$. The frequency dependence of (2.146) is illustrated in figure 2.9 for the case of the III–V semiconductor GaAs.

It was mentioned in section 2.4.2 that in semiconductors the plasma frequency lies in the far infrared. With suitable doping, it can be made close in value to the TO frequency ω_T, so the dielectric function contains both plasma and optic-phonon contributions. The total dielectric function can be found by a combination of the derivations we have given. For completeness, we record the form quoted by Palik *et al* (1976) for the dielectric tensor of a doped polar semiconductor in the presence of a static magnetic field $\mathbf{B}_0$ in the z direction. The tensor has the form of (2.133), with

$$\varepsilon_1 = \varepsilon_\infty \left(1 + \frac{\omega_L^2 - \omega_T^2}{\omega_T^2 - \omega^2 - i\omega\Gamma} + \frac{\omega_p^2(\omega + i/\tau)}{\omega[\omega_c^2 - (\omega + i/\tau)^2]} \right) \qquad (2.149)$$

ε_2 in the form of (2.135) and

$$\varepsilon_3 = \varepsilon_\infty \left(1 + \frac{\omega_L^2 - \omega_T^2}{\omega_T^2 - \omega^2 - i\omega\Gamma} - \frac{\omega_p^2}{\omega(\omega + i/\tau)} \right). \qquad (2.150)$$

Following the implications of this relatively complicated expression would take us too far afield, however, and we present it here only for reference.

2.5   Polaritons

2.5.1   Dispersion relations

The propagation of light through a nonmagnetic bulk medium characterized by a frequency-dependent dielectric function is governed by Maxwell's equations. For a plane wave in an isotropic or cubic medium the equation $\nabla \cdot \mathbf{D} = 0$ gives

$$\varepsilon(\omega)\, \mathbf{q} \cdot \mathbf{E} = 0 \qquad (2.151)$$

where all fields are proportional to $\exp(i\mathbf{q} \cdot \mathbf{r} - i\omega t)$ and we neglect spatial dispersion (the $\mathbf{q}$ dependence of ε) which is adequate in most cases. This equation has the solutions

$$\varepsilon(\omega) = 0 \qquad (2.152)$$

or

$$\mathbf{q} \cdot \mathbf{E} = 0. \qquad (2.153)$$

In the absence of damping, the first of these gives $\omega = \omega_L$ for an ionic medium, with the dielectric function given by (2.146), and $\omega = \omega_p$ for the plasma with the dielectric function of (2.132). These are the longitudinal modes; they have no surface counterpart so will not be discussed further. The second solution, (2.153), is the transversality condition ($\mathbf{E}$ transverse to $\mathbf{q}$). For this solution, the $\nabla \times \mathbf{E}$ and $\nabla \times \mathbf{H}$ equations yield in the usual way, as discussed in any text on electromagnetic theory (e.g. Jackson 1998),

$$q^2 = \varepsilon(\omega)\omega^2/c^2 \qquad (2.154)$$

which is the propagation equation for the transverse mode.

    If the damping term in $\varepsilon(\omega)$ is neglected, then it is an easy matter to draw the dispersion curve corresponding to (2.154). For the phonon case, with

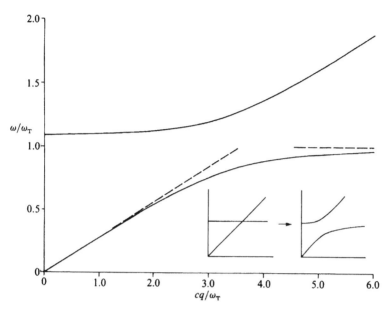

Figure 2.10. Dispersion curve for the bulk phonon–polariton in GaAs, taking numerical values quoted for figure 2.9. Asymptotic lines $q \approx [\varepsilon(0)]^{1/2}\omega/c$ and $\omega = \omega_T$ are shown. Inset shows qualitative origin in mode mixing.

$\varepsilon(\omega)$ given by (2.146), the asymptotic forms are as follows and are illustrated in figure 2.10:

$$
\begin{aligned}
q^2 &\approx \varepsilon(0)\omega^2/c^2 & \omega &\ll \omega_T \\
q^2 &\to \infty & \omega &\to \omega_T \\
q^2 &< 0 & \omega_T &< \omega < \omega_L \qquad (2.155) \\
q &= 0 & \omega &= \omega_L \\
q^2 &\approx \varepsilon_\infty \omega^2/c^2 & \omega &\to \infty.
\end{aligned}
$$

Note that q is pure imaginary for $\omega_T < \omega < \omega_L$, and therefore this frequency interval appears as a stop band. It will be seen in chapter 6 that it is this interval in which a surface mode appears.

The dispersion curve drawn in figure 2.10 may be seen as resulting from the crossing of the phonon line $\omega = \omega_T$ and the photon curve $q = \varepsilon^{1/2}\omega/c$, where ε is imagined to change slowly from $\varepsilon(0)$ to ε_∞. These modes interact strongly, and the crossover is therefore eliminated with repulsion of the curves, as shown schematically in the inset to figure 2.10. Thus the full dispersion curve describes a mode of mixed phonon–photon character. Many years ago, the word polariton was coined to describe such a mixed

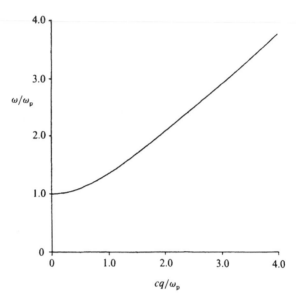

Figure 2.11. Bulk plasmon–polariton dispersion curve.

mode, so one may say that figure 2.10 depicts the bulk polariton dispersion curve. Nowadays *polariton* is used for any mixed mode involving a photon.

The plasma dielectric function of (2.132) is formally the $\omega_T \to 0$ limit of the phonon function. The dielectric function, drawn in figure 2.8 for $\tau \to \infty$, and the plasma–polariton dispersion curve, drawn as figure 2.11, can clearly be derived from the corresponding phonon curves by letting $\omega_T \to 0$.

For a doped polar semiconductor (in the absence of a magnetic field) $\varepsilon(\omega)$ is given by an expression of the form of (2.149) with $\omega_c = 0$. Consequently $\varepsilon(\omega)$ has two resonances, one plasma-like at $\omega = 0$ and one phonon-like at $\omega = \omega_T$ and there are two branches in the polariton dispersion curve rather than the one of figure 2.11.

The dispersion curves of figures 2.10 and 2.11 are drawn for the case when damping is neglected in the dielectric functions. Equation (2.154) is then a relation between two real variables q and ω, and the dispersion curve is a natural representation. However, if damping is included, (2.154) relates two variables q and ω that in principle are complex. There is then not much sense in discussing (2.154) in isolation, since the conditions of a given experiment will determine what trajectory is followed in the space of the two complex variables q and ω. For example, in inelastic light (Raman or Brillouin) scattering in a transparent bulk medium, the experimental constraint is that q is real, since it is the magnitude of the wavevector difference between the incident and scattered light. The real part of ω then appears as a line position, and the imaginary part as a linewidth. Conversely, it is usually the case in attenuated-total-reflection (ATR) experiments, which will be discussed in chapter 6, that ω is constrained to be real. In this case, the imaginary part of q is interpreted as the inverse of the decay length of the surface mode excited in the ATR experiment. With damping present, then, discussion of the dispersion relation requires reference to the experimental conditions.

We have given a very brief account of bulk polaritons in the simple cases of an isotropic medium whose dielectric response is dominated by one or two resonances and in which spatial dispersion can be neglected. These bulk modes were the subject of very active investigation, particularly by Raman scattering, from the mid-1960s. A good account of that work is found in various papers in the conference proceedings edited by Burstein and DeMartini (1974). Several extensions of the calculations outlined here were necessary. The theory of bulk polaritons in anisotropic crystals with multiple resonances is given by Merten in Burstein and DeMartini (1974), and several articles describe the related experimental results. As seen in section 1.4.1, for a theory of light-scattering intensities it is necessary to go beyond the solution of the homogeneous equations of motion, which in this case are Maxwell's equations together with the appropriate dielectric functions. Several articles in Burstein and DeMartini (1974) give accounts of light-scattering theory; one approach is to use the linear-response method, which is summarized in section 1.3.2 and discussed further in the Appendix.

2.5.2 Reflection of light

The properties of the bulk polaritons are basic to the reflection of light incident on the medium since transmitted radiation travels in the form of a polariton. We illustrate this with the simplest example, reflection of light in normal incidence. As shown in any optics text (e.g. Born and Wolf 1980, Lipson *et al* 1995) the reflectivity (power reflection coefficient) is given by

$$R = \left| \frac{q_1 - q}{q_1 + q} \right|^2 = \left| \frac{\varepsilon_1^{1/2} - \varepsilon^{1/2}}{\varepsilon_1^{1/2} + \varepsilon^{1/2}} \right|^2 \tag{2.156}$$

where q_1 and ε_1 are the wavevector and dielectric constant in the medium of incidence, and q and ε are the same things in the reflecting medium. Suppose ε_1 is a real constant, e.g. 1 for vacuum, and ε is given by (2.146). If we omit damping, $\Gamma = 0$, then as seen from figures 2.9 and 2.10, ε is negative and q (or $\varepsilon^{1/2}$) is pure imaginary for $\omega_T < \omega < \omega_L$. A simple sketch in the complex plane shows that for a pure imaginary value, $\varepsilon^{1/2} = i\alpha$ say, $|\varepsilon_1^{1/2} - i\alpha| = |\varepsilon_1^{1/2} + i\alpha|$ so that $R = 1$ in this interval. Otherwise, where $\varepsilon^{1/2}$ is real, $0 < R < 1$ so that some light is reflected and some transmitted. In terms of figure 2.10, for $\omega_T < \omega < \omega_L$, there is no bulk polariton to transmit

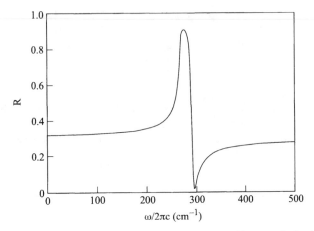

Figure 2.12. Reflectivity (2.156) versus frequency for GaAs with numerical values given in figure 2.10 and incidence in vacuum, $\varepsilon_1 = 1$.

radiation, so all the incident power is reflected. The interval $\omega_T < \omega < \omega_L$ is called the *reststrahl* (residual ray) region because radiation in this narrow band is totally reflected by a bulk ionic crystal.

The reflectivity of GaAs over a range including the reststrahl region is illustrated in figure 2.12 for the same numerical values used in figure 2.9. The inclusion of damping ($\Gamma \neq 0$) slightly rounds the shoulders of the region with $R \approx 1$ and reduces R below 1 through the reststrahl region, because of absorption in the medium, but our qualitative discussion based on (2.156) is still relevant. It will be noted in figure 2.12 that $R \approx 0$ at a frequency slightly above ω_L. This is the point at which $\mathrm{Re}(\varepsilon) = 1$ (see figure 2.9), or in terms of figure 2.10 it is the point where the upper polariton branch crosses the photon line $q = \omega/c$. Here in the absence of damping the two media are impedance matched and all the incident power is transmitted.

The reflection formula (2.156) can be applied with ε taking the plasma form (2.132) with the polariton dispersion curve illustrated in figure 2.11. As mentioned, formally the plasma dielectric function is the $\omega_T \rightarrow 0$ limit of the phonon function. Consequently the reflectivity curve resembles figure 2.12 but with the $R \approx 1$ region stretching down to zero frequency.

We shall subsequently discuss a number of reflectivity spectra taken in oblique incidence, and it will be convenient to define notation at this point. As seen in figure 2.13, in oblique incidence the direction of the incident beam and the normal to the surface define a plane, called the *plane of incidence*. As in section 2.2.2 for acoustic waves, the polarization (direction of **E** field) in a plane-polarized incident beam can be defined with reference to the plane of incidence. The conventional terminology, as indicated in figure 2.13, is p-polarization for **E** in the plane of incidence and s-polarization for **E** transverse. Even for non-dispersive media, the reflectivities for the two

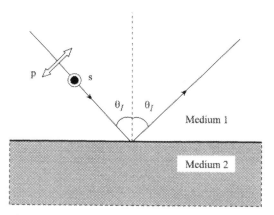

Figure 2.13. Notation and definitions for oblique-incidence reflectivity. The angle of incidence θ_1 and the directions of the optical **E** field in s- and p-polarization are marked.

polarizations are quite different. In particular, the p-reflectivity is zero when the angle of incidence is equal to the Brewster angle $\theta_B = \tan^{-1}(\varepsilon^{1/2}/\varepsilon_1^{1/2})$ but the s-reflectivity is never zero. For dispersive media, though, a high-reflectivity region is generally seen in the reststrahl band because there is no mode to carry transmitted power. With caution, therefore, our discussion of normal incidence can be applied to help give a general view of oblique-incidence spectra.

2.5.3 Exciton–polaritons

As mentioned in section 2.4.1, the exciton–polariton is the most important case in which spatial dispersion of the dielectric function must be taken into account. *Excitons* are bound electron–hole pairs occurring in semi-conductors; introductory discussions are given by Kittel (1986) and Seeger (2002), while Rashba and Sturge (1982) give a comprehensive account. The type of exciton of interest here arises in a direct-gap, polar semiconductor, such as GaAs or ZnSe. It is formed from an electron near the bottom of the conduction band and a hole near the top of the valence band. Since the valence band usually comprises a heavy-hole band, a light-hole band and a band split off by the spin–orbit coupling, and the bands are not strictly isotropic, the full theory of the exciton is quite complicated. We consider the simplest model, in which an electron of effective mass m_e is bound by the Coulomb interaction to a hole of effective mass m_h.

If the exciton moves with wavevector **q** it is found to have energy

$$E(q) = \hbar\omega(q) = \hbar\omega_e + \hbar^2 q^2/2M \qquad (2.157)$$

where $M = m_e + m_h$ and $\hbar\omega_e = E_g - E_0$ is the energy of the exciton at rest (the band-gap energy E_g reduced by the exciton binding energy E_0). Since the exciton couples strongly to light it gives a contribution to the dielectric function, which is found to be

$$\varepsilon(q,\omega) = \varepsilon_\infty + [S/(\omega_e^2 + Dq^2 - \omega^2 - i\omega\Gamma)] \qquad (2.158)$$

where $D = 1/2M$, S is the dipole strength of the exciton resonance and ε_∞ is the background dielectric constant.

The exciton–polariton dispersion relation is found by solving Maxwell's equations with the use of the dielectric function of (2.158). The result is an obvious generalization of (2.154):

$$\varepsilon(q,\omega)[q^2 - \varepsilon(q,\omega)\omega^2/c^2] = 0. \qquad (2.159)$$

This gives a longitudinal mode if $\varepsilon(q,\omega) = 0$ and a transverse mode if the second factor vanishes. For the moment we concentrate on the latter. As explained in section 2.5.1, the dispersion relation is most meaningful when damping is neglected. The vanishing of the second factor in (2.159) then gives a quadratic equation in q^2. It is easily shown that there is one real solution for q for $\omega < \omega_L$ but two for $\omega > \omega_L$, where $\omega_L(>\omega_e)$ is the generalization of the corresponding phonon–polariton parameter. The dispersion curve, as sketched in figure 2.14, resembles that for a phonon–polariton, figure 2.10, except that the lower branch bends up for large q. As shown in the inset to figure 2.14, the curve may be viewed as resulting from the 'crossing' of the photon dispersion curve with the exciton dispersion of (2.157). An important practical difference is that exciton–polaritons occur near the

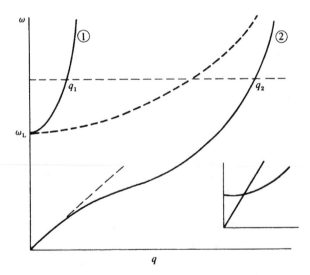

Figure 2.14. Sketch of exciton–polariton dispersion curves (full lines). Inset shows origin of curve in crossing of photon and exciton curves. The two transverse modes propagating at frequency $\omega > \omega_L$ are indicated. The longitudinal mode is also shown (broken line).

semiconductor gap frequency, typically in the visible or near infrared region, whereas phonon–polaritons occur in the far infrared.

Figure 2.14 brings out a major point of interest. Suppose s-polarized light of frequency $\omega > \omega_L$ is incident from vacuum on the surface of an excitonic medium. As indicated, two transverse modes of wavevectors q_1 and q_2 can propagate in the medium at this frequency. It should therefore be possible to evaluate two transmission coefficients T_1 and T_2 and one reflection coefficient R to give the amplitudes transmitted into the two propagating modes and reflected into the vacuum. However, Maxwell's equations yield only two boundary conditions. Thus, just as for the dipole-exchange modes discussed later in sections 5.4 and 5.5, an *additional boundary condition* (ABC) is required. For the dipole-exchange modes the ABC can be found from the microscopic torque equations of motion, since the whole continuum theory may be derived from the microscopic theory. Similarly in the present case the ABC can in principle be derived from the quantum-mechanical equation of motion of the exciton in the presence of the surface. However, since the exciton is an electron–hole pair, what is required is in effect the solution of a quantum-mechanical three-body problem. The exact theory is rather intractable, and various phenomenological forms are often postulated, such as

$$\partial \mathbf{P}/\partial z + \xi \mathbf{P} = 0 \qquad (2.160)$$

where $\mathbf{P}$ is the excitonic polarization. This is analogous to the macroscopic dipole-exchange theory (see section 5.4.1). Expressions for the reflection and transmission coefficients with the above ABC are given by Tilley (1980), for example. An alternative description of the surface is called the *dead-layer model*, in which it is assumed that $\mathbf{P} = 0$ in a layer of thickness $2a_B$ at the surface (a_B denotes the Bohr radius of the exciton). The boundary conditions that are applied are the usual electromagnetic ones at the two boundaries of the dead layer, together with $\mathbf{P} = 0$ at the interface between the dead layer and the rest of the medium.

For p-polarized incident light the situation is somewhat more complicated. In addition to the two transverse modes, a longitudinal mode is excited. As seen from (2.159) its wavevector $\mathbf{q}_L$ is given by $\varepsilon(\mathbf{q}_L, \omega) = 0$. The ABC, (2.160) or whatever is used, then has two components, so that altogether there are four boundary conditions. These are sufficient to determine the reflection coefficient R, two transverse transmission coefficients T_1 and T_2 and a longitudinal transmission coefficient T_L. It might be thought that ordinary reflectivity measurements would discriminate between different ABCs. In practice, however, because of the number of free parameters and variability of specimens, experimental data can be fitted by a range of ABCs. Some discussion of reflectivity experiments is given by Weisbuch and Ulbrich (1982).

As was suggested originally by Brenig *et al* (1972), the properties of exciton–polaritons have been investigated by resonant Brillouin scattering,

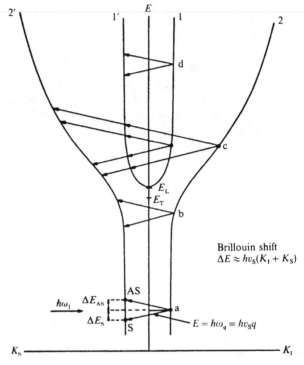

Figure 2.15. Illustration of the kinematics of Brillouin scattering via the transverse exciton–polariton modes. The lines denoting scattering processes have slopes equal to $+v_A$ (Stokes) and $-v_A$ (anit-Stokes), where v_A is the acoustic-phonon velocity. For incident frequency $\omega < \omega_L$ only one Stokes and one anti-Stokes process are possible, but for $\omega > \omega_L$ four of each occur. After Weisbuch and Ulbrich (1982).

resonant meaning here that the incident frequency ω_I is close to ω_L. As shown in figure 2.15, light of frequency $\omega > \omega_L$ incident on the medium generates two transmitted waves at wavevectors q_1 and q_2. As a simplification, let us first ignore complications due to the absorption of light in the medium and reflection of phonons at the surface. The Brillouin scattering may then be described by the kinematics of simple bulk scattering, as described in section 1.4.1. For normal-incidence backscattering, the Brillouin process must scatter a polariton from one of the $+q$ branches to one of the $-q$ branches. Conservation of frequency and wavevector shows that the frequency shift for Stokes scattering is found from a line of slope v_s equal to the relevant acoustic phonon velocity. Likewise, anti-Stokes scattering requires a line of slope $-v_s$. These are illustrated in figure 2.15. It is seen that for $\omega > \omega_L$, both Stokes and anti-Stokes lines split into four, corresponding to the four scattering possibilities $1 \to 1'$, $1 \to 2'$, $2 \to 1'$ and $2 \to 2'$.

A description of the scattering that includes light absorption and phonon reflection can be given by an extension of the calculation for normal-incidence Brillouin scattering to be outlined in section 3.3.2; the details are found in Tilley (1980). Experimental results on a range of polar semiconductors are reviewed by Weisbuch and Ulbrich (1982).

2.6 Magnetic susceptibilities, polaritons and magnetostatic modes

The dielectric function is the natural quantity for describing the interaction of a nonmagnetic system with electromagnetic radiation. We now turn to the corresponding techniques for magnetic systems. Some discussion has already been given in section 2.3.4 on magnetic dipole–dipole interactions. Here we study the alternative formulation by means of a frequency-dependent magnetic susceptibility; it will be seen that some of the previous results can be rederived from this point of view.

The correct description of magnetic response depends on the frequency of the interacting radiation (Landau and Lifshitz 1971). For frequencies in the visible or near-infrared range, it is necessary in general to use a gyrotropic dielectric tensor $\ddot{\varepsilon}$ of the form of (2.133). However, for frequencies around the magnetic resonance, that is, in the microwave region for ferromagnets and in the far infrared for antiferromagnets, the analogous quantity is the frequency-dependent, *gyromagnetic susceptibility tensor* $\ddot{\chi}$. This is what we now discuss.

2.6.1 Susceptibilities

We consider first a ferromagnet in the presence of a magnetic field containing a static part (which we take along z) and an electromagnetic part at frequency ω arising, for example, from incident radiation:

$$\mu_0 \mathbf{H} = B_0 \hat{\mathbf{z}} + \mu_0 \mathbf{h}(\mathbf{r}) \exp(-i\omega t). \tag{2.161}$$

We are concerned with the linear response, so we take the magnetization in the form

$$\mathbf{M} = M_0 \hat{\mathbf{z}} + \mathbf{m}(\mathbf{r}) \exp(-i\omega t) \tag{2.162}$$

where M_0 is the static magnetization. To relate $\mathbf{m}(\mathbf{r})$ to $\mathbf{h}(\mathbf{r})$ we use the continuum form of the equation of motion (2.74), namely

$$d\mathbf{M}/dt = \gamma(\mathbf{M} \times \mathbf{B}) = \gamma\mu_0(\mathbf{M} \times \mathbf{H}). \tag{2.163}$$

Substituting (2.161) and (2.162) and then linearizing (2.163) gives

$$-i\omega\mathbf{m}(\mathbf{r}) = \gamma\hat{\mathbf{z}} \times [\mu_0 M_0 \mathbf{h}(\mathbf{r}) - B_0 \mathbf{m}(\mathbf{r})] \tag{2.164}$$

where it has been assumed that $T \ll T_C$ so that $|\mathbf{m}| \ll M_0$. Equation (2.164) can be re-expressed as a linear susceptibility relation involving the x and y components of the fluctuating fields:

$$\begin{pmatrix} m_x \\ m_y \end{pmatrix} = \begin{pmatrix} \chi_a & i\chi_b \\ -i\chi_b & \chi_a \end{pmatrix} \begin{pmatrix} h_x \\ h_y \end{pmatrix} \tag{2.165}$$

where

$$\chi_a = \omega_m \omega_0 / (\omega_0^2 - \omega^2) \tag{2.166}$$

$$\chi_b = \omega_m \omega / (\omega_0^2 - \omega^2) \tag{2.167}$$

and we have defined the frequencies

$$\omega_0 = \gamma B_0, \qquad \omega_m = \gamma \mu_0 M_0. \tag{2.168}$$

It is seen from (2.166) and (2.167) that both χ_a and χ_b have poles at the ferromagnetic resonance frequency ω_0, which appeared previously in (2.70) as the long-wavelength limit of the magnon frequency. The electromagnetic wavelength is much greater than the interatomic spacing and it is natural that the response should be resonant at the relevant frequency of the system. It may be verified that $\bar{\chi}$ is diagonalized by the transformation $m^{\pm} = m_x \pm i m_y$, $h^{\pm} = h_x \pm i h_y$ to give

$$\begin{pmatrix} m^+ \\ m^- \end{pmatrix} = \begin{pmatrix} \omega_m / (\omega_0 - \omega) & 0 \\ 0 & \omega_m / (\omega_0 + \omega) \end{pmatrix} \begin{pmatrix} h^+ \\ h^- \end{pmatrix} = \begin{pmatrix} \chi^+ & 0 \\ 0 & \chi^- \end{pmatrix} \begin{pmatrix} h^+ \\ h^- \end{pmatrix} \tag{2.169}$$

where $\chi^{\pm} = \chi_a \pm \chi_b$.

A similar derivation to that just given holds for a two-sublattice anti-ferromagnet. We outline the derivation for the case when the static field is along the ordering direction $\hat{\mathbf{z}}$. For each sublattice (labelled by $p = 1, 2$) we write instead of (2.163)

$$\frac{d\mathbf{M}_p}{dt} = \gamma \mu_0 (\mathbf{M}_p \times \mathbf{H}_p). \tag{2.170}$$

This follows by the same continuum limiting process as described for the ferromagnetic case. The effective sublattice magnetizations and fields are expressed as sums of static and r.f. terms:

$$\mathbf{M}_p = \pm M_0 \hat{\mathbf{z}} + \mathbf{m}_p(\mathbf{r}) \exp(-i\omega t) \tag{2.171}$$

$$\mu_0 \mathbf{H}_p = [B_0 \pm (B_A + B_E)]\hat{\mathbf{z}} + \mu_0 \mathbf{h}_p \exp(-i\omega t). \tag{2.172}$$

Here the upper and lower signs correspond to $p = 1$ and $p = 2$ respectively. Following the notation in section 2.3.3, B_0 and B_A are the static applied field and the effective anisotropy field, respectively. Also B_E is the static effective exchange field and M_0 is the static sublattice magnetization (both having the same magnitude, but opposite sign, for each sublattice). We take the same r.f.

field on each sublattice (putting $\mathbf{h}_1 = \mathbf{h}_2 \equiv \mathbf{h}$), since this is the case in practice, and linearize (2.170) in $\mathbf{h}$ and $\mathbf{m}_p$. The susceptibility that corresponds to interaction with electromagnetic fields is the response of the total magnetization $\mathbf{m}(\mathbf{r}) = \mathbf{m}_1(\mathbf{r}) + \mathbf{m}_2(\mathbf{r})$ to $\mathbf{h}$. It may be shown (e.g. Sarmento and Tilley 1982, Mills 1984) that (2.165) applies with

$$\chi_a = \tfrac{1}{2}(\chi^+ + \chi^-), \qquad \chi_b = \tfrac{1}{2}(\chi^+ - \chi^-) \tag{2.173}$$

where, for the antiferromagnetic case, we have

$$\chi^{\pm} = \frac{2\omega_A \omega_m}{\omega_A(2\omega_E + \omega_A) - (\omega \mp \omega_0)^2}. \tag{2.174}$$

The frequencies ω_0 and ω_m are defined as in (2.168); additionally $\omega_E = \gamma\mu_0 H_E$ and $\omega_A = \gamma\mu_0 H_A$ consistent with section 2.3.3.

Some properties of (2.174) are worth noting. In contrast to the dielectric functions for a plasma and an ionic medium, (2.132) and (2.146), which are completely determined by properties of the medium, $\ddot{\chi}$ here depends on the applied field; the same is true of the ferromagnet, (2.166) and (2.167). In the absence of a static field, $\omega_0 = 0$, the susceptibilities in (2.174) have a pole at $\omega_{AF} = [\omega_A(2\omega_E + \omega_A)]^{1/2}$. This is the antiferromagnetic resonance frequency, which previously appeared in (2.88) as the long-wavelength magnon frequency. In the absence of a field, furthermore, $\chi^+ = \chi^-$, so that $\ddot{\chi}$ is diagonal ($\chi_b = 0$). In the presence of a field, matters become more complicated. As mentioned, (2.174) hold when the static field is applied along $\hat{\mathbf{z}}$, the easy axis for the sublattice magnetizations. In this case, it is seen that $\ddot{\chi}$ is non-diagonal and the resonances in χ^+ and χ^- are Zeeman split to $\omega_{AF} \pm \omega_0$ (provided $\omega_0 < \omega_{AF}$). If the applied field is sufficiently large that $\omega_0 = \omega_{AF}$, it implies that one of the frequencies goes to zero. This signals the onset of a phase transition to a state (the so-called *spin–flop phase*) in which the sublattice magnetizations are no longer along $\hat{\mathbf{z}}$ and $-\hat{\mathbf{z}}$, but instead are canted relative to the field direction; details can be found, e.g. in White (1983).

Even when the field is smaller but is applied at an angle to $\hat{\mathbf{z}}$, the equilibrium directions of $\mathbf{M}_1$ and $\mathbf{M}_2$ change since they are pulled towards the field direction. A calculation of $\ddot{\chi}$ can still be carried out using linearization of the equations of motion (Almeida and Mills 1988) but the resulting form is much more complicated than that shown here. We note in passing that even the calculation of the bulk magnon dispersion curve, along the lines of section 2.3.3, for general field direction has only been carried out quite recently (Osman *et al* 1996).

Apart from the linearization, two simplifications have been made in these derivations. First, no allowance has been made for spatial dispersion, that is, q dependence of $\ddot{\chi}$. The lowest-order q dependence can be included by retaining ∇^2 terms of the kind appearing in (2.81); we return to this in sections 5.4 and 5.5 when discussing dipole-exchange modes (but see also

Lim *et al* 1998). Second, we have omitted the damping terms that could be included in a fuller treatment.

Given the susceptibility tensor for either a ferromagnet or an antiferromagnet, we are now in a position to describe the coupling of the magnet to electromagnetic radiation. The general method is to solve Maxwell's equations incorporating $\bar{\chi}$, which will be done in section 2.6.2. The resulting modes are magnetic polaritons, and occupy the 'retarded mode' region $q \approx \omega/c$ identified in figure 2.5. They are of interest for antiferromagnets since ω_{AF} typically lies in the far infrared. For ferromagnets, however, with the resonance frequency ω_0 in the microwave part of the spectrum, the corresponding wavelength $2\pi/q$ is much larger than typical sample dimensions so simple retarded bulk modes are not of interest. However, for ferromagnets, the magnetostatic modes, occurring for $q \gg \omega/c$ are of real importance. They have already been discussed briefly from one point of view in section 2.3.4 and we look at them from the present perspective in section 2.6.3.

2.6.2 Magnetic polaritons

The bulk polariton modes were first discussed by Auld (1960) for a ferromagnet; later authors dealt with antiferromagnets (Manohar and Venkataraman 1972, Bose *et al* 1975, Sarmento and Tilley 1977) and ferrimagnets (Oliveira *et al* 1979). Maxwell's equations lead to

$$q^2 \mathbf{h} - \mathbf{q}(\mathbf{q} \cdot \mathbf{h}) = (\varepsilon \omega^2/c^2)(\mathbf{h} + \mathbf{m}) \tag{2.175}$$

for a plane wave $\exp(\mathrm{i}\mathbf{q} \cdot \mathbf{r} - \mathrm{i}\omega t)$. With choice of axes such that the static magnetic field is in the z direction, there is no loss of generality in choosing $q_y = 0$. Equation (2.175) then yields

$$h^+ = \xi_1 m^+ - \xi_2 m^- \tag{2.176}$$

$$h^- = -\xi_2 m^+ + \xi_1 m^- \tag{2.177}$$

where $h^\pm$ and $m^\pm$ were introduced in section 2.6.1 and

$$\xi_1 = \tfrac{1}{2}(2\varepsilon\omega^2/c^2 - q_x^2)/(q^2 - \varepsilon\omega^2/c^2) \tag{2.178}$$

$$\xi_2 = \tfrac{1}{2}q_x^2/(q^2 - \varepsilon\omega^2/c^2). \tag{2.179}$$

When combined with the susceptibility equations $m^\pm = \chi^\pm h^\pm$ these yield the polariton dispersion relation in a general form:

$$(1 - \xi_1\chi^+)(1 - \xi_1\chi^-) = \xi_2^2\chi^+\chi^-. \tag{2.180}$$

It is clear from (2.178) and (2.179) that, as might be expected, the dispersion relation depends on the angle θ between the propagation vector $\mathbf{q}$ and the z axis. The special cases $\theta = 0$ (propagation along the static field) and $\theta = \pi/2$ (propagation transverse to the static field) are worth singling out; they are

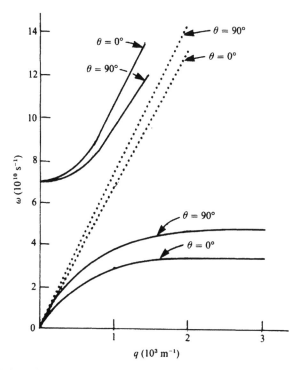

Figure 2.16. Dispersion curves for bulk polaritons propagating along the static magnetic field ($\theta = 0$) and perpendicular to it ($\theta = 90°$) in a bulk ferromagnet: $B_0 = \mu_0 M_0 = 0.175\,\text{T}$. After Auld (1960).

usually called the *Faraday* and *Voigt* geometries respectively. For the former, $q_x = 0$ and so $\xi_2 = 0$, it follows from (2.180) that two modes propagate given by $\xi_1 \chi^{\pm} = 1$. The susceptibilities $\chi^{\pm}$ correspond to the two signs of circular polarization so that these are the eigenmodes for propagation in the Faraday geometry. Furthermore, it follows from the forms of $\chi^{\pm}$ for a ferromagnet that χ^{+} is resonant for $\omega = \omega_0$, that is, in a positive static field, while χ^{-} is resonant for $\omega = -\omega_0$, that is, in a negative static field. Thus in a positive field, the $+$ mode has a typical polariton form and the $-$ mode does not couple strongly to the electro-magnetic field. These properties are seen in the $\theta = 0$ dispersion curve of figure 2.16.

For the Voigt geometry in a ferromagnet, $q_x = q$. It can then be shown that the eigenmodes are the states of plane polarization with **h** along y and along z. The former couples to the magnetic resonance but the latter does not, so as in the Faraday geometry there is one 'polariton-type' dispersion curve and one 'photon-type'. This is also shown in figure 2.16.

Equation (2.180) applies equally to antiferromagnets, and as in the case of ferromagnets the dispersion relation depends on the angle θ. Here again, the eigenmodes are the circular polarization states in the Faraday geometry

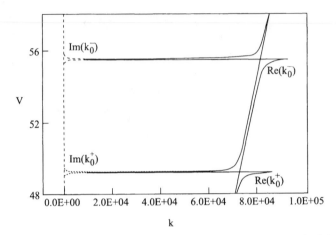

Figure 2.17. Faraday-geometry polariton dispersion curves for the uniaxial antiferromagnet FeF_2 in a magnetic field of 3 T. Vertical axis is $\omega/2\pi c$ in cm^{-1} and horizontal axis is $k \equiv q$ in m^{-1}. Parameters are (Brown *et al* 1994) $\mu_0 M_0 = 0.056$ T, $B_E = 53.3$ T, $B_A = 19.7$ T and dielectric constant $\varepsilon = 5.5$. The superscripts $+$ and $-$ denote the two states of circular polarization. After Lim *et al* (1998).

$(\theta = 0)$ and plane polarization states in the Voigt geometry $(\theta = \pi/2)$. For simplicity, we show in figure 2.17 just the former. The curves are drawn for a small value of damping consistent with the reflectivity spectra of Brown *et al* (1994). Because of the nonzero applied field, the dispersion curves for the $+$ and $-$ modes are Zeeman split in accordance with (2.174). Each resembles the phonon–polariton dispersion curve, figure 2.10, but the reststrahl region where q is essentially imaginary occupies a very small interval.

Far infrared (FIR) reflectivity spectra related to these dispersion curves are shown in figure 2.18. It should be said that this form of spectroscopy on antiferromagnets is very demanding and a sophisticated instrumental design is required to obtain the necessary resolution. As indicated, the spectra are taken in oblique incidence in s-polarization; this means that the FIR **h** field is in the plane of incidence and therefore couples to both resonant modes (see the denominator in (2.174) in the susceptibility elements). The spectra show two narrow reststrahl-like features with peaks in R separated by the Zeeman splitting. It is characteristic of magnetic spectra that the minimum in R lies on the low-frequency side of the line rather than the high-frequency side as in a dielectric, figure 2.12. The theoretical curves in figure 2.18 are based on a generalization of the standard electromagnetic analysis of reflectivity (Lipson *et al* 1995) to include the susceptibility (2.174). Within the experimental arrangement used for figure 2.18 it is possible to rotate the sample about a horizontal axis so that the field is at an angle to the easy axis. The corresponding spectra together with an analysis using the

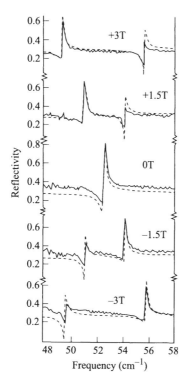

Figure 2.18. 4.2 K reflectivity spectra of the uniaxial antiferromagnet FeF_2 in different fields. The applied field is along the easy (z) axis (vertical) and the plane of incidence is horizontal with angle of incidence $\theta = 45°$. Resolution is $0.125\,cm^{-1}$. Spectra are for s-polarization, i.e. FIR **E** field vertical. Continuous curves, experiment; dashed curves, theory. After Brown *et al* (1994).

relevant expression for $\bar{\bar{\chi}}$ (Almeida and Mills 1988) are given by Abraha *et al* (1994) and a full review of the theory is in Abraha and Tilley (1996).

Perhaps the most striking property of the spectra in figure 2.18 is that they exhibit *non-reciprocity*, that is, the reflectivity changes when the field is reversed, $R(-H_0) \neq R(H_0)$. Non-reciprocity derives ultimately from the lack of time reversal symmetry in a magnetic field, but it requires in addition the presence of a surface to remove translational invariance; further examples will be encountered in chapter 5. The symmetry arguments are given in the review of this topic by Camley (1987) and there is a more recent discussion by Dumelow *et al* (1998).

2.6.3 Magnetostatic modes

We have already, in section 2.3.4, introduced the magnetostatic modes of a bulk ferromagnet starting from the Hamiltonian with dipole–dipole interactions included. They may also, more conventionally, be discussed from the viewpoint of macroscopic electrodynamics with use of the susceptibility functions of section 2.6.1. The polaritons of the previous section are sometimes called *retarded modes* since the derivation retains a finite value for the velocity of light so that the interaction between distant spins is not instantaneous. For magnetostatic modes we take the approximation $c \to \infty$ and the modes are also called unretarded. With $c \to \infty$ the coupling between the magnetic and electric fields is removed and the relevant Maxwell's equations take the form

$$\nabla \times \mathbf{h}(\mathbf{r}) = 0, \qquad \nabla \cdot [\mathbf{h}(\mathbf{r}) + \mathbf{m}(\mathbf{r})] = 0. \tag{2.181}$$

The first of these means that a scalar potential ψ, the magnetostatic potential, can be introduced:

$$\mathbf{h} = \nabla \psi. \tag{2.182}$$

With the use of (2.165) to express $\mathbf{m}$ in terms of $\mathbf{h}$, the divergence equation in (2.181) leads to

$$(1 + \chi_a)\left(\frac{\partial^2 \psi}{\partial x^2} + \frac{\partial^2 \psi}{\partial y^2}\right) + \frac{\partial^2 \psi}{\partial z^2} = 0 \tag{2.183}$$

and for ψ expressed in a plane-wave form $\exp(i\mathbf{q} \cdot \mathbf{r} - i\omega t)$ this gives, with $q^2 = q_x^2 + q_y^2 + q_z^2$,

$$(1 + \chi_a)q^2 - \chi_a q_z^2 = 0. \tag{2.184}$$

Equation (2.184) can be shown to be identical to the earlier expression (2.96) for the magnetostatic mode frequency of a ferromagnet without exchange and for zero demagnetization. In (2.96) θ is the angle between $\mathbf{q}$ and the z axis, so that we can put $q_z = q\cos\theta$ and (2.184) reads

$$1 + \chi_a \sin^2 \theta = 1 + \frac{\omega_m \omega_0}{\omega_0^2 - \omega^2}\sin^2 \theta = 0 \tag{2.185}$$

where the form of χ_a has been substituted from (2.166). Given the definitions of ω_0 and ω_m in (2.168) it is easy to show that (2.185) reduces to (2.96). It may be asked why the two different-looking derivations lead to the same result. The point is that the dipole–dipole interaction (2.58) is derived from Maxwell's equations without retardation. Both derivations therefore have the same starting point: the Hamiltonian for the magnetic system, approximated in continuum form, and the magnetostatic equations. The difference is simply the order in which the steps of the derivation are taken.

The derivation above applies equally to antiferromagnets, so that with the understanding that $\hat{\mathbf{z}}$ is the direction of the static field and of the sublattice orderings, the magnetostatic modes are given by (2.184) with χ_a substituted from (2.173). In this case there are two modes split in frequency, and the equivalent of (2.96) is found to be

$$\omega^2(\mathbf{q}) = [\omega_A(2\omega_E + \omega_A) + \omega_A\omega_m \sin^2 \theta]$$
$$\pm \{\omega_A^2\omega_m^2 \sin^4 \theta + 4\omega_0^2[\omega_A(2\omega_E + \omega_A) + \omega_A\omega_m \sin^2 \theta]\}^{1/2}. \quad (2.186)$$

The above result simplifies in the case of zero field ($\omega_0 = 0$) when the lower of the two modes occurs at the AFMR frequency (2.88). Further discussion is given in chapter 5.

For a third, and final, view of bulk magnetostatic modes we remark that the limit $c \to \infty$ used in this section is equivalent, as expected, to the limit $q \gg \omega/c$ of the retarded modes of section 2.6.2. In that limit the two quantities ξ_1 and ξ_2 in (2.178) and (2.179) become

$$\xi_1 \to -q_x^2/2q^2 = -\tfrac{1}{2}\sin^2 \theta, \qquad \xi_2 \to q_x^2/2q^2 = \tfrac{1}{2}\sin^2 \theta. \quad (2.187)$$

Equation (2.180) then reduces to

$$1 - \tfrac{1}{2}(\chi^+ + \chi^-)\sin^2 \theta = 0 \quad (2.188)$$

which in view of (2.173) is the same as (2.184).

References

Abraha K, Brown D E, Dumelow T, Parker T J and Tilley D R 1994 *Phys. Rev. B* **50** 6808
Abraha K and Tilley D R 1996 *Surf. Sci. Reports* **24** 125
Abrikosov A A, Gorkov L P and Dzyaloshinski I E 1963 *Methods of Quantum Field Theory in Statistical Physics* (New Jersey: Prentice-Hall)
Almeida N S and Mills D L 1988 *Phys. Rev. B* **37** 3400
Anderson P W 1963 in *Magnetism* vol 1 ed Rado G T and Suhl H (New York: Academic) p 25
Ashcroft N W and Mermin N D 1976 *Solid State Physics* (New York: Holt, Rinehart and Winston)
Auld B A 1960 *J. Appl. Phys.* **31** 1642
Bland J A C and Heinrich B (eds) 1994 *Ultrathin Magnetic Structures I* (Berlin: Springer)
Blinc R and Zeks B 1974 *Soft Modes in Ferroelectrics and Antiferroelectrics* (Amsterdam: North-Holland)
Bloch F 1930 *Z. Phys.* **61** 206
Boardman A D 1982 in *Electromagnetic Surface Modes* ed A D Boardman (Chichester: Wiley)
Born M and Huang K 1985 *Dynamical Theory of Crystal Lattices* (Oxford: Clarendon)
Born M and Wolf E 1980 *Principles of Optics* 6th edition (Oxford: Pergamon)
Bose M S, Foo E N and Zuniga M A 1975 *Phys. Rev. B* **12** 3885
Bransden B H and Joachain C J 2000 *Quantum Mechanics* 2nd edition (Englewood Cliffs, NJ: Prentice-Hall)
Brenig W, Zeyher R and Birman J L 1972 *Phys. Rev. B* **6** 4617
Brown D E, Dumelow T, Parker T J, Abraha K and Tilley D R 1994 *Phys. Rev. B* **49** 12266
Burstein E and DeMartini F 1974 *Polaritons* (New York: Pergamon)

Camley R E 1987 *Surf. Sci. Reports* **7** 103

Cohen M H and Keffer F 1955 *Phys. Rev.* **99** 1128

Cottam M G and Maradudin A A 1984 in *Surface Excitations* ed V M Agranovich and R Loudon (Amsterdam: North-Holland) p 1

Dumelow T, Camley R E, Abraha K and Tilley D R 1998 *Phys. Rev. B* **58** 897

Fetter A L and Walecka J D 1971 *Quantum Theory of Many-Particle Systems* (New York: McGraw Hill)

Heisenberg W 1928 *Z. Phys.* **49** 619

Herring C 1966 in *Magnetism* vol 4 ed G Rado T and H Suhl (New York: Academic)

Jackson J D 1998 *Classical Electrodynamics* 3rd edition (New York: Wiley)

Keffer F 1966 *Handbuch der Physik* **18**(2) 1

Keldysh L V, Kirzhnitz D A and Maradudin A A 1989 *The Dielectric Function of Condensed Systems* (Amsterdam: North-Holland)

Kittel C 1986 *Introduction to Solid State Physics* 6th edition (New York: Wiley)

Landau L D and Lifshitz E M 1970 *Theory of Elasticity* (Oxford: Pergamon)

Landau L D and Lifshitz E M 1971 *Electrodynamics of Continuous Media* (Oxford: Pergamon)

Levstik A, Tilley D R and Zeks B 1984 *J. Phys. C* **17** 3793

Lim S C, Osman J and Tilley D R 1998 *J. Phys.: Condens. Matter* **10** 1891

Lindhard J 1954 *Kgl. Danske Mat.-fys. Medd.* **28** 8

Lines M E and Glass A M 1979 *Principles and Applications of Ferroelectrics and Related Materials* (Oxford: Oxford University Press)

Lipson S G, Lipson H and Tannhauser D S 1995 *Optical Physics* 3rd edition (Cambridge: Cambridge University Press)

Lovett D L 1999 *Tensor Properties of Crystals* 2nd edition (Bristol: Adam Hilger)

Manohar C and Venkataraman G 1972 *Phys. Rev. B* **5** 1993

Maradudin A A, Montroll E W, Weiss G H and Ipatova I P 1971 *Theory of Lattice Dynamics in the Harmonic Approximation* (New York: Academic)

Mattis D C 1981 *The Theory of Magnetism I* (Berlin: Springer)

Mills D L 1984 in *Surface Excitations* ed V M Agranovich and R Loudon (Amsterdam: North-Holland) p 379

Oliveira F A, Khater A F, Sarmento E F and Tilley D R 1979 *J. Phys. C* **12** 4021

Osman J, Almeida N S and Tilley D R 1996 *J. Phys.: Condens. Matter* **8** 11181

Palik E D, Kaplan R, Gammon R W, Kaplan H, Wallis R F and Quinn J J 1976 *Phys. Rev. B* **13** 2497

Rashba E I and Sturge M D 1982 *Excitons* (Amsterdam: North-Holland)

Rickayzen G D 1980 *Green Functions and Condensed Matter* (London: Academic)

Sarmento E F and Tilley D R 1977 *J. Phys. C* **10** 795

Sarmento E F and Tilley D R 1982 in *Electromagnetic Surface Modes* ed A D Boardman (Chichester: Wiley)

Seeger K 2002 *Semiconductor Physics* 8th edition (Berlin: Springer)

Tilley D R 1980 *J. Phys. C* **13** 781

Weiss P 1907 *J. Phys.* **6** 661

Weisbuch C and Ulbrich R G 1982 in *Topics in Applied Physics* **51**: *Light Scattering in Solids III* ed M Cardona and G Guntherodt (Berlin: Springer) p 207

White R M 1983 *Quantum Theory of Magnetism* (Berlin: Springer)

PART TWO

SINGLE-SURFACE, FILM AND RELATED EXCITATIONS

Chapter 3

Surface phonons and elastic modes

We start the study of surfaces with the modes that can propagate parallel to the surface of an elastic solid and along an elastic film (or slab). These can be described by the discrete model of lattice dynamics or by the continuum model of elasticity theory; the corresponding results for a bulk medium were given in sections 2.1 and 2.2. Modifications to the bulk modes are discussed in addition to the properties of the surface modes.

3.1 Lattice dynamics

As mentioned in section 2.1.2, the simplest example of a mode that is localized at a surface occurs for the diatomic lattice. The arrangement of interest is shown in figure 3.1, which is like figure 2.1(c) with the 'crystal' terminating on a mass m_2. The bulk modes were discussed in section 2.1.2 and the dispersion graph is figure 2.2; we recall in particular that in the stop band $(2C/m_1)^{1/2} < \omega < (2C/m_2)^{1/2}$ the Bloch wavenumber q has the form $\pi/2a + iy$, where y is real. One particular such value of q is associated with a surface excitation, as we shall now see.

All the masses, except the end one, have the equation of motion (2.12) or (2.13). For the end mass the equivalent is

$$m_2 \frac{d^2 u_0}{dt^2} = C(u_1 - u_0). \tag{3.1}$$

Suppose we now substitute the trial solutions (2.14) and (2.15) into the equations of motion (2.12), (2.13) and (3.1). All the equations except (3.1) are satisfied provided the ratio U_1/U_2 has the same form as in the bulk system. The condition for (3.1) to hold is

$$-m_2\omega^2 U_2 = C[U_1 \exp(iqa) - U_2]. \tag{3.2}$$

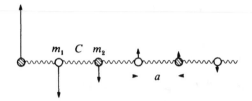

Figure 3.1. 'Surface' of a diatomic lattice (on left-hand side). The longitudinal displacement of each mass in the surface mode is proportional to the arrow shown.

Upon substitution of U_1/U_2 this becomes

$$2(C - m_2\omega^2)\cos(qa) = (2C - m_2\omega^2)\exp(iqa). \tag{3.3}$$

This equation and (2.16) are simultaneous equations for ω^2 and q. It is clear from (3.3) that they cannot be satisfied for real q. There is, however, a solution for real ω and complex q. Elimination of $\exp(iqa)$ between (3.3) and (2.16) gives

$$\omega^2(m_1 m_2\omega^2 - Cm_1 - Cm_2) = 0. \tag{3.4}$$

The first solution is $\omega^2 = 0$; implying $q = 0$ with all the u_n equal. That is, this solution corresponds to uniform translation of the lattice. The second solution is $\omega = \pm\omega_s$, where

$$\omega_s^2 = C(m_1^{-1} + m_2^{-1}). \tag{3.5}$$

This frequency lies in the stop band, as already shown in figure 2.2.

To understand the nature of the mode at frequency ω_s, we recall that within the stop band q is complex, $q = \pi/2a + iy$. Substitution into (2.16) with use of (3.5) gives

$$\sinh^2(ya) = (m_1 - m_2)^2/4m_1 m_2. \tag{3.6}$$

It then follows from the equations of motion that

$$\frac{u_{2n+1}}{u_{2n}} = -\frac{(2C - m_2\omega^2)\exp(-ya)}{2C\sinh(ya)} \tag{3.7}$$

and

$$\frac{u_{2n+2}}{u_{2n+1}} = \frac{2C\sinh(ya)\exp(-ya)}{2C - m_2\omega^2}. \tag{3.8}$$

These relations exhibit the relative phases of the motions of successive masses, and they show that the amplitude of the motion decreases exponentially with distance away from the surface. The amplitude and phase relations are shown schematically in figure 3.1. Because of the decreasing amplitudes it is natural to refer to the mode under discussion as a *surface mode*, in accordance with section 1.1.2.

The above calculation illustrates how a surface mode can arise within a stop band of a bulk dispersion equation. The surface mode we have found occurs when the lighter mass m_2 is the end atom and a repetition of the calculation shows that there is no surface mode when the system terminates at the heavier mass m_1. It can be seen also that, as mentioned before, there is no surface mode in a monatomic crystal. In fact, the diatomic crystal becomes monatomic if the masses are put equal, $m_1 = m_2$. In that case, it is seen from (3.6) that $y = 0$, so that the wavevector q is purely real and the mode at frequency ω_s is not localized at the surface. In terms of the bulk dispersion curve, figure 2.2, there is no stop band at $q = \pi/2a$ for $m_1 = m_2$. This follows from (2.16), which gives $\omega^2 = \omega_s^2$ for both branches in this case. Thus for $m_1 = m_2$ the surface mode degenerates into one of the bulk modes.

As it stands, the calculation we have outlined leaves one major question unanswered. Suppose the system also terminates at the right-hand end, and consists of N masses in all. There must be N normal modes, but in (3.4) we have identified only two of them. Thus we have 'lost' $N - 2$ modes, which looks like carelessness. A discussion of the N-mass problem is given by Wallis (1957). What happens is that for the finite system of N masses there are solutions of the complete set of equations of motion with real values of q. As can be seen from the discussion of the continuum limit in section 2.1.1, the equations of motion of the masses at each end of the line of masses play the same part as the boundary conditions in a continuum theory. Their effect here is that only $N - 2$ discrete real values q_n describe normal modes. Since these values are real they occur in the acoustic and optic mode regions of figure 2.2. It may be said that the effect of restricting the system to a finite length $(N - 1)a$ is to *discretize* or *quantize* the acoustic and optic mode continua. The complete set of normal modes consists of the $N - 2$ quantized bulk modes, with real q $(= q_n)$, and the two surface modes with complex q. For large N these two modes are localized, one at each end of the line of atoms. As $N \to \infty$, the spacing in frequency between the quantized modes becomes very small, and in the limit they become the bulk modes. We shall encounter a closely analogous situation with magnons in thin films in section 4.2.

It was explained in section 1.1.2 that because of translational invariance parallel to the surface, a surface mode is characterized by a wavevector $\mathbf{q}_\parallel$ in the surface plane. The 1D models that we have been concerned with so far can only describe modes that have no variation in the surface plane, that is, modes for which $\mathbf{q}_\parallel = 0$. We could in principle extend our discussion by considering the lattice dynamics of a surface on a 3D crystal. This was done for the first time by Wallis (1959), and later theoretical literature can be traced from Mazur and Maradudin (1981). As might be expected, the calculations become fairly laborious, since one is dealing with 3D displacements of masses in a 3D crystal structure. We do not pursue the matter further here.

    As will be seen below, surface modes with $q_{\parallel}$ near the centre of the 2D Brillouin zone have been studied experimentally by inelastic light scattering for some time. However, the investigation of surface modes for all values of $q_{\parallel}$ in the Brillouin zone was a purely theoretical matter until about 1980. After that date, the experimental techniques of inelastic neutral-atom scattering and inelastic electron scattering became sufficiently well developed to yield data on surface modes. The experimental techniques were described in section 1.4.2; we defer our discussion of the results until section 3.6, since it is convenient to deal first with the surface modes predicted by elasticity theory.

3.2   Surface elastic waves

3.2.1   Rayleigh waves

As mentioned in section 1.1.2, the free surface of an isotropic elastic medium supports a surface mode known as the *Rayleigh wave*. It results from a mixture of waves with p-transverse and longitudinal polarizations and its properties follow from extension of results given in sections 2.2.1 and 2.2.2. We look for a solution of the wave equation (2.26) and the corresponding equation for the transverse component, together with the boundary conditions (2.30) and (2.32). The displacement is written as a sum of longitudinal and transverse parts, $\mathbf{u} = \mathbf{u}_L + \mathbf{u}_T$, each part satisfying the wave equation with the appropriate velocity. In order to find a surface mode, we draw on the form of the pseudo-surface wave and assume that both parts are localized at the surface:

$$\mathbf{u}_{L,T} \propto \exp[i(q_x x - \omega t)]\exp(\kappa_{L,T} z). \tag{3.9}$$

Here we are assuming, as in section 2.2.2, that the sample occupies the half space $z < 0$ so that the real exponential factors in (3.9), with $\kappa_{L,T} > 0$, describe decay of the mode amplitude away from the surface. The complex exponential term is in accordance with the 2D form of Bloch's theorem, as in (1.25). It corresponds to propagation parallel to the surface plane along the x direction; since the medium is isotropic there is no loss of generality. We must assume that both the longitudinal and transverse components have the same value of q_x so that the boundary conditions can be satisfied for all values of x. Since the components separately satisfy the corresponding wave equation, the constants κ_L and κ_T are given by

$$\kappa_{L,T} = (q_x^2 - \omega^2/v_{L,T}^2)^{1/2}. \tag{3.10}$$

The x and z components of $\mathbf{u}_L$ and $\mathbf{u}_T$ are related through the equations $\nabla \times \mathbf{u}_L = 0$ and $\nabla \cdot \mathbf{u}_T = 0$, which imply

$$iq_x u_{Lz} - \kappa_L u_{Lx} = 0 \tag{3.11}$$

$$iq_x u_{Tx} + \kappa_T u_{Tz} = 0. \tag{3.12}$$

These can be satisfied by introducing two amplitude factors a and b:

$$\mathbf{u_L} = a(q_x, 0, -i\kappa_L) \exp[i(q_x x - \omega t)] \exp(\kappa_L z) \tag{3.13}$$

$$\mathbf{u_L} = b(\kappa_T, 0, -iq_x) \exp[i(q_x x - \omega t)] \exp(\kappa_T z). \tag{3.14}$$

With these forms substituted, the first boundary condition, (2.30), gives

$$2\kappa_L q_x a + (q_x^2 + \kappa_T^2)b = 0. \tag{3.15}$$

It is convenient to write the other boundary condition, (2.32), in terms of v_L and v_T. It is seen from (2.27) and (2.28) that

$$\sigma/(1 - \sigma) = 1 - 2v_T^2/v_L^2. \tag{3.16}$$

Then (2.32) yields

$$2v_T^2 \kappa_T q_x b + [v_L^2(\kappa_L^2 - q_x^2) + 2v_T^2 q_x^2]a = 0. \tag{3.17}$$

Here κ_L^2 can be eliminated in favour of κ_T^2 with the help of (3.10) to give

$$2\kappa_T q_x b + (q_x^2 + \kappa_T^2)a = 0. \tag{3.18}$$

Equations (3.15) and (3.18) give two independent expressions for the ratio a/b; equating these, we obtain the dispersion equation for the Rayleigh wave:

$$4\kappa_L \kappa_T q_x^2 = (q_x^2 + \kappa_T^2)^2. \tag{3.19}$$

Squaring this and substituting for κ_L^2 and κ_T^2 we find the more explicit form

$$16q_x^4(q_x^2 - \omega^2/v_L^2)(q_x^2 - \omega^2/v_T^2) = (2q_x^2 - \omega^2/v_T^2)^4. \tag{3.20}$$

This equation is homogeneous in ω and q_x, so the solution must have the property that ω is proportional to q_x, and we write

$$\omega = \xi v_T q_x. \tag{3.21}$$

It is clear from (3.20) that ξ is determined by the ratio v_T^2/v_L^2, so in view of (3.16), ξ depends on Poisson's ratio σ and not on Young's modulus E. The dependence of ξ on σ is shown in figure 3.2; as is seen it varies only a little around the value 0.9.

Equation (3.21) shows that the Rayleigh wave is like an ordinary acoustic wave, but with a velocity ξv_T that is slower than either of the bulk velocities v_L and v_T. It is sometimes convenient to represent the dispersion relation as in figure 3.3. This answers the question: What modes propagate for a given value of the in-plane wavevector component q_x? The lowest frequency mode is the Rayleigh wave. At a somewhat higher frequency transverse bulk modes given by $\omega = v_T(q_x^2 + q_z^2)^{1/2}$ can occur. However, on the diagram only q_x is given and q_z can take any value. Thus the transverse modes occupy a *bulk continuum* defined by $\omega \geq v_T q_x$. At higher frequencies still is the longitudinal bulk continuum, $\omega \geq v_L q_x$. These bulk continua are represented by shaded regions in figure 3.3.

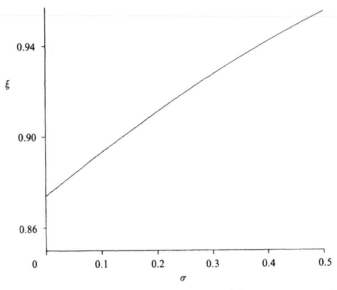

Figure 3.2. Dependence on Poisson's ratio σ of the Rayleigh-wave parameter ξ of (3.21).

It is of interest to ask how the above calculation is modified in an elastic film with a second free surface, say at a distance L from the first. It may be expected intuitively, and it is true, that for large L, which means $\kappa_L L \gg 1$ and $\kappa_T L \gg 1$, there is a Rayleigh wave on each surface. (The analogous

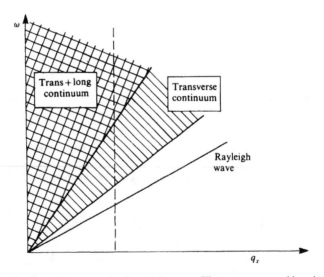

Figure 3.3. The dispersion curve for Rayleigh waves. The transverse and longitudinal bulk continua are shown shaded. Also shown (broken line) for later reference is an approximation to the frequency scan in a Brillouin-scattering experiment.

property is not, however, true for all types of surface mode. For example, it does not hold for the surface magnetostatic mode of a ferromagnetic film, to be discussed in section 5.1). For a given q_x these two surface modes are degenerate in frequency. If, as a thought experiment, we let L decrease, the amplitudes of the Rayleigh waves overlap and the degeneracy is lifted. In order to make this idea precise, we solve the elastic equations of motion with the boundary conditions (2.30) and (2.32) applied at $z = -L$ as well as at $z = 0$. In place of (3.13) and (3.14) we take trial solutions

$$\mathbf{u}_L = [a_1(q_x, 0, -i\kappa_L) \exp(\kappa_L z) + a_2(q_x, 0, i\kappa_L) \exp(-\kappa_L z)]$$
$$\times \exp[i(q_x x - \omega t)] \tag{3.22}$$

$$\mathbf{u}_T = [b_1(\kappa_T, 0, -iq_x) \exp(\kappa_T z) + b_2(-\kappa_T, 0, -iq_x) \exp(-\kappa_T z)]$$
$$\times \exp[i(q_x x - \omega t)]. \tag{3.23}$$

The terms in $\exp(-\kappa_L z)$ and $\exp(-\kappa_T z)$ do not appear in (3.13) and (3.14) because they would diverge as $z \to -\infty$. However, the sample now extends only to $z = -L$ so they must be included. Equations (3.22) and (3.23) introduce four amplitudes a_1, a_2, b_1 and b_2. The boundary conditions give four homogeneous linear equations in these amplitudes so that there is a solvability condition in the form that a 4×4 determinant, Δ (say), must be zero, and this provides the dispersion equation. The general form is quite complicated, but for large thickness Δ can be expanded as a series in $\exp(-\kappa_L L)$ and $\exp(-\kappa_T L)$. The lowest-order terms are

$$\Delta = [4q_x^2 \kappa_L \kappa_T - (q_x^2 + \kappa_T^2)^2]^2 - [4q_x^2 \kappa_L \kappa_T + (q_x^2 + \kappa_T^2)^2]^2 \exp(-2\kappa_T L). \tag{3.24}$$

As $L \to \infty$ the equation $\Delta = 0$ gives (3.19) twice, once for the uncoupled Rayleigh wave on each surface. For finite L, the second term mixes the Rayleigh waves and lifts the degeneracy, as stated earlier.

3.2.2 Other film, surface and interface modes

The coupled Rayleigh waves just described are not the only modes that can travel along an elastic film. For a localized mode on a semi-infinite medium occupying $z < 0$ it is necessary to have a value of q_z in the lower half of the complex plane, for example $q_z = -i\kappa$ as in (3.9) or $q_z = \pi/2a - iy$ by analogy with the surface mode discussed in section 3.1. For a film, however, (3.22) and (3.23), with imaginary q_z, are not the only possible solutions of the wave equations and boundary conditions. There are also solutions with q_z real; in acoustics these are often referred to as *slab modes* while the corresponding modes in optics are called *guided waves*. They are the analogue of the discretized bulk modes of the 1D mass chain that were discussed in section 3.1. To see how the slab modes arise, we note that we could write down solutions like (3.22) and (3.23) with factors $\exp(\pm iq_{Lz})$ and $\exp(\pm iq_{Tz})$ replacing $\exp(\pm\kappa_L z)$ and $\exp(\pm\kappa_T z)$. As can be seen from our discussion of the pseudo-surface mode in section 2.2.2, there are also combinations of ω and q_x for which q_{Tz} is real while q_{Lz} is imaginary. Whether the q_z values are real or imaginary, there are four amplitudes involved, and the solvability condition for the boundary conditions gives the dispersion equation of the mode in question.

The above discussion was concerned with the coupled p-transverse and longitudinal modes of an elastic film. As we have implied, there are no surface modes in s-polarization. However, there are s-polarized slab modes with real q_z. The calculation of their properties is easily outlined. The appropriate trial function is

$$u = [a_1 \exp(iq_z z) + a_2 \exp(-iq_z z)] \exp[i(q_x x - \omega t)] \tag{3.25}$$

with $q_x^2 + q_z^2 = \omega^2/v_T^2$. Application of the boundary condition (2.32) at each surface gives two homogeneous equations in a_1 and a_2 which can be solved for the dispersion equation.

All the elastic modes of a film are collectively called *Lamb waves*. We have not given details of the calculations; the interested reader is referred to the definitive account by Meeker and Meitzler (1964). Two further ramifications are worth mentioning. The introduction of one or more interfaces modifies the mode spectrum and therefore, at least in principle, changes the specific heat. Burt (1973) uses a very neat Green-function method to derive the specific heat of a semi-infinite medium. Second, for some problems it is necessary to have a quantum theory of the acoustic field. The quantization procedure for a semi-infinite medium is given by Oliveros and Tilley (1983).

The Rayleigh wave and the Lamb waves are connected with stress-free surfaces. In the presence of interfaces between different elastic media other modes may occur. The analogue of the Rayleigh wave for a single interface between two media is called the *Stoneley wave*; this exists, however, only provided the elastic constants of the two media satisfy certain conditions (Scholte 1947). When it does exist, the Stoneley wave has the dispersion equation

$$\omega = v_S q_x \tag{3.26}$$

where v_S is given by an equation similar to, but more complicated than, (3.20) for the Rayleigh-wave velocity; this equation is quoted by Albuquerque (1980), for example.

The simplest interface-related wave in s-polarization is the *Love wave*, which occurs in a film of thickness L on a semi-infinite substrate provided that the transverse acoustic velocity v_T' of the substrate exceeds the value v_T for the film. For a wave travelling in the x direction, the transverse (y)

displacement is

$$u = [A\exp(iq_z z) + B\exp(-iq_z z)]\exp[i(q_x x - \omega t)] \tag{3.27}$$

in the film, and

$$u = C\exp(\kappa_z z)\exp[i(q_x x - \omega t)] \tag{3.28}$$

in the substrate, which is assumed to occupy the region $z < -L$. Here

$$q_z^2 = (\omega^2/v_T^2) - q_x^2 \tag{3.29}$$

and

$$\kappa_z^2 = q_x^2 - (\omega^2/v_T'^2). \tag{3.30}$$

In order for Love waves to exist, both q_z and κ_z must be real, so that the displacement is an oscillatory function of z in the film and a decaying function in the substrate, which is the reason for the requirement $v_T' > v_T$. The combination of oscillatory and decaying dependence is similar to the behaviour of a guided optical wave in a film on a substrate, which will be discussed briefly in section 6.2.1.

Equations (3.27) and (3.28) must satisfy three boundary conditions: zero stress at the top surface of the film, plus equal displacements and stress at the film–substrate interface. These three boundary conditions give three homogeneous equations for the amplitudes A, B and C and the solvability condition is the dispersion equation. As shown by Albuquerque *et al* (1980), for example, it is

$$\tan(q_z L) = \mu' \kappa_z / \mu q_z \tag{3.31}$$

where μ and μ' are the Lamé parameters of the film and substrate respectively. Equation (3.31) is similar to the one for the guided optical waves mentioned and also to the equation for the bound states in a quantum-mechanical square well of finite depth. As is the case for all guided waves, (3.31) determines a discrete set of values of q_z. A graphical analysis similar to that applied in quantum mechanics (see e.g. Bransden and Joachain 2000) shows that as the frequency increases the number of Love waves also increases, the higher waves oscillating more rapidly within the film. An example of the dispersion curves is shown in figure 3.4.

3.2.3 Green functions

The Green function for the 1D wave equation in an infinite medium was derived in section 2.2.3. We now see how the calculation is generalized for a semi-infinite medium and a film. We still seek a solution of (2.48), the wave equation in the presence of a point driving force, but the boundary conditions are different. Whereas (2.50) is the solution of (2.48) that is bounded as $z \to \infty$ and as $z \to -\infty$, we now require in the case of a

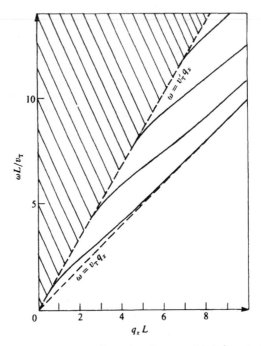

Figure 3.4. Love wave (full curves) dispersion for $v_T = 0.575v_T'$ and $\mu' = 5\mu$. The bulk continuum is shaded. After Albuquerque *et al* (1980).

semi-infinite medium that the solution is bounded as $z \to -\infty$ and satisfies the condition of zero stress across the plane $z = 0$. It is seen from (2.31) that the latter condition is simply

$$\frac{\mathrm{d}u}{\mathrm{d}z} = 0 \quad \text{at } z = 0. \tag{3.32}$$

In order to satisfy (3.32) we add to (2.50) the complementary function $C\exp(-\mathrm{i}qz)$, which is the solution of the homogeneous wave equation that is bounded as $z \to -\infty$. The amplitude C is determined by the requirement that the complete solution satisfies (3.32); the resulting expression for u is

$$u = (\mathrm{i}qf/2\rho\omega^2 A)\{\exp(\mathrm{i}q|z - z'|) + \exp[-\mathrm{i}q(z + z')]\} \tag{3.33}$$

where, as in section 2.2.3, weak damping ($\omega\tau \gg 1$) has been assumed. From (2.46) the surface-related Green function $\langle\langle u(z); u(z')^*\rangle\rangle_\omega$ is the coefficient of f on the right-hand side of this equation.

The result just derived, that the Green function has a term in $z - z'$ and a term in $z + z'$, is typical of Green functions in a semi-infinite medium. These dependences can be viewed pictorially as illustrated in figure 3.5: the disturbance originating at the source point z' can propagate to the observation point z either directly or via a reflection off the surface.

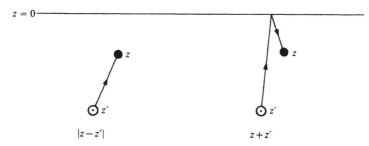

Figure 3.5. To illustrate the exponential factors in the acoustic Green function for a semi-infinite medium. The source point z' and the observation point z are horizontally displaced for ease of drawing.

The calculation of the Green function for a semi-infinite medium can be extended without undue difficulty to a film of thickness L. In that case the zero-stress boundary condition, (2.32), is applied at both $z = 0$ and $z = -L$, and the complementary function contains an additional term $D \exp(\mathrm{i} q z)$. The details are left as an exercise for the reader; the exponential factors that appear may be illustrated in a similar way to figure 3.5.

Fluctuation spectra can be derived from the Green function of (3.33) in the same way that (2.52) was derived from (2.51) for an infinite medium. Note, however, that because the surface at $z = 0$ destroys translational invariance in the z direction, the fluctuation spectra depend on z and z' separately, not simply on $z - z'$. Examples of fluctuation spectra $\langle |u(z)|^2 \rangle_\omega$, with $z = z'$, are given by Loudon (1978a) for a film of thickness L.

The wavevector Fourier transform of the Green function may be expected to be of physical interest, and indeed it was seen in section 1.4.1 that the Brillouin-scattering cross section is closely related to this Fourier transform. To be precise, for Brillouin scattering we shall need the Green function involving the displacement derivatives $u_{zz} = \partial u_z(z)/\partial z$, where the subscript in $u_z(z)$ is inserted to emphasize that we are dealing with longitudinal modes whose displacement is in the z direction. We define

$$\langle\langle u_{zz}(z); u_{zz}(z')^* \rangle\rangle_\omega = \frac{\partial}{\partial z}\frac{\partial}{\partial z'}\langle\langle u_z(z); u_z(z')^* \rangle\rangle_\omega. \tag{3.34}$$

Differentiating the Green function part in (3.33) yields

$$\langle\langle u_{zz}(z); u_{zz}(z')^* \rangle\rangle_\omega = (\mathrm{i}q^3/2\rho\omega^2 A)\{\exp(\mathrm{i}q|z - z'|) - \exp[-\mathrm{i}q(z + z')]\}. \tag{3.35}$$

Because of the loss of translational invariance, the Fourier transform involves two wavevectors Q_z and Q'_z separately; it is found from (3.35) to be

$$\langle\langle u_{zz}(Q); u_{zz}(Q') \rangle\rangle_\omega = \frac{\mathrm{i}q^4}{\rho\omega^2 A L_0^2}\frac{1}{(Q'_z - q)(Q_z + q)(Q_z - Q'_z)}. \tag{3.36}$$

Here a macroscopically large thickness L_0 has been introduced for the evaluation of the integrals defining the Fourier transform and the limit $L_0 \to \infty$ has been taken in the exponential factors that arise. Equation (3.36) will be used in the discussion of Brillouin scattering in the next section. In particular it will be seen there that the factor L_0^2 in (3.36) eventually cancels out from the expression for the cross section.

The results derived so far in this section are for the 1D wave equation in which the amplitude of the wave depends only on the coordinate z normal to the surface. Consequently these Green functions contain no information about surface modes, which have a plane-wave dependence upon distance along the xy, or surface, plane. In order to incorporate surface modes, it is necessary to generalize the Green functions to include this dependence on x and y. As usual, the x axis is chosen to lie along the propagation direction. The natural generalization of a Green function such as that of (3.34) is then $\langle\langle u_{ij}(z); u_{kl}(z')^* \rangle\rangle_{Q_x,\omega}$. Here i, j, k and l can be x, y or z, and the Green function depends on the in-plane wavevector Q_x as well as on ω. The quantities $u_{ij}(z)$ are displacement derivatives $\partial u_i / \partial x_j$ as distinct from the strain components $\bar{u}_{ij}$ defined in (2.18). These Green functions are found by solving the wave equation in the presence of a force

$$F = (f/A)\exp(\mathrm{i}Q_x x - \mathrm{i}\omega t)\delta(z - z') \tag{3.37}$$

which is an obvious generalization of (2.44). The mathematical steps are similar to those that have been described for the 1D case.

The most important property of these more general Green functions is that appropriate ones have poles corresponding to the presence of surface modes. For example, for a semi-infinite medium all the Green functions $\langle\langle u_{ij}(z); u_{kl}(z')^* \rangle\rangle_{Q_x,\omega}$ with i, j, k, and l all equal to x or z contain the denominator

$$D_R = 4v_T^4 Q_x^2 q_{Lz} q_{Tz} + (\omega^2 - 2v_T^2 Q_x^2)^2 \tag{3.38}$$

where $Q_x^2 + q_{Nz}^2 = \omega^2/v_N^2$ with $N = L$ or T. This result was first derived by Loudon (1978b). It is easily verified that the equation $D_R = 0$ is satisfied only when q_{Lz} and q_{Tz} are both imaginary, and that it is then equivalent to the dispersion equation (3.20) for the Rayleigh wave. Thus the Green functions have a pole when the Rayleigh-wave condition is satisfied. On the other hand, when i or k is equal to y the Green function has no pole; this corresponds to the absence of s-polarized surface modes.

We have given an elementary method for calculating Green functions for systems with planar interfaces. A more advanced method, called *surface Green function matching* (SGFM) has been developed by Garcia-Moliner and collaborators (for details see Garcia-Moliner 1977, Garcia-Moliner and Flores 1979). The method has been applied in a range of problems, including in particular the theory of electronic surface states.

3.3 Normal-incidence Brillouin scattering

3.3.1 Some experimental results

As explained in section 1.4.1, the application of Brillouin scattering to acoustic (and other) surface modes stems from the development by Sandercock of the multipass Fabry–Pérot interferometer as a practical spectroscopic instrument. The results first published (Sandercock 1972a,b) were obtained with a single multipass interferometer, while many of the later results involved the use of a tandem instrument, which has the advantage of a larger free spectral range. The experimental techniques and results are reviewed in Sandercock (1982).

It is convenient to start with the first results, shown in figure 3.6, which were for Brillouin scattering from silicon and germanium surfaces. At all the wavelengths used, the refractive indices of Ge and Si are high. Thus, although outside the sample, the incident and scattered light propagate in directions at right angles, inside both propagate very nearly along the normal to the surface. It is therefore possible to simplify the analysis by using the theory for normal-incidence backscattering.

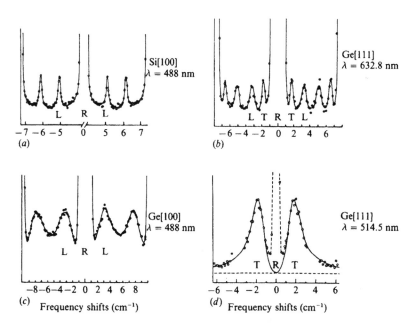

Figure 3.6. Brillouin backscattering from Si and Ge surfaces. The optical absorption coefficients are about $10^6\,\mathrm{m}^{-1}$ in (a), $2 \times 10^7\,\mathrm{m}^{-1}$ in (b) and $6 \times 10^7\,\mathrm{m}^{-1}$ in (c) and (d). Spectra (a) to (c) were taken with a single multipass interferometer, while (d) was taken with a tandem interferometer which has a larger free spectral range. After Sandercock (1982).

The main features of interest in the spectra are these. First, although the Stokes and anti-Stokes peaks are at approximately the positions that would be predicted by simple 'bulk-type' kinematics, they are substantially broadened compared with typical Brillouin spectra of transparent media. Second, as the frequency of the incident light increases, approaching the band-gap frequency, the optical absorption coefficient increases and with it the line broadening. The line broadening is indeed due to absorption of light in the sample, and for that reason it is referred to as *opacity broadening*. The third feature of interest is that the broadening is asymmetric.

3.3.2   Theoretical formulation and discussion

The first question is the nature of the coupling between the incident light and the excitations that are responsible for the scattering. In the present case, two mechanisms may be involved. First, thermally excited acoustic fluctuations within the sample frequency-modulate the optical dielectric function. This coupling, which governs Brillouin scattering in a bulk sample, is described by the acousto-optic (or Pöckels) tensor. In the second mechanism, the acoustic fluctuations modulate the surface of the sample, and light is scattered off the moving surface with a frequency shift. This surface-ripple mechanism is generally dominant in highly reflective samples, whereas the acousto-optic mechanism dominates in relatively transparent samples, like those used for the results of figure 3.6. Here we give the theory in detail for acousto-optic coupling, and the effects of surface-ripple coupling will be described later. We use essentially the method of Loudon (1978a), which is a development of the macroscopic formulation of light scattering by bulk samples (Hayes and Loudon 1978).

The notation is defined in figure 3.7. We discuss Stokes scattering, and so $\omega_S < \omega_I$ and a phonon is created by the scattering process. The incident-light wavevectors are given by

$$k_{I1}^2 = \varepsilon_1 \omega^2 / c^2 \tag{3.39}$$

$$k_{I2}^2 = \varepsilon_2(\omega_I)\omega^2 / c^2 \tag{3.40}$$

where ε_1 and $\varepsilon_2(\omega_I)$ are the dielectric constants at frequency ω_I of the upper and lower media; usually the upper medium is vacuum and $\varepsilon_1 = 1$. It is emphasized in (3.40) that ε_2 is generally frequency-dependent and, as seen in section 2.4.1, it may be complex. For most cases, ε_2 varies sufficiently slowly with frequency that we may take $\varepsilon_2(\omega_S) = \varepsilon_2(\omega_I)$. The exception is when the sample is a semiconductor with ω_I close to the band-gap frequency. This is the case of *resonant Brillouin scattering* via exciton–polaritons that was mentioned in section 2.5.3. The theory can be given as an extension of the present formalism (Tilley 1980) but it will not be discussed here. With the sign convention of figure 3.7, the incident light in the medium has a

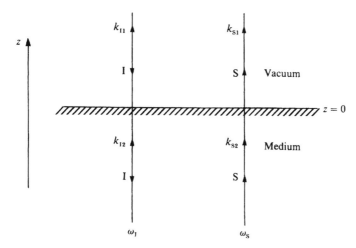

Figure 3.7. Notation for wavevectors of incident light, with frequency ω_I and scattered light, with frequency ω_S, for normal-incidence backscattering.

factor $\exp(-ik_{I2}z)$ so that it is necessary to take the solution of (3.40) with $\mathrm{Re}(k_{I2}) > 0$ and $\mathrm{Im}(k_{I2}) > 0$. The incident-light amplitude inside the medium is related to that outside by the usual optics formula derived by application of standard electromagnetic boundary conditions:

$$E_{I2} = 2k_{I1}E_{I1}/(k_{I1} + k_{I2}) \equiv fE_{I1} \qquad (3.41)$$

which defines the Fresnel factor f for later use.

As mentioned in section 3.2.3, coupling to thermally-excited acoustic excitations within the medium is conventionally described in terms of displacement derivatives u_{ij} and it is convenient to start by considering a single Fourier component:

$$u_{ij} = u_{ij}(Q'_z)\exp(iQ'_z z - i\omega t). \qquad (3.42)$$

The coupling between this and the incident light is described by the Pöckels tensor $p^{\alpha\beta\eta}$. Its effect is to produce a polarization P_S^{α} within the medium at frequency $\omega_S = \omega_I - \omega$:

$$P_S^{\alpha} = P_{S0}^{\alpha}\exp(iK_0 z - i\omega_S t) \qquad (3.43)$$

where

$$P_{S0}^{\alpha} = -\varepsilon_0\varepsilon_2^2 p^{\alpha\beta\eta}E_{I2}^{\beta}u^{\eta}(Q'_z)^* \qquad (3.44)$$

$$K_0 = -k_{I2} - Q'_z. \qquad (3.45)$$

In (3.44) the summation convention applies and the prefactors result from the definition of the Pöckels tensor.

Equation (3.44) already contains the polarization selection rules for the scattering process. For example, with plane-polarized incident light on an isotropic sample, we may take the x axis along the $\mathbf{E}$ vector of the incident beam, so that $\beta = x$. The only non-vanishing component of the Pöckels tensor is then p^{xxzz}. Thus $\mathbf{P}_{S0}$, and ultimately the scattered light $\mathbf{E}_{S1}$, are also x-polarized, and scattering is solely by the longitudinal component u^{zz}. For simplicity, we assume that we are dealing with such scattering; the derivation is easily generalized.

The polarization $\mathbf{P}_S$ of (3.43) extends from the surface some distance into the sample. Its variation with z depends partly on the z-dependence of the thermally-excited acoustic field u^{zz}, but the appearance of the factor $\exp(iK_0 z)$ ensures that $\mathbf{P}_S$ is confined to a thickness of the order of the optical skin depth. It is a straightforward exercise using Maxwell's equations to show that $\mathbf{P}_S$ appears as a driving term on the wave equation:

$$d^2 E_{S2}^x/dz^2 + k_{S2}^2 E_{S2}^x = -(\omega_S^2/\varepsilon_0 c^2) P_{S0}^x \exp(iK_0 z). \tag{3.46}$$

This shows that $\mathbf{P}_S$ acts as a source within the medium for radiation of frequency ω_S. The radiated light propagates to the surface at $z = 0$, where some of it is reflected and some is transmitted as a field $\mathbf{E}_{S1}$. It is this field $\mathbf{E}_{S1}$ of frequency ω_S that is ultimately detected in the spectrometer. In order to find an expression for it, we solve (3.46) together with the electromagnetic boundary conditions at $z = 0$. In $z < 0$ the solution of (3.46) consists of a particular integral plus a complementary function:

$$E_{S2}^x = A_0 P_{S0}^x \exp(iK_0 z) + A_2 \exp(-ik_{S2} z) \tag{3.47}$$

while for $z > 0$ there is only the complementary function:

$$E_{S1}^x = A_1 \exp(ik_{S1} z). \tag{3.48}$$

Since $\mathrm{Im}(k_{S2}) > 0$ and $\mathrm{Im}(k_{S1}) > 0$ the complementary functions in (3.47) and (3.48) tend to zero as $z \to -\infty$ and $z \to \infty$ respectively.

The amplitude A_0 in (3.47) is found by direct substitution in (3.46), while A_1 and A_2 are found from the boundary conditions of continuous E^x and H^y. The expression for A_2, together with (3.44) for P_{S0}^x and (3.41), enables one to find the scattered amplitude E_{S1} in terms of the incident amplitude E_{I1}:

$$E_S^*(Q_z') = (\omega_S \varepsilon_2^2/c) gpf u^{zz}(Q_z')^* E_{I1}/(Q_z' + k_{I2} + k_{S2}) \equiv h(Q_z') u^{zz}(Q_z') E_{I1} \tag{3.49}$$

where p is shorthand for p^{xxzz} and f is defined in (3.41). The factor g is

$$g = k_{S1}/(k_{S1} + k_{S2}) \tag{3.50}$$

and in (3.49) the notation $h(Q_z')$ is introduced for later use.

Where have we got to? We are viewing the light scattering as a three-stage process. In the first stage incident light of frequency ω_I is transmitted across the interface into the medium. This is described by (3.41). In the second stage, the incident light within the medium mixes with the thermally-excited acoustic field u^{zz} to produce a driving polarization $\mathbf{P}_S$ at frequency

ω_S, as described by (3.43) and (3.44). In the third stage, $\mathbf{P}_S$ radiates light at frequency ω_S whose amplitude in the upper medium is given by (3.49).

The experimentally measured quantity is the differential cross section. As shown in section 1.4.1, this is given by

$$\frac{d^2\sigma}{d\Omega\,d\omega_S} = \frac{\omega_I\omega_S A\bar{A}}{4\pi^2 c^2 |E_{11}|^2} \sum_{Q_z,Q'_z} \langle E^*_{S1}(Q_z)E_{S1}(Q'_z)\rangle_{\omega_S} \tag{3.51}$$

where A is the surface area of the sample and $\bar{A}$ is the area illuminated by the incident light. In comparison with (1.63), we take the medium of incidence as vacuum, $\varepsilon_I = 1$, and since we are discussing normal-incidence backscattering $\theta_S = 0$. Equation (3.49) relates E_{S1} to u^{zz} so that the power spectrum inside the sum in (3.51) is proportional to the power spectrum of u^{zz}. The latter is related, via the fluctuation–dissipation theorem, to the imaginary part of the corresponding Green function given in (3.36). Using these relations, we can eventually write

$$\frac{d^2\sigma}{d\Omega\,d\omega_S} = \frac{\omega_I\omega_S^3 A\bar{A}\varepsilon_2^4 k_B T}{4\pi^3 c^4 \omega}|gpf|^2 \,\mathrm{Im}\sum_{Q_z,Q'_z} \frac{\langle\langle u^{zz}(Q_z); u^{zz} Q'_z\rangle\rangle}{(Q_z + K^*)(Q'_z + K)} \tag{3.52}$$

where

$$K = k_{I2} + k_{S2} \tag{3.53}$$

and the classical form of the fluctuation–dissipation theorem has been used, since $k_B T \gg \hbar\omega$ in the experiments.

It remains to evaluate the sums over Q_z and Q'_z appearing in (3.52). By converting the sums to integrals in the usual way and then using contour integration (see e.g. Oliveira *et al* 1980) the final result is

$$\frac{d^2\sigma}{d\Omega\,d\omega_S} = \frac{\omega_I\omega_S^3 \bar{A}\varepsilon_2^4 k_B T}{4\pi^3 c^4 \rho v_T^3}|gpf|^2 \,\mathrm{Im}\frac{iq}{(K+q)(K^*-q)(K-K^*)} \tag{3.54}$$

where the symbols g, p, f, q and K have been defined previously.

Very often, as for the Si results of figure 3.6, the phonon damping is negligible; in that case q is the purely real quantity ω/v_L. It is then easy to show that

$$\mathrm{Im}\frac{iq}{(K+q)(K^*-q)(K-K^*)} = \frac{q^2}{(q^2 - K_1^2 + K_2^2)^2 + 4K_1^2 K_2^2} \tag{3.55}$$

where $K = K_1 + iK_2$. Since K is independent of frequency, the Brillouin line-shape is given by the q dependence of (3.55); this result was first derived by Dervisch and Loudon (1976). The function is shown in figure 3.8; as indicated there its peak is at $q_m = (K_1^2 + K_2^2)^{1/2}$ and it has width $\Delta q = 2K_2$. It is clear from (3.55) and figure 3.8 that the Dervisch–Loudon lineshape is in qualitative agreement with the experimental results of figure 3.6; furthermore the quantitative agreement with the Ge data is good (Dervisch and Loudon 1976).

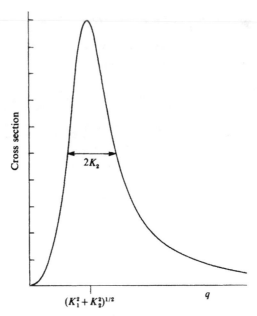

Figure 3.8. Lineshape of equation (3.55).

As mentioned, the theory described so far has been for scattering by the acousto-optic mechanism, whereas in highly reflective samples, particularly metals, the surface-ripple mechanism makes an important contribution and can be dominant. The scattering amounts to a Doppler shift of the incident light as it is reflected from the moving surface. At time t, the position of the surface is shifted from its equilibrium position $z = 0$ to $z = u_z(z = 0, t)$. The theory involves expanding the reflected light amplitude up to first order in the small quantity $u_z(z = 0, t)$ so that $E_{S1} \propto u_z(z = 0, t)$. A formula similar to (3.51) applies for the cross section, and since E_{S1} is linear in $u_z(z = 0)$ it transpires that the cross section is proportional to the power spectrum $\langle |u_z(z = 0)|^2 \rangle_\omega$. This is readily obtained by means of the fluctuation–dissipation theorem from the Green function of (3.35). This Green function is the fundamental quantity for surface-ripple scattering as well as for acousto-optic scattering. The theory of surface-ripple scattering was first given by Loudon (1978c), Rowell and Stegeman (1978a) and Bortolani *et al* (1979).

We have given the detailed theory of normal-incidence scattering by the acousto-optic mechanism because it is the simplest example and it contains the essential ideas of all such calculations. For more complicated geometries, such as oblique-incidence scattering, we give a largely qualitative discussion since the theory follows similar lines to what has been done here. In addition to the references cited, accounts of the theory, including in some cases both surface-ripple and acousto-optic mechanisms, are given by Subbaswamy and Maradudin (1978), Rowell and Stegeman (1978b) and Marvin *et al*

(1980a,b). The most complete theoretical account of Brillouin scattering by phonons is that given by Bortolani *et al* (1983); it is sufficiently general to describe scattering by unsupported films and by films on substrates as well as scattering by semi-infinite media, and it includes both scattering mechanisms.

3.4 Oblique-incidence Brillouin scattering

We now review some results obtained with more general scattering geometries. By 'oblique-incidence' we mean the geometry and/or materials are such that the excitations responsible for the scattering have a significant wavevector component q_x parallel to the surface. Reviews are given by Sandercock (1982), Stegeman and Nizzoli (1984) and Nizzoli (1986).

The experimental scan over scattered frequency is closely approximated by a scan over ω at fixed q_x in the dispersion curve. This is shown for the Rayleigh-wave spectrum in figure 3.3. Since a surface wave has a definite frequency for given q_x, there should therefore be a sharp peak in the cross section at this frequency; the peak is broadened only by acoustic damping. By contrast, where the scan line passes through a bulk continuum, the cross section shows opacity-broadened structure like that discussed in the previous section. This is clearly illustrated by the results for a Si surface shown in figure 3.9. The sharp lines marked R are the Rayleigh-wave peaks, while the lines T and L are due to opacity-broadened scattering by transverse and longitudinal bulk phonons. The scattering is dominated by the acousto-optic mechanism (Loudon 1978b).

In the analysis of oblique-incidence scattering, the cross section is found from the imaginary part of an appropriate Green function, typically $\langle\langle u_{ij}(Q_z); u_{kl}(Q_z') \rangle\rangle_{Q_x,\omega}$. As mentioned in section 3.2.3, this has a pole in the complex ω plane when ω and Q_x are related by the dispersion relation for the Rayleigh wave. If damping is small, this pole is very near the real axis, and in the vicinity of the pole the Green function may be written

$$G(Q_x,\omega) = G_0(\omega)[\omega - \omega_R(Q_x) - i\eta]^{-1} \tag{3.56}$$

where the Rayleigh-wave frequency is written $\omega_R(Q_x) + i\eta$, and η in the small imaginary part represents the damping. $G_0(\omega)$ is a function which has no singularities near $\omega_R(Q_x)$. With the use of the symbolic identity (A.27) of the Appendix, (3.56) gives for the imaginary part

$$\text{Im}\, G(Q_x,\omega) = \pi G_0(\omega_R)\delta[\omega - \omega_R(Q_x)]. \tag{3.57}$$

It is this delta function that corresponds to the sharp Rayleigh peak observed in the experiments.

Scattering from metal surfaces is governed by the surface-ripple mechanism. Typical spectra are shown in figure 3.10. As for Si (see figure 3.9), the Rayleigh wave shows as a sharp peak, while at higher frequency

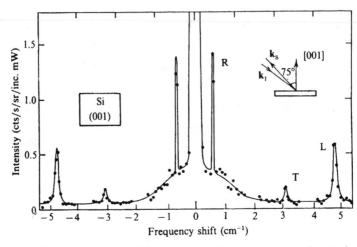

Figure 3.9. Brillouin spectrum of Si surface in oblique-incidence backscattering. After Sandercock (1978).

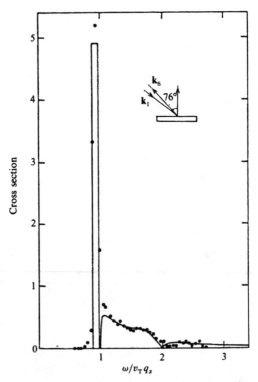

Figure 3.10. Oblique-incidence Brillouin backscattering from the surface of polycrystalline Al. The points are the experimental data, and the line the theory for surface-ripple scattering. After Loudon and Sandercock (1980).

shifts there is a broadened structure. The theoretical expressions involve the optical and acoustic parameters, which are well known, but not the acousto-optic coefficients. The theoretical curve is drawn with the approximation that the sample is an isotropic elastic medium with $\sigma = \frac{1}{3}$ but there are no adjustable parameters.

For many semiconductors, both acousto-optic and ripple mechanisms contribute to the cross section. Examples of spectra and their analysis may be found in the references already cited. These articles also deal with light scattering by thin films on substrates.

Subsequent to the work reviewed here, interest shifted to Brillouin scattering by more complicated materials, particularly superlattices and other layered structures. Two early examples are the work of Harley and Fleury (1979) on layered transition-metal dichalcogenides like $TaSe_2$ and the study of Rayleigh waves on Cu/Nb superlattices by Kueny *et al* (1982). Further discussion will be given in section 8.2.2.

3.5   Acoustic waves on gratings

Various techniques have been developed for depositing, or inscribing, a diffraction-type grating on a surface. The commonest is to prepare a grating pattern holographically in an overlayer of photoresist, then, for example, to apply an etch in such a way that the grating pattern is transferred to the underlying surface. On general grounds, it can be seen that a grating of period D introduces Brillouin-zone boundaries at wavenumber component $q_x = n\pi/D$, where n takes integer values and x is the direction perpendicular to the grating lines. Thus gaps in the dispersion curves of surface waves should appear at the zone boundaries. One reason for the study of acoustic waves on such surfaces is because of the basic importance of these gaps. A second reason is because of the need to understand rough surfaces since a rough surface can be modelled as a random Fourier series of gratings. It will be seen in sections 6.4 and 6.5 that gratings have an important part to play in modifying the properties of electromagnetic surface waves and therefore in some applications. In the present section we review some of the corresponding work that has been carried out on acoustic surface waves.

A detailed theoretical study of propagation perpendicular to the grating lines was carried out by Glass *et al* (1981); a further discussion of propagation and attenuation was given by Glass and Maradudin (1983) with the extension to a general direction of propagation by Mayer *et al* (1991). The application of these methods to the propagation of a Rayleigh wave on a rough surface was given by Eguiluz and Maradudin (1983).

We outline the method of approach by reference to figure 3.11. The corrugated surface is defined by a function $\zeta(x)$ so that the elastic solid occupies the region $z < \zeta(x)$. The amplitude of $\zeta(x)$ is denoted ζ_0 and the

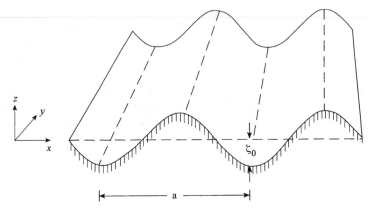

Figure 3.11. Notation for a surface in the form of a periodic grating. After Mayer *et al* (1991).

region $-\zeta_0 < z < \zeta_0$ is referred to as the *selvedge* (this is one of the rare references to dressmaking in the physics literature). In a sense, the theoretical problem can be stated easily: the acoustic wave equation (2.25) is to be solved in the medium, then the boundary conditions of zero stress are applied at the surface $z = \zeta(x)$. The difficulty, obviously, is that this surface

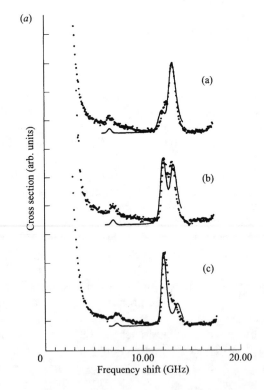

is not flat. One standard approach, and the first to be applied by Glass *et al* (1981), makes use of what is known as the *Rayleigh hypothesis* (Rayleigh 1907). For $z < -\zeta_0$ the acoustic-wave amplitude can be written exactly as a superposition of the bulk solutions of (2.25), as was done in section 3.2.1 for the Rayleigh wave on a flat surface. The Rayleigh hypothesis, which is expected to hold only for small values of ζ_0, is that this expansion can be continued into the selvedge region $-\zeta_0 < z < \zeta(x)$. The boundary conditions then determine the expansion coefficients and then the dispersion curve. The calculation of the dispersion equation (3.19) is an example of this process. Glass *et al* (1981) derive the dispersion equation first by means of the Rayleigh hypothesis and then by a Green function method that should apply for larger ζ_0. They find that in fact the two methods give identical results provided $\zeta(x)$ can be written as an even function of x.

An oblique-incidence Brillouin-scattering study of the grating-induced modification of the acoustic modes on a Si surface was reported by Dutcher *et al* (1992) and the theory was then given by Giovannini *et al* (1992). As is the case for other oblique-incidence measurements, for example those shown in figure 3.10, it is possible to vary the in-plane wave-vector component Q of the scattering mode by varying the angles of incidence and scattering. It is convenient to present the experimental and theoretical results together, in figure 3.12. The first main feature on the

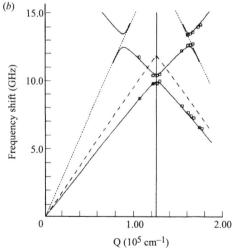

Figure 3.12. (a) (left) Calculated (lines) and experimental (dots) Brillouin spectra for a Si grating of period $D = 250$ nm and depth 17.5 nm. The parallel wavevector Q is (a) 1.58×10^5 cm^{-1}, (b) 1.61×10^5 cm^{-1} and (c) 1.64×10^5 cm^{-1}. (b) Extended-zone plot of scattering peaks from Brillouin spectra (open squares) compared with theoretical dispersion curves. Solid curve, Rayleigh wave. Dashed line, $\omega = v_T Q$ (transverse continuum boundary). Dotted line, $\omega = v_L Q$ (longitudinal continuum boundary). Vertical line at $Q = 1.26 \times 10^5$ cm^{-1} is the zone boundary. After Giovannini *et al* (1992).

theoretical dispersion curve, figure 3.12b, is the expected energy gap in the Rayleigh-wave dispersion curve at the first zone boundary. The second gap, at around $Q = 1.6 \times 10^5 \, \text{cm}^{-1}$ and $\omega/2\pi = 13 \, \text{GHz}$ is due to hybridization between the Rayleigh wave (RW) and what is called the *longitudinal resonance* (LR). The latter is a mode with large amplitude near the surface but just inside the longitudinal continuum so that it has some coupling to bulk longitudinal modes; its dispersion curve follows the continuum boundary closely. Since its polarization is longitudinal, it does not produce a surface ripple. The spectra are shown in figure 3.12(a) for values of Q in the neighbourhood of the RW/LR hybridization and it can be seen that they have three main features. At around 7 GHz there is a small peak due to the RW folded into the second zone. At around 13 GHz there is a strong doublet; the relative intensities of the two parts vary rapidly with Q. This doublet is due to the RW/LR hybridization. As is seen from figure 3.12(a), the agreement between the measured peak positions and the dispersion curves (calculated without any adjustable parameters) is very good.

The outline of the theoretical calculation of the cross section is as follows. The Rayleigh hypothesis is applied, and since the grating depth, denoted by δ is small the calculations are carried out to first order in δ. The calculation is equivalent to that of section 3.3.2 and only the surface-ripple contribution to scattering is included since this is dominant for the silicon surface. One consequence that might be expected is that where the LR is not hybridized with the RW it does not give any ripple and is therefore not seen in the Brillouin spectrum. This is largely true, although as discussed by Giovannini *et al* (1992) and by Lee *et al* (1994) the full story is more complicated. Nevertheless it explains why the doublet appears only close to $Q = 1.6 \times 10^5 \, \text{cm}^{-1}$. As can be seen from figure 3.12(a), the general agreement between the calculations and the experimental spectra is excellent.

These experimental and theoretical studies have been extended to higher folded modes on a corrugated Si surface by Lee *et al* (1994).

3.6   Inelastic particle scattering

3.6.1   Helium atom scattering

As in the case for bulk samples, inelastic light scattering has the advantage of high precision, but it has the disadvantage that it can only probe excitations near the centre of the Brillouin zone. Thus the modes investigated by Brillouin scattering, as described in previous sections, are of long wavelength, and an account can be given entirely within continuum elasticity theory. In order to carry out experiments on surface modes of shorter wavelength, which are described by the microscopic phonon dynamics to which an introduction was given in section 3.1, it is necessary to use probe particles carrying greater momentum than photons. Neutron scattering, which has been extensively applied in the study of bulk phonons, is of very limited use for surface phonons for the reasons mentioned in section 1.4.2. The two techniques that have been applied with success are inelastic scattering of ^{4}He atoms and inelastic electron scattering; the former is discussed below and the latter in the next section. A review of the earlier work is given by Tonnies (1987) and comparisons of the two techniques are given by Bunjes *et al* (1994) and Ibach (1991). Briefly, ^{4}He scattering probes the surface electronic density while inelastic electron scattering results from interaction with the atomic core electrons so that it is governed by the motion of the nuclei. The energy resolution of ^{4}He scattering is around 10^{-4} eV while that for electron scattering is around 10^{-3} eV.

Some experimental details of ^{4}He scattering, for which the incident-particle energy is typically about 20 meV, were given in section 1.4.2. It was shown there that a time-of-flight (TOF) spectrum gives a scan along the parabolic path in $(\omega, \mathbf{q}_\parallel)$ space described by (1.68). These scans can be projected on to the $(\omega, q_\parallel)$ plane to give phonon dispersion curves. The first results published were for ionic crystals, particularly LiF (Brusdeylins *et al* 1980; later references are given by Brusdeylins *et al* 1985), and for metals (references are given by Doak *et al* 1983). As an example, figure 3.13 shows TOF spectra for an Ag(111) surface. The angle between the incident and scattered beams was fixed at 90°, so that $\theta_1 = 45°$ corresponds to specular reflection. The data are therefore paired in angles of incidence equidistant from 45°, and the farther the angle is from 45° the greater the corresponding wavenumber. The peak positions can be plotted in the Brillouin zone, as in figure 3.14. The lower set of data points and the theoretical curve in figure 3.14 are the extension across the zone of the Rayleigh-wave dispersion curve shown in figure 3.3. Although theory and experiment are in good agreement for the Rayleigh wave, this is not the case for the upper set of data points, which fall away from the corresponding theoretical curve. Bortolani *et al* (1984) gave the full dynamical theory of the scattering, and in particular by choice of surface force constants they were able to fit the data points and also the relative magnitudes of the peaks in the TOF spectra. The full theory involves both the force constants of the phonon dynamics and the interaction potential between the incident atomic beam and the surface; a review is given by Bortolani *et al* (1986). More recent results for scattering from the Ag(001) surface are given by Bunjes *et al* (1994).

Like those shown in figure 3.13, the early results concerned scattering by acoustic modes. The cross section for scattering of atoms by optic modes is smaller, but even so Brusdeylins *et al* (1985) were able to detect optic-mode scattering in NaF.

The earlier work on helium scattering, as in the examples above, dealt with scattering from single surfaces. Later, attention was given to surfaces with adsorbed or deposited overlayers. For example, Bracco *et al* (1990)

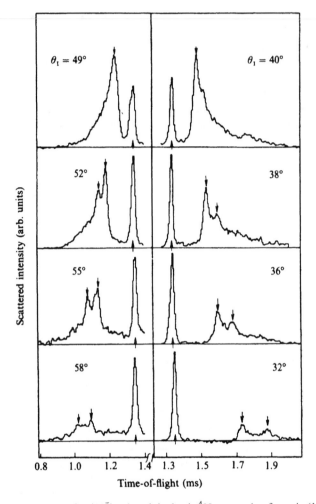

Figure 3.13. TOF spectra for $\langle 11\bar{2} \rangle$ azimuth inelastic ^{4}He scattering from Ag(111) surface. Values of the angle of incidence θ_1 are marked. The incoherent elastic peak attributed to crystal imperfections is shown by an arrow at the bottom and the inelastic peaks are shown by arrows at the top. After Doak *et al* (1983).

studied the Ag(110) surface with chemisorbed oxygen; an understanding of such faces is important for catalysis. The dispersion curves resulting from their data, figure 3.15, consist of an acoustic mode and two surface resonances together with a weak higher-energy structure. They carried out detailed lattice-dynamical calculations for various models and the best fit was obtained with a combination of reconstruction of the Ag surface layers together with (3×1) adsorption of the O atoms. The reconstruction

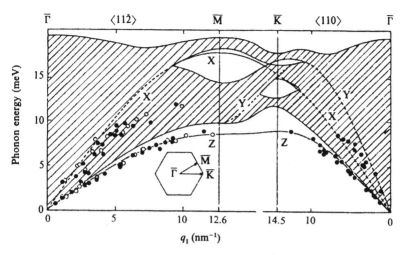

Figure 3.14. Reduced-zone plot of data points in inelastic ^{4}He scattering from Ag(111). Full circles, bulk crystal; open circles, epitaxial crystal. The bulk continuum is shaded. Surface dispersion curves calculated by Armand and by Bortolani *et al.* The broken curves are pseudo-surface modes. After Doak *et al* (1983).

(see the inset) consists of a 'missing-row' structure on the top face of the Ag. As a second example of a study of a technically important surface, we mention the determination by Lange and Tonnies (1996) of the Rayleigh-wave dispersion curve throughout the surface Brillouin zone of the (diamond) C(111) surface with adsorbed H(1 × 1). They used the very high incident energy of 80.5 meV so as to map energy-loss features up to 70 meV.

Epitaxially deposited overlayers have been the subject of many studies; a useful list of references is given by Gillman *et al* (1996). These authors used an *in-situ* system to investigate ^{4}He scattering from ultrathin (one, two and three monolayers) MBE-deposited epilayers of KBr on RbCl, and they compared their results with detailed shell-model calculations.

Finally we mention a recent instrumental development of significance. Doak *et al* (1999) have successfully focused a ^{4}He beam with the use of a nano-engineered Fresnel zone plate with an outermost zone width of 100 nm. The focused spot had a central maximum of width 25 μm. This development opens the way for the use of ^{4}He beams in scanning microscopy.

3.6.2 Electron scattering

Inelastic electron scattering was established as a viable technique for the observation of surface phonons by Lehwald *et al* (1983). The essential improvement in experimental technique was the application of a high-resolution spectrometer for relatively large incident energies of the order of

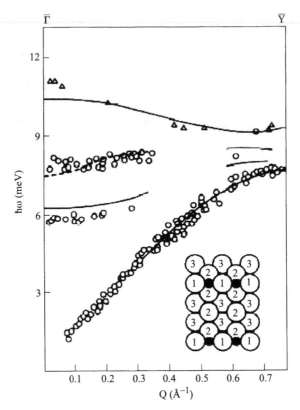

Figure 3.15. Experimental data (points; circles for strong features, triangles for weak) and calculated dispersion curves for inelastic ^{4}He scattering from Ag(110) with chemisorbed O atoms. Inset shows surface structure assumed for calculated curves. After Bracco *et al* (1990).

some hundreds of electron volts. Scattering spectra obtained in that work for two different values of surface-wavevector transfer in scattering off the Ni(100) surface are shown in figure 3.16(a). As for other forms of scattering, the kinematic analysis, given by (1.64) and (1.65) in section 1.4.2, permits translation of the peak positions into a dispersion curve in the $(\omega, q_{\parallel})$ plane which is shown in figure 3.16(b). As for the results obtained by ^{4}He scattering, the dispersion curve is the extension across the Brillouin zone of the Rayleigh wave of elasticity theory.

The theoretical curves in figure 3.16(a) are spectral densities, and do not represent a full theory since the interaction between the electron beam and the surface ions is not included. The full theory is given by Xu *et al* (1985), and compared with experimental data in which further surface-phonon branches are observed on Ni(001).

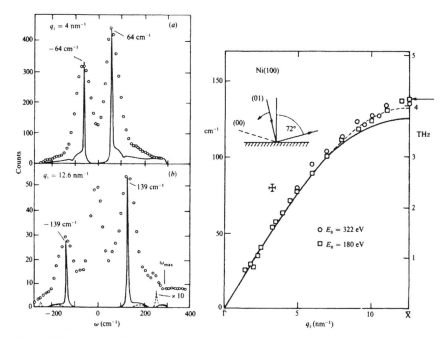

Figure 3.16. (a) Inelastic electron-loss spectra for Ni(100) surface. Beam energy 180 eV. The scattered beam was collected at fixed angle and the angle of incidence was varied to vary $q_{\parallel}$. The full lines are calculated spectral densities for the vertical motion of atoms in the first layer. (b) Loss-spectrum peak positions versus $q_{\parallel}$. Full line: dispersion curve from nearest-neighbour central-force model, with force constant chosen to fit bulk phonon spectrum. Broken line: 20% increase in force constant between first and second layers. After Lehwald *et al* (1983).

The first observation of a surface optic phonon by inelastic electron scattering was reported by Oshima *et al* (1984) on the TaC(100) surface. This is a metallic compound, so the conductivity screens out the Coulomb interaction, which would otherwise lead to a surface polariton (see section 6.1) that might obscure the microscopic optic phonon. As figure 3.17 shows, the scattering peaks lead to a dispersion curve in which ω is almost independent of $q_{\parallel}$. TaC has the simple rock-salt structure, and in a nearest-neighbour central-interaction model the theory of the (100) surface mode reduces exactly to the theory of the diatomic linear chain, given in section 3.1. The surface-mode frequency of (3.5) is marked on figure 3.17. The difference between this and the experimental value may be due to modified force constants near the surface (surface relaxation) or to the effect of second- and third-neighbour forces.

As occurred for ^{4}He scattering, after the work on clean surfaces attention was given to surfaces with adsorbates. A review and discussion

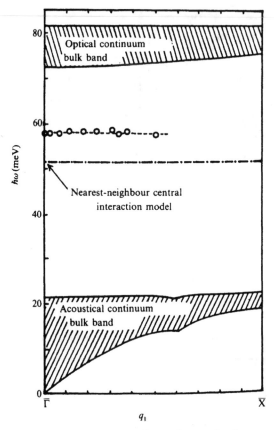

Figure 3.17. Experimental dispersion curve for surface optic phonons on TaC(100) as determined by inelastic electron scattering. The theoretical value given by (3.5) is marked. After Oshima *et al* (1984).

of studies of such surfaces are given by Ibach (1991). An instructive example is the case of hydrogen and deuterium adsorbed on the Ni(110) surface, for which the spectra are shown in figure 3.18(a). The two main scattering peaks are due to transitions between the quantized vibrational states of the adsorbed species. The points of interest stressed by Ibach are first that the frequency ratio 640/510 ($\approx$1.25) for the lower peaks is not $\sqrt{2}$, as would be expected for simple classical vibrations of H and D, and second that the H lines are substantially broadened compared with the D lines. These effects are attributed to the quantum nature of the H vibrational levels arising from the light mass. As illustrated in figure 3.18(b), the excited vibrational states are broadened into a band and the selection rule

$$q_{\parallel}^{(i)} - q_{\parallel}^{(f)} = k_{\parallel,H}^{(i)} - k_{\parallel,H}^{(f)} \tag{3.58}$$

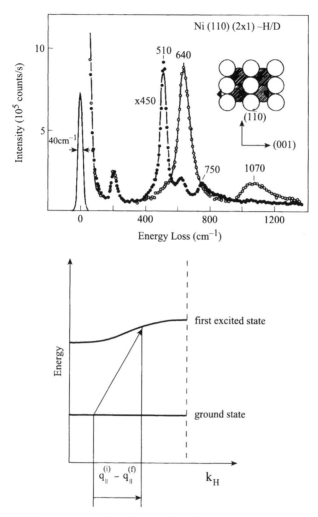

Figure 3.18. (a) Electron energy loss spectra for hydrogen (open circles) and deuterium (full circles) adsorbed on Ni(110). Instrumental resolution and surface structure (inset) shown. (b) Line broadening mechanism for H adsorbate (schematic). The excited vibrational states are broadened into a band. After Ibach (1991).

between the initial and final electron ($q_{||}$) and H ($k_{||,H}$) in-plane momentum components leads to a band of energies rather than a single energy for the transition. This does not occur for D (or any other heavier atom) because the vibrational states are too strongly localized for the energy broadening by band formation to be significant. It is the band of energies that is seen as the linewidth in figure 3.18(a).

3.7 Surface waves on liquids

There are a number of similarities between surface waves on liquids and elastic surface waves on solids. Here, therefore, we briefly review their properties. However, there are several distinctive features. First, there are two restoring forces: gravity and surface tension. The former is dominant for long waves on the ocean, for example, while the latter determines ripples on a cup of tea. Second, liquids do not support shear stresses so only longitudinal effects occur. Third, many liquids, including water, are incompressible.

Despite these differences, the derivation of the dispersion relation for surface waves is similar in spirit to the discussion of the Rayleigh wave in section 3.2.1. The starting point is the tensor formulation of hydrodynamics (see Landau and Lifshitz 1959, for example) and the application to surface waves is reviewed by Loudon (1984). With neglect of the damping due to viscosity and for an incompressible fluid, the dispersion relation is

$$\rho\omega^2 = (q^2 - (\omega^2/v_L^2))^{1/2}(\alpha q^2 + \rho g) \tag{3.59}$$

where ρ is the density, v_L is the (longitudinal) bulk velocity, α is the surface tension (surface energy per unit area) and g is the acceleration due to gravity. In the limit $q \gg \omega/v_L$, which is often satisfied, this is approximated as

$$\rho\omega^2 = \alpha q^3 + \rho g q. \tag{3.60}$$

The first term in (3.59) or (3.60) is due to surface tension and, as implied, it is dominant for large q (short wavelength), while the second is due to gravity and is dominant for small q (long wavelength). Substitution of the numerical values for water shows that the two terms in (3.60) are equal for $q/2\pi = 58.4 \, \text{m}^{-1}$ or wavelength $\lambda = 1.71 \, \text{cm}$, so this may be taken as the intermediate point between the long-wavelength gravity waves and the short-wavelength surface-tension ripples.

The frequencies involved in the surface ripples are too low to be detectable in Brillouin scattering. However, a number of studies, cited by Loudon (1984), have applied other spectroscopic techniques. Byrne and Earnshaw (1979a) used photon-correlation spectroscopy to measure the frequency shifts involved in inelastic scattering of obliquely incident light. In this technique the time dependence of the photon–photon correlation function, after correction for instrumental effects, gives the evolution with time of the scattering excitation at the wavevector q determined by the scattering geometry. The time dependence corresponding to a Lorentzian in the frequency domain is a simple damped oscillation:

$$f(t) = f_0 \cos(\omega_0 t) \exp(-\Gamma t). \tag{3.61}$$

The results are presented in the form of the q dependence of ω_0 and Γ, and later work (Earnshaw and McGivern 1987) gives results that are in excellent agreement with those predicted from bulk viscosity values.

One reason for the interest in the surface of water is that it serves as a suitable substrate for layers that provide models of biological membranes whose elastic properties are of considerable interest. The application of photon-correlation spectroscopy to monolayers of fatty acids is described by Byrne and Earnshaw (1979b).

References

Albuquerque E L 1980 *J. Phys. C* **13** 2623
Albuquerque E L, Loudon R and Tilley D R 1980 *J. Phys. C* **13** 1775
Bortolani V, Nizzoli F, Santoro G, Marvin A and Sandercock J R 1979 *Phys. Rev. Lett.* **43** 224
Bortolani V, Marvin A M, Nizzoli F and Santoro G 1983 *J. Phys. C* **16** 1757
Bortolani V, Franchini A, Nizzoli F and Santoro G 1984 *Phys. Rev. Lett.* **52** 429
Bortolani V, Franchini A, Nizzoli F and Santoro G 1986 in *Electromagnetic Surface Waves* ed G I Stegeman and R F Wallis (Berlin: Springer)
Bracco G, Tatarek R and Vandoni G 1990 *Phys. Rev. B* **42** 1852
Bransden B H and Joachain C J 2000 *Quantum Mechanics* 2nd edition (Englewood Cliffs, NJ: Prentice-Hall)
Brusdeylins G, Doak R P and Tonnies J P 1980 *Phys. Rev. Lett.* **44** 1417
Brusdeylins G, Rechsteiner R, Skofronick J G, Tonnies J P, Benedek G and Miglio L 1985 *Phys. Rev. Lett.* **54** 466
Bunjes N, Luo N S, Ruggerone P, Tonnies J P and Witte G 1994 *Phys. Rev. B* **50** 8897
Burt M G 1973 *J. Phys. C* **6** 855
Byrne D and Earnshaw J C 1979a *J. Phys. D* **12** 1133
Byrne D and Earnshaw J C 1979b *J. Phys. D* **12** 1145
Dervisch A and Loudon R 1976 *J. Phys. C* **9** L669
Doak R B, Harten U and Tonnies J P 1983 *Phys. Rev. Lett.* **51** 578
Doak R B, Grisenti R E, Rehbein S, Schmahl G, Tonnies J P and Wöll Ch 1999 *Phys. Rev. Lett.* **83** 4229
Dutcher J R, Lee S, Hillebrands B, McLaughlin G J, Nickel B G and Stegeman G I 1992 *Phys. Rev. Lett.* **68** 2464
Earnshaw J C and McGivern R C 1987 *J. Phys. D* **20** 82
Eguiluz A G and Maradudin A A 1983 *Phys. Rev. B* **28** 728
Garcia-Moliner F 1977 *Ann. Phys. Paris* **2** 179
Garcia-Moliner F and Flores F 1979 *Introduction to the Theory of Solid Surfaces* (Cambridge: Cambridge University Press)
Gillman E S, Baker J, Hernandez J J, Bishop G G, Li J A, Safron S A, Skofronick J G, Bonart D and Schröder U 1996 *Physica B* **219–220** 428
Giovannini L, Nizzoli F and Marvin A M 1992 *Phys. Rev. Lett.* **69** 1572
Glass N E, Loudon R and Maradudin A A 1981 *Phys. Rev. B* **24** 6843
Glass N E and Maradudin A A 1983 *J. Appl. Phys.* **54** 796
Harley R T and Fleury P A 1979 *J. Phys. C* **12** L863
Hayes W and Loudon R 1978 *Scattering of Light by Crystals* (New York: Wiley)
Ibach H 1991 *Physica Scripta* **T39** 323

Kueny A, Grimsditch M, Miyano K, Banerjee I, Falco C M and Schuller J K 1982 *Phys. Rev. Lett.* **48** 166

Landau L D and Lifshitz E M 1959 *Fluid Mechanics* (Oxford: Pergamon)

Lange G and Tonnies J P 1996 *Phys. Rev. B* **53** 9614

Lee S, Giovannini L, Dutcher J R, Nizzoli F, Stegeman G I, Marvin A M, Wang Z, Ross J D, Amoddeo A and Caputi L S 1994 *Phys. Rev. B* **49** 2273

Lehwald S, Szeftel J M, Ibach H, Rahman T S and Mills D L 1983 *Phys. Rev. Lett.* **50** 518

Loudon R 1978a *J. Phys. C* **11** 403

Loudon R 1978b *J. Phys. C* **11** 2623

Loudon R 1978c *Phys. Rev. Lett.* **40** 581

Loudon R 1984 in *Surface Excitations* ed V M Agranovich and R Loudon (Amsterdam: North-Holland) p 589

Loudon R and Sandercock J R 1980 *J. Phys. C* **13** 2609

Marvin A M, Bortolani V and Nizzoli F 1980a *J. Phys. C* **13** 229

Marvin A M, Bortolani V, Nizzoli F and Santoro G 1980b *J. Phys. C* **13** 1607

Mayer A P, Zierau W and Maradudin A A 1991 *J. Appl. Phys.* **69** 1942

Mazur P and Maradudin A A 1981 *Phys. Rev. B* **24** 2996

Meeker T R and Meitzler A H 1964 in *Physical Acoustics* vol 1A ed W P Mason (New York: Academic) p 112

Nizzoli F 1986 in *Electromagnetic Surface Excitations* ed R F Wallis and G I Stegeman (Berlin: Springer) p 138

Oliveira F A, Cottam M G and Tilley D R 1980 *Phys. Stat. Sol. (b)* **107** 737

Oliveros M C and Tilley D R 1983 *Phys. Stat. Sol. (b)* **119** 675

Oshima C, Souda R, Aono M, Otani S and Ishizawa Y 1984 *Phys. Rev. B* **30** 5361

Rayleigh, Lord 1907 *Phil. Mag.* **14** 70

Rowell N and Stegeman G I 1978a *Solid State Commun.* **26** 809

Rowell N and Stegeman G I 1978b *Phys. Rev. B* **18** 2598

Sandercock J R 1972a *Phys. Rev. Lett.* **28** 237

Sandercock J R 1972b *Phys. Rev. Lett.* **29** 1735

Sandercock J R 1978 *Solid State Commun.* **26** 547

Sandercock J R 1982 in *Light Scattering in Solids III* ed M Cardona and G Güntherodt (Berlin: Springer)

Scholte J G 1947 *Mon. Not. R. Astron. Soc.: Geophys. Suppl.* **5** 120

Stegeman G I and Nizzoli F 1984 in *Surface Excitations* ed V M Agranovich and R Loudon (Amsterdam: North-Holland) p 195

Subbaswamy K R and Maradudin A A 1978 *Phys. Rev. B* **18** 4181

Tilley D R 1980 *J. Phys. C* **13** 781

Tonnies J P 1987 in *Surface Phonons* ed W Kress (Berlin: Springer)

Wallis R F 1957 *Phys. Rev.* **105** 540

Wallis R F 1959 *Phys. Rev.* **116** 302

Xu M L, Hall B M, Tong S Y, Rocca M, Ibach H, Lehwald S and Black J E 1985 *Phys. Rev. Lett.* **54** 1171

Chapter 4

Surface magnons

In chapter 2 an introductory account was given of the theoretical models required to describe magnons in simple ferromagnets and antiferromagnets. Results were obtained for the dispersion relations of the bulk magnons in infinite samples of these materials. In the present chapter we concentrate on the exchange-dominated region, where the interactions between different magnetic moments (or spins) are short range. It will be shown how the previous theoretical results in sections 2.3.2 and 2.3.3 generalize to semi-infinite materials and to thin films. In appropriate cases it is found that surface magnons may occur localized near the surface(s) and that the bulk-magnon properties are modified.

First, using the Heisenberg model, we present the theory of bulk and surface magnons in semi-infinite ferromagnets and ferromagnetic films. Then the theory is extended to bulk and surface magnons in Heisenberg antiferromagnets, emphasizing important effects due the lattice structure. Next we discuss surface magnons for the Ising model (where the exchange interactions are anisotropic, by contrast with the Heisenberg model), taking the case of an Ising ferromagnet in a transverse applied magnetic field. We then make some generalizations to exchange-coupled magnetic bilayers, since this type of film geometry has been of considerable interest for spin-dependent transport phenomena such as giant magnetoresistance and has led to a wealth of experimental data. This is followed by an account of other experimental results for surface magnetic effects in Heisenberg or exchange-dominated magnetic materials.

The additional effects occurring due to the long-range magnetic dipole–dipole interactions in surface geometries will be the topic of chapter 5.

4.1 Semi-infinite Heisenberg ferromagnets

We now generalize the theory of section 2.3.2 to the case of a semi-infinite Heisenberg ferromagnet occupying the half-space $z < 0$. The system is assumed to be effectively infinite in the x and y directions, so that there is only one surface (the plane $z = 0$) to consider. The Hamiltonian will be taken to have the same form as in (2.59), except that the exchange J_{ij} near the surface may differ in general from the value in the bulk of the material. This modification can occur because the exchange interaction is related to overlap integrals between electronic wave functions, as explained in section 2.3.1, and these wave functions and/or the atomic spacing may be perturbed near a surface. As before, we make use of both microscopic and macroscopic (or continuum) methods, and finally some Green-function applications are described.

4.1.1   Microscopic theory

As a specific example, we consider a simple cubic lattice with a (001) surface. The exchange interactions J_{ij} will be assumed to couple nearest neighbours only, having the value J_S if both i and j are in the surface layer and the bulk value J otherwise. This is a model first considered by Fillipov (1967) and it is represented schematically in figure 4.1. The direction of average spin alignment will be taken along the z axis, which is also the direction of the applied magnetic field B_0, and to determine the equilibrium configuration we again use mean-field theory as in section 2.3.2. Equations (2.71) and (2.72) still apply, but the effective exchange field $B_E(i)$ will now depend on the location of site i relative to the surface. We introduce an index $n(= 1, 2, 3, \ldots)$ to label the layers parallel to the surface; n is related to the coordinate z by $z = -(n-1)a$. From symmetry considerations it is clear that $B_E(i)$ and

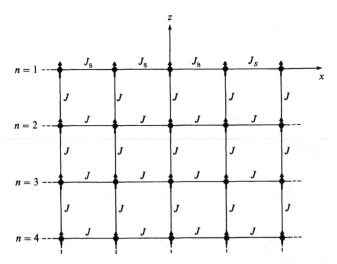

Figure 4.1. The (001) surface of a semi-infinite ferromagnet with simple-cubic structure, indicating the nearest-neighbour exchange constants J and J_S.

$\langle S_i^z \rangle$ depend on position only through the layer index n. For $n = 1$, each spin has four neighbours in the surface and one in layer 2, and so

$$g\mu_B B_E(1) = 4J_S \langle S^z \rangle_1 + J \langle S^z \rangle_2. \tag{4.1}$$

Similarly for $n > 1$, there are six neighbours and

$$g\mu_B B_E(n) = J \langle S^z \rangle_{n-1} + 4J \langle S^z \rangle_n + J \langle S^z \rangle_{n+1}. \tag{4.2}$$

In the case of $S = \frac{1}{2}$ the mean-field equations are

$$\langle S^z \rangle_n = \frac{1}{2}\tanh\{g\mu_B[B_0 + B_E(n)]/2k_B T\} \qquad (n \geq 1) \tag{4.3}$$

by analogy with (2.73). When (4.1) and (4.2) are substituted into (4.3) we have a set of recurrence relationships satisfied by the spin averages $\langle S^z \rangle_n$. In general they have to be solved numerically, but we note that for $T \to 0$ there is the analytic solution that $\langle S_n^z \rangle \to \frac{1}{2}$ for all n. An example of the numerical solution of (4.3) using an iterative approach is given in figure 4.2, taking $B_0 = 0$ and $J_S/J = 0.5$.

We next evaluate the frequencies of the magnetic excitations, showing that they consist of localized surface magnons in addition to bulk magnons. We restrict attention to the low-temperature region $T \ll T_C$ in order to avoid complications due to magnetic surface reconstruction, i.e. the modified values of $\langle S^z \rangle_n$ near the surface. This enables each $\langle S^z \rangle_n$ to be replaced by the spin quantum number S. The linearized equation of

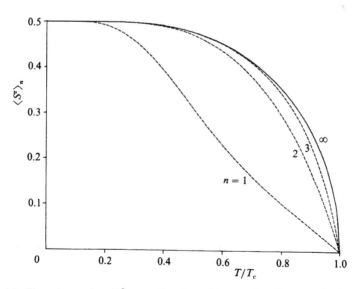

Figure 4.2. The spin average $\langle S^z \rangle_n$ as a function of temperature for several values of the layer index n in a semi-infinite Heisenberg ferromagnet with simple-cubic structure, taking $B_0 = 0$ and $J_S/J = 0.5$.

motion, obtained using (2.63) and (2.64), is

$$i\frac{d}{dt}S_j^+ = g\mu_B B_0 + S\sum_i J_{ij}(S_j^+ - S_i^+). \tag{4.4}$$

The dispersion relations for bulk and surface magnons can now be deduced from this equation, as follows.

The magnons correspond to wave-like solutions of the form

$$S_j^+ = s_n(\mathbf{q}_\parallel)\exp(i\mathbf{q}_\parallel \cdot \mathbf{r}_{\parallel,j})\exp(-i\omega t) \tag{4.5}$$

for any operator S_j^+ in layer n, where we are considering a Fourier component with angular frequency ω and 2D wave vector $\mathbf{q}_\parallel = (q_x, q_y)$ parallel to the surface. The factor $\exp(i\mathbf{q}_\parallel \cdot \mathbf{r}_{\parallel,j})$ in (4.5), where $\mathbf{r}_\parallel = (x, y)$, is in accordance with Bloch's theorem and the translational symmetry of the system in the xy plane. Because of the surface, there is no such factor for the z direction and $s_n(\mathbf{q}_\parallel)$ depends on z through the layer index n. From (4.4) and (4.5) we obtain the following infinite series of coupled equations for $s_n(\mathbf{q}_\parallel)$:

$$\{\omega - g\mu_B B_0 - SJ - 4SJ_S[1 - \gamma(\mathbf{q}_\parallel)]\}s_1 + SJs_2 = 0 \tag{4.6}$$

$$\{\omega - g\mu_B B_0 - 2SJ - 4SJ[1 - \gamma(\mathbf{q}_\parallel)]\}s_n + SJ(s_{n-1} + s_{n+1}) = 0 \quad (n > 1) \tag{4.7}$$

where

$$\gamma(\mathbf{q}_\parallel) = \tfrac{1}{2}[\cos(q_x a) + \cos(q_y a)]. \tag{4.8}$$

Equations (4.6) and (4.7) can be solved by a variety of techniques to obtain the frequencies ω of the various magnon modes, e.g. as discussed in the review article by Wolfram and Dewames (1972). However, a simple approach (which parallels that used in section 3.2 for elastic waves in semi-infinite media) is to note that the bulk magnons correspond to solutions for $s_n(\mathbf{q}_\parallel)$ made up of two waves (one incident and one reflected):

$$s_n(\mathbf{q}_\parallel) = A(\mathbf{q}_\parallel)\exp[-iq_z(n-1)a] + B(\mathbf{q}_\parallel)\exp[iq_z(n-1)a]. \tag{4.9}$$

When this is substituted into (4.7) we obtain the dispersion relation that $\omega = \omega_B(\mathbf{q})$, where $\mathbf{q} = (\mathbf{q}_\parallel, q_z)$ is a 3D wave vector and

$$\omega_B(\mathbf{q}) = g\mu_B B_0 + 2SJ[3 - 2\gamma(\mathbf{q}_\parallel) - \cos(q_z a)]. \tag{4.10}$$

This is equivalent to the bulk-magnon dispersion relation in an *infinite* simple-cubic ferromagnet at $T \ll T_C$, as can be seen from (2.67) and (2.69). The ratio B/A, which is a reflection coefficient for the bulk magnon at the surface $z = 0$, can be determined using (4.6). It is easily shown that B/A is a complex number with modulus unity and a phase angle that depends on the surface parameter J_S.

   The surface magnons may be found from the ansatz (as discussed in section 1.1.2) of seeking attenuated solutions for $s_n(\mathbf{q}_{||})$:

$$s_n(\mathbf{q}_{||}) = C(\mathbf{q}_{||})\exp(-\kappa na). \tag{4.11}$$

From (4.6) and (4.7) this leads to $\omega = \omega_S(\mathbf{q}_{||})$, where

$$\omega_S(\mathbf{q}_{||}) = g\mu_B B_0 + 4SJ[1 - \gamma(\mathbf{q}_{||})] - SJ(\Delta - 1)^2/\Delta \tag{4.12}$$

and

$$\Delta \equiv \exp(-\kappa a) = \left\{1 + 4\left(1 - \frac{J_S}{J}\right)[1 - \gamma(\mathbf{q}_{||})]\right\}^{-1} \tag{4.13}$$

is the fractional decrease per layer in the magnon amplitude. Equation (4.12) represents a *localized* surface mode only if

$$|\Delta| < 1 \tag{4.14}$$

or, equivalently, $\mathrm{Re}(\kappa) > 0$. This condition may be satisfied in two ways. One possibility is $0 < \Delta < 1$, which occurs for $J_S < J$ and any $\mathbf{q}_{||} \neq 0$. The spin deviations on adjacent layers are then in phase, and the mode is called an *acoustic surface magnon* (see figure 4.3(a) for a semi-classical representation): its frequency is less than that of the corresponding bulk magnon, i.e. $\omega_S(\mathbf{q}_{||}) < \omega_B(\mathbf{q}_{||}, q_z)$ for any q_z. The other case is $-1 < \Delta < 0$, which implies a phase change of $180°$ between the spin deviations on adjacent

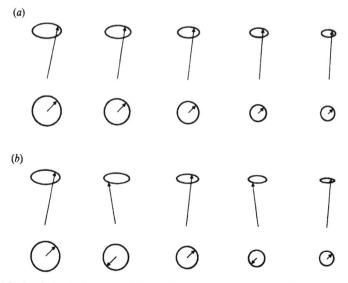

Figure 4.3. Semi-classical representation of surface magnons in a ferromagnet: (a) an acoustic surface magnon and (b) an optic surface magnon. Comparison should be made with figure 2.4 for a bulk magnon.

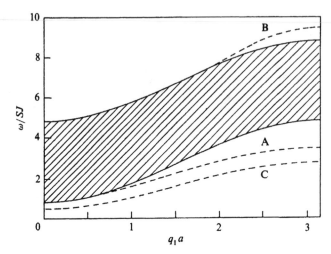

Figure 4.4. A plot of magnon frequency ω (in units of SJ) against $|\mathbf{q}_{\parallel}|a$, in a semi-infinite Heisenberg ferromagnet for propagation wave vector $\mathbf{q}_{\parallel} = (q_{\parallel}, 0)$. The bulk magnon continuum is shown, together with three surface magnon branches corresponding to: A, $J_S/J = 0.5$, $B_{AS} = 0$; B, $J_S/J = 1.8$, $B_{AS} = 0$; C, $J_S/J = 0.5$, $g\mu_B B_{AS}/SJ = -0.8$. Also $g\mu_B B_0/SJ = 0.8$.

layers (see figure 4.3(b)). This mode is known as an *optic surface magnon*, and its frequency is such that $\omega_S(\mathbf{q}_{\parallel}) > \omega_B(\mathbf{q}_{\parallel}, q_z)$. It can be shown from (4.13) that a necessary condition for optic surface magnons to exist in some part of the 2D Brillouin zone for $\mathbf{q}_{\parallel}$ is $J_S > \frac{5}{4}J$ and that these modes occur only for $|\mathbf{q}_{\parallel}|$ above some nonzero critical value.

In figure 4.4 we give some numerical examples of the magnon dispersion relations as predicted by this simple model. We plot frequency ω against $\mathbf{q}_{\parallel}$ for wave vector $\mathbf{q}_{\parallel}$ in the (100) direction. The bulk magnons appear as a continuum, with lower and upper edges corresponding to $q_z = 0$ and $q_z = \pm\pi/a$ respectively. Curves A and B are examples of acoustic and optic surface magnons, respectively, as given by (4.12).

Further discussions of the surface-magnon dispersion relations have been given by various authors. The early work was extensively reviewed by Wolfram and Dewames (1972), while more recent developments are described by Mills (1984, 1994) and Cottam and Slavin (1994). We briefly quote here some extensions and generalizations to the results derived above for the surface-magnon frequencies. An additional effect to be taken into account is *surface anisotropy* (also known as *pinning*). This can often be represented phenomenologically by adding an extra term to the Hamiltonian (2.59) of the form

$$-g\mu_B \sum_i B_A(i)S_i^z. \tag{4.15}$$

The anisotropy may arise because the crystalline fields at the surface have a lower symmetry than in the bulk or because of surface impurities (e.g. see Heinrich and Cochran 1993). In the simplest case the effective anisotropy field may be taken to have a constant value B_{AS} for spins in the surface layer and to be zero otherwise. When the theory is modified to take this into account, one finds that the surface-magnon frequency is still given formally by (4.12) but the definition of Δ becomes

$$\Delta \equiv \exp(-\kappa a) = \left\{ 1 - \frac{g\mu_B B_{AS}}{SJ} + 4\left(1 - \frac{J_S}{J}\right)[1 - \gamma(\mathbf{q}_{\parallel})] \right\}^{-1}. \qquad (4.16)$$

If $B_{AS} < 0$ (i.e. the anisotropy field is in the opposite direction to the applied magnetic field), the existence condition (4.14) for a surface magnon can be satisfied for certain $\mathbf{q}_{\parallel}$ values even if $J_S = J$, unlike in the previous situation of $B_{AS} = 0$. An example of an acoustic surface magnon in the presence of surface anisotropy has been included in figure 4.4 (see curve C). Note that even at $\mathbf{q}_{\parallel} = 0$ it is split off below the lower edge of the bulk-magnon continuum.

Calculations of the surface-magnon spectrum for a variety of other models are to be found in the literature already cited. A relatively minor modification of the preceding theory occurs when the applied field, the anisotropy field and the average spin alignment are all *parallel* to the surface. It is easily shown that this leaves the formal results for the magnon frequencies unaltered. The invariance is a consequence of the symmetry of the Heisenberg Hamiltonian under rotations, and we shall later (in sections 5.1 and 5.4) find a different behaviour when the dipole–dipole interactions are included, giving a lowering of the symmetry. In general, the applied field, the effective anisotropy field and the direction of average spin alignment may not necessarily be collinear, and this leads to more complicated results. Furthermore, modifying the assumptions concerning the exchange interactions also has an important effect. For example, another simple choice would be to take all exchange interactions (including those involving spins at the surface) to have their bulk values, denoted by J for nearest neighbours and J' for next-nearest neighbours (e.g. see Wallis *et al* 1967). For a simple cubic lattice with a (001) surface and in the absence of pinning, this leads to an acoustic surface magnon for all $\mathbf{q}_{\parallel} \neq 0$ provided $J > 4J' > 0$. The more general situation in which there may be nearest and next-nearest exchange interactions, together with several exchange parameters near the surface different from their bulk values, is extremely complicated. Several branches may be predicted for the surface-magnon spectrum. In certain cases the surface spins may even orient in a different direction from those in the bulk, thus providing an example of surface reconstruction.

The existence conditions and frequencies of the surface magnons are sensitive to the lattice structure of the material and/or to the crystallographic orientation of the surface. It is relatively straightforward to generalize the calculations given earlier in this section for the sc lattice with (001) surface to fcc and bcc lattices with (001) surface and also to all three lattices with (011) surface orientation.

4.1.2   Macroscopic theory

By analogy with discussion in section 2.3.2 for bulk ferromagnets, it is useful to consider also the approach based on using a *continuum approximation* for the ferromagnetic medium. It is applicable here provided the wavelength of any excitation is large compared with the lattice spacing a, i.e. we require $q_{\|}a \ll 1$ and $qa \ll 1$. In addition, the formalism as described here also requires that $|\kappa a| \ll 1$, and consequently it is restricted to acoustic-type surface modes.

We employ the torque equation of motion (2.74) with the instantaneous effective magnetic field calculated from (2.75). It is easily seen that (2.78) and (2.79) still apply for our model of a semi-infinite ferromagnet with sc structure and nearest-neighbour exchange. In going over to the continuum approximation we must again make a Taylor series expansion of (2.78) for the fluctuating field $\mathbf{b}_i$ at site i. For any site that is *not* in the surface layer, there are six nearest neighbours and the exchange constant is J. In the continuum limit it leads to (2.81) and (2.82) as previously; these results apply provided $z < 0$. However, for a spin in the surface layer $z = 0$, there are four nearest neighbours in the same layer (with exchange J_S) and one nearest neighbour in layer 2 (with exchange J). This gives approximately

$$\mathbf{b}(\mathbf{r})_{z=0} = \left[(4J_S + J) - aJ\frac{\partial}{\partial z} + a^2 J_S^2 \left(\frac{\partial^2}{\partial x^2} + \frac{\partial^2}{\partial y^2} \right) + \frac{1}{2}a^2 J^2 \frac{\partial^2}{\partial z^2} \right] \boldsymbol{\mu}(\mathbf{r})_{z=0}.$$

(4.17)

When this is substituted into (2.79) in the continuum limit and a comparison is made with (2.82) for $z < 0$, we arrive at the following boundary condition to be satisfied at the surface:

$$\partial \boldsymbol{\mu}(\mathbf{r})/\partial z - \xi \boldsymbol{\mu}(\mathbf{r}) = 0 \qquad (z = 0). \tag{4.18}$$

Here the quantity ξ depends on the surface characteristics:

$$\xi = -\frac{g\mu_B B_{AS}}{aSJ} + \left(1 - \frac{J_S}{J} \right) aq_{\|}^2 \tag{4.19}$$

and we have introduced a surface anisotropy field B_{AS} as before. The boundary condition (4.19) has the form introduced by Rado and Weertman (1959). Often it is postulated phenomenologically, but here it has been derived from the continuum limit. Note that it has the same form as (2.160) postulated for exciton–polaritons; this analogy will become clearer when we discuss dipole–exchange modes in section 5.4.

We now use the property of translational invariance parallel to the surface to write

$$\boldsymbol{\mu}(\mathbf{r}) = \boldsymbol{\mu}(z)\exp(i\mathbf{q}_{||} \cdot \mathbf{r}_{||}) \tag{4.20}$$

and by seeking travelling-wave and attenuated solutions of (2.82) and (4.18) we deduce the bulk-magnon and surface-magnon frequencies, respectively. The results are

$$\omega_{B}(\mathbf{q}) = g\mu_{B}B_{0} + SJa^{2}(q_{||}^{2} + q_{z}^{2}) \tag{4.21}$$

$$\omega_{S}(\mathbf{q}_{||}) = g\mu_{B}B_{0} + SJa^{2}(q_{||}^{2} - \xi^{2}). \tag{4.22}$$

These are, as expected, just the long-wavelength (small $|\mathbf{q}|$ and $|\mathbf{q}_{||}|$) approximations of the previous results (4.10) and (4.12). An attenuated solution is possible only if $\xi > 0$, and this becomes the existence condition for a surface magnon. In fact, by using (4.19) and approximating (4.13) for the case of $aq_{||} \ll 1$, it can be checked that ξ is equivalent to κ of the microscopic theory in this limit. The frequency $\omega_{S}(\mathbf{q}_{||})$ in (4.22) corresponds to an acoustic surface-magnon branch below the bulk magnon continuum. Because the present analysis has been carried out for long wavelengths, no optic surface magnons are predicted.

In summary, the continuum approach can be a useful simplification when we are concerned only with long-wavelength excitations. For example, the low-temperature contributions of the bulk magnons and acoustic surface magnons to the magnetization are dominated by the long-wavelength region, where the excitations have lower frequencies and are therefore more readily excited thermally. Similarly it is these excitations that participate in a one-magnon light-scattering process or in magnetic resonance, due to the kinematic restrictions mentioned in section 1.4.

4.1.3   Green functions and applications

So far we have concentrated on the *frequencies* of the bulk and surface magnons, but more generally we may wish to know the *intensities* associated with the excitations. This information can be obtained from spin-dependent Green functions of the form $\langle\langle S^{+}; S^{-}\rangle\rangle_{\omega}$, since magnons are the excitations associated with transverse spin components. These Green functions contain a full description of the dynamic behaviour of the magnetic system, and can be employed to evaluate the thermodynamic properties of the magnons or the intensities of the magnons as measured, for example, by light scattering.

Explicit forms of the Green functions for a semi-infinite Heisenberg ferromagnet were derived by Cottam (1976), using a microscopic theory, and by Moul and Cottam (1979), using a macroscopic approach. We shall discuss only the latter case, for which linear-response theory can be employed in an analogous manner to the treatment of elastic waves (long-wavelength acoustic phonons) in section 3.2.3. Briefly, the calculation involves taking an additional term, representing an external driving field $\mathbf{b}_{\text{ext}}(\mathbf{r})$, to be included on the right-hand side of (2.77) for the total field. This produces an additional term in the linearized equation of motion (2.79), and by calculating the response of $\boldsymbol{\mu}$ to $\mathbf{b}_{\text{ext}}$ we may deduce the required Green functions. This is in accordance with the general formalism described in the Appendix for applying a stimulus (force) at one point and calculating its linear response at another point. The final result for the Green function can be expressed as

$$\langle\langle S^+(z); S^-(z')\rangle\rangle_\omega = \frac{i}{AJaQ}\left\{\exp(iQ|z-z'|) + \left(\frac{Q+i\xi}{Q-i\xi}\right)\exp(-iQ(z+z'))\right\}$$
(4.23)

where A is the (macroscopically large) surface area of the ferromagnet and Q is a complex quantity which satisfies

$$\omega = g\mu_B B_0 + SJa^2(q_\parallel^2 + Q^2)$$
(4.24)

with $\text{Im}(Q) \geq 0$. The other notation is as before. The bulk magnons correspond to Q taking the value of a real wave vector component q_z in the z direction, and it can be seen that (4.24) then reproduces the bulk-magnon dispersion relation (4.21). The vanishing of the denominator in the second term of (4.23) occurs for $Q = i\xi$, provided $\xi > 0$, and the substitution of this into (4.24) gives the surface-magnon dispersion relation (4.22). The first term in (4.23) depends on the position labels z and z' only through $|z-z'|$ and may be identified as the Green function for the infinite crystal, while the second term, which depends on $(z+z')$, contains the effects due to the surface. The occurrence of two exponential terms with these position dependences is a typical feature of Green functions for excitations in semi-infinite media, and was seen also in (3.35) for a semi-infinite elastic medium.

It is now a straightforward application of the fluctuation–dissipation theorem (see the Appendix) to deduce from (4.23) the correlation functions $\langle S^-(z')S^+(z)\rangle_\omega$ between any two spin operators. As an example, we take the case of $z' = z$ to evaluate

$$F(\omega, z) \equiv \langle|S^+(z)|^2\rangle_\omega = \langle|S^x(z)|^2 + |S^y(z)|^2\rangle_\omega$$
(4.25)

as a function of ω and z for a fixed value of the wave vector $\mathbf{q}_\parallel$. It follows from (4.24) that the bulk magnons correspond to $\omega > \Omega_0 \equiv g\mu_B B_0 + SJa^2 q_\parallel^2$. In this frequency range we obtain using (A.29):

$$F(\omega, z) = \frac{[n(\omega)+1]}{\pi AJaq_z}\left\{1 + \frac{(q_z^2 - \xi^2)\cos(2q_z z) + 2q_z\xi\sin(2q_z z)}{(q_z^2 + \xi^2)}\right\}.$$
(4.26)

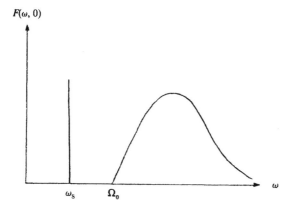

$F(\omega, 0)$

ω_S Ω_0 ω

Figure 4.5. Schematic plot of the correlation function $F(\omega,0)$ for a semi-infinite Heisenberg ferromagnet as a function of frequency ω. After Cottam and Maradudin (1984).

This is nonzero at the surface of the ferromagnet ($z = 0$) where it takes the value

$$F(\omega, 0) = \frac{2[n(\omega) + 1]q_z}{\pi AJa(q_z^2 + \xi^2)} \qquad (4.27)$$

and away from the surface it oscillates as a function of z. In (4.26) and (4.27) ω is given by (4.24) with Q real and equal to q_z, and $n(\omega)$ is again the Bose–Einstein factor introduced in (1.51). In addition there is a surface-magnon contribution to $F(\omega, z)$, occurring when $\omega < \Omega_0$. It corresponds to the pole at $Q = i\xi$ in (4.23), and from (A.27) and (A.29) we obtain in this region

$$F(\omega, z) = \frac{4[n(\omega_S) + 1]a\xi}{SA} \exp(2\xi z)\delta(\omega - \omega_S). \qquad (4.28)$$

The strength of this delta-function contribution at the surface-magnon frequency decreases with distance $-z$ from the surface, as expected. The behaviour of $F(\omega, 0)$ as a function of ω, as given by (4.27) and (4.28), is represented schematically in figure 4.5. The integrated intensities $I_B(\omega)$ and $I_S(\omega)$, associated with the bulk and surface magnons respectively, can be obtained by integrating (4.27) and (4.28) over the appropriate ranges of ω. For the surface-mode contribution this gives $I_S(\omega)$ as simply the strength of the delta-function in (4.28), but the evaluation of the bulk-mode contribution is much more complicated (see Cottam and Maradudin 1984).

The Green function $\langle\langle S^+(z); S^-(z')\rangle\rangle_\omega$ may also be employed in a number of other physical applications, and we give two examples here. One application is to calculate the average spin deviation $S - \langle S^z(z)\rangle$ in the semi-infinite ferromagnet. This quantity will be nonzero due to thermal effects of the bulk and surface magnons, providing corrections to mean-field theory even at $T \ll T_C$. The connection with the Green-function formalism can be achieved by using the spin commutation property (e.g. see (2.61)) that $S^+S^- - S^-S^+ = 2S^z$ for operators at the same site. We also take for simplicity $S = \frac{1}{2}$, in which case the identity $S^+S^- + S^-S^+ = 1$ follows from the properties of the Pauli spin matrices. It is then easy to show that the average spin deviation is given by

$$\delta S(z) \equiv \tfrac{1}{2} - \langle S^z(z) \rangle = \langle S^-(z)S^+(z) \rangle. \qquad (4.29)$$

The correlation function on the right-hand side can be obtained from the Green function (4.23) by using the fluctuation–dissipation theorem and integrating over ω and $\mathbf{q}_\parallel$ (see e.g. Cottam and Maradudin (1984) for details). A partial cancellation takes place between the bulk- and surface-magnon contributions, leading to the overall approximate result:

$$\delta S(z) = \left(\frac{a}{2\pi}\right)^3 \int \mathrm{d}^3\mathbf{q}[1 + \cos(2q_z z)]n[\omega_\mathrm{B}(\mathbf{q})]. \qquad (4.30)$$

This result was first derived in Mills and Maradudin (1967) by using a different method. It follows that, for small enough $|z|$ (such that $\cos(2q_z z) \approx 1$ throughout the dominant region of integration), $\delta S(z)$ is *twice* the corresponding value in an infinite crystal. The condition is $|z| \ll z_0$, where $z_0 = aT_\mathrm{C}/T$ when $B_0 = B_\mathrm{A} = 0$. For $|z| \gg z_0$ it can be shown that $\delta S(z)$ is equal to the infinite-crystal result plus a small correction term of order $z_0/|z|$ or $aT_\mathrm{C}/|z|T$.

Another application of the Green function (4.23) is to light scattering, where the calculation closely parallels that in section 3.3.2 for Brillouin scattering from acoustic phonons. It was established in section 1.4.1 that the light-scattering cross section is related to a correlation function involving components of the scattered electric field. This can be related to a polarization in the scattering medium, which can then be expanded in terms of the appropriate variables describing the excitation. For light scattering from vibrational modes (see section 3.3.2) this led to a description in terms of atomic displacement derivatives. For light scattering from magnons the polarization must be expanded in powers of the spin operators $\mathbf{S}$ or, equivalently, the fluctuating magnetization (e.g. see Hayes and Loudon 1978). In this way, and with the aid of the fluctuation–dissipation theorem, one may relate the differential cross section $\mathrm{d}^2\sigma/\mathrm{d}\Omega\,\mathrm{d}\omega_\mathrm{S}$ for scattering from the magnons in a Heisenberg ferromagnet to $\langle\langle S^+(z); S^-(z') \rangle\rangle_\omega$ in lowest order. Details of the formalism and results are to be found elsewhere (see Cottam 1978b, Moul and Cottam 1979). For example, it is found that the contribution to the cross section for Stokes scattering from the bulk magnons in an absorptive medium has the form

$$\left(\frac{\mathrm{d}^2\sigma}{\mathrm{d}\Omega\,\mathrm{d}\omega_\mathrm{S}}\right) = \frac{F[n(\omega) + 1]}{q_z}\left|\frac{1}{(K - q_z)} + \frac{\phi(q_z)}{(K + q_z)}\right|^2 \qquad (4.31)$$

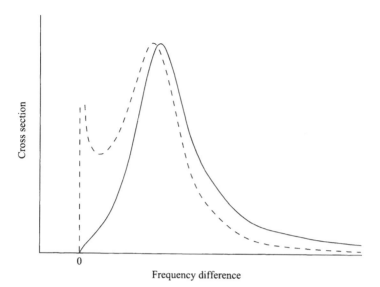

Figure 4.6. Calculated lineshapes for the cross section versus frequency difference $(\omega - \Omega_0)$ for light scattering at normal incidence from bulk magnons in an optically absorptive ferromagnet. Raman scattering in a low-temperature limit has been assumed, together with $\mathrm{Im}(K)/\mathrm{Re}(K) = 0.2$ for the optical parameters and the cases of $\phi = -1$ (solid curve) and $\phi = 1$ (broken curve).

where ω is the frequency shift of the scattered light, K is related to the wave vector components of the incident and scattered light perpendicular to the surface as defined in (3.53), and

$$\phi(q_z) = (q_z + i\xi)/(q_z - i\xi). \qquad (4.32)$$

The overall factor F depends on the scattering geometry and on the polarizations of the incident and scattered light; it is effectively a constant across a given spectrum. The quantity $\phi(q_z)$, which also depends on the 'pinning' parameter ξ introduced in (4.18), can be regarded as a complex reflection coefficient for bulk magnons at the surface. The predicted bulk-magnon spectrum at $\omega > \Omega_B$ is illustrated in figure 4.6 for two limiting cases: $\phi(q_z) \approx 1$ (zero pinning) and $\phi(q_z) \approx -1$ (large pinning). In both cases the asymmetric main peak corresponds *approximately* to the frequency at which $q_z = \mathrm{Re}(K)$, while the width is proportional to $\mathrm{Im}(K)$. For $\phi(q_z) \approx 1$ there is a subsidiary peak at $\omega \approx \Omega_0$, which is associated with a singularity in the magnon density of states. Details are given in the references cited above. Later, in section 4.6, we shall describe some subsequent Brillouin scattering experiments that provide good agreement with this theory.

4.2 Heisenberg ferromagnetic films

The semi-infinite limit treated in section 4.1 represents a satisfactory approximation if the thickness of the ferromagnet is large compared with the surface mode attenuation length and any other characteristic length that may be involved (e.g. the optical penetration depth in a light-scattering experiment). If this condition is *not* fulfilled, then it becomes necessary to treat the ferromagnet as a film (or slab) of finite thickness and to take account explicitly of the *two* surfaces. The same general methods of calculation can be employed as before, but the imposition of two sets of boundary conditions makes the results algebraically more complicated. We shall show that the main effects of finite film thickness on the magnons in Heisenberg ferromagnets may typically include (a) a splitting of the surface-magnon branch, usually into two parts, for appropriate film thickness and (b) an effective 'quantization' of the bulk-magnon frequencies. The values of film thickness of interest may vary considerably depending on the particular ferromagnetic material and on the wave vector, but typically they might lie within the range of a few nanometres up to a few micrometres. Review accounts of the bulk and surface magnons in Heisenberg ferromagnetic thin films have been given, for example, by Puszkarskii (1979) and Mills (1994).

4.2.1 Microscopic theory

We first consider a microscopic approach, generalizing directly the calculation given in section 4.1.1 for a semi-infinite ferromagnet. As before we assume an sc structure with nearest-neighbour exchange interactions that may be modified in the surface layer(s). However, there are now N atomic layers and we allow for the possibility that nearest-neighbour exchange parameters J_S and J'_S, acting between spins in the upper and lower surfaces respectively, are different from one another and from the bulk exchange J (see figure 4.7). In analysing the magnon frequencies we again restrict attention to the low-temperature region $T \ll T_C$ so that $\langle S^z \rangle_n \approx S$ in each layer. Instead of obtaining the infinite series of coupled equations represented by (4.6) and (4.7), we now have the following finite set:

$$\{\omega - g\mu_B B_0 - SJ - 4SJ_S[1 - \gamma(\mathbf{q}_{\|})]\}s_1 + SJs_2 = 0 \qquad (4.33)$$

$$\{\omega - g\mu_B B_0 - 2SJ - 4SJ[1 - \gamma(\mathbf{q}_{\|})]\}s_n$$
$$+ SJ(s_{n-1} + s_{n+1}) = 0 \qquad (1 < n < N - 1) \qquad (4.34)$$

$$\{\omega - g\mu_B B_0 - SJ - 4SJ'_S[1 - \gamma(\mathbf{q}_{\|})]\}s_N + SJs_{N-1} = 0. \qquad (4.35)$$

The frequencies of the magnons can be deduced as before. Therefore for the bulk modes we seek travelling-wave solutions for the coefficients s_n as in (4.9). It is easily verified from (4.34) that the dispersion relation (4.10) is

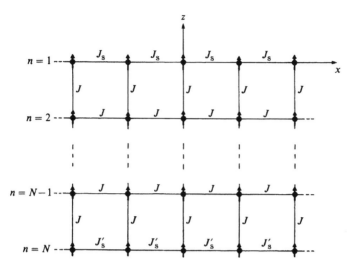

Figure 4.7. Model for the finite-thickness ferromagnetic film in the case of a simple-cubic structure and (001) surfaces. The nearest-neighbour exchange constants J_S, J'_S and J are indicated.

still applicable in the thin-film case. However, the boundary conditions (4.33) and (4.35) each lead to restrictions on the ratio B/A of the coefficients in (4.9), and a consistency condition then follows that

$$\tan(q_z L)$$

$$= \frac{i[\Delta - \exp(iq_z a)][\Delta' - \exp(iq_z a)] - i[\Delta - \exp(-iq_z a)][\Delta' - \exp(-iq_z a)]}{[\Delta - \exp(iq_z a)][\Delta' - \exp(iq_z a)] + [\Delta - \exp(-iq_z a)][\Delta' - \exp(-iq_z a)]}$$

(4.36)

where $L = (N - 1)a$ is the thickness of the film. The quantity Δ' is defined as for Δ in (4.13), but with J_S replaced by J'_S. The above result also holds if there are anisotropy fields B_{AS} and B'_{AS} acting in the surface layers $z = 0$ and $z = -L$ respectively, provided the definitions of Δ and Δ' are generalized as in (4.16). Equation (4.36) will be satisfied only by certain discrete values of q_z. This represents a 'quantization' of the wave-vector component and hence of the bulk-magnon frequency, as we discuss shortly. We recall that an analogous quantization effect was discussed in the lattice dynamics context in section 3.1.

The surface-magnon dispersion relation can be found by taking

$$s_n(\mathbf{q}_\parallel) = C(\mathbf{q}_\parallel) \exp(-\kappa na) + C'(\mathbf{q}_\parallel) \exp(\kappa na)$$

(4.37)

where $\mathrm{Re}(\kappa) > 0$. The term proportional to $C'(\mathbf{q}_\parallel)$ did not appear in (4.11) for the semi-infinite limit, because the corresponding exponential term would increase in magnitude without bound as $n \to \infty$. In the present case

n is always finite, and so both terms occur. On substituting (4.37) into (4.33)–(4.35), we eventually obtain $\omega = \omega_S(\mathbf{q}_{\parallel})$, where

$$\omega_S(\mathbf{q}_{\parallel}) = g\mu_B B_0 + 4SJ[1 - \gamma(\mathbf{q}_{\parallel})] + SJ[2 - \exp(\kappa a) - \exp(-\kappa a)] \quad (4.38)$$

and κ satisfies

$$\tanh(\kappa L) = \frac{[\Delta - \exp(\kappa a)][\Delta' - \exp(\kappa a)] - [\Delta - \exp(-\kappa a)][\Delta' - \exp(-\kappa a)]}{[\Delta - \exp(\kappa a)][\Delta' - \exp(\kappa a)] + [\Delta - \exp(-\kappa a)][\Delta' - \exp(-\kappa a)]}.$$
$$(4.39)$$

In the limit of large film thickness ($L \to \infty$) we have $\tanh(\kappa L) \to 1$, and (4.39) has two solutions for κ corresponding to $\exp(-\kappa a) = \Delta$ and $\exp(-\kappa a) = \Delta'$, provided $|\Delta| < 1$ and $|\Delta'| < 1$. These correspond to surface magnons of the type discussed in section 4.1, one localized at the upper surface ($z = 0$) and one at the lower surface ($z = -L$). More generally it can be shown that there are either zero, one or two surface-magnon solutions, depending on the values of L, Δ and Δ' (see Puszkarskii 1979, Cottam and Kontos 1980).

A numerical example of the magnon dispersion relations for an ultrathin film, with $N = 6$ and $\mathbf{q}_{\parallel} = (q_{\parallel}, 0)$, is shown in figure 4.8. For all values of the in-plane wavevector there are six discrete modes in total, of which one is a surface mode at small $q_{\parallel}$ and two are surface modes at larger $q_{\parallel}$ (eventually becoming degenerate with one another).

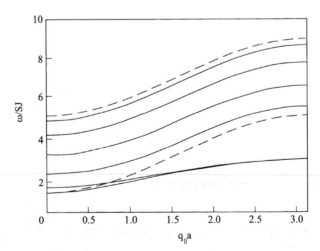

Figure 4.8. Plot of magnon frequency ω (in units of SJ) against $|\mathbf{q}_{\parallel}|a$ in a Heisenberg ferromagnetic film with 6 layers, taking $\mathbf{q}_{\parallel} = (q_{\parallel}, 0)$. The broken lines indicate the edges of the bulk-mode region, so the lines inside and outside correspond to quantized bulk and surface modes respectively. The parameters correspond to $J_S/J = 0.25$, $B_{AS} = 0$ and $g\mu_B B_0/SJ = 1$.

4.2.2 Macroscopic theory

For a more detailed discussion of the bulk and surface magnons in a thin film we restrict attention to the case of long wavelengths (small wavevectors), since the results simplify to some extent. The dispersion relations are obtained *either* directly from a macroscopic continuum approach, using the torque equation of motion as in section 4.1.2, *or* by taking $|q_{\parallel}a| \ll 1$, $|q_z a| \ll 1$ and $|\kappa a| \ll 1$ in (4.36), (4.38) and (4.39), and approximating accordingly. For the bulk magnons the frequencies are given by (4.21) where q_z must now satisfy

$$\tan(q_z L) = q_z(\xi + \xi')/(\xi\xi' - q_z^2). \tag{4.40}$$

The surface parameter ξ referring to the surface at $z = 0$ has been defined in (4.19), and ξ' for the other surface at $z = -L$ is given by a similar expression. In the special case of $\xi = \xi' = 0$ the above condition becomes simply $\tan(q_z L) = 0$, and hence

$$q_z = m\pi/L \qquad \text{with } m = 0, 1, 2, \ldots . \tag{4.41}$$

On the other hand, if ξ and ξ' are sufficiently large that $|\xi\xi'| \gg q_z^2$ for the wavevectors of interest, then (4.40) approximates to $\tan(q_z L) = q_z L_0$ where L_0 is a characteristic length defined by

$$L_0 = (1/\xi) + (1/\xi'). \tag{4.42}$$

Provided ξ and ξ' are of comparable magnitude, the above inequality implies $|q_z L_0| \ll 1$ and so $\tan(q_z L) \approx 0$ as before. Thus for large pinning (large ξ and ξ') we again obtain the approximate quantized solutions for q_z as in (4.41). Each of the above limiting cases corresponds to fitting an integral number of half-wavelengths into the film thickness L, with antinodes at the surfaces in the zero-pinning case and nodes at the surface in the large-pinning case, as illustrated in figure 4.9. For general values of ξ and ξ' the solutions of (4.40) can be investigated graphically by plotting each side of the equation as a function of q_z and looking for the intersections. The quantization of the

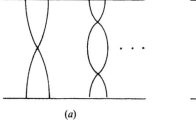 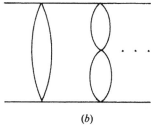

         (a)   (b)

Figure 4.9. The quantization of the bulk-magnon wave vectors in a Heisenberg ferromagnetic film for (a) zero pinning and (b) large pinning.

bulk magnons has been discussed quite extensively in the literature (e.g. see Puszkarskii 1979).

For the surface magnons the long-wavelength limits of (4.38) and (4.39) yield

$$\omega_S(\mathbf{q}_{\parallel}) = g\mu_B B_0 + SJa^2(q_{\parallel}^2 - \kappa^2) \qquad (4.43)$$

and

$$\tanh(\kappa L) = \kappa(\xi + \xi')/(\xi\xi' + \kappa^2) \qquad (4.44)$$

where the physical solutions correspond to $\kappa > 0$. Clearly, in the limit of $L \to \infty$, we have $\tanh(\kappa L) \to 1$ and (4.44) may be rearranged as a quadratic equation in κ with the two solutions $\kappa = \xi$ and $\kappa = \xi'$. Hence we recover the dispersion relation (4.22) of the semi-infinite case for a surface mode localized at $z = 0$, whilst the corresponding result involving ξ' refers to the surface mode at $z = -L$. The existence conditions for these modes are $\xi > 0$ and $\xi' > 0$. In the case of film thickness L large but finite (such that $\kappa L \gg 1$) the approximate solutions of (4.44) for κ may be obtained by a simple iterative procedure (see Moul and Cottam 1983). For example, when $\xi = \xi'$ the two solutions are

$$\kappa = \xi[1 \pm 2\exp(-\xi L)] \qquad (4.45)$$

provided $L \gg L_0$, where L_0 is defined in (4.42).

For general values of L the nature and existence of the surface magnons may also be investigated by graphical methods (see Puszkarskii 1979, Moul and Cottam 1983). For example, in the special case of symmetric boundary conditions ($\xi = \xi'$) we note first that (4.44) may be rearranged and factorized into two parts to give either

$$\kappa = \xi\tanh(\kappa L/2) \quad \text{or} \quad \kappa = \xi\coth(\kappa L/2). \qquad (4.46)$$

The graphical solution of these equations then leads to the following conclusions. First, if $\xi \leq 0$ there are no surface-magnon solutions. Second, if $\xi > 0$ there are two solutions (one from each expression in (4.46)) when $L > 2/\xi$ and only one solution (from the second expression) when $L \leq 2/\xi$. Note that $2/\xi$ is just the value of L_0 for the case of $\xi = \xi'$. In figure 4.10 a numerical example is given for the variation of the surface-magnon frequencies with film thickness. For large L the two branches converge to the semi-infinite result.

Calculations of Green functions of the form $\langle\langle S^+(z); S^-(z')\rangle\rangle_\omega$ for a film enable the spectral intensities of the magnons to be discussed, just as in the semi-infinite limit; see Cottam and Kontos (1980) for a microscopic approach and Moul and Cottam (1983) for a macroscopic theory. The results are algebraically more complicated than for the preceding semi-infinite limit in section 4.1.3. For example, the spectral function $F(\omega, z)$, defined as in (4.25), takes a different form from figure 4.5. There may be up to two surface-magnon peaks (occurring at $\omega < \Omega_0$), and the

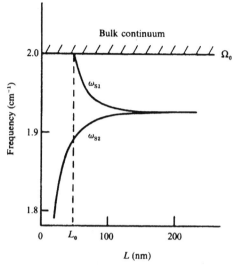

Figure 4.10. Calculated dependence of the acoustic surface-magnon frequencies (labelled ω_{S1} and ω_{S2}) on thickness L for a symmetric ferromagnetic film. The assumed parameters are $\Omega_0 = 2\ \text{cm}^{-1}$, $SJa^2 = 3 \times 10^{-15}\ \text{cm}^{-1}\ \text{m}^2$ and $\xi = 4 \times 10^7\ \text{m}^{-1}$.

bulk-magnon spectrum no longer forms a smooth continuum but consists of a series of sharp resonance peaks corresponding to the 'quantized' bulk magnons. The contribution of the bulk magnons to $F(\omega, z)$ at $z = 0$ is illustrated in figure 4.11 for two different values of L. The Green-function

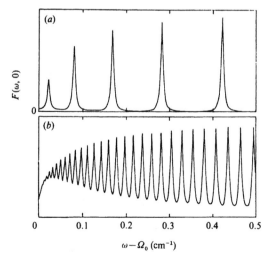

Figure 4.11. Calculated frequency dependence of the bulk-magnon contribution to $F(\omega, 0)$ for Heisenberg ferromagnetic films of different thickness L: (a) 190 nm and (b) 1000 nm. The assumed parameters are $SJa^2 = 3.7 \times 10^{-15}\ \text{cm}^{-1}\ \text{m}^2$ and $\xi = \xi' = 4 \times 10^7\ \text{m}^{-1}$. After Moul and Cottam (1983).

results have also been applied to calculate the spectra for light scattering and spin wave resonance. We shall discuss some of those results in section 4.6.

4.3   Heisenberg antiferromagnets

The magnons in bulk Heisenberg antiferromagnets were described in section 2.3.3. It was shown there that, as a consequence of the two sublattices of 'spin-up' and 'spin-down' sites, there are two branches to the magnon spectrum, which become degenerate in magnitude in the case of zero applied field $B_0 = 0$. We now generalize these results to consider the spectrum of surface and bulk magnons, mainly for a semi-infinite geometry. The calculations can be carried out in a similar manner to those of section 4.1 for semi-infinite ferromagnets and can be generalized, if required, to finite-thickness films as in section 4.2.

We take the antiferromagnet to occupy the half-space $z < 0$ and the directions of average spin alignment to be parallel and antiparallel to the z axis for sublattices 1 and 2 respectively. For simplicity we assume that the exchange couples only the nearest-neighbour sites on opposite sublattices and that all exchange interactions are the same as in the bulk. Likewise we assume that the anisotropy is the same at the surface as in the bulk. It is found that the results for the surface magnons in antiferromagnets depend sensitively on the lattice structure and on the crystallographic orientation of the surface(s); we shall consider two cases here, namely body-centred tetragonal antiferromagnets and simple-cubic antiferromagnets.

4.3.1   Body-centred tetragonal antiferromagnets

Antiferromagnets such as FeF_2 and MnF_2 have a rutile crystal structure in which the magnetic ions lie on a body-centred tetragonal lattice. A view of a semi-infinite sample with a (001) surface is shown schematically in figure 4.12. Note that each layer (labelled with index n) parallel to the surface contains magnetic sites of one sublattice type only and the nearest-neighbour exchange coupling is to sites in *adjacent* layers only. We see that the surface layer in figure 4.12 is occupied only by spins belonging to sublattice 1 (with spins 'up'), and the surface therefore has the effect of removing the equivalence between the two sublattices that would exist in an infinite crystal when $B_0 = 0$.

The linearized equations of motion satisfied by the spin operators in each layer may be found just as in the ferromagnetic case of section 4.1.1, except that we now employ the Hamiltonian (2.84). In the case of zero applied field the infinite set of coupled equations for the spin amplitudes

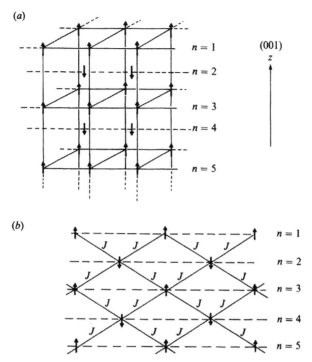

Figure 4.12. The (001) surface of a semi-infinite antiferromagnet with body-centred tetragonal structure: (a) perspective view and (b) side view.

$s_n(\mathbf{q}_{\parallel})$, defined as in (4.15), is

$$(\omega - g\mu_B B_A - 4SJ)s_1 - 4SJ\gamma_0(\mathbf{q}_{\parallel})s_2 = 0 \tag{4.47}$$

$$(\omega + g\mu_B B_A + 8SJ)s_{2m} + 4SJ\gamma_0(\mathbf{q}_{\parallel})(s_{2m-1} + s_{2m+1}) = 0 \quad (m \geq 1) \tag{4.48}$$

$$(\omega - g\mu_B B_A - 8SJ)s_{2m+1} - 4SJ\gamma_0(\mathbf{q}_{\parallel})(s_{2m} + s_{2m+2}) = 0 \quad (m \geq 1) \tag{4.49}$$

where we have denoted

$$\gamma_0(\mathbf{q}_{\parallel}) = \cos(\tfrac{1}{2}q_x a)\cos(\tfrac{1}{2}q_y a). \tag{4.50}$$

It was assumed for simplicity that the spin averages $\langle S^z \rangle$ may be approximated by the values S and $-S$ for sites on sublattices 1 and 2, respectively, at low temperatures $T \ll T_N$. The above finite difference equations may be solved to obtain the dispersion relations of bulk and surface magnons, as we now indicate.

Following the same approach as for the semi-infinite ferromagnet, we investigate the bulk magnons by seeking travelling-wave solutions for s_n of the form (4.9), but with different coefficients for the two sublattices (corresponding to odd and even layer indices). This leads to $\omega = \pm\omega_B(\mathbf{q})$

where

$$\omega_B(\mathbf{q}) = \{[g\mu_B B_A + 8SJ]^2 - [8SJ\gamma_0(\mathbf{q}_\|)\cos(\tfrac{1}{2}q_z c)]^2\}^{1/2} \qquad (4.51)$$

and c is the lattice parameter corresponding to the z direction in the body-centred tetragonal lattice. This is equivalent to the bulk-magnon dispersion relation in an *infinite* body-centred tetragonal antiferromagnet at $T \ll T_N$, as can be deduced from (2.87). Also for wave vector $\mathbf{q} = 0$ the result can be written as in (2.88), the bulk AFMR frequency, provided we now define $\omega_A = g\mu_B B_A$ and $\omega_E = 8SJ$.

The surface magnons correspond to attenuated solutions in which s_n takes a form analogous to (4.11) but with different amplitude factors for each sublattice. It can be shown from (4.47)–(4.49) that the attenuation factor κ of any surface magnon must satisfy

$$\exp(-\kappa c) = (2\omega_A + \omega_E - 2\omega)/\omega_E. \qquad (4.52)$$

The only solution for the surface-magnon frequency that also satisfies the localization condition $\mathrm{Re}(\kappa) > 0$ is found to be $\omega = \omega_S(\mathbf{q}_\|)$, where

$$\omega_S(\mathbf{q}_\|) = -\tfrac{1}{4}\omega_E[1 - \gamma_0^2(\mathbf{q}_\|)]$$
$$+ \{(\omega_E + \omega_A)(\omega_A + \tfrac{1}{2}\omega_E[1 - \gamma_0^2(\mathbf{q}_\|)]) + \tfrac{1}{16}\omega_E^2[1 - \gamma_0^2(\mathbf{q}_\|)]^2\}^{1/2}.$$
$$(4.53)$$

This is the surface-magnon branch first predicted by Mills and Saslow (1968). It is an acoustic mode, i.e. it occurs at frequencies below those of the corresponding bulk magnons with wave vectors $(\mathbf{q}_\|, q_z)$, as illustrated in figure 4.13. When $\mathbf{q}_\| = 0$ the above expression reduces to

$$\omega_S(0) = [\omega_A(\omega_E + \omega_A)]^{1/2}. \qquad (4.54)$$

Since $\omega_A/\omega_E \ll 1$ in many rutile-structure antiferromagnets (e.g. it is approximately 0.015 for MnF_2), it follows that $\omega_S(0)/\omega_B(0)$ in such cases is close to $1/\sqrt{2}$. This difference between the surface and bulk frequencies is advantageous for experimental studies of surface modes by techniques that involve small-wavevector excitations (such as antiferromagnetic resonance and inelastic light scattering).

The effect of the anisotropy field and the exchange parameter having different values at the surface compared with their bulk values has been studied by Dewames and Wolfram (1969). They concluded that, even when the surface parameters are changed significantly from the bulk values, the ratio $\omega_S(0)/\omega_B(0)$ is still close to $1/\sqrt{2}$ for $\omega_A/\omega_E \ll 1$. However, at nonzero $\mathbf{q}_\|$ these surface parameters play a larger role leading to the prediction, in certain cases, of *two* surface-magnon branches that may occur either below or above the bulk-magnon continuum.

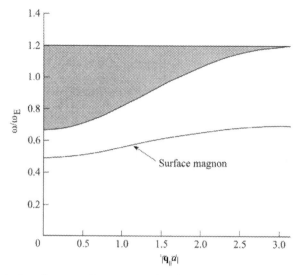

Figure 4.13. A plot of magnon frequency against $|q_{\parallel}a|$ in a semi-infinite Heisenberg anti-ferromagnet with body-centred tetragonal structure, taking $\mathbf{q}_{\parallel} = (q_{\parallel}, 0)$ and $\omega_A/\omega_E = 0.2$. The bulk-magnon continuum and an acoustic surface-magnon branch are shown.

4.3.2 Simple-cubic antiferromagnets

Antiferromagnets such as $KNiF_3$ and $RbMnF_3$ have a crystal structure (the perovskite structure) in which the magnetic ions lie on a simple-cubic lattice, as sketched in figure 4.14 for a semi-infinite sample with a (001) surface. Notice in this case that each layer (labelled by index n as before) contains equal numbers of magnetic sites from both sublattices and that the nearest-neighbour exchange coupling is to sites in the same layer as well as to the adjacent layers. Hence, by contrast with a body-centred tetragonal antiferromagnet, the occurrence of a (001) surface in a simple cubic anti-ferromagnet does *not* remove the equivalence between the two sublattices. It is primarily this symmetry property that gives rise to differences between the surface-magnon characteristics in the two structures considered here.

The coupled equations satisfied by the spin amplitudes s_n at $T \ll T_N$ can be derived by analogy with (4.47)–(4.49). They are (see Wolfram and Dewames 1972)

$$(\omega - g\mu_B B_A - 5SJ)s_1^{(1)} - 4SJ\gamma(\mathbf{q}_{\parallel})s_1^{(2)} - SJs_2^{(2)} = 0 \qquad (4.55)$$

$$(\omega + g\mu_B B_A + 5SJ)s_1^{(2)} + 4SJ\gamma(\mathbf{q}_{\parallel})s_1^{(1)} + SJs_2^{(1)} = 0 \qquad (4.56)$$

$$(\omega - g\mu_B B_A - 6SJ)s_n^{(1)} - 4SJ\gamma(\mathbf{q}_{\parallel})s_n^{(2)} - SJ(s_{n-1}^{(2)} + s_{n+1}^{(2)}) = 0 \qquad (n > 1)$$

$$(4.57)$$

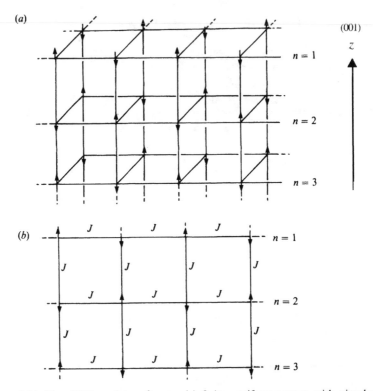

(a)

(001)

z

$n = 1$

$n = 2$

$n = 3$

(b)

$n = 1$

$n = 2$

$n = 3$

Figure 4.14. The (001) surface of a semi-infinite antiferromagnet with simple-cubic structure: (a) perspective view and (b) side view.

$$(\omega + g\mu_B B_A + 6SJ)s_n^{(2)} + 4SJ\gamma(\mathbf{q}_{\parallel})s_n^{(1)} + SJ(s_{n-1}^{(1)} + s_{n+1}^{(1)}) = 0 \qquad (n > 1). \tag{4.58}$$

The superscripts 1 and 2 in the spin wave amplitudes refer to the sublattice type within a layer, and $\gamma(\mathbf{q}_{\parallel})$ is the geometrical factor defined in (4.8). The bulk-magnon dispersion relation can easily be found from the above by seeking travelling-wave solutions as before. The result is $\omega = \pm\omega_B(\mathbf{q})$ where

$$\omega_B(\mathbf{q}) = \{[g\mu_B B_A + 6SJ]^2 - 4S^2 J^2 [2\gamma(\mathbf{q}_{\parallel}) + \cos(q_z a)]^2\}^{1/2}. \tag{4.59}$$

As expected, this is in agreement with the bulk-magnon dispersion relation (2.87) for an infinite antiferromagnet with this structure. The bulk-magnon frequency at $\mathbf{q} = 0$ can again be expressed formally as (2.88) provided we now use the appropriate expression $\omega_E = 6SJ$ for the exchange frequency.

The surface-magnon solutions of (4.55)–(4.58) are more complicated to obtain than for the examples discussed earlier in this chapter. This is because, except in some limiting cases, the surface magnons in sc antiferromagnets cannot be described in terms of a *single* attenuation factor (Wolfram and Dewames 1972). In particular, the simple ansatz of (4.11) for $s_n(\mathbf{q}_\parallel)$ will not be applicable in general and this is a consequence of the fact that each layer n in the semi-infinite sc structure contains sites from both sublattices. For the case considered here, with surface exchange and anisotropy parameters the same as in the bulk, Wolfram and Dewames (1972) calculate a single surface-magnon branch of the acoustic type with frequency $\omega_S(\mathbf{q}_\parallel)$ given by

$$\omega_S(\mathbf{q}_\parallel) = \left\{ \frac{(\omega_A + \omega_E)\{4\omega_E^2[1 - \gamma^2(\mathbf{q}_\parallel)] + 12\omega_A\omega_E + 9\omega_A^2\}}{3(2\omega_A + 3\omega_E)} \right\}^{1/2}. \quad (4.60)$$

When $\mathbf{q}_\parallel = 0$ and $\omega_A/\omega_E \ll 1$ this approximates to

$$\omega_S(0) \approx (2\omega_A\omega_E + \tfrac{1}{2}\omega_A^2)^{1/2}. \quad (4.61)$$

Comparing this with (2.88) we see that $\omega_S(0)/\omega_B(0)$ is close to 1 when $\omega_A/\omega_E \ll 1$ (compared with $1/\sqrt{2}$ found for a body-centred tetragonal antiferromagnet).

The extension of the above theory to include cases where the anisotropy fields and exchange parameters at the surface may differ from their bulk values has been made by Wolfram and Dewames (1969). The results for the excitation spectrum become very complicated and include the possibilities of various truncated acoustic and optic surface-magnon branches.

4.3.3 Some generalizations

The effect of an applied magnetic field B_0 on the surface magnons can also be studied. First, in the case of an infinite antiferromagnet with B_0 applied along the z direction, we recall the result in section 2.3.3 that the degeneracy in the bulk-magnon frequencies is lifted. The two branches have frequencies equal in magnitude to $\omega_B(\mathbf{q}) \pm g\mu_B B_0$ at $T \ll T_N$, where $\omega_B(\mathbf{q})$ is given by (2.87). As B_0 is increased, the lower branch becomes zero for some value of the wavevector $\mathbf{q}$ (normally $\mathbf{q} = 0$) and this represents an instability in the system. The spins reorient from alignment parallel and antiparallel to the z axis to new directions that are almost at right angles to B_0. This is the *spin–flop phase transition* that has been observed in a number of antiferromagnets and was briefly mentioned in section 2.6.1. In the semi-infinite case there may be acoustic-type surface magnons with a lower frequency at $\mathbf{q}_\parallel = 0$ than that of the $\mathbf{q} = 0$ bulk magnons, as we have already seen in section 4.3.1 for the rutile-structure MnF_2. Consequently the spin–flop phase transition may occur at the surface of a semi-infinite antiferromagnet at a smaller applied field B_{sf} than for the bulk, corresponding to $B_{sf}(\text{surface})/B_{sf}(\text{bulk}) \simeq 1/\sqrt{2}$ (Mills 1968). Some more discussion related to the above prediction will be given in section 4.6 on experimental studies.

Apart from studies of the magnon dispersion relations, there have been some calculations of thermodynamic properties. In particular, Mills and Saslow (1968) have evaluated the sublattice magnetization and the specific heat for a semi-infinite antiferromagnet with body-centred tetragonal structure. It is the long-wavelength magnons that influence most strongly the thermodynamic properties at $T \ll T_N$ and, as shown earlier in this section, the surface magnons in antiferromagnets of this structure are well split off in frequency below the bulk magnons. In turn this leads to an enhancement in the surface-magnon contribution to the thermodynamic properties, by contrast with the situation for ferromagnets or simple-cubic antiferromagnets. Further theoretical work for semi-infinite antiferromagnets has been concerned with evaluating the spin-dependent Green functions (see e.g. Cottam 1978a). These results were applied to predict the lineshapes and integrated intensities for Raman scattering from the surface and bulk magnons.

The magnons in Heisenberg antiferromagnetic thin films at magnetic field values below the spin–flop transition have also be calculated (see e.g. Diep 1991, Pereira and Cottam 2000), extending the methods of section 4.2. Qualitatively, the effects on the magnon spectrum are similar to the ferromagnetic case in that there may be additional surface branches and the bulk spectrum consists of discrete (or 'quantized') modes. The results are particularly interesting in the case of body-centred tetragonal antiferromagnets because there is a symmetry difference depending on whether the number of layers, N, in the film is odd or even. If N is odd, the top and bottom surface layers are made up from spins of the same sublattice type. On the other hand, they are made up of spins from opposite sublattices if N is even. This has consequences for the degeneracy and splitting of the modes. By contrast, this odd and even property for N does not occur for simple-cubic antiferromagnetic films, because each layer has equal numbers of spins from each sublattice.

A thorough analysis of the phase behaviour of thin antiferromagnetic films has been given by Carriço *et al* (1994). Applying their calculations to FeF_2 and MnF_2, they showed that spatially nonuniform canted states could intermediate the transition between the antiferromagnetic and spin–flop phases.

4.4   Surface magnons in Ising ferromagnets

The Ising model in a transverse magnetic field was discussed in section 2.3.5 in terms of its bulk properties. As noted, it provides a good description of some real anisotropic magnetic materials when subjected to a transverse field, as well as being of much wider applicability as a pseudo-spin model for hydrogen-bonded ferroelectrics.

We now provide the generalization (see Cottam *et al* 1984) of the previous bulk-magnon theory to the case of a semi-infinite medium described by the Hamiltonian (2.97). The first step, as in section 2.3.5, is to find the thermal average $\langle \mathbf{S}_i \rangle \equiv \mathbf{R}_i$ at any site i using mean field theory. As before, only components R_i^x and R_i^z will be nonzero, but they now depend on the site i because of surface reconstruction. Equation (2.99) is replaced by

$$\mathbf{R}_i = (\mathbf{B}_i^0/|\mathbf{B}_i^0|) \tanh(|\mathbf{B}_i^0|/2k_\mathrm{B}T) \tag{4.62}$$

with $\mathbf{B}_i^0$ denoting the static part of the instantaneous field $\mathbf{B}_i$:

$$\mathbf{B}_i^0 = \left(\Omega_i, 0, \sum_j J_{ij} R_j^z \right). \tag{4.63}$$

By analogy with section 2.3.5 we obtain the linearized equations of motions by putting

$$\mathbf{S}_i = \mathbf{R}_i + \boldsymbol{\mu}_i \exp(-\mathrm{i}\omega t). \tag{4.64}$$

After substituting into the torque equation (2.74) and expanding to first order in $\boldsymbol{\mu}_i$, we obtain the coupled equations:

$$-\mathrm{i}\omega\mu_i^x = \mu_i^y \sum_j J_{ij} R_j^z \tag{4.65}$$

$$-\mathrm{i}\omega\mu_i^y = \Omega_i \mu_i^z - R_i^x \sum_j J_{ij}\mu_j^z - \mu_i^x \sum_j J_{ij} R_j^z \tag{4.66}$$

$$-\mathrm{i}\omega\mu_i^z = \Omega_i \mu_i^y. \tag{4.67}$$

Solving the above equations is not straightforward because R_i^x and R_i^z depend on i in a complicated manner given by (4.62). The general solutions for R_i^x and R_i^z (and hence ω) may be investigated numerically (see Cottam *et al* 1984). However, we examine here a case in which an analytic solution is obtainable. We take temperature $T > T_\mathrm{C}$, whereupon (4.62) and (4.63) simplify to give

$$R_i^x = \tfrac{1}{2}\tanh(\Omega_i/2k_\mathrm{B}T), \qquad R_i^z = 0. \tag{4.68}$$

Equation (4.65) then has $\omega = 0$ as a solution, while (4.66) and (4.67) combine to give

$$(\omega^2 - \Omega_i^2)\mu_i^z + \Omega_i R_i^x \sum_j J_{ij}\mu_j^z = 0. \tag{4.69}$$

To illustrate the solution of (4.69), which yields the bulk- and surface-mode frequencies, we take the case of the (001) surface of a simple cubic lattice. To allow for variations of Ω_i and J_{ij} at the surface we make the following assumptions. We take $\Omega_i \equiv \Omega_\mathrm{S}$ if i is in the surface layer ($z = 0$) and $\Omega_i \equiv \Omega$ (the bulk value) otherwise. Also we take J_{ij} to couple nearest neighbours only, with $J_{ij} \equiv J_\mathrm{S}$ if both spins are in the surface layer and $J_{ij} \equiv J$

otherwise. From (4.68) the two values of R_i^x in this model are $\frac{1}{2}\tanh(\Omega_S/2k_BT) \equiv R_S^x$ if i is in the surface layer and $\frac{1}{2}\tanh(\Omega/2k_BT) \equiv R^x$ otherwise. Writing $\mu_i^- = \mu_n(\mathbf{q}_{||})\exp(i\mathbf{q}_{||}\cdot\mathbf{r}_{||})$, where n is a layer index as before with $n=1$ being the surface and $\mathbf{r}_{||} = (x,y)$, we re-express (4.69) as

$$[\omega^2 - \Omega_S^2 + 4\Omega_S R_S^x J_S\gamma(\mathbf{q}_{||})]\mu_1 + \Omega_S R_S^x J\mu_2 = 0 \qquad (4.70)$$

$$[\omega^2 - \Omega^2 + 4\Omega R^x J\gamma(\mathbf{q}_{||})]\mu_n + \Omega R^x J(\mu_{n-1} + \mu_{n+1}) = 0 \qquad (n>1) \quad (4.71)$$

where $\gamma(\mathbf{q}_{||})$ is defined in (4.8). The above finite-difference equations can be solved by seeking travelling-wave and attenuated solutions in an analogous manner to the solution of (4.6) and (4.7) for the semi-infinite Heisenberg ferromagnet. The bulk modes are found to have the dispersion relation

$$\omega_B(\mathbf{q}) = \{\Omega^2 - 2\Omega R^x J[2\gamma(\mathbf{q}_{||}) + \cos(q_z a)]\}^{1/2} \qquad (4.72)$$

which is equivalent to (2.104) in the high-temperature phase (where $R^z = 0$) for a simple cubic structure. The frequency ω of any surface mode must satisfy

$$\omega^2 - \Omega^2 + 4\Omega R^x J\{4\gamma(\mathbf{q}_{||}) + \exp[-\kappa(\omega)a] + \exp[\kappa(\omega)a]\} = 0 \qquad (4.73)$$

provided $\mathrm{Re}[\kappa(\omega)] > 0$ for the frequency-dependent attenuation factor, where

$$\exp[-\kappa(\omega)a] = \left(\frac{\Omega_S^2 - \omega^2}{\Omega_S R_S^x J}\right) - 4\gamma(\mathbf{q}_{||})\left(\frac{J_S}{J}\right). \qquad (4.74)$$

For the particular case of $\Omega_S = \Omega$ (and therefore $R_S^x = R^x$) the solutions of (4.73) are $\omega = \pm\omega_S(\mathbf{q}_{||})$, where

$$\omega_S(\mathbf{q}_{||}) = \left\{\Omega^2 - \Omega R^x\left[4J_S\gamma(\mathbf{q}_{||}) + \frac{J}{4\sigma\gamma(\mathbf{q}_{||})}\right]\right\}^{1/2} \qquad (4.75)$$

and we have denoted $\sigma = (J_S/J) - 1$. The attenuation factor κ is now given by

$$\exp(-\kappa a) = [4\sigma\gamma(\mathbf{q}_{||})]^{-1}. \qquad (4.76)$$

The surface mode exists only if σ and $\mathbf{q}_{||}$ are such that $\mathrm{Re}(\kappa) > 0$. This leads to the possibility of 'acoustic' and 'optic' surface modes, just as for the semi-infinite Heisenberg ferromagnet in section 4.1. The conditions for acoustic and optic modes to exist in some part of the 2D Brillouin zone for $\mathbf{q}_{||}$ are $J_S > \frac{5}{4}J$ and $J_S < \frac{3}{4}J$ respectively. A numerical illustration of the dispersion relations is given in figure 4.15.

The above basic theory for the surface magnons in a semi-infinite medium has been extended to include the case of $T < T_C$ (Cottam *et al* 1984) and the case of a thin film on a substrate (Salman and Cottam 1994). In the former paper the authors also examine other theoretical

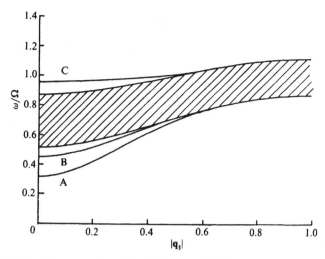

Figure 4.15. Plot of frequency (in units of Ω) versus $|\mathbf{q}_{||}|$ (in units such that $|\mathbf{q}_{||}| = 1$ at the 2D Brillouin zone boundary) for the excitations of the Ising model in a transverse field. We have taken $\mathbf{q}_{||} = (q_{||}, 0)$, $T = 1.5T_C$ and $J/\Omega = 0.5$. The bulk-magnon continuum is shown together with three surface branches for J_S/J values of: A, 1.75; B, 1.5; C, 0.25. After Cottam *et al* (1984).

models for surface modes in ferroelectrics, apart from the pseudo-spin approach. These are based on the Landau theory of phase transitions and on polariton theory. The Landau theory had previously been applied by Binder (1981) to discuss surface reconstruction in ferroelectrics; it involves an expansion of the free energy in terms of the order parameter, in this case the polarization of the medium. It is a macroscopic continuum method, which has some analogy with the continuum approach employed in section 4.1.2 for the semi-infinite Heisenberg model. The Landau-theory calculation of surface reconstruction was extended to ferroelectric films by Tilley and Zeks (1984); a later discussion may be found in Ong *et al* (2001). The second of the above methods, namely that based on polariton theory, is mentioned in chapter 6.

4.5   Exchange-coupled magnetic bilayers

It is appropriate at this stage to discuss some exciting properties in surface magnetism associated with two ferromagnetic layers that are weakly coupled by exchange interactions across a very thin spacer layer. Typically the magnetic materials of interest here are ferromagnetic metals (e.g. Fe) separated by a nonmagnetic metal layer (e.g. Cr) with thickness of the order of 10 nm or less. The usual description of exchange interactions, such as outlined in section 2.3.1, is based on the overlapping of relatively-localized electronic wavefunctions to provide coupling that is significant between nearest and maybe next-nearest neighbours typically. In metals, however, there are conduction electrons providing additional effects of exchange coupling that may extend over greater distances. An example is the well-known Ruderman–Kittel–Kasuya–Yosida (or RKKY) mechanism (see e.g. Kittel 1986, Hathaway 1994), which was originally developed to describe indirect exchange between the $4f$ ion cores in rare-earth metals via mutual coupling to surrounding conduction (s) electrons. In its simplest form, assuming a free-electron model, the effective exchange depends on the spatial separation r between the magnetic ions as

$$J(r) \propto [2k_F r \cos(2k_F r) - \sin(2k_F r)]/(2k_F r)^4 \qquad (4.77)$$

where k_F is the Fermi wavevector for the electron gas. It is apparent that this function oscillates in sign with an amplitude that decreases with distance; effectively it becomes zero after four or five oscillations of period π/k_F.

Another mechanism for coupling between magnetic films separated by a spacer (including non-metals and even a vacuum gap) is provided by the long-range magnetic dipole–dipole interactions. This is a separate topic that we shall consider in sections 5.2 and 5.4.

4.5.1 Oscillatory exchange and giant magnetoresistance

In pioneering work Grünberg *et al* (1986) reported experimental evidence, obtained from Brillouin scattering and magneto-optical measurements, for an *antiferromagnetic* exchange coupling between two ferromagnetic Fe layers via an intermediate layer of non-ferromagnetic Cr with thickness 0.9 nm. This and similar work on other trilayer metallic systems soon led to two important developments. First, in further studies on Fe/Cr/Fe structures, Baibich *et al* (1988) observed a sharp decrease in the electrical resistance of the system when an external magnetic field was applied. This effect was so pronounced that it was termed *giant magnetoresistance* (GMR): an example of their data is shown in figure 4.16 for different layer thicknesses, as marked, for multilayer samples. GMR has subsequently become of technological importance for applications to magnetoresistive recording heads and information storage (see e.g. Grünberg 2000a, Hartmann 2000). Second, in separate measurements for Fe/Cr/Fe systems, Parkin *et al* (1990) found that as the spacer thickness was increased the effective exchange between the magnetic layers oscillated and eventually decayed. Other studies soon followed and a complicated behaviour became apparent. Some systems exhibited oscillations with a period of about 1 nm or more, while others seemed to have short-period oscillations ($\sim$0.2 nm, which would be expected on the RKKY model for a free-electron gas) superimposed on the long-period oscillations (see e.g. Grünberg 2000b).

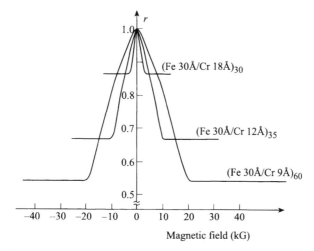

Figure 4.16. Measured magnetoresistance ratio $r \equiv R(B_0)/R(B_0 = 0)$ of three Fe/Cr multilayer structures versus applied magnetic field. The current and the applied field are parallel (or antiparallel) and are in the layer plane. After Baibich *et al* (1988).

One convenient way to study the oscillatory exchange is to use a spacer of variable thickness (e.g. in the form of a small-angle wedge as illustrated schematically in figure 4.17) between the magnetic films. Domains of alternating magnetization direction can then be observed (see Unguris *et al* 1991), as we shall discuss in section 4.6. Good reviews of both oscillatory exchange and GMR, covering both theory and experiment, are to be found in the books edited by Heinrich and Bland (1994) and Hartmann (2000).

To explain why oscillatory exchange coupling of ferromagnetic layers across metallic nonmagnetic spacers is observed in many systems, but is absent in others, we might start with the RKKY model. This was first extended to a trilayer geometry of Gd/Y/Gd by Yafet (1987), and there have been numerous elaborations subsequently (see e.g. Hathaway 1994). While it is capable of explaining oscillations, the approach has some serious shortcomings. For example, it is derived from perturbation theory that assumes the ferromagnetic layers have only a small effect on the electrons in the spacer metal, the predicted oscillations for free electrons do not reproduce the long-period oscillations seen in many systems, and the non-oscillatory ferromagnetic or antiferromagnetic exchange coupling in some materials is not explained. Other approaches that have produced promising results are reviewed by Hathaway (1994) and Grünberg (2000b); they include non-perturbative many-body theory for strongly hybridized systems and band-structure theory. This is an ongoing field of study as more experimental data become available.

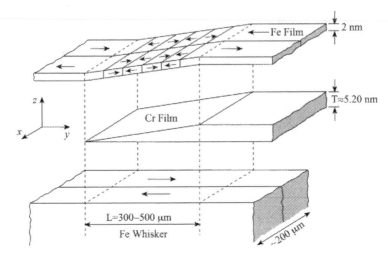

Figure 4.17. Schematic exploded view of a Fe film separated from a Fe substrate by a Cr wedge (with angle of order 10^{-3} deg). The arrows in the Fe show the magnetization direction in each domain. After Unguris *et al* (1991).

In cases where there is weak antiferromagnetic coupling across the spacer between two ferromagnetic metals, it is easy to understand in general terms how a magnetoresistive effect (with dependence on applied magnetic field) can arise. If there is no applied field the magnetizations in the two ferromagnets will be antiparallel to one another, but the application of a relatively small field (e.g. of order 0.1 T or less for a single trilayer) will switch the magnetizations to the parallel configuration. A consequent change in electrical resistance will occur if there is, for example, a spin-dependent scattering of the current-carrying electrons at the interfaces. What was surprising was the magnitude of the effect, and hence the label 'giant' in GMR. The resistance is usually larger when the magnetizations are anti-parallel, and values with $(R_{\downarrow\uparrow} - R_{\uparrow\uparrow})/R_{\uparrow\uparrow} \approx 1$ were obtained in some of the early measurements by Baibich *et al* (1988). The effect is enhanced at low temperatures, and occurs both when the current is perpendicular to the plane and in the plane (as in figure 4.16), although the latter configuration is favoured for applications to date. A theory that accounted for most of the observed features of GMR was proposed by Camley and Barnas (1989). Their semiclassical model involved solving the Boltzmann transport equation with spin-dependent scattering assumed at the interfaces. The strength of the predicted GMR depends on the ratios of layer thickness to the mean free path for each of the layers and on the asymmetry in scattering from spin-up and spin-down electrons. Quantum-mechanical models of GMR were developed by Lévy *et al* (1990), among others. Both types of theory introduce spin dependence for the scattering at the interfaces; it would be of interest to have a more thorough understanding of its microscopic origin and for the role of interface roughness and imperfections.

In all of the above discussion the antiparallel alignment of magnetizations in the two ferromagnetic layers was due to antiferromagnetic coupling across a spacer. While this is the usual situation for GMR, the antiparallel alignment can be produced in other ways. These include having different coercivities for the ferromagnetic layers or 'pinning' the magnetization of one ferromagnetic layer by *exchange biasing* (i.e. putting it in direct contact with an antiferromagnet). If GMR is obtained by one of the latter methods, the arrangement is sometimes referred to as a *spin-valve system* (see Hartmann 2000).

4.5.2   Effects of biquadratic exchange

Other interesting effects came to the forefront in experimental studies on layers of metallic ferromagnets that are exchange coupled through non-magnetic metals. The arguments outlined in section 2.3.1 for the origin of exchange terms of the form $S_1 \cdot S_2$ between spin operators at neighbouring sites may, in principle, be generalized to give terms proportional to $(S_1 \cdot S_2)^n$ where n is a positive integer. This can be shown, for example, using higher-order perturbation theory as described in the book by Yosida (1996). The usual Heisenberg exchange (also known as *bilinear exchange*) corresponds to the simplest case of $n = 1$, while the case of $n = 2$ is called *biquadratic exchange*.

It has been known for a long time that biquadratic exchange could occur as a small effect in some bulk materials such as MnO (see Harris and Owen 1963). However, experiments on Fe/Cr/Fe systems by Rührig *et al* (1991), and later extended by Azevedo *et al* (1996), showed that the bilinear and quadratic exchange across the spacer could be comparable and that, in some circumstances, the biquadratic effect could dominate. To consider this further, we write the overall exchange coupling between two interface spins S_1 and S_2 across the spacer layer in the form

$$-J_{BL}(S_1 \cdot S_2) - J_{BQ}(S_1 \cdot S_2)^2 \qquad (4.78)$$

where J_{BL} and J_{BQ} are the bilinear and biquadratic exchange constants respectively. As we have seen, J_{BL} can be either positive or negative favouring parallel and antiparallel alignment, respectively, for the magnetizations in the two magnetic layers. It is found generally that J_{BQ} is negative, and so the second term in (4.78) is a minimum when $(S_1 \cdot S_2)^2$ is zero, i.e. when S_1 and S_2 are at $\pm 90^0$ to one another. Hence the biquadratic term favours the magnetization directions in the two magnetic layers being perpendicular. Rührig *et al* (1991) made their measurements for a case in which the oscillatory J_{BL} term was very small (and so $|J_{BL}| \ll |J_{BQ}|$); they observed a perpendicular orientation of magnetization directions. Subsequent studies in Fe/Cr/Fe and similar systems identified cases where $|J_{BL}|$ and J_{BQ} could be comparable. Various different explanations have been put forward to explain the microscopic origins of such a large biquadratic component to the interface exchange coupling. These range from effects due to interface roughness (Slonczewski 1991, Demokritov *et al* 1994), magnetic impurities in the spacer layer (Slonczewski 1993, Schäfer *et al* 1995), and higher harmonics in the oscillatory exchange coupling (Edwards *et al* 1993). These may occur in combination and it is not clear which mechanism is dominant overall.

More recently the biquadratic exchange has been studied mainly for its effect on the magnetization profile and magnetoresistance in *multilayers*, both periodic and quasiperiodic (see e.g. the review by Albuquerque and Cottam 2003). We shall therefore return to this topic in sections 9.3 and 9.5.

We conclude this section by mentioning briefly some related theory for the magnon spectrum in exchange-coupled magnetic bilayers. Suppose we consider an interface formed between two identical semi-infinite media. Specifically, each half of the combined structure is as represented in figure 4.1 (with exchange constants J and J_S) while additionally there is a bilinear exchange coupling I between nearest sites across the interface. This is a special case of a theory worked out by Chen *et al* (1995), who obtained the localized interface–magnon dispersion relations and Green functions when the two media could, in general, be different. Their results simplify for the above model (with I representing the coupling across the spacer), and we find that there may be *two* localized modes. The corresponding frequencies can be written in the same form as (4.12) with Δ taking two possible values, provided the localization condition (4.14) is satisfied. One value of Δ is just identical to (4.13), and thus the frequency is independent of I/J, while the other is

$$\Delta = \left\{ 1 + 4\left(1 - \frac{J_S}{J}\right)[1 - \gamma(\mathbf{q}_{\parallel})] - 2\frac{I}{J} \right\}^{-1}. \tag{4.79}$$

For each mode the spectral intensity (as deduced from the Green's functions) depends on the ratio I/J. Bezerra and Cottam (2002) considered the additional effects on the magnon frequencies of having biquadratic (as well as bilinear) exchange across magnetic interfaces, mainly in the context of multilayers.

4.6   Experimental studies

The first clear example in which exchange-dominated surface magnons were observed directly came from spin wave resonance (SWR) in measurements by Yu *et al* (1975) for films of the ferrimagnet yttrium iron garnet (YIG) deposited on a nonmagnetic substrate. A *ferrimagnet* is an ordered magnetic material that has two (or more) sublattices of spins, some of which are oppositely aligned. However, unlike an antiferromagnet where the magnetic moments of the two sublattices cancel out in zero applied magnetic field, ferrimagnets have an overall nonzero magnetization. Thus their low-temperature behaviour in many respects is similar to that of ferromagnets, including the bulk spin–wave properties (see e.g. Keffer 1966).

We have already referred briefly to SWR in section 1.4.3. To elaborate further, r.f. radiation is applied to the magnetic sample such that typically the fluctuating magnetic-field component is perpendicular to the spin-ordering direction. Resonance absorption occurs when the frequency of the r.f. field matches that of a magnon excitation. In thin-film samples the technique can be used to excite selectively the magnons with small but nonzero wave vector q. For bulk magnons this corresponds to having a standing-wave pattern of magnons across the thickness of the film, as was illustrated schematically in figure 4.9. If the film thickness is equal to an *odd* number of half-wavelengths of a $q \neq 0$ bulk magnon, then there will be a net magnetic moment to couple with the applied r.f. field, and resonance absorption may be observed. Although the technique was originally applied to bulk magnons (see e.g. Frait and Fraitova (1988) and Dutcher (1994) for reviews), it was used in the experiments of Yu *et al* (1975) to observe a surface magnon bound to the YIG/substrate interface. It was split off in frequency below the bulk-magnon region, which is the expected behaviour if YIG behaves similarly to a ferromagnetic film (i.e. as in section 4.2) at low temperatures $T \ll T_C$. A full theoretical analysis has been given by Harada *et al* (1977).

Recent magnetic resonance studies, made by Suran *et al* (1998), on thin films of bcc structure Ni showed good evidence for highly localized, exchange-dependent surface magnons. The film thickness L varied in the range from 2.4 nm to 12 nm, and the films were either grown epitaxially on a W substrate or sandwiched between two thick W layers. In the experiments the surfaces were deliberately roughened, and up to four localized surface modes were observed in appropriate cases. These modes might be surface magnons of the type discussed in section 4.2, but corresponding to the (111) growth geometry of the Ni films and an effective surface exchange (due to the roughening). Alternatively the observed modes might be localized magnetic modes associated with W impurities in the surfaces. This is still a topic for further theoretical study.

The light-scattering experiments to date have not provided evidence of the exchange-dominated *surface* magnons. This is because, in most magnetic materials, the long wavelengths probed by light scattering (see section 1.4.1) correspond to a regime where the magnetic dipole–dipole interactions are either dominant or comparable with exchange effects. In such cases the surface magnons have a quite different character, and light scattering has indeed proved to be an excellent probe of their properties, as we shall discuss in chapter 5. Nevertheless, light scattering from exchange-dominated *bulk* magnons at surfaces has provided verification for the type of theory in

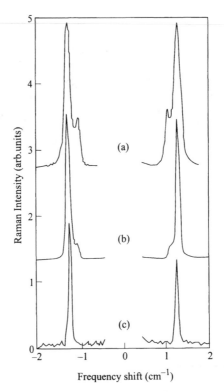

Figure 4.18. Measured Stokes and anti-Stokes spectra for one-magnon Raman scattering from Bi-YIG at 295 K with an applied magnetic field of 1.108 T. The excitation wavelength was (a) 457.9 nm, (b) 496.5 nm and (c) 524.5 nm. The spectral resolutions were (a) $0.072\,\mathrm{cm}^{-1}$, (b) $0.066\,\mathrm{cm}^{-1}$ and (c) $0.055\,\mathrm{cm}^{-1}$. After Cottam *et al* (1994).

section 4.1. An experimental and theoretical analysis of high-resolution Raman scattering data was reported by Cottam *et al* (1994) for a sample of yttrium iron garnet (YIG) doped with Bi. The ferrimagnet YIG (or $Y_3Fe_5O_{12}$) has strong exchange interactions and, as mentioned, at $T \ll T_C$ it behaves in many respects like a ferromagnet. The doping with Bi (to give $Bi_xY_{3-x}Fe_5O_{12}$ with $x = 0.41$) makes it relatively opaque at certain wavelengths of the incident light, so that the scattering takes place only near to the surface of the thick crystal sample. An interesting feature of the measured spectra is that a pronounced lineshape asymmetry and a subsidiary peak appear at higher optical absorption. An example is shown in figure 4.18, where the spectra labelled (a) to (c) represent increasing excitation wavelength, implying *decreasing* optical absorption. The type of theory developed in section 4.1.3, which resulted in a predicted lineshape given by (4.31), leads to excellent agreement with the experimental data when the instrumental resolution and the temperature are taken into account, as can be seen from

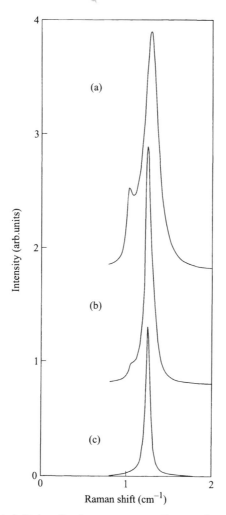

Figure 4.19. Calculated Stokes lineshapes corresponding to the one-magnon Raman scattering data shown in figure 4.18 for Bi-YIG at 295 K, assuming the pinning term $\xi = 10^5 \, \text{cm}^{-1}$. After Cottam *et al* (1994).

figure 4.19. This analysis also enabled the pinning parameter ξ of the Bi-YIG sample to be determined, since this quantity affects the reflection coefficient ϕ (see also figure 4.6).

In the case of ferromagnetic layers coupled by exchange through a metallic spacer, we have already mentioned in section 4.5 some references to experiments. These have mainly involved light scattering, the magneto-optical Kerr effect (or MOKE) and scanning electron microscopy with polarization analysis (or SEMPA). We have already described Raman and Brillouin scattering of light in earlier sections (starting in section 1.4.1), and we now comment on the other two techniques, both of which are specific to magnetic materials. They are extensively reviewed in the book edited by Heinrich and Bland (1994). MOKE is a linear optical technique that is based on changes in the linear susceptibility as a function of the applied magnetic field. Its origin lies in the spin–orbit coupling that acts like a magnetic field on the current induced by the electromagnetic field of an incident light beam. As this results in a small rotation of the polarization of light travelling through a magnetic material, it yields a probe for its magnetization. The use of modulation techniques can lead to a high sensitivity; a surface-sensitive variation of the technique is known as SMOKE. On the other hand, the SEMPA technique relies on the fact that, when electrons are incident from a scanning electron microscope, the secondary electrons emitted from a ferromagnet have a spin polarization which is related to the net spin density in the material and hence to the magnetization. It requires a ultra-high vacuum environment. SEMPA is particularly suitable for imaging the magnetization variations in the top few atomic layers of samples. As an example, we mention its application to the wedge-shaped-spacer geometry of magnetic bilayers depicted in figure 4.17 (see also section 4.5.1). SEMPA images measured by Unguris *et al* (1991) of the in-plane magnetization in the Fe film and the Fe substrate are shown in figure 4.20. The pattern of domains, which appear black or white according to the magnetization direction, along the Cr wedge, is clearly seen. Since the spacer thickness

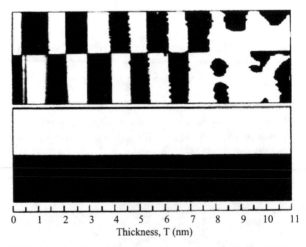

Figure 4.20. SEMPA image of the in plane magnetization for the Fe/Cr/Fe sample with a Cr wedge, as depicted in figure 4.17. The patterns of domains in the Fe film and Fe substrate are shown in the upper and lower panels, respectively. The thickness of the Cr wedge, which varies linearly across the image, is shown on the bottom scale. Each image represents a region 490 μm by 280 μm. After Unguris *et al* (1991).

varies linearly along the wedge, the spatial period of the oscillatory exchange can be deduced.

In the context of exchange-coupled magnetic bilayers, it is relevant to mention further work by Wang *et al* (1994). They studied effects of an applied magnetic field on Fe multilayers coupled antiferromagnetically through Cr spacers using magneto-optical and magnetization measurements. The conditions of the experiments were such that they confirmed the long-standing prediction that the surface spin–flop phase transition in an exchange-coupled antiferromagnet occurs at a lower value, by a factor of $\sqrt{2}$, than in the bulk (see section 4.3.3).

Two other techniques that have been successfully used to study the surface magnetization and other magnetic structure in films are spin-polarized electron photoemission and spin-polarized LEED; see e.g. Pierce (1986) and Hopster (1994) for reviews. In spin-polarized photoemission, the incident electromagnetic radiation (for example, from a synchrotron source) produces photoelectrons whose dependence on angle, energy and spin polarization can be measured to deduce the surface magnetic structure. The application to europium chalcogenides (such as the ferromagnets EuO and EuS) shows that in certain circumstances there may be surface spin instabilities corresponding to surface reconstruction in the outermost one or two atomic layers (see Campagna *et al* 1974).

Spin-polarized LEED is similar to the LEED technique described in section 1.2.3 but involves using a spin-polarized incident electron beam and/or measuring the polarization of the diffracted beam. In this way Celotta *et al* (1979) were able to measure the temperature dependence of the surface magnetization in a Ni crystal with a (110) surface. A more thorough investigation (Pierce 1986) has been carried out for the ferromagnetic glass $Ni_{0.4}Fe_{0.4}B_{0.2}$ at low temperatures $T \ll T_C$, and some data are shown in figure 4.21. The reduced magnetization $M(T)/M(0)$ is found to decrease with increasing temperature according to the power law

$$M(T)/M(0) = 1 - bT^{3/2}. \qquad (4.80)$$

The coefficient b is greater (by a factor of about 3) for the surface layer compared with the bulk. On the basis of the theory for a simple-cubic semi-infinite ferromagnet with nearest-neighbour exchange coupling (see section 4.1.3) we would predict $b_{surface}/b_{bulk} = 2$, as well as the $T^{3/2}$ dependence. However, since $Ni_{0.4}Fe_{0.4}B_{0.2}$ needs to be described by a more complicated theoretical model, a value of about 3 for the ratio of the b coefficients is not unreasonable. Similar experiments on $Ni_{0.78}Fe_{0.22}$, but extending to higher values of T/T_C, were reported by Mauri *et al* (1989).

Measurements of the surface magnetization in the antiferromagnet NiO have been reported using conventional LEED by Palmberg *et al* (1968) and Hayakawa *et al* (1971). In this case the magnetic unit cell of the NiO is larger than the spatial unit cell (because of the antiferromagnetic ordering), so there

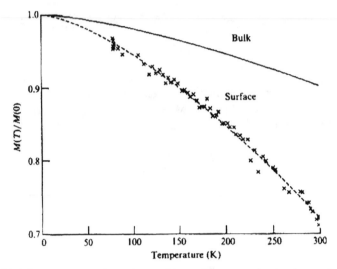

Figure 4.21. The temperature dependence of the surface magnetization (crosses) measured by spin-polarized LEED in $Ni_{0.4}Fe_{0.4}B_{0.2}$ compared with that of the bulk (full line) determined by conventional methods. The broken line is a best fit to the experimental data for the surface layer using (4.80). After Pierce (1986).

are specific *magnetic* Bragg reflections and it is unnecessary to employ beams that are spin-polarized.

Finally we mention that there have been measurements of the surface specific heat in various magnetic materials due to Henderson *et al* (1969, 1971). However, there are considerable difficulties in comparing this type of data with theory, as was discussed by Mills (1984).

References

Albuquerque E L and Cottam M G 2003 *Phys. Rep.* **376** 225
Azevedo A, Chesman C, Rezende S M, Aguiar F M, Bian X and Parkin S S P 1996 *Phys. Rev. Lett.* **76** 4837
Baibich M N, Broto J M, Fert A, Nguyen Van Dau F, Petroff F, Etienne P, Creuzet G, Friederich A and Chazelas J 1988 *Phys. Rev. Lett.* **61** 2472
Bezerra C G and Cottam M G 2002 *J. Appl. Phys.* **91** 7221
Binder K 1981 *Ferroelectrics* 35 99
Camley R E and Barnas J 1989 *Phys. Rev. Lett.* **63** 664
Campagna M, Sattler K and Siegmann H C 1974 *AIP Conf. Proc.* **18** 1388
Carriço A S, Camley R E and Stamps R L 1994 *Phys. Rev. B* **50** 13453
Celotta R J, Pierce D J, Wang G C, Bader S D and Fetcher G P 1979 *Phys. Rev. Lett.* **43** 728
Chen N N, Cottam M G and Khater A F 1995 *Phys. Rev. B* **51** 1003
Cottam M G 1976 *J. Phys. C* **9** 2121

Cottam M G 1978a *J. Phys. C* **11** 151

Cottam M G 1978b *J. Phys. C* **11** 165

Cottam M G and Kontos D 1980 *J. Phys. C* **13** 2945

Cottam M G and Maradudin A A 1984 in *Surface Excitations* ed V M Agranovich and R Loudon (Amsterdam: North-Holland) p 1

Cottam M G and Slavin A N 1994 in *Linear and Nonlinear Spin Waves in Magnetic Films and Superlattices* ed M G Cottam (Singapore: World Scientific) p 1

Cottam M G, Tilley D R and Zeks B 1984 *J. Phys. C* **17** 1793

Cottam M G, Zhang P X and Lockwood D J 1994 *Solid State Commun.* **92** 967

Demokritov S, Tsymbal E, Grünberg P, Zinn W and Schuller I K 1994 *Phys. Rev. B* **49** 720

Dewames R E and Wolfram T 1969 *Phys. Rev.* **185** 720

Diep H T 1991 *Phys. Rev. B* **43** 8509

Dutcher J R 1994 in *Linear and Nonlinear Spin Waves in Magnetic Films and Superlattices* ed M G Cottam (Singapore: World Scientific) p 287

Edwards D M, Mathon J, Muniz R B, Villeret M and Ward J M 1993 in *Magnetism and Structure in Systems of Reduced Dimension* ed R F C Farrow *et al* (New York: Plenum) p 401

Fillipov B N 1967 *Sov. Phys.* **9** 1048

Frait Z and Fraitova D 1988 in *Spin Waves and Magnetic Excitations 2* ed A S Borovik-Romanov and S K Sinha (Amsterdam: Elsevier Science) p 1

Grünberg P 2000a *J. Mag. Mag. Mat.* **226–230** 1688

Grünberg P 2000b in *Magnetic Multilayers and Giant Magnetoresistance* ed U Hartmann (Berlin: Springer) p 49

Grünberg P, Schreiber R, Pang Y, Brodsky M B and Sowers H 1986 *Phys. Rev. Lett.* **57** 2442

Harada I, Nagai O and Nagamiya T 1977 *Phys. Rev. B* **16** 4882

Harris E A and Owen J 1963 *Phys. Rev. Lett.* **11** 9

Hartmann U (ed) 2000 *Magnetic Multilayers and Giant Magnetoresistance* (Berlin: Springer)

Hathaway K B 1994 in *Ultrathin Magnetic Structures II* ed B Heinrich and J A C Bland (Berlin: Springer) p 45

Hayakawa K, Namikawa K and Miyaka S 1971 *J. Phys. Soc. Japan* **31** 1408

Hayes W and Loudon R 1978 *Scattering of Light by Crystals* (New York: Wiley)

Heinrich B and Bland J A C (ed.) 1994 *Ultrathin Magnetic Structures II* (Berlin: Springer)

Heinrich B and Cochran J F 1993 *Adv. Phys.* **42** 523

Henderson A J, Meyer H and Guggenheim H J 1969 *Phys. Rev.* **185** 128

Henderson A J, Meyer H and Guggenheim H J 1971 *J. Phys. Chem. Solids* **32** 1047

Hopster H 1994 in *Ultrathin Magnetic Structures I* ed J A C Bland and B Heinrich (Berlin: Springer) p 123

Keffer F 1966 *Handbuch der Physik* **18** 1

Kittel C 1986 *Introduction to Solid State Physics* 6th edition (New York: Wiley)

Lévy P M, Zhang S and Fert A 1990 *Phys. Rev. Lett.* **65** 1643

Mauri D, Scholl D, Siegman H C and Kay E 1989 *Phys. Rev. Lett.* **62** 1900

Mills D L 1968 *Phys. Rev. Lett.* **20** 18

Mills D L 1984 in *Surface Excitations* ed V M Agranovich and R Loudon (Amsterdam: North-Holland) p 379

Mills D L 1994 in *Ultrathin Magnetic Structures I* ed J A C Bland and B Heinrich (Berlin: Springer) p 91

Mills D L and Maradudin A A 1967 *J. Phys. Chem. Solids* **28** 1855
Mills D L and Saslow W M 1968 *Phys. Rev.* **171** 488
Moul R C and Cottam M G 1979 *J. Phys. C* **12** 5191
Moul R C and Cottam M G 1983 *J. Phys. C* **16** 1307
Ong L H, Osman J and Tilley D R 2001 *Phys. Rev. B* **63** 144109
Palmberg P W, Dewames R E and Viedevoe L A 1968 *Phys. Rev. Lett.* **21** 682
Parkin S S P, More N and Roche K P 1990 *Phys. Rev. Lett.* **64** 2304
Pereira J M and Cottam M G 2000 *J. Appl. Phys.* **87** 5941
Pierce D T 1986 in *Magnetic Properties of Low-Dimensional Systems* ed L M Falicov and
    J L Moran-Lopez (Heidelberg: Springer) p 58
Puszkarskii H 1979 *Prog. Surf. Sci.* **9** 191
Rado G T and Weertman J R 1959 *J. Phys. Chem. Solids* **11** 315
Rührig M, Schäfer R, Hubert A, Mosler R, Wolf J A, Demokritov S and Grünberg P 1991
    Phys. Stat. Solidi (a) **125** 635
Salman A and Cottam M G 1994 *Surf. Rev. Lett.* **1** 23
Schäfer M, Demokritov S, Müller-Pfeiffer F, Schäfer R, Schneider M, Grünberg P and
    Zinn W 1995 *J. Appl. Phys.* **77** 1
Slonczewski J C 1991 *Phys. Rev. Lett.* **67** 3172
Slonczewski J C 1993 *J. Appl. Phys.* **73** 5957
Suran G, Rothman J and Meyer C 1998 *IEEE Trans. Magnetics* **34** 849
Tilley D R and Zeks B 1984 *Solid State Commun.* **49** 823
Unguris J, Celotta R J and Pierce D T 1991 *Phys. Rev. Lett.* **67** 140
Wallis R F, Maradudin A A, Ipatova I P and Klochikhim A A 1967 *Solid State Commun.*
    5 89
Wang R W, Mills D L, Fullerton E E, Mattson J E and Bader S D 1994 *Phys. Rev. Lett.*
    72 920
Wolfram T and Dewames R E 1969 *Phys. Rev.* **185** 762
Wolfram T and Dewames R E 1972 *Prog. Surf. Sci.* **2** 233
Yafet Y 1987 *J. Appl. Phys.* **61** 4058
Yosida K 1996 *Theory of Magnetism.* (Berlin: Springer)
Yu J T, Turk R A and Wigen P E 1975 *Phys. Rev. B* **11** 420

Chapter 5

Surface magnetostatic and dipole-exchange modes

This chapter deals with surface magnetic excitations in situations where the magnetic dipole–dipole interactions are no longer negligible, as they were assumed to be in chapter 4 for surface modes in exchange-dominated magnetic systems. We have already discussed in a general way that the relative importance of the exchange and dipolar terms, as regards the *dynamical properties* of the excitations, is influenced by the wavevector $\mathbf{q}$ (see section 2.3 and figure 2.5).

To summarize, it is typically the case that the dipolar terms become significant when $|\mathbf{q}| < 10^8\,\mathrm{m}^{-1}$ (about $1/100$ of the value for a Brillouin zone boundary wavevector). For an intermediate wavevector range $10^7\,\mathrm{m}^{-1} < |\mathbf{q}| < 10^8\,\mathrm{m}^{-1}$, the exchange and dipolar terms may be comparable, and this is known as the *dipole-exchange region*. For wavelengths that are sufficiently long that $|\mathbf{q}| < 10^7\,\mathrm{m}^{-1}$, the dipolar terms are dominant. In general, the dipole–dipole effects may either be treated microscopically using the Hamiltonian term (2.58) or, more conveniently, macroscopically within a continuum theory using Maxwell's equations of electromagnetizm. Examples of these two methods for the case of bulk excitations were given in sections 2.3.4 and 2.6.3 respectively. A useful simplification occurs if $|\mathbf{q}| \gg \omega/c$, where ω is the frequency and c is the velocity of light. This is the so-called *magnetostatic region*, which was described for bulk excitations in section 2.6.3. The corresponding range of values of $|\mathbf{q}|$ is typically from about $3 \times 10^3\,\mathrm{m}^{-1}$ to $10^7\,\mathrm{m}^{-1}$ for ferromagnets. At even smaller $|\mathbf{q}|$ we have the electromagnetic region where it is necessary to employ the full form of Maxwell's equations including retardation, as outlined in section 2.6.2 for bulk magnetic polaritons. For surface geometries the electromagnetic region will be dealt with in chapter 6, where there are sections covering theory and experiment for surface magnetic polaritons.

We begin this chapter by considering surface magnetostatic modes in ferromagnets, first for films (including the semi-infinite limit) and then for double-layer systems. A similar, but briefer, treatment for antiferromagnets follows. These modes were first predicted by Mercereau and Feynman (1956), Walker (1957) and Damon and Eshbach (1961); they were subsequently observed by microwave techniques (see Brundle and Freedman 1968). Next we discuss the more complex problem of surface modes in the dipole-exchange region for ferromagnets and antiferromagnets, using both the macroscopic and microscopic methods. Finally we describe the experimental results for magnetostatic modes and dipole-exchange modes in various surface geometries. These studies are now very extensive and include light scattering, microwave techniques, far infrared reflectivity and spin wave resonance.

5.1 Magnetostatic modes for ferromagnetic films

We now obtain the surface and bulk magnetostatic modes for a ferromagnetic film or slab. The usual configuration for surface magnetostatic modes is to have the static magnetization M_0 parallel to the surfaces, and we consider this case first before examining what happens for M_0 perpendicular to the surfaces.

5.1.1 Magnetization parallel to the film surfaces

For this case we adopt the geometry shown in figure 5.1, with the z axis along M_0 and the x axis perpendicular to the slab surface. The 2D in-plane wave vector $\mathbf{q}_{||}$ has components

$$\mathbf{q}_{||} = (q_y, q_z) = q_{||} \, (\sin \phi, \, \cos \phi) \tag{5.1}$$

where $q_{||} = (q_y^2 + q_z^2)^{1/2}$ and ϕ is the angle between $\mathbf{q}_{||}$ and M_0. The theory for this geometry was first presented by Damon and Eshbach (1961). Also there

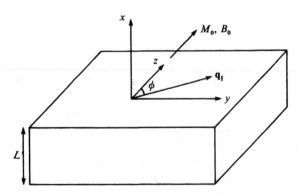

Figure 5.1. The assumed geometry and coordinate axes for calculating magnetostatic modes in a ferromagnetic film with M_0 parallel to the surfaces.

have been many review accounts of magnetostatic theory and we mention, for example, Mills (1984), Tilley (1986), Cottam and Slavin (1994) and Kabos (1995). The calculations will be described for general film thickness L, so the surfaces are at $x = 0$ and $x = -L$. The results for a semi-infinite medium $(L \to \infty)$ with just one surface at $x = 0$ will be obtained as a special limit.

In the magnetostatic region the wavelength of any excitation is very large compared with the lattice spacing, and so a continuum approximation for the ferromagnet should be valid. We follow the same approach using macroscopic field variables as in section 2.6.3 for the bulk case. A scalar potential ψ is introduced, related to the fluctuating magnetic field $\mathbf{h}$ as in (2.182). For the region *inside* the ferromagnetic film (i.e. for $0 < x < -L$) ψ satisfies the differential equation (2.183), where χ_a is the susceptibility component defined in (2.166). For the nonmagnetic region outside the ferromagnet we have simply

$$\nabla^2 \psi = 0. \tag{5.2}$$

From the property of translational invariance in the y and z directions it follows that $\psi(\mathbf{r})$ must be of the form $\psi_1(x) \exp(i\mathbf{q}_{||} \cdot \mathbf{r}_{||})$, where $\mathbf{r}_{||} = (y, z)$. For (2.183) and (5.2) to be satisfied and for ψ to vanish at $x = \pm\infty$, we have the general solution

$$\psi_1(x) = \begin{cases} a_1 \exp(-q_{||}x) & x > 0 \\ a_2 \exp(iq_x x) + a_3 \exp(-iq_x x) & 0 > x > -L \\ a_4 \exp(q_{||}x) & x < -L. \end{cases} \tag{5.3}$$

Because of (2.183), the above quantity q_x, which can be real or imaginary, is given by

$$(1 + \chi_a)(q_x^2 + q_{||}^2) - \chi_a q_z^2 = 0. \tag{5.4}$$

The amplitude coefficients a_j $(j = 1, 2, 3, 4)$ in (5.3) can now be determined by applying the standard electromagnetic boundary conditions at $x = 0$ and $x = -L$. In the magnetostatic case these boundary conditions are (i) that ψ must be continuous across a boundary, and (ii) that $(h_x + m_x)$ inside the ferromagnet at $x = 0$ and $x = -L$ must be equal to h_x immediately outside. They lead to four homogeneous linear equations for the four coefficients, and the condition for a solution is found to be

$$q_{||}^2 + 2q_{||}q_x(1 + \chi_a)\cot(q_x L) - q_x^2(1 + \chi_a)^2 - q_y^2 \chi_b^2 = 0. \tag{5.5}$$

When q_x is substituted from (5.4), the above equation determines the dispersion relations.

The solutions for the mode frequencies are particularly simple when $q_z = 0$, i.e. angle $\phi = 90°$ for the propagation direction. This is referred to as the *Voigt configuration*. First, we see that (5.4) can be satisfied if

$\chi_a = -1$. From (2.166) this leads to $\omega = \pm\omega_B$, where

$$\omega_B(q_x, \mathbf{q}_\parallel) = [\omega_0(\omega_0 + \omega_m)]^{1/2}. \tag{5.6}$$

This represents the frequency of the bulk magnetostatic modes in the slab, and it is in fact equivalent to the result in (2.96) with $\theta = 90°$. We recall the definitions in section 2.6 that ω_0 and ω_m are related to the applied field and the magnetization by $\omega_0 = \gamma B_0$ and $\omega_m = \gamma\mu_0 M_0$. Note that (5.6) is independent of the wave-vector components q_x and q_y. The second way in which equation (5.4) can be satisfied when $q_z = 0$ is if $q_x = \pm iq_\parallel$, and these two possibilities correspond to surface modes in the film. This follows since (5.3) for the case $0 > x > -L$ then has localized solutions rather than wave-like solutions. The surface solutions have a common frequency $\omega = \omega_S(\mathbf{q}_\parallel)$, which is found by substituting for q_x in (5.5). On using (2.166) the result can be rearranged as

$$\omega_S(\mathbf{q}_\parallel) = [(\omega_0 + \tfrac{1}{2}\omega_m)^2 - \tfrac{1}{4}\omega_m^2 \exp(-2q_\parallel L)]^{1/2}. \tag{5.7}$$

It is important to note that the mode solutions, as given in terms of the amplitude coefficients a_j in (5.3), are different for the two surface states. If $q_y > 0$ it can be shown that the two cases of $q_x = iq_\parallel$ and $q_x = -iq_\parallel$ correspond to surface states localized near the lower surface $(x = -L)$ and the upper surface $(x = 0)$, respectively, and vice versa if $q_y < 0$. This property is another aspect of *non-reciprocal propagation* (see also section 2.6.2), and we shall discuss it in more detail shortly. From (5.6) and (5.7) it follows that ω_S is greater than ω_B (by contrast to the exchange-dominant case), and the limiting surface mode frequencies are

$$\omega_S(\mathbf{q}_\parallel) = \begin{cases} [\omega_0(\omega_0 + \omega_m)]^{1/2} & (q_\parallel L \ll 1) \\ (\omega_0 + \tfrac{1}{2}\omega_m) & (q_\parallel L \gg 1). \end{cases} \tag{5.8}$$

The latter case corresponds to the surface mode for a semi-infinite ferromagnet. The behaviour of ω_S and ω_B as functions of $q_\parallel L$ is sketched in figure 5.2.

When $q_z \neq 0$ the surface and bulk magnetostatic modes can still be investigated using (5.4) and (5.5). First, let us consider $q_y = 0$, implying angle $\phi = 0°$ and $q_\parallel = |q_z|$. The waves propagate parallel (or antiparallel) to the applied magnetic field, and in this special case it may be shown that (5.5) can be re-expressed as two separate equations for modes that are symmetric or antisymmetric with respect to the middle of the film:

$$q_x \cot(q_x L/2) = -q_\parallel \quad \text{symmetric modes} \tag{5.9}$$

$$q_x \tan(q_x L/2) = q_\parallel \quad \text{antisymmetric modes.} \tag{5.10}$$

These equations have solutions only for real values of q_x. Therefore there are no surface modes in this case, and the solutions of (5.9) and (5.10) lead to

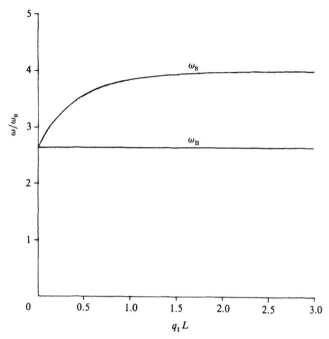

Figure 5.2. The dependence of the surface and bulk magnetostatic mode frequencies on $q_{\parallel}L$ for a ferromagnetic film, assuming the Voigt configuration ($q_z = 0$) and $\omega_m/\omega_0 = 6$.

quantized bulk modes. For $L \rightarrow \infty$ they form a continuum with dispersion relation given by

$$\omega_B(q_x, \mathbf{q}_{\parallel}) = \{\omega_0^2 + [\omega_0\omega_m q_x^2/(q_x^2 + q_{\parallel}^2)]\}^{1/2} \tag{5.11}$$

which is consistent with the bulk-mode expression in (2.96). Note that there is an explicit dependence on wavevector, in contrast to (5.6) for the Voigt configuration. The frequencies extend from ω_0 (when $q_x \approx 0$) to $[\omega_0(\omega_0 + \omega_m)]^{1/2}$ (when $q_x \gg q_{\parallel}$), for any fixed in-plane wavevector. It is clear from (5.11) that the frequency of these $\phi = 0$ bulk modes decreases with increasing $q_{\parallel}$, and so these modes have a negative group velocity $\partial\omega_B/\partial q_{\parallel}$ parallel to the surface. For this reason they are often referred to as *magnetostatic backward bulk modes*. In the Voigt configuration the group velocity is zero.

For general values of the propagation angle ϕ between 0 and 90° the results are more complicated. Some of the features of the dispersion relations are represented schematically in figure 5.3, where the frequencies are plotted against the components q_y and q_z of the in-plane wave vector $\mathbf{q}_{\parallel}$. The spectrum of bulk magnetostatic modes consists of discrete sheets that become degenerate at $\omega = [\omega_0(\omega_0 + \omega_m)]^{1/2}$ in the limit of $q_z \rightarrow 0$. These sheets correspond to solutions with q_x real in (5.4), as in the special case of $\phi = 0$ already discussed. Above the bulk region in figure 5.3 there is a

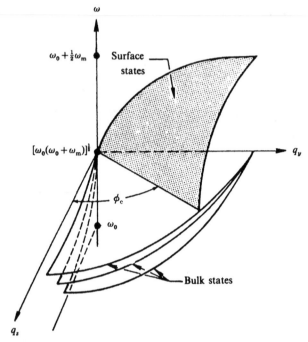

Figure 5.3. Schematic plot of the dispersion curves for the magnetostatic modes in a ferromagnetic film (with M_0 parallel to the surfaces) as a function of q_y and q_z. After Damon and Eshbach (1961).

single sheet corresponding to a surface modes (see the shaded region). The imaginary values of q_x are found from (5.4) to be $q_x = \pm i\beta$ where

$$\beta = \left[q_y^2 + \frac{q_z^2}{(1 + \chi_a)} \right]^{1/2} \tag{5.12}$$

provided the quantity inside the square brackets is positive. Hence the attenuation constant is changed from $|q_y|$ (when $q_z = 0$) to β. From (5.5) the surface-mode frequency satisfies

$$q_{||}^2 + 2q_{||}\beta(1 + \chi_a) \coth(\beta L) + \beta^2(1 + \chi_a)^2 - q_y^2 \chi_b^2 = 0. \tag{5.13}$$

The surface mode frequency decreases as the propagation wavevector is rotated towards the q_z axis: for example, if $q_{||}L \gg 1$ the surface mode frequency is

$$\omega_S(\mathbf{q}_{||}) = \tfrac{1}{2}[(\omega_0/\sin\phi) + (\omega_0 + \omega_m)\sin\phi]. \tag{5.14}$$

As $\mathbf{q}_{||}$ moves away from the q_y axis (or, in other words, as ϕ varies from $90°$) $\omega_S(\mathbf{q}_{||})$ decreases until at a critical angle it reaches the value $[\omega_0(\omega_0 + \omega_m)]^{1/2}$ corresponding to the top of the bulk magnon region (see figure 5.3). In fact,

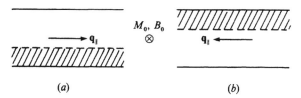

(a) (b)

Figure 5.4. Illustration of nonreciprocal propagation of surface magnetostatic modes for a ferromagnetic film in the Voigt configuration. When $\mathbf{q}_{\|}$ is reversed the localization (see amplitude envelope for ψ) of the surface mode switches from one film surface to the other as in cases (a) and (b).

(5.14) holds only for $\phi_c < \phi < 180° - \phi_c$, where ϕ_c is given by

$$\sin \phi_c = [\omega_0/(\omega_0 + \omega_m)]^{1/2}. \qquad (5.15)$$

Hence the surface magnetostatic modes are predicted to have non-reciprocal propagation properties, in the sense that $\omega_S(\mathbf{q}_{\|}) \neq \omega_S(-\mathbf{q}_{\|})$. Indeed no surface mode is possible, localized at that *same* surface, when $\mathbf{q}_{\|}$ is reversed. The behaviour on reversing $\mathbf{q}_{\|}$ is illustrated in figure 5.4 for the case of the Voigt configuration ($\phi = 90°$). We suppose that for one direction of $\mathbf{q}_{\|}$ there is a surface mode localized at the lower surface of the slab, as in figure 5.4(a). Then if $\mathbf{q}_{\|}$ is reversed in direction, keeping B_0 and M_0 fixed, the new surface mode is localized near the upper surface, as in figure 5.4(b). The two modes are degenerate in frequency and have the same attenuation length $q_{\|}^{-1}$. A thorough explanation of why the surface magneto-static modes have this non-reciprocity property was given by Mills (1984), who concluded that it requires consideration of the space-group symmetry of the dipolar system as well as the lack of time-reversal symmetry.

Surface magnetostatic modes in ferromagnets typically have frequencies that correspond to the microwave region, and there are applications for signal processing (see e.g. Marcelli and Nikitov 1996). The frequencies are also very suitable for experimental studies by Brillouin scattering. These techniques, amongst others, and the extent to which they provide verification of the foregoing theory, are discussed later in this chapter. For a comparison of theory with Brillouin scattering data it is also necessary to have results for the magnetic Green functions; these quantities have been derived in the magnetostatic limit by Cottam (1979).

5.1.2 Magnetization perpendicular to the film surfaces

Next we consider the case when the applied field B_0 and the static magnetiza-tion M_0 are perpendicular to the surfaces. It is appropriate now to choose this direction as the z axis, with the film occupying the space between $z = 0$ and $z = -L$. The x and y axes are parallel to the surfaces, and without loss of generality we may take $\mathbf{q}_{\|} = (q_x, 0)$ for the 2D in-plane wave vector.

The magnetostatic calculation proceeds in a similar manner to that in section 5.1.1, including the introduction of a magnetostatic potential function ψ. Writing $\psi(\mathbf{r}) = \psi_1(z) \exp(i\mathbf{q}_\| \cdot \mathbf{r}_\|)$, where now $\mathbf{r}_\| = (x, y)$, the general solution for ψ_1 takes the same form as in (5.3) except that x and q_x are replaced by z and q_z. The condition satisfied by q_z is that

$$(1 + \chi_a)q_\|^2 + q_z^2 = 0. \tag{5.16}$$

Another condition on q_z is obtained by applying the usual electromagnetic boundary conditions at the film surfaces. In the present geometry these imply that ψ and $\partial\psi/\partial z$ must be continuous at $z = 0$ and $z = -L$, leading to either

$$q_z \tan(q_z L/2) = q_\| \quad \text{symmetric modes} \tag{5.17}$$

or

$$q_z \cot(q_z L/2) = -q_\| \quad \text{antisymmetric modes.} \tag{5.18}$$

Since $q_\| = |q_x|$ is positive, it follows that the only solutions for q_z are real. We therefore conclude that, when M_0 is perpendicular to the film surfaces, no surface magnetostatic modes occur. The discrete q_z values obtained from (5.17) and (5.18) correspond to bulk magnetostatic modes whose amplitudes are symmetric and antisymmetric about the middle of the film. On substituting for χ_a in (5.16) and rearranging, the bulk-mode frequency is

$$\omega_B(\mathbf{q}_\|, q_z) = [\omega_0(\omega_0 + \omega_m) - \omega_0\omega_m q_z^2/(q_\|^2 + q_z^2)]^{1/2}. \tag{5.19}$$

The group velocity $\partial\omega_B/\partial q_\|$ is positive for these modes, and hence they are often referred to as *magnetostatic forward bulk modes* (in contrast to the backward bulk modes discussed in the previous section).

5.2 Magnetostatic modes for double-layer ferromagnets

We next consider the magnetostatic theory for a ferromagnetic double-layer system as represented in figure 5.5. Here there are two magnetic layers of thickness L_1 and L_2 separated by a spacer layer of thickness d. Such systems are of interest because they give rise to coupled modes, which have been studied by Brillouin scattering. For simplicity we take the magnetic layers to be of the same material and to be magnetized in the same direction (the z axis) parallel to the film surfaces. Also we assume the spacer layer to be nonmagnetic, so that the coupling between the two magnetic layers is due to the long-range dipolar fields in the magnetostatic theory. Improvements in the techniques for thin film preparation and deposition have made feasible the fabrication of such double-layer ferromagnets. In the Brillouin scattering experiments the layer thickness (for the ferromagnets and for the spacer) might typically lie in the range of a few nanometres (nm) up to a few

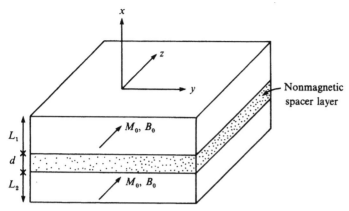

Figure 5.5. A ferromagnetic double layer separated by a nonmagnetic spacer layer. The static magnetization M_0 is assumed to be the same in both ferromagnets and to be parallel to the film surfaces.

hundred nm. Suitable ferromagnets would be metals such as Fe and Ni or semiconductors such as EuO and EuS.

Unlike the metallic trilayer systems considered in the context of GMR in section 4.4, the present case does not require a metal spacer to provide the exchange coupling. Also the spacer thickness can be much larger than in section 4.4.

The magnetostatic theory described in section 5.1.1 for a single film can straightforwardly be generalized to double-layer systems of the type shown in figure 5.5 (see Grünberg 1980 for details). The differential equations (2.183) and (5.2) are still applicable for the magnetostatic scalar potential ψ inside and outside, respectively, of the ferromagnetic media. The solutions for ψ in the different media can be written down by analogy to (5.3), and application of the boundary conditions at the interfaces $x = 0$, $x = -L_1$, $x = -(L_1 + d)$ and $x = -(L_1 + L_2 + d)$ leads to expressions that can be solved for the mode frequencies.

We shall quote the results only for the special case of $q_z = 0$ (i.e. the Voigt configuration where the propagation wave vector $\mathbf{q}_\parallel$ is perpendicular to M_0), when it is found that

$$\omega = [(\omega_0 + \tfrac{1}{2}\omega_m)^2 - \tfrac{1}{4}\omega_m^2\alpha]^{1/2}. \tag{5.20}$$

The quantity α is equal to 1 for a bulk mode, giving the same result as in (5.6) for ω_B in a single film, while for a surface mode α satisfies

$$[\alpha - \exp(-2q_\parallel L_1)][\alpha - \exp(-2q_\parallel L_2) - \alpha\exp(-2q_\parallel d)$$
$$\times [1 - \exp(-2q_\parallel L_1)][1 - \exp(-2q_\parallel L_2)] = 0. \tag{5.21}$$

This is a quadratic equation for α, giving two solutions as might be expected for coupled surface modes in a double-layer system. In the case of a large separation of the magnetic layers ($d \rightarrow \infty$) the solutions for α from (5.21) simplify to become $\exp(-2q_\parallel L_1)$ and $\exp(-2q_\parallel L_2)$. The corresponding frequencies given by (5.20) are just the surface magnetostatic mode frequencies for single isolated films of thickness L_1 and L_2, as can be seen by comparison with (5.7). In the opposite limit of separation $d \rightarrow 0$ the solutions of (5.21) become $\alpha = 1$ and $\alpha = \exp[-2q_\parallel(L_1 + L_1)]$. The former of these solutions represents a mode that has become degenerate with the bulk magnons, while the latter corresponds to the surface mode for a slab of overall thickness $L_1 + L_2$.

The general predictions of (5.20) and (5.21) for the mode frequencies are illustrated by numerical examples in figure 5.6. The frequencies are plotted against the spacer thickness d for the two cases of $L_1 = L_2 = 15$ nm (full curves) and $L_1 = 15$ nm and $L_2 = 150$ nm (broken curves). It has been assumed that $B_0 = 0.05$ T and $\mu_0 M_0 = 2.2$ T, appropriate to Fe.

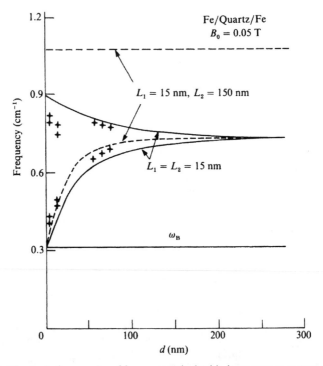

Figure 5.6. The mode frequencies of ferromagnetic double-layer systems versus the spacer-layer thickness d for two combination of the thickness L_1 and L_2 of the magnetic layers. The crosses denote experimental points obtained by Brillouin scattering from Fe films with $L_1 = L_2 = 15$ nm separated by quartz. After Grünberg *et al* (1982).

The theory has been extended to situations where the static magnetizations in the two magnetic films may be different. For example, Grünberg (1981) took the magnetizations to be of different magnitudes and to be in the directions either parallel or antiparallel to one another, while being perpendicular to the wavevector $q_{\parallel}$. The situation of magnetizations being antiparallel but of the same magnitude (which can be achieved experimentally by tailoring the coercive forces of the magnetic films in a proper way) is particularly interesting in that it has different symmetries under time reversal compared with the case of parallel alignment that we discussed earlier in this section. This has consequences for the non-reciprocal behaviour of the modes; see Grünberg (1981) for details. Also Camley and Maradudin (1982) gave the magnetostatic theory of interface modes between two semi-infinite media with different magnetizations (i.e. the case of $d = 0$ and infinitely large slab thicknesses).

We shall mention the double-layer magnetic systems again in section 5.6 in the context of Brillouin scattering experiments. The generalization from double-layer systems to magnetostatic modes in superlattices is described in sections 9.1 and 9.5.

5.3   Magnetostatic modes for antiferromagnets

A theory of the long-wavelength surface magnetostatic modes in semi-infinite antiferromagnets was first given by Tarasenko and Kharitonov (1971) and extended by Camley (1980). A further analysis, correcting some inconsistencies in the earlier papers, was made by Luthi *et al* (1983). The magnetostatic theory was then extended to antiferromagnetic films by Stamps and Camley (1984).

The calculations can be carried out in a manner that closely parallels the treatment for the ferromagnetic case as described in section 5.1. Therefore, in this section it will suffice to summarize only the main steps and to discuss the implications of the results; the interested reader is referred to the original papers for details. As well as the similarities, it should be noted that there are important differences between the ferromagnetic and antiferromagnetic cases. On the practical side, the frequencies of the long-wavelength modes in antiferromagnets are typically in the far infrared region of the spectrum, rather than in the microwave region as for ferromagnets. This has consequences for their experimental study, as already pointed out in chapter 2, and we will discuss this further in sections 5.6 and 5.7. On the theory side, another difference is that antiferromagnets have two sublattices of spins and consequently two branches are expected for the bulk-mode spectrum, as we have already seen from the discussion in section 2.6.3.

By analogy with the theory for the ferromagnetic case, we have a torque equation of motion for *each* of the sublattice magnetizations $\mathbf{M}_p (p = 1, 2)$, as in (2.170) for the derivation of the susceptibilities. Further, the fluctuating fields $\mathbf{m}(\mathbf{r})$ and $\mathbf{h}(\mathbf{r})$ in the antiferromagnet must satisfy the unretarded Maxwell's equations (2.181). The theory then proceeds along the same lines as for the ferromagnetic film in section 5.1 (see also section 2.6.3) with the only essential difference being the modified result for the susceptibility χ_a, which is given for a uniaxial antiferromagnet by (2.173) and (2.174).

We assume a film geometry and choice of coordinate axes similar to those in figure 5.1. In this case the sublattice magnetizations and anisotropy fields are taken to be in the surface plane parallel and antiparallel to the z axis. For simplicity, we present first the results in the semi-infinite limit $L \to \infty$, and then we describe the effects of finite film thickness L.

The bulk magnetostatic modes for the semi-infinite case are found, as expected, to be the same as worked out in section 2.6.3 for an infinite bulk antiferromagnet (see also Loudon and Pincus 1963). The two frequencies are given in general by (2.186), from which we note that the dependence on the 3D wave vector $\mathbf{q}$ occurs only through the angle θ, defined there as the angle between $\mathbf{q} = (q_x, \mathbf{q}_{\parallel})$ and the z axis. In the case of zero applied magnetic field (when $\omega_0 = 0$) the previous expression for the bulk frequencies simplify to give $\omega_B^{\pm}(\mathbf{q})$ where

$$\omega_B^+(\mathbf{q}) = [\omega_A(2\omega_E + \omega_A) + 2\omega_A\omega_m \sin^2\theta]^{1/2} \tag{5.22}$$

$$\omega_B^-(\mathbf{q}) = [\omega_A(2\omega_E + \omega_A)]^{1/2}. \tag{5.23}$$

It should be noted that $\omega_B^+ \neq \omega_B^-$, unlike the situation for bulk Heisenberg antiferromagnets with $\omega_0 = 0$ (see section 2.3.3), and this is a consequence of having now included the lower-symmetry dipolar effects in the calculations. One of the bulk modes ω_B^- is independent of $\mathbf{q}$ and is equal to the bulk antiferromagnetic resonance (AFMR) frequency, while the other mode ω_B^+ depends on both $\mathbf{q}$ (through the angle θ) and the sublattice magnetization (through ω_m).

The surface magnetostatic mode frequency ω_S in the case of zero applied magnetic field is found to be (e.g. see Camley 1980)

$$\omega_S(\mathbf{q}_{\parallel}) = \left[\omega_A(2\omega_E + \omega_A) + \frac{2\omega_A\omega_m \sin^2\phi}{1 + \sin^2\phi} \right]^{1/2} \tag{5.24}$$

where ϕ is the angle between the in-plane 2D wave vector $\mathbf{q}_{\parallel}$ and the direction of M_0 (the z axis). When the frequencies of the surface and bulk magnetostatic modes are plotted against ϕ, it is found that the surface branch lies between the two bulk-mode regions (see figure 5.7). The operation of changing $\mathbf{q}_{\parallel} \to -\mathbf{q}_{\parallel}$ (or, equivalently, $\phi \to \phi + 180°$) can be seen to leave the right-hand side of (5.24) invariant. Hence $\omega_S(\mathbf{q}_{\parallel}) = \omega_S(-\mathbf{q}_{\parallel})$, and so the non-reciprocal propagation property of the surface magnetostatic

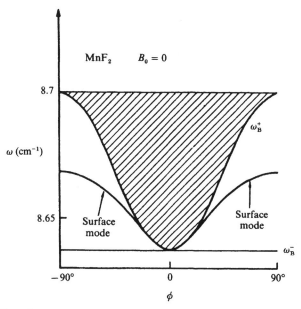

Figure 5.7. The calculated frequencies of bulk and surface magnetostatic modes plotted against ϕ for a semi-infinite antiferromagnet in zero applied field. Parameters appropriate to MnF$_2$ at low temperatures $T \ll T_N$ have been assumed. The bulk continuum is shown shaded. After Luthi *et al* (1983).

modes in ferromagnets (see section 5.1.1) does not occur for uniaxial anti-ferromagnets *in zero applied field*.

However, in the presence of an applied field B_0, the propagation of the surface magnetostatic mode in a semi-infinite antiferromagnet does become non-reciprocal (see Camley 1980). For example, if B_0 is along the z direction and $\mathbf{q}_{\parallel}$ is parallel or antiparallel to the y direction (corresponding to $\phi = \pm 90°$), the surface-mode frequency is

$$\omega_S(\mathbf{q}_{\parallel}) = (\omega_0 q_y/|q_y|) + [\omega_A(2\omega_E + \omega_A) + \omega_A \omega_m]^{1/2}. \quad (5.25)$$

Thus a wave travelling in the $+y$ direction has its frequency increased by an amount ω_0 whereas a wave in the $-y$ direction has its frequency decreased by ω_0. This type of non-reciprocity is different from that in a semi-infinite ferromagnet where there is no surface wave propagation when $\mathbf{q}_{\parallel}$ is reversed. The results for the case of general propagation angle ϕ when $B_0 \neq 0$ are illustrated in figure 5.8. The surface modes do not propagate in all directions for ϕ (contrasting with the situation in figure 5.7 where only $\phi = 0$ is excluded). Instead, there are critical angles ϕ_c^+ and ϕ_c^- where the surface branches terminate when they intersect the bulk-mode regions.

The generalization of the magnetostatic theory to an antiferromagnetic film of finite thickness L was made by Stamps and Camley (1984). The

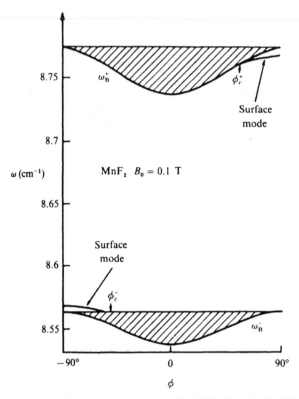

Figure 5.8. As in figure 5.7 but for nonzero applied field $B_0 = 0.1$ T. After Luthi *et al* (1983).

expressions for the mode frequencies are complicated except when $\phi = \pm 90°$. In this case it is found that the bulk-mode frequencies are independent of both thickness L and wavevector component q_x, taking the values $[\omega_A(2\omega_E + \omega_A) + 2\omega_A\omega_m]^{1/2}$ and $[\omega_A(2\omega_E + \omega_A)]^{1/2}$ when $B_0 = 0$, consistent with (5.22) and (5.23). Additionally, when $B_0 = 0$, there are two surface modes with frequencies given by

$$\omega_S^\pm(\mathbf{q}_{\parallel}) = \{\omega_A(2\omega_E + \omega_A) + \omega_A\omega_m[1 \pm \exp(-|q_y|L)]\}^{1/2}. \tag{5.26}$$

For $|q_y|L \gg 1$ the above frequencies reduce to (5.24) in the case of $\phi = \pm 90°$, corresponding to two uncoupled surface modes on each surface of a thick slab. If L is reduced the two modes become coupled and split in frequency. For $|q_y|L \ll 1$ the values for $\omega_S^\pm(\mathbf{q}_{\parallel})$ given by (5.26) become approximately equal to the bulk-mode frequencies.

The dependence of the bulk and surface magnetostatic modes on the propagation angle ϕ is shown in figure 5.9 for a film with $q_{\parallel}L = 2$ and $B_0 = 0$. The bulk modes (solid lines) are now quantized, and for $|\phi|$ greater

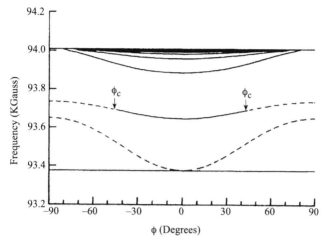

Figure 5.9. The calculated frequencies of bulk and surface magnetostatic modes plotted against ϕ for an antiferromagnetic film in zero applied field with $q_{\parallel} L = 2$. Parameters appropriate to MnF$_2$ at low temperatures $T \ll T_N$ have been assumed. Here the quantized bulk modes and the surface modes are shown by solid and broken lines respectively. After Stamps and Camley (1985).

than a critical value ϕ_c there are two surface modes (broken lines). One of the surface modes becomes a bulk mode for $|\phi| < \phi_c$, leaving just one surface branch.

For a film the dispersion relations for bulk and surface modes are found to be reciprocal, but it turns out that the localization of the modes is non-reciprocal. This means that reversing the direction of propagation switches the amplitude distributions describing the localization of the surface modes (see Stamps and Camley 1984).

5.4 Dipole-exchange modes for ferromagnets

When discussing the bulk properties of ferromagnets in section 2.3.4 we treated the exchange interactions and the dipole–dipole interactions together to obtain the bulk-magnon dispersion relation. It was also noted earlier that the exchange and dipole–dipole terms are typically comparable, as regards their contribution to the frequencies of the magnetic modes, for wavevectors in the range 10^7 to 10^8 m^{-1}. In films (or in other systems of limited geometry) the problem of solving for the dipole-exchange modes is much more complicated than in the bulk case, mainly because of the boundary conditions at the surfaces and the role of the surfaces in truncating the long-range dipolar interactions. Two approaches can be followed. One is to incorporate exchange into the magnetostatic theory of section 5.1 by again using macroscopic field variables. The other is a microscopic approach analogous to that in section 2.3.4, where a Hamiltonian representation is used for the dipole–dipole and Heisenberg exchange interactions. It is necessary to introduce 2D dipole–dipole sums consistent with the in-plane symmetry for a film and these must be evaluated separately. Both methods are described in the following paragraphs.

5.4.1 Macroscopic theory

Accounts of the macroscopic (or continuum) dipole-exchange theory are to be found in several review articles, for example by Wolfram and Dewames (1972), Mills (1984, 1994), Kalinikos (1994) and Kabos (1995).

We recall section 2.6.1 where the linear response was obtained, in the magnetostatic limit, for the fluctuating part of the magnetization $\mathbf{m}(\mathbf{r})$ inside a magnetic material to a fluctuating electromagnetic field $\mathbf{h}(\mathbf{r})$. In the present case (2.161) for the total field will be generalized to

$$\mu_0\mathbf{H} = B_0\hat{\mathbf{z}} + \mathbf{b}_{ex}(\mathbf{r}) + \mu_0\mathbf{h}(\mathbf{r})\exp(-\mathrm{i}\omega t) \qquad (5.27)$$

where the additional term $\mathbf{b}_{ex}(\mathbf{r})$ is the contribution due to the exchange: it has a static part and a fluctuating part, as derived in (2.81) for a simple cubic ferromagnet. When (5.27) and (2.162) are substituted into the torque equation (2.163) and linearization approximations are made as before, it is found that the previous macroscopic equation of motion (2.164) becomes

$$-\mathrm{i}\omega\mathbf{m}(\mathbf{r}) = \gamma\mathbf{i}_z \times \{\mu_0 M_0\mathbf{h}(\mathbf{r}) - [B_0 - (D/\gamma)\nabla^2]\mathbf{m}(\mathbf{r})\} \qquad (5.28)$$

where the static part of $\mathbf{b}_{ex}(\mathbf{r})$ has cancelled out. Here D is the exchange-dependent parameter introduced in (2.70), e.g. at $T \ll T_C$ it is equal to SJa^2 in a simple cubic ferromagnet with nearest-neighbour exchange J. Note that the differential operator $[B_0 - (D/\gamma)\nabla^2]$ represents the effective field acting on the spins in the long-wavelength continuum limit, as may also be seen by reference to (2.82).

The theory can now be developed in a broadly similar fashion to the magnetostatic case of section 5.1, except that it becomes necessary to solve higher-order differential equations than encountered previously. We adopt once again the film geometry in figure 5.1 with the static magnetization M_0 parallel to the surfaces, since this is the case of interest for surface modes. The differential equation for the scalar potential ψ within the film can be obtained from (2.183) for the magnetostatic case by making the formal replacement $B_0 \rightarrow [B_0 - (D/\gamma)\nabla^2]$ to give

$$(\hat{\theta}^2 + \omega_m\hat{\theta} - \omega^2)\left(\frac{\partial^2\psi}{\partial x^2} + \frac{\partial^2\psi}{\partial y^2}\right) + (\hat{\theta}^2 - \omega^2)\frac{\partial^2\psi}{\partial z^2} = 0 \qquad (5.29)$$

where the differential operator $\hat{\theta}$ is defined by

$$\hat{\theta} = \omega_0 - D\left(\frac{\partial^2}{\partial x^2} + \frac{\partial^2}{\partial y^2} + \frac{\partial^2}{\partial z^2}\right). \tag{5.30}$$

This holds for $0 > x > -L$, whereas outside the film $\nabla^2 \psi = 0$ as before.

For the regions outside the film ($x > 0$ and $x < -L$) the characteristic solutions for $\psi = \psi_1(x) \exp(i\mathbf{q}_\| \cdot \mathbf{r}_\|)$ take the same form as in the magnetostatic case of (5.3). However, within the film the previous expression in (5.3) generalizes to

$$\psi_1(x) = \sum_{j=1}^{3} [a_2^{(j)} \exp(iq_x^{(j)} x) + a_3^{(j)} \exp(-iq_x^{(j)} x)] \qquad (0 > x > -L) \tag{5.31}$$

representing the solution of the sixth-order differential equation (5.29). Here $q_x^{(j)}$ (with $j = 1, 2, 3$) are the roots of

$$(q_x^2 + q_\|^2)[(\omega_0 + Dq_\|^2 + Dq_x^2)^2 + \omega_m(\omega_0 + Dq_\|^2 + Dq_x^2) - \omega^2]$$
$$- q_z^2(\omega_0 + Dq_\|^2 + Dq_x^2) = 0. \tag{5.32}$$

Notice that (5.32) is a cubic equation in q_x^2, so solutions can always be expressed analytically.

We illustrate the nature of these solutions by taking the case of $q_z = 0$, which is the Voigt configuration ($\mathbf{q}_\| \perp M_0$). From (5.32) it follows that

$$q_x^{(1)} = \pm iq_\| \tag{5.33}$$

$$q_x^{(2)} = \pm D^{-1/2}[(\omega^2 + \tfrac{1}{4}\omega_m^2)^{1/2} - \tfrac{1}{2}\omega_m - \omega_0 - Dq_\|^2]^{1/2} \tag{5.34}$$

$$q_x^{(3)} = \pm iD^{-1/2}[(\omega^2 + \tfrac{1}{4}\omega_m^2)^{1/2} + \tfrac{1}{2}\omega_m + \omega_0 + Dq_\|^2]^{1/2}. \tag{5.35}$$

The root $q_x^{(1)}$ is pure imaginary and has the same value as that found for the surface magnetostatic mode (the Damon–Eshbach or DE mode) in section 5.1.1 for a ferromagnetic slab. The other two sets of roots depend on the exchange parameter D. The root $q_x^{(2)}$ is typically real (for the values of the frequency ω of interest) in which case it corresponds to a bulk-like mode, while $q_x^{(3)}$ is pure imaginary corresponding to a highly localized surface mode that has no analogue in the magnetostatic theory. The resulting *dipole-exchange modes* have eigenstates that are mixtures of the three types of waves in accordance with (5.31).

The degree of mixing is found by application of the boundary conditions. There will be the usual magnetostatic boundary conditions as employed in section 5.1, but these are insufficient. Clearly we also require some *additional boundary conditions* (or ABCs) in order to solve uniquely for all the coefficients $a_2^{(j)}$ and $a_3^{(j)}$ in (5.31). We may note that this situation is formally analogous to that encountered in spatial dispersion theory for exciton–polaritons (see sections 2.5.3 and 6.1.4). In the present case the ABCs can be found by writing down the linearized torque equation of motion for a spin in the surface layer and then taking the continuum limit. The general form of these 'exchange' boundary conditions can be rather complicated, as discussed by Wolfram and Dewames (1972), but a simplified form that is often used is (e.g. see Mills 1984)

$$\left[\frac{\partial m^{\mu}(\mathbf{r})}{\partial x} - \xi m^{\mu}(\mathbf{r})\right]_{x=0} = 0, \qquad \left[\frac{\partial m^{\mu}(\mathbf{r})}{\partial x} + \xi' m^{\mu}(\mathbf{r})\right]_{x=-L} = 0 \qquad (5.36)$$

where superscript μ denotes x or y, and ξ and ξ' are the effective surface anisotropy (or 'pinning') parameters at $x = 0$ and $x = -L$ respectively. We may remark that the above equations are analogous to (4.18) derived for a Heisenberg ferromagnet in the continuum approximation.

Wolfram and Dewames (1972) concluded that the mixing of the three types of waves in (5.31) is small when $q_{\parallel} < 10^6 \, \text{m}^{-1}$ typically, but becomes large when $q_{\parallel} > 10^8 \, \text{m}^{-1}$. For $q_z = 0$ and small $q_{\parallel}$ the solutions are approximately:

(i) a magnetostatic surface mode with frequency given to a very good approximation by (5.7), and
(ii) standing bulk-magnon modes with $q_x^{(2)}$ real and taking a series of discrete (or 'quantized') values.

The latter case is analogous to the quantization of the wavevector component for Heisenberg ferromagnetic films as discussed in section 4.2. A simplified treatment gives the quantized $q_x^{(2)}$ values as $n\pi/L$ with $n = 1, 2, 3, \ldots$, and it is then simple to show using (5.34) that the corresponding frequencies are

$$\omega_{\text{B}} = \left[\left(\omega_0 + Dq_{\parallel}^2 + \frac{Dn^2\pi^2}{L^2}\right)\left(\omega_0 + \omega_m + Dq_{\parallel}^2 + \frac{Dn^2\pi^2}{L^2}\right)\right]^{1/2}. \qquad (5.37)$$

The admixture of the $q_x^{(3)}$ wave is small for small $q_{\parallel}$ and it seems that its main role is in enabling the boundary conditions at $x = 0$ and $x = -L$ to be satisfied.

We further illustrate these results for the case of $q_z = 0$ by means of figures 5.10 and 5.11, which both show the frequency versus L dependence for Fe films. For small L the frequencies of the standing bulk modes become large, as predicted from (5.37), and will eventually exceed the frequency of the surface mode. This leads to the behaviour shown in figure 5.10. Away from a 'crossover' region there is little interaction between the surface and bulk branches, but near a 'crossover' there is appreciable mixing of the modes and the occurrence of mode repulsion as shown by the more detailed plot in figure 5.11. For large film thickness the number of standing bulk magnons intersecting the surface mode branch becomes very large. Hence in the limit of $L \to \infty$ the surface branch is a virtual

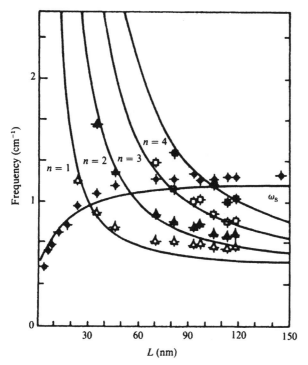

Figure 5.10. Calculated frequencies (full lines) of the dipole-exchange surface mode and the lowest standing bulk modes versus thickness L for Fe films with exchange included, taking $B_0 = 0.1$ T and $q_{\parallel} = 1.82 \times 10^7 \text{ m}^{-1}$. Experimental points obtained from Brillouin scattering are shown ($\bullet$, ω_S; $\triangle$, ω_B for $n = 1$; $\blacktriangle$, ω_B for $n = 2$; $\square$, ω_B for $n = 3$; $\blacksquare$, ω_B for $n = 4$). After Grünberg *et al* (1982).

branch in the quasi-continuum of bulk-magnon states. This gives a damping of the surface modes, i.e. there is a radiative decay of the surface modes into bulk magnons of the same frequency. The term 'leaky' is sometimes used to describe this aspect of the surface-mode behaviour. Wolfram and Dewames (1972) calculate for $q_z = 0$ and $L \to \infty$ that the radiative damping of the surface dipole-exchange mode is proportional to $q_{\parallel}^3$ while its frequency is modified to

$$\omega_S(\mathbf{q}_{\parallel}) = \omega_0 + \tfrac{1}{2}\omega_m + D'q_{\parallel}^2. \tag{5.38}$$

This result provides a correction to (5.8) for the case of $q_{\parallel}L \gg 1$. Here D' is an effective exchange parameter, which is expected to be of the same magnitude as the bulk exchange factor D. The magnitude of the damping term (when $q_z = 0$) is such that the effect is negligible for $q_{\parallel} < 10^7 \text{ m}^{-1}$. However, the damping increases rapidly with $q_{\parallel}$ so that no well-defined surface mode of the magnetostatic type exists for $q_{\parallel} > 10^8 \text{ m}^{-1}$.

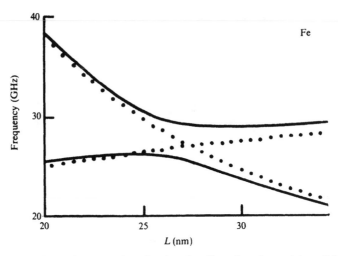

Figure 5.11. Similar to figure 5.10 but showing the effect of mode repulsion of the surface mode and the $n = 1$ bulk mode. In this case $B_0 = 0.02\,\mathrm{T}$ and $q_{\parallel} = 2.42 \times 10^7\,\mathrm{m}^{-1}$. After Grünberg *et al* (1982).

By an extension of the above method, the case of $q_z \neq 0$ for the dipole-exchange modes has been analysed in detail by Camley and Mills (1978) for a semi-infinite ferromagnet and by Camley *et al* (1981) for a ferromagnetic film. In particular, they developed a linear-response formalism to calculate the spin-dependent Green functions, which were then applied to light scattering as will be discussed in section 5.6. They studied the effect of the exchange interactions on the non-reciprocal properties of the surface mode as the propagation angle ϕ (see figure 5.1 for its definition) is varied. For example, in the case of a semi-infinite ferromagnet, the authors found that, as ϕ decreases towards the critical angle ϕ_c defined in (5.15), the damping of the mode increases dramatically until the surface mode is no longer well defined. This occurs over a finite interval of angle ϕ near to ϕ_c, with this interval being broader if $Dq_{\parallel}^2$ is large. This is illustrated in figure 5.12 by some numerical calculations for the spectral function $F_y(\mathbf{q}_{\parallel}, \omega)$, defined as the Fourier transform with respect to $\mathbf{q}_{\parallel}$ and ω of the spin correlation function $\langle S^y(\mathbf{r}, t) S^y(\mathbf{r}', t') \rangle$ evaluated at the surface layer $x = 0$. Plots of $F_y(\mathbf{q}_{\parallel}, \omega)$ against ω for various values of ϕ are shown, taking three cases of the ratio $Dq_{\parallel}^2/\omega_0$ covering a range from weak to strong exchange effects. In figure 5.12(a) with $Dq_{\parallel}^2/\omega_0 = 0.01$ the broadening of the spectrum is very small for most values of ϕ, but it sets in fairly abruptly when ϕ decreases below $\phi_c \approx 22°$. The broadening is stronger in figure 5.12(b) with $Dq_{\parallel}^2/\omega_0 = 0.1$, but nevertheless the peaks are well defined for most ϕ and frequencies are close to those given by the magnetostatic theory. However, in figure 5.12(c) with $Dq_{\parallel}^2/\omega_0 = 1.0$ the radiative damping is severe even for ϕ appreciably larger than ϕ_c.

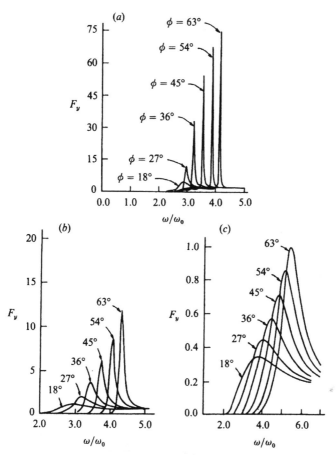

Figure 5.12. The calculated frequency spectrum, $F_y(\mathbf{q}_{||}, \omega)$ versus ω/ω_0, of spin fluctuations in the surface layer of a transversely-magnetized semi-infinite ferromagnet. Curves are shown for different values of the propagation angle ϕ (as marked) for fixed values of $Dq_{||}^2/\omega_0$: (a) 0.01, (b) 0.1 and (c) 1.0. It has been assumed that $\omega_m = 7\omega_0$, which implies $\phi_c \approx 22°$. After Camley and Mills (1978).

Other calculations for the dipole-exchange modes, including their spectral intensities and light-scattering intensities, have been given by Cochran and Dutcher (1988a,b), who used a normal-mode expansion instead of a Green function approach. They included the effects of magnetic damping and metallic conductivity, as well as dipolar, exchange and pinning terms. A similar theory, but ignoring damping, was developed by Rado and Hicken (1988).

An extension of the dipole-exchange theory to thin films where the magnetization direction has an out-of-plane component was made by Stamps and Hillebrands (1991a). They considered a ferromagnetic film with a strong out-of-plane anisotropy that is sufficient to overcome the demagnetizing field and lead to a perpendicular magnetization in zero applied field. In this geometry there is no surface magnetic mode of the Damon–Eshbach (DE) type, just as in the magnetostatic limit. However, Stamps and Hillebrands considered the application of a magnetic field parallel to the film surfaces. Depending on the magnitude of the applied field, the magnetization could have components both parallel and perpendicular to the film surfaces. The authors calculated the frequency of the surface dipole-exchange mode, showing that it goes through a minimum at a critical value of the field related to switching of the magnetization to an in-plane alignment.

Finally we mention an alternative formulation of macroscopic dipole-exchange theory that avoids using the magnetostatic potential. It was developed by Kalinikos and Slavin (1986, 1989), based on the tensorial Green-function approach of Vendik and Chartorizhskii (1970). Basically it is a perturbation method involving matrix elements of the dipole–dipole interactions (in a representation of the eigenstates of the fluctuating magnetization across the film). While it is mathematically more complex, it has the distinct advantage of yielding approximate analytical solutions for the magnetic modes and the linear-response functions for arbitrary directions of the static magnetization, while still giving sufficient accuracy for practical applications, such as SWR and Brillouin scattering (Cottam *et al* 1995).

5.4.2 Microscopic theory

As a consequence of the small wavevectors that are typically involved, the macroscopic theory is generally a good approximation for the dipole-exchange modes. However, a microscopic theory becomes necessary *either* at large wavevectors (for any value of the film thickness L) *or* for ultrathin films with only a few atomic layers. In both cases the preceding continuum approximation would break down. Microscopic descriptions of the dipole-exchange magnons in ferromagnetic films have been used, for example, by Benson and Mills (1969), Erickson and Mills (1991a,b), and Stamps and Hillebrands (1991b). A generalization of the method to describe both linear and nonlinear magnons within a perturbation theory has also been developed (see Costa Filho *et al* 1998, 2000); this will be discussed in section 12.1.

The microscopic theory follows the same approach as in section 4.2 for a Heisenberg ferromagnetic film, except that the dipole–dipole terms are added to the Hamiltonian. When (2.58) and (2.59) are combined we have

$$H = -\tfrac{1}{2}\sum_{i,j} J_{ij}\mathbf{S}_i\cdot\mathbf{S}_j - g\mu_B B_0 \sum_i S_i^z + \tfrac{1}{2}(g\mu_B)^2 \sum_{i,j}\sum_{\alpha,\beta} D_{ij}^{\alpha\beta} S_i^\alpha S_j^\beta \qquad (5.39)$$

where $\alpha,\beta = x, y$ or z, and the dipole–dipole coupling is

$$D_{ij}^{\alpha\beta} = \{|\mathbf{r}_{ij}|^2\delta_{\alpha\beta} - 3r_{ij}^\alpha r_{ij}^\beta\}/|\mathbf{r}_{ij}^5|, \qquad (\mathbf{r}_{ij} = \mathbf{r}_j - \mathbf{r}_i). \qquad (5.40)$$

The coupled equations of motion for the S^+ operators in the layers of the film can be constructed as in sections 4.1 and 4.2. However, the dipolar terms now provide coupling to the S^- operators, as in (2.89) for a bulk medium, and to *all* other layers. The result, which generalizes (4.33)–(4.35) for a film with N layers, is

$$\omega s_n(\mathbf{q}_\parallel) = \sum_{n'=1}^N \{A_{nn'}(\mathbf{q}_\parallel)s_{n'}(\mathbf{q}_\parallel) + [B_{nn'}(\mathbf{q}_\parallel) + B_{n'n}(-\mathbf{q}_\parallel)]\, t_{n'}(-\mathbf{q}_\parallel)\}. \quad (5.41)$$

Here $s_n(\mathbf{q}_\parallel)$ is an amplitude factor for the S^+ operators, and $t_n(\mathbf{q}_\parallel)$ is a similarly defined factor for the S^- operators. When the static magnetization M_0 is parallel to the surface the A and B coefficients are

$$A_{nn'}(\mathbf{q}_\parallel) = C(n)\delta_{n,n'} - SJ\delta_{n+1,n'} - SJ\delta_{n-1,n'} - S(g\mu_B)^2 D_{nn'}^{zz}(\mathbf{q}_\parallel) \quad (5.42)$$

$$B_{nn'}(\mathbf{q}_\parallel) = \tfrac{1}{4}S(g\mu_B)^2[D_{nn'}^{xx}(\mathbf{q}_\parallel) - D_{nn'}^{yy}(\mathbf{q}_\parallel) - 2iD_{nn'}^{xy}(\mathbf{q}_\parallel)] \quad (5.43)$$

with

$$C(n) = g\mu_B B_0 + SJ(2 - \delta_{n,1} - \delta_{n,N}) + 4SJ[1 - \gamma(\mathbf{q}_\parallel)] - (g\mu_B)^2 \sum_{n''=1}^N D_{nn''}^{zz}(0).$$
$$(5.44)$$

The notation for the exchange terms is the same as in chapter 4, except that we have, for simplicity, ignored modifications to the exchange parameter in the surface layers. The quantities $D_{nn'}^{\alpha\beta}(\mathbf{q}_\parallel)$ are Fourier transforms of the dipole terms $D_{ij}^{\alpha\beta}$ with respect to the 2D wave vector $\mathbf{q}_\parallel$; they are slowly convergent, due to the long-range nature of the dipole interactions, if evaluated directly. Instead, they can be conveniently evaluated by a transformation to Bessel functions, yielding a rapidly convergent series (Benson and Mills 1969).

The corresponding equation of motion for the $t_n(\mathbf{q}_\parallel)$ operators is

$$-\omega t_n(-\mathbf{q}_\parallel) = \sum_{n'=1}^N \{A_{nn'}(-\mathbf{q}_\parallel)\, t_{n'}(-\mathbf{q}_\parallel) + [B_{nn'}(\mathbf{q}_\parallel) + B_{n'n}(-\mathbf{q}_\parallel)]\, s_{n'}(\mathbf{q}_\parallel)\}.$$
$$(5.45)$$

It then follows that the condition for (5.41) and (5.45) to have a nontrivial solution can be expressed in a determinantal form as

$$\det \begin{bmatrix} \mathbf{A}(\mathbf{q}_\parallel) - \omega\mathbf{I}_N & 2\tilde{\mathbf{B}}(\mathbf{q}_\parallel) \\ 2\mathbf{B}(\mathbf{q}_\parallel) & \mathbf{A}(\mathbf{q}_\parallel) + \omega\mathbf{I}_N \end{bmatrix} = 0 \quad (5.46)$$

where we have used some symmetry properties of the dipole sums. Here $\mathbf{A}(\mathbf{q}_\parallel)$ and $\mathbf{B}(\mathbf{q}_\parallel)$ are $N \times N$ matrices with elements defined by (5.42) and (5.43) respectively, the tilde denotes a matrix transpose, and $\mathbf{I}_N$ is the $N \times N$ unit matrix. The N positive solutions of (5.46) for ω represent the

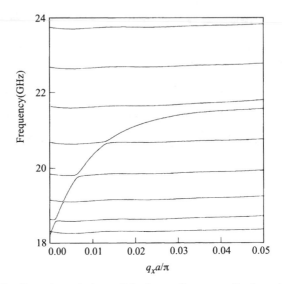

Figure 5.13. The dispersion relations of the lowest-frequency dipole-exchange magnons for a GdCl$_3$ film with $N = 16$. The Voigt configuration is assumed, and the parameter values are given in the text. After Costa Filho *et al* (2000).

frequencies of the dipole-exchange magnons (they are degenerate in magnitude with the negative solutions). They can be obtained numerically, and an example is given in figure 5.13. This refers to the Voigt configuration ($q_z = 0$), and we have taken $N = 16$, $\omega_0 = 10\,\mathrm{GHz}$, $\omega_m = 23\,\mathrm{GHz}$, and $\omega_{\mathrm{ex}} \equiv 6SJ/g\mu_B = 15\,\mathrm{GHz}$. These values for ω_m and ω_{ex} would be appropriate for a weak ferromagnet such as GdCl$_3$. Only the lowest nine branches for small wave vector q_x are shown. It can be seen that there is a mode similar to the purely-magnetostatic Damon–Eshbach (DE) surface mode starting at $q_x = 0$ at the frequency 18.2 GHz. There are hybridizations (mode mixing) until the sixth mode, and the DE modes tends to flatten out at the approximate value 21.5 GHz as q_x increases. These numerical values are close to the limiting frequencies in (5.8) obtained from the magnetostatic theory.

5.4.3 Double-layer systems

The magnetostatic theory for ferromagnetic double layers, as described in section 5.2, can in principle be generalized to include exchange effects following the same steps as in the previous two sections. Consideration needs to be given to exchange effects both within each magnetic layer and across the spacer layer, so the complete analysis is rather complex and additional assumptions or simplifications are often made in the macroscopic theories.

Review accounts, covering theory and experiment, have been given by Grünberg (1989) and Dutcher (1994). As might be expected qualitatively for thin film thickness, the mode spectrum typically consists of two coupled DE-type surface modes that are modified by the exchange, plus some standing (or quantized) bulk modes across the layer thicknesses. In particular, Cochran and Dutcher (1988c) extended their single-film dipole-exchange theory (see section 5.4.1) to the double-layer systems, finding good agreement with Brillouin scattering data as we shall discuss later in section 5.6. Other theories, using slightly different assumptions and model systems, were developed by Vayhinger and Kronmüller (1988), Barnas and Grünberg (1989) and Vohl *et al* (1989). An alignment of the film magnetizations either parallel or antiparallel was assumed in all these calculations. Subsequently Cochran *et al* (1990) allowed for the magnetization directions to take other orientations, as might be induced by a suitable in-plane applied magnetic field or a biquadratic component to the interlayer exchange.

Microscopic formalisms may also be applied to the double-layer systems with exchange. For example, Stamps *et al* (1996) employed a simple model in which an ultrathin ferromagnetic film was coupled by an effective exchange to another ultrathin film that was taken to be either ferromagnetic or antiferromagnetic. The spin-operator equations of motion were written down for the atomic layers in each film (e.g. as an extension of the approach in sections 4.2 and 4.3 for the exchange-dominated case). The dipole–dipole terms, however, were approximated in terms of their appropriate demagnetizing fields, which is equivalent to including just the static (or mean-field) part of the full dipole–dipole interaction. A fuller analysis would involve extending the Hamiltonian formalism described in section 5.4.2 to the double-layer geometry. Some preliminary results, assuming dipolar and exchange coupling within each film but only the long-range dipolar coupling between films, have been obtained by Pereira *et al* (2004).

5.5 Dipole-exchange modes for antiferromagnets

Dipole-exchange theories for antiferromagnetic films have been developed using both the macroscopic and microscopic methods. Because of qualitative similarities with the approaches used in the ferromagnetic case, we describe the results only briefly here and provide references to the original papers. The main differences between the dipole-exchange modes for the two types of magnetic materials are those already encountered in the Heisenberg limit (see sections 4.2 and 4.3), namely the two-sublattice structure in antiferromagnetic films leads to two sets of bulk bands and a strong dependence on structure.

The macroscopic, or continuum, theory was first developed by Stamps and Camley (1987) and applied to antiferromagnetic films with bcc structure and (001) surfaces. It represented an extension of their earlier magnetostatic theory for antiferromagnetic films (see section 5.3) by including in the torque equations of motion for each sublattice the effective fields that represent the dynamical effects of the exchange. These additional terms, along with their inclusion in the boundary conditions, are readily deduced by adapting the method described in section 5.4.1 for ferromagnets in the continuum approximation.

We assume the same geometry as used in section 5.3 when calculating the magnetostatic modes for an antiferromagnetic film, i.e. the sublattice magnetization directions are in the surface plane parallel and antiparallel to the z axis. By analogy with the ferromagnetic case in section 5.4.1, the first step is to modify the torque equation of motion for each sublattice to include the exchange terms. An expression for the dependence of the scalar potential on the coordinate x perpendicular to the film surfaces can be constructed similarly to (5.29) and (5.30) for dipole-exchange modes in ferromagnetic films. It is again found that the inclusion of effective exchange fields leads to the applied field term ω_0 being replaced by a differential operator. However, if on one sublattice the effective exchange field acts in the same direction as the applied field, then on the other sublattice it must act in the opposite direction. Hence the overall result is more complicated than for a ferromagnet, and it was found by Stamps and Camley (1987) that a tenth-order differential equation resulted (instead of the sixth-order equation in section 5.4.1). The characteristic solutions are formed by a superposition (mixing) of modes as in (5.31), except that now five terms are involved. Equation (5.32) is replaced by a fifth-order polynomial in q_x^2, where q_x behaves as a wavenumber (real for bulk modes and imaginary for surface modes) in the direction perpendicular to the film surfaces. Hence the main conclusion is that the dipole-exchange modes of bcc antiferromagnets are mixtures of *five* types of waves. The degree of mixing depends sensitively on the in-plane wavevector $\mathbf{q}_{\parallel}$ (its magnitude and direction) and on the surface pinning conditions.

Numerical examples were given by Stamps and Camley (1987) for MnF_2 films, typically with several hundred atomic layers. The frequencies of the bulk modes in zero applied magnetic field are approximated reasonably well for these thicknesses by assuming they are standing bulk magnons with $q_x = n\pi L$ (where L is the film thickness and n is an integer). Thus (5.22) is replaced by

$$\omega_B^+(\mathbf{q}) = \{\omega_A(2\omega_E + \omega_A) + 2\omega_A\omega_m[(n\pi/L)^2 + q_y^2]/[(n\pi/L)^2 + q_{\parallel}^2]\}^{1/2} \quad (5.47)$$

after noting that $\sin^2\theta = (q_x^2 + q_y^2)/(q_x^2 + q_{\parallel}^2)$. The two surface modes found in the Voigt geometry (i.e. when $q_z = 0$) and for zero applied magnetic field are qualitatively very similar to (5.26) for the magnetostatic case, but there are some frequency shifts. Examples were identified of bulk modes and surface modes showing crossover effects, either with or without repulsion (depending on the symmetries). Crossover with repulsion was discussed in section 5.4.1 for the ferromagnetic case, while crossover without repulsion can occur in antiferromagnets if the modes have opposite symmetries.

A microscopic theory of dipole-exchange modes in antiferromagnets was developed by Pereira and Cottam (1999). A Hamiltonian approach analogous to that of section 5.3.2 was followed, and the method is convenient for films with less than about one hundred atomic layers. Numerical examples were given for ultrathin films with three different structures, all with (001) surfaces, namely MnF_2 (a body-centred tetragonal antiferromagnet), $RbMnF_3$ (a simple-cubic antiferromagnet), and K_2NiF_4 (a layered antiferromagnet). In the latter two cases each atomic layer has an equal number of sites from each sublattice. In the first case, however, each atomic layer is made up of sites from one sublattice only, in an alternating fashion (see figure 4.12). This leads to a simplification because the specification of a layer index n in the film also keeps track of which sublattice is involved. The Hamiltonian is taken to be the same as (2.84) for a Heisenberg antiferromagnet with anisotropy terms plus extra dipole–dipole interactions as in (5.39) and (5.40) for the ferromagnetic case. Equations analogous to (5.41)–(5.46) can then be written down in the linear spin–wave approximation. In particular, it is found that the frequencies of the dipole-exchange modes are found from a determinantal equation of the same form as (5.46) provided the elements of the $N \times N$ matrix $\mathbf{A}(\mathbf{q}_{\parallel})$ and $\mathbf{B}(\mathbf{q}_{\parallel})$ are appropriately redefined; see Pereira and Cottam (1999) for details.

5.6 Brillouin scattering

Brillouin scattering of light has proved to be a sensitive and successful experimental technique for studying the magnetostatic modes and dipole-exchange modes in layered ferromagnetic media. To a remarkable extent it has provided confirmation of the theoretical results described earlier in this chapter. Some reviews of Brillouin scattering from surface and bulk modes in ferromagnetic films have been given by Grünberg (1985, 1989), Cottam and Lockwood (1986), and Dutcher (1994). Generally the experiments have employed relatively opaque ferromagnetic materials (such as semiconductors and metals) in order to increase the importance of light scattering from the surface modes, since the light then penetrates only near the surface. The high-contrast multipass Fabry–Pérot interferometer, which made these measurements feasible, was mentioned in section 1.4.1.

The first observation of surface magnetic modes by light scattering was in the ferromagnetic semiconductor EuO (Grünberg and Metawe 1977). The experiment was carried out in a backscattering geometry, so that the incident light wavevector $\mathbf{k}_I$ and scattered light wavevector $\mathbf{k}_S$ were antiparallel to one

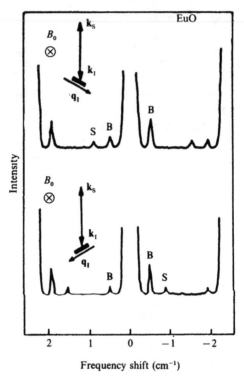

Figure 5.14. Brillouin spectra for EuO for two different orientations of the crystal, showing scattering from the surface (S) and bulk (B) magnons in a thick sample. The applied field is $B_0 = 0.3$ T. After Grünberg and Metawe (1977).

another and at oblique incidence to the surface of the sample. The applied field B_0 and the magnetization M_0 were both in a direction parallel to the surface and at right angles to the in-plane wave vector $\mathbf{q}_{||}$ (i.e. $\phi = 90°$). Some spectra for two different orientations of the sample, which can be regarded as semi-infinite in thickness, are shown in figure 5.14. The bulk-magnon peaks (labelled B) are seen to occur on both sides of the spectrum, but the surface mode peak (labelled S) occurs on one side of the spectrum only. Depending on the scattering geometry, it is observed either on the Stokes side or on the anti-Stokes side but never on both. The surface peak can be switched from one side to the other either by keeping B_0 fixed and reversing the tilt of the sample (as in figure 5.14) or, equivalently, by reversing B_0 for a fixed scattering geometry. The exchange interaction in EuO is relatively weak, and under the experimental conditions the assumptions of the magnetostatic limit are well realized. The frequency shifts associated with the bulk and surface modes agree closely with the theoretical expressions (5.6) and (5.8), respectively, for $q_{||}L \gg 1$. Also the property of the surface mode peak appearing on only one side of the spectrum is a consequence of the non-reciprocal propagation. In going from a Stokes to anti-Stokes scattering contribution, one effectively reverses the wavevector $\mathbf{q}_\parallel$, and it has been shown in section 5.1.1 that the surface magnetostatic mode with wavevector $\mathbf{q}_\parallel$ has no counterpart *at the same surface* with opposite wavevector $-\mathbf{q}_\parallel$.

Similar Brillouin scattering measurements have been made on thick samples of other materials, notably Fe and Ni (see e.g. Sandercock and Wettling 1979). The optical penetration depth in these metals is much shorter than in EuO and also the effect of exchange interactions is more important. These two factors make the surface mode scattering relatively more intense and also give rise to a greater spread of the wavevector component perpendicular to the surface for the bulk magnons participating in the light scattering. Consequently the bulk-magnon peaks are broadened and become asymmetric in lineshape. This 'opacity broadening' is analogous to the behaviour for the phonon case discussed in sections 3.3 and 3.4. It is shown for the case of Fe in figure 5.15 (e.g. see the bulk peaks B in the spectrum for $\phi = 90°$). The effect of varying the propagation angle ϕ is also illustrated in figure 5.15. As ϕ is decreased from $90°$ the surface mode

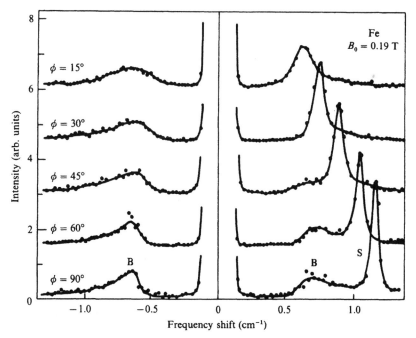

Figure 5.15. Brillouin spectra for Fe with $B_0 = 0.19\,\mathrm{T}$, showing scattering from the surface and bulk magnons in a thick sample for different values of the propagation angle ϕ. The bulk and surface peaks in the spectrum for $\phi = 90°$ are indicated by B and S respectively. After Sandercock and Wettling (1979).

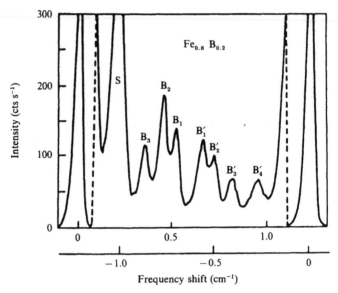

Figure 5.16. Brillouin spectrum for a film of amorphous $Fe_{0.8}B_{0.2}$ of thickness 106 nm, showing a surface mode (S) and several standing bulk magnons (B_1–B_3 and B'_1–B'_3). After Grimsditch *et al* (1979).

peak shifts to lower frequency and eventually merges with the bulk. This is in accordance with the theory of section 5.1.1, and from (5.15) we estimate that $\phi_c \cong 20°$ consistent with the experimental data. A more complete analysis of the light-scattering results has been given by Camley and Mills (1978) using their Green function theory, which is applicable to semi-infinite ferromagnets in the dipole-exchange region. The agreement with experiment is generally good.

Next we consider Brillouin scattering from ferromagnetic thin films. There is now an approximate quantization of the q_x wave-vector component of the bulk magnons (as explained in section 5.4.1 on dipole-exchange theory). This leads to the observation in the Brillouin spectrum of a series of standing bulk mode peaks, which replace the broadened continuum described above for the case of $L \to \infty$. This effect is clearly demonstrated in amorphous magnetic films (see e.g. Grimsditch *et al* 1979), and we show an example in figure 5.16. Here the peak labelled S is the Stokes surface mode, and B_1–B_3 and B'_1–B'_3 label the Stokes and anti-Stokes standing bulk magnons, respectively. Similar results have been obtained in films of ferromagnetic metals (see e.g. Grünberg *et al* 1982, Vernon *et al* 1984, Dutcher 1994). As L is decreased the standing bulk mode peaks are observed to move farther apart and to higher frequency shifts, as predicted using (5.37). Hence, for the series of spectra shown for Fe films in figure 5.17,

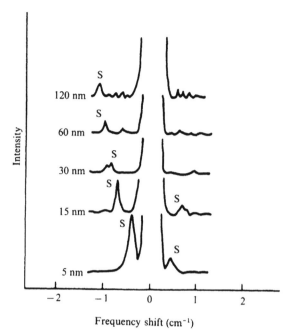

Figure 5.17. Brillouin spectra for Fe films of various thickness L deposited on a sapphire substrate. The applied field is $B_0 = 0.1$ T. After Grünberg *et al* (1982).

there are several standing bulk mode peaks as well as the surface mode peak S when $L = 120$ nm, but when $L = 5$ nm the bulk modes all have frequencies outside the experimental range. The corresponding plots of the mode frequencies versus L have been given in figure 5.10, demonstrating good agreement between theory and experiment. Another important feature of figure 5.17 is that for sufficiently small values of L a surface mode peak is observed on both the Stokes and anti-Stokes side of the spectrum, by contrast with the case of a thick film. This effect can be understood in terms of the non-reciprocal propagation of the surface mode as illustrated schematically in figure 5.4. If the direction of propagation in figure 5.4(a) corresponds to Stokes scattering, then the situation in figure 5.4(b) (where $\mathbf{q}_{\parallel}$ has been reversed) will correspond to anti-Stokes scattering. If the film thickness L is large compared with the optical penetration depth, it follows that no surface mode peak will be observed for Stokes scattering off the top surface because the light will not penetrate to the lower surface where the mode is localized. In this case there will only be anti-Stokes scattering from the surface. However, if L becomes comparable with or smaller than the optical penetration depth, the light will reach both surface regions and a surface peak will occur on both sides of the spectrum. The optical penetration depth in Fe is of order 15–20 nm. A careful analysis of the residual

Stokes/anti-Stokes asymmetry in thin metallic films was made by Camley *et al* (1982). They concluded that it is due to the contribution of the off-diagonal spin–spin correlation functions $\langle S^x S^y \rangle$ to the light scattering intensity, and their theory provided a good fit to experimental data on Fe, Ni and perm-alloy films with thickness of order 10 nm or less.

The mode repulsion effect that was mentioned in section 5.4.1 (see also figure 5.11) between the surface mode and the standing bulk modes has been studied in detail experimentally by Brillouin scattering in Fe films (see e.g. Kabos *et al* 1984) and theoretically by Cottam *et al* (1995). The overall agreement is good.

An unexpected feature of the data in figure 5.17 for Fe films on a sapphire substrate concerns the *intensity* of the surface mode: the anti-Stokes scattering from the surface mode is much more intense in very thin films than in thicker films. On the other hand, this enhancement does not occur for thin films of Fe on a silicon substrate (see Grünberg *et al* 1982). A similar type of behaviour has been observed with other materials and was explained by Cottam (1983) in terms of multiple internal reflections of the light within the film. The reflection coefficients are influenced by the optical parameters of the substrate material, as well as those of the ferro-magnet, and this can lead to an enhancement in appropriate cases, as in figure 5.18. Brillouin scattering data for ultrathin Fe films (with thickness L of the order of a nm or less) are discussed by Dutcher (1994).

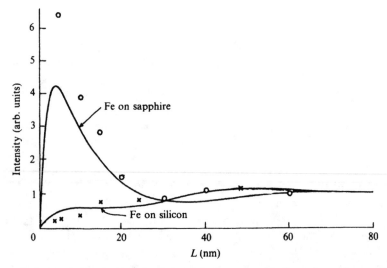

Figure 5.18. The Brillouin intensity for anti-Stokes scattering from the surface mode in Fe films for two choices of substrate, sapphire and Si. The experimental points (O, Fe on sapphire; ×, Fe on Si) are due to P. Grünberg and coworkers. The lines are from the theory by Cottam (1983). After Cottam and Lockwood (1986).

Finally we mention Brillouin light scattering measurements on double-layer magnetic systems, where numerous references were given in section 5.4.3. Some of the earlier data obtained by Grünberg *et al* (1982) for Fe films separated by a quartz spacer layer are shown in figure 5.6, where there is fair agreement with the predicted frequencies from magnetostatic theory (section 5.2). The discrepancies that become apparent at small spacer-layer thickness are probably due to the exchange-coupling effects ignored in the simple theory; see section 5.4.3 and references quoted there, particularly Grünberg (1989). The exchange coupling across thin spacers between ferromagnetic transition metals has attracted much interest due to an enhancement in the magnetoresistance (the so-called 'giant' magneto-resistance effect discussed in section 4.4). Reviews of this topic are to be found in the book edited by Heinrich and Bland (1994). As already noted, Brillouin scattering has been a useful technique in studying this effect, e.g. see the work on Fe/Cr/Fe structures by Demokritov *et al* (1991).

5.7 Other experimental studies

Although Brillouin scattering has provided a vast amount of experimental data concerning the magnetostatic modes and dipole-exchange modes in ferromagnets, other experimental techniques have also been applied to study the surface modes in ferromagnets and (more recently) in anti-ferromagnets.

The surface magnetostatic modes in ferromagnets were first investigated by direct microwave excitation, and we briefly describe some early experiments by Brundle and Freedman (1968) on a rectangular slab of yttrium iron garnet (YIG). The sample dimensions and the experimental arrangement are indicated in figure 5.19. The microwave source is a coaxial line carrying a r.f. current at one end of the slab ($y = 0$), and the detector is a similar line at the other end ($y = d$). One type of experiment involved inducing a surface wave by a short pulse in the source wire. The surface wave propagates in the y direction, and the delay time for arrival at the detector is measured. The delay is simply d/v_g, where v_g is the group velocity of the surface wave. The group velocity can be calculated from the magnetostatic dispersion relation (5.7) using $v_g = \partial \omega_S / \partial q_{\parallel}$, and the result is easily shown to be

$$v_g = (L\omega_m^2 / 4\omega_S) \exp(-2q_{\parallel}L). \tag{5.48}$$

Brundle and Freedman employed a fixed microwave frequency, but varied the applied field B_0, and they compared the delay times with those calculated from (5.48). The agreement was not particularly good, possibly because of the effect of finite slab dimensions in the y and z directions. Better results were obtained in a second type of experiment in which a continuous r.f.

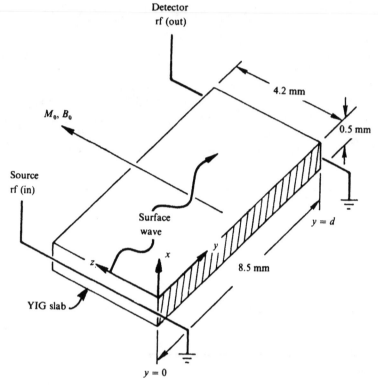

Figure 5.19. The geometry for the microwave experiments of Brundle and Freedman (1968) on a YIG slab. After Wolfram and Dewames (1972).

current was maintained in the source wire. The surface wave is assumed to travel on the top surface of the slab to the end $y = d$ and then to return to $y = 0$ along the lower surface. When the total path length $2d$ (neglecting the slab thickness L compared with d) is an integral multiple of the wave-length, there is constructive interference, which can be observed as a series of geometrical resonances as B_0 is varied. This enabled ω_S to be deduced at a series of discrete $q_{||}$ values, namely $q_{||}^{(n)} = n\pi/d$ with $n = 1, 2, \ldots$. The resulting dispersion relation was found to be in reasonable agreement with (5.7).

Subsequently other microwave experiments were made for the time delay and for the surface wave dispersion. Generally these employed much thinner films (e.g. $L \approx 10\,\mu\text{m}$) and led to better agreement with the magneto-static theory. Another aspect that has been studied experimentally concerns the directional effects for propagation of the nonreciprocal surface modes. By appropriate rotation of the applied magnetic field one can essentially 'steer' a magnetic surface wave. References are given in the review article by Wolfram and Dewames (1972). Magnetostatic waves in films have several convenient features (e.g. tunability by varying the external field, low group velocity, comparatively low propagation losses etc.) that make them suitable for the design of compact controllable and non-reciprocal microwave devices, such as delay lines, filters, resonators (see e.g. Adam 1988, Ishak 1988), and for nonlinear applications (see chapter 12).

We have already included discussion and examples of the technique of spin wave resonance (SWR) in section 4.4. However, the same general considerations occur in systems where the dipolar effects play a role. There is a considerable amount of data from SWR and other magnetic resonance techniques for ultrathin Fe films on various types of substrates. Reviews of this work have been given by Dutcher (1994) and Heinrich (1994).

The existence of surface magnetic modes in the antiferromagnet FeF_2 has been established by optical reflectivity measurements, utilizing the non-reciprocity of these modes in an applied magnetic field. The experimental conditions were such that retardation effects were of importance, and so the results will be discussed later in chapter 6 on surface polaritons.

References

Adam J D 1988 *Proc. IEEE* **76** 159
Barnas J and Grünberg P 1989 *J. Mag. Mag. Mat.* **82** 186
Benson H and Mills D L 1969 *Phys. Rev.* **178** 839
Brundle L K and Freedman N J 1968 *Electron. Lett.* **4** 132
Camley R E 1980 *Phys. Rev. Lett.* **45** 283
Camley R E and Maradudin A A 1982 *Solid State Commun.* **41** 585
Camley R E and Mills D L 1978 *Phys. Rev. B* **18** 4821
Camley R E, Rahman T S and Mills D L 1981 *Phys. Rev. B* **23** 1226
Camley R E, Grünberg P and Mayr C M 1982 *Phys. Rev. B* **26** 2609
Cochran J F and Dutcher J R 1988a *J. Appl. Phys.* **63** 3814
Cochran J F and Dutcher J R 1988b *J. Mag. Mag. Mat.* **73** 299
Cochran J F and Dutcher J R 1988c *J. Appl. Phys.* **64** 6092
Cochran J F, Rudd J, Muir W B, Heinrich B and Celinski Z 1990 *Phys. Rev. B* **42** 508
Costa Filho R N, Cottam M G and Farias G A 1998 *Solid State Commun.* **108** 439
Costa Filho R N, Cottam M G and Farias G A 2000 *Phys. Rev. B* **62** 6545
Cottam M G 1979 *J. Phys. C* **12** 1709
Cottam M G 1983 *J. Phys. C* **16** 1573
Cottam M G and Lockwood D J 1986 *Light Scattering in Magnetic Solids* (New York: Wiley)
Cottam M G and Slavin A N 1994 in *Linear and Nonlinear Spin Waves in Magnetic Films and Superlattices* ed M G Cottam (Singapore: World Scientific) p 1
Cottam M G, Rojdestvenski I V and Slavin A N 1995 in *High Frequency Processes in Magnetic Materials* ed G Srinivasan and A N Slavin (Singapore: World Scientific) p 394
Damon R W and Eshbach J R 1961 *J. Phys. Chem. Solids* **19** 308

Demokritov S, Wolf J A and Grünberg P 1991 *Europhys. Lett.* **15** 881

Dutcher J R 1994 in *Linear and Nonlinear Spin Waves in Magnetic Films and Superlattices* ed M G Cottam (Singapore: World Scientific) p 287

Erickson R P and Mills D L 1991a *Phys. Rev. B* **43** 10715

Erickson R P and Mills D L 1991b *Phys. Rev. B* **44** 11825

Grimsditch M, Malozemoff A P and Brunsch A 1979 *Phys. Rev. Lett.* **43** 711

Grünberg P 1980 *J. Appl. Phys.* **51** 4338

Grünberg P 1981 *J. Appl. Phys.* **52** 6824

Grünberg P 1985 *Prog. Surf. Sci.* **18** 1

Grünberg P 1989 in *Light Scattering in Solids V* ed M Cardona and G Güntherodt (Berlin: Springer) p 303

Grünberg P and Metawe F 1977 *Phys. Rev. Lett.* **39** 1561

Grünberg P, Cottam M G, Vach W, Mayr C M and Camley R E 1982 *J. Appl. Phys.* **53** 2078

Heinrich B 1994 in *Ultrathin Magnetic Structures II* ed B Heinrich and J A C Bland (Berlin: Springer) p 195

Heinrich B and Bland J A C (eds) 1994 *Ultrathin Magnetic Structures I and II* (Berlin: Springer)

Ishak W S 1988 *Proc. IEEE* **76** 171

Kabos P 1995 in *High Frequency Processes in Magnetic Materials* ed G Srinivasan and A N Slavin (Singapore: World Scientific) p 3

Kabos P, Wilber W D, Patton C E and Grünberg P 1984 *Phys. Rev. B* **29** 6396

Kalinikos B A 1994 in *Linear and Nonlinear Spin Waves in Magnetic Films and Superlattices* ed M G Cottam (Singapore: World Scientific) p 89

Kalinikos B A and Slavin A N 1986 *J. Phys. C* **19** 7013

Kalinikos B A and Slavin A N 1989 *Acta Phys. Pol. A* **75** 541

Loudon R and Pincus P 1963 *Phys. Rev.* **132** 673

Luthi B, Mills D L and Camley R E 1983 *Phys. Rev. B* **28** 1475

Marcelli R and Nikitov S A (eds) 1996 *Nonlinear Microwave Signal Processing: Towards a New Range of Devices* (Dordrecht: Kluwer)

Mercereau J and Feynman R P 1956 *Phys. Rev.* **104** 63

Mills D L 1984 in *Surface Excitations* ed V M Agranovich and R Loudon (Amsterdam: North-Holland) p 379

Mills D L 1994 in *Ultrathin Magnetic Structures II* ed J A C Bland and B Heinrich (Berlin: Springer) p 91

Pereira J M and Cottam M G 1999 *J. Appl. Phys.* **85** 4949

Pereira J M, Costa Filho R N and Cottam M G 2004 *J. Mag. Mag. Mat.* **272** 1235

Rado G T and Hicken R J 1988 *J. Appl. Phys.* **63** 3885

Sandercock J R and Wettling W 1979 *J. Appl. Phys.* **50** 7784

Stamps R L and Camley R E 1984 *J. Appl. Phys.* **56** 3497

Stamps R L and Camley R E 1987 *Phys. Rev. B* **35** 1919

Stamps R L and Hillebrands B 1991a *Phys. Rev. B* **43** 3532

Stamps R L and Hillebrands B 1991b *Phys. Rev. B* **44** 12417

Stamps R L, Camley R E and Hicken R J 1996 *Phys. Rev. B* **54** 4159

Tarasenko V V and Kharitonov V D 1971 *Zh. Eksp Teor. Fiz.* **60** 2321 [1971 *Sov. Phys.–JETP* **33** 1246]

Tilley D R 1986 in *Electromagnetic Surface Excitations* ed R F Wallis and G I Stegeman (Heidelberg: Springer) p 30

Vayhinger K and Kronmüller H 1988 *J. Mag. Mag. Mat.* **72** 307

Vendik O G and Chartorizhskii D N 1970 *Fiz. Tverd. Tela* **12** 1538 [1970 *Sov. Phys.–Solid State* **12** 2357]

Vernon S P, Lindsay S M and Stearns M B 1984 *Phys. Rev. B* **29** 4439

Vohl M, Barnas J and Grünberg P 1989 *Phys. Rev. B* **39** 12003

Walker L R 1957 *Phys. Rev.* **105** 390

Wolfram T and Dewames R E 1972 *Prog. Surf. Sci.* **2** 233

Chapter 6

Surface polaritons

We now turn to the surface and film modes that are related to the bulk ordinary and magnetic polaritons discussed in sections 2.5 and 2.6 respectively. In each case, we first give the theory, for single interfaces and then generalized to two interfaces, followed by a discussion of experimental results. A number of these modes, particularly the surface plasmon–polaritons, have been used in applications in spectroscopy and as sensors, and some account of these applications is included.

6.1 Single interface modes

6.1.1 Basic properties and electrostatic approximation

It has been known for many years that an electromagnetic wave can propagate along an interface between two media of which at least one is dispersive. In fact the calculation to be presented in this section for a mode at the interface between two isotropic media appears as a problem in Landau and Lifshitz (1971). Interest increased greatly from the late 1960s, when attenuated total reflection (ATR) was established as a viable technique for experimental study of surface polaritons. Detailed reviews of that work are given by Otto (1974, 1976) as well as in several articles in Burstein and DeMartini (1974) and in Agranovich and Mills (1982).

We consider the plane interface between two semi-infinite dielectric media and, as shown in figure 6.1, we take the z axis normal to the interface and the x axis as the direction of propagation of the mode. Explicit calculation shows that there is no surface mode with the $\mathbf{E}$ field in the y direction, i.e. s-polarization, so we take $\mathbf{E}$ in the xz plane (p-polarization):

$$\mathbf{E} = (E_{1x}, 0, E_{1z}) \exp(iq_x x - i\omega t) \exp(iq_{1z} z) \qquad (6.1)$$

in medium 1, with a similar form in medium 2. We take a single frequency ω, and since boundary conditions will be applied on the whole plane $z = 0$ the

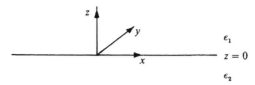

Figure 6.1. Notation for surface-polariton calculation. Medium 1 occupies the half-space $z > 0$ and medium 2 occupies $z < 0$.

wavevector component q_x must be the same in both media. Clearly q_x is the propagation vector, and one of our aims is to find the dispersion equation $\omega(q_x)$.

For the mode to be localized, $\mathbf{E}$ must decrease in magnitude with distance from the interface. Thus $\mathrm{Im}(q_{1z}) > 0$ and $\mathrm{Im}(q_{2z}) < 0$. In order for (6.1) to satisfy Maxwell's equations in both media, the wavevectors must satisfy

$$q_x^2 + q_{iz}^2 = \varepsilon_i \omega^2 / c^2, \qquad i = 1, 2. \tag{6.2}$$

Thus, when ε_1 and ε_2 are real, the localization requirement is $q_x^2 > \varepsilon_i \omega^2 / c^2$. The equation $\nabla \cdot \mathbf{D} = 0$ gives the ratios of the field amplitudes in (6.1):

$$q_x E_{ix} + q_{iz} E_{iz} = 0, \qquad i = 1, 2. \tag{6.3}$$

The amplitudes in the two media are related by continuity of the tangential component of $\mathbf{E}$ and the normal component of $\mathbf{D}$:

$$E_{1x} = E_{2x} \tag{6.4}$$

$$\varepsilon_1 E_{1z} = \varepsilon_2 E_{2z}. \tag{6.5}$$

Equations (6.3)–(6.5) are four homogeneous equations in the four field amplitudes, and the solvability condition is the required dispersion relation. Equations (6.3) and (6.4) give E_{iz} in terms of E_{1x}, and then substitution in (6.5) gives

$$q_{1z}/\varepsilon_1 = q_{2z}/\varepsilon_2. \tag{6.6}$$

Substitution from (6.2) leads to the explicit result for the dispersion equation:

$$q_x^2 = \frac{\omega^2}{c^2} \left(\frac{\varepsilon_1 \varepsilon_2}{\varepsilon_1 + \varepsilon_2} \right). \tag{6.7}$$

As for bulk polaritons in section 2.5, we start by neglecting damping, so that ε_1 and ε_2 are both real. From (6.2) and the localization requirement it then follows that q_{1z} and q_{2z} are both pure imaginary, and we write $q_{1z} = i\kappa_1$ and $q_{2z} = -i\kappa_2$, where the signs incorporate the localization condition. Equation (6.6) now shows that for a surface polariton to exist, ε_1 and ε_2 must have opposite signs:

$$\varepsilon_1 \varepsilon_2 < 0. \tag{6.8}$$

Since the right-hand side of (6.7) must be positive, we have the further condition

$$\varepsilon_1 + \varepsilon_2 < 0. \tag{6.9}$$

Equations (6.8) and (6.9) determine the frequency intervals in which the surface polaritons may occur. In many experimental situations, one of the dielectric functions is frequency-independent and positive, as is the case if medium 1, say, is vacuum. In that case, (6.8) and (6.9) show that the other dielectric constant, ε_2 say, has to be negative, indeed $\varepsilon_2 < -\varepsilon_1$. Medium 2 is then called the *surface-active* medium.

As an example, we consider the interface between vacuum, $\varepsilon_1 = 1$ and a polar medium described by the dielectric function of (2.146). Equation (6.9) then shows that the surface polariton occupies the frequency interval $\omega_T < \omega < \omega_S$ where

$$\varepsilon_2(\omega_S) = -1 \tag{6.10}$$

or explicitly

$$\omega_S^2 = (\varepsilon_\infty \omega_L^2 + \omega_T^2)/(\varepsilon_\infty + 1). \tag{6.11}$$

It can be seen, by reference to figure 2.9 for example, that $\omega_S < \omega_L$ so that the surface polariton occupies part of the frequency interval $\omega_T < \omega < \omega_L$ in which, as seen from figure 2.10, no bulk-polariton mode propagates. This interval is often called the *surface-mode window*.

The dispersion curve calculated from (6.7) is illustrated in figure 6.2; the asymptotic values are

$$q_x \to \omega/c, \qquad |q_{iz}| \to 0 \qquad \text{as } \omega \to \omega_T \tag{6.12}$$

$$q_x \to \infty, \qquad |q_{iz}| \to \infty \qquad \text{as } \omega \to \omega_S. \tag{6.13}$$

These limiting values for $|q_{iz}|$ show that the surface polariton is weakly localized at the low-frequency end, and strongly localized at the high-frequency end. In a similar way to figures 3.3 and 3.4 for surface elastic waves, the bulk continuum regions are shown shaded. At a point (q_x, ω) within one of these regions a bulk mode propagates with that value of q_x and some real value of q_z.

For a second example, we take $\varepsilon_1 = 1$ and $\varepsilon_2(\omega)$ in the simple plasma form of (2.119), again with damping neglected. The surface mode, usually called the *surface plasmon–polariton* (SPP), occupies the interval $0 < \omega < \omega_S$ where now

$$\omega_S^2 = \omega_p^2/2. \tag{6.14}$$

The dispersion curve is illustrated in figure 6.3. Equations (6.12) and (6.13) for the limiting forms still apply on the understanding that ω_T is replaced by zero.

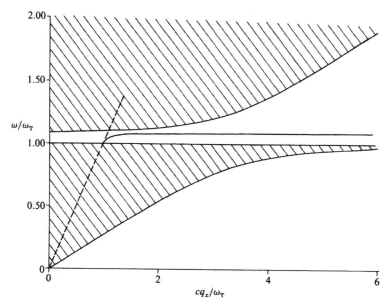

Figure 6.2. Surface polariton dispersion curve for a GaAs/vacuum interface. Numerical values for GaAs are quoted in figure 2.9 with $\omega_S/\omega_T = 1.081$. The bulk continuum is shown shaded.

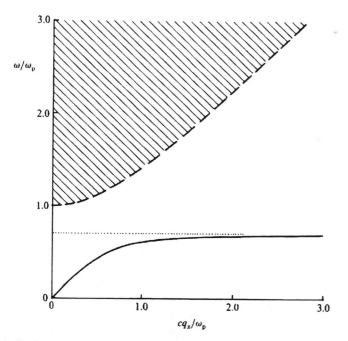

Figure 6.3. Surface plasmon–polariton dispersion curve. The bulk continuum is shaded.

It will be recalled that the plasma form of the dielectric function applies both to a metal, for which ω_p is typically in the near ultraviolet, and to a doped semiconductor, in which case ω_p is typically in the far infrared with a value that depends on the carrier concentration and therefore on the doping. For a polar semiconductor, $\varepsilon_2(\omega)$ may have one or more poles resulting from optic phonons in addition to the frequency dependence due to the plasma response. We mention two special cases that may be studied as exercises. The first concerns undoped $Ga_{1-x}Al_xAs$. This random-alloy semiconductor has two optically active phonons, one near the GaAs frequency and the other near the AlAs frequency, so we may take

$$\varepsilon_2(\omega) = \varepsilon_\infty\left(1 + \frac{S_1\omega_1^2}{\omega_1^2 - \omega^2 - i\omega\Gamma_1} + \frac{S_2\omega_2^2}{\omega_2^2 - \omega^2 - i\omega\Gamma_2}\right) \qquad (6.15)$$

where the parameters are known from a fit to infrared reflectivity data (Kim and Spitzer 1979). It may be verified in this case that the surface-polariton dispersion curve has two branches. In the second case, doped InSb is described by the dielectric function of (2.150), with terms arising from both plasma and phonon interactions. It is again straightforward to obtain the surface polariton dispersion relation.

Finally it will be recalled that, in the presence of a magnetic field, $\varepsilon_2(\omega)$ becomes a gyrotropic tensor of the form of (2.133). The surface-mode spectrum can be found by an extension of the method used here. It is discussed by Palik *et al* (1976) for the tensor form of $\varepsilon_2(\omega)$ including a single phonon resonance as given in (2.149) and (2.150).

Equations (6.3)–(6.5) may be used to find the relative amplitudes and phases of the electric-field components in the two media, and the magnetic-field components can be derived from Maxwell's equations in the usual way. Furthermore, Poynting's theorem can be applied to find the directions and magnitudes of the power flow in the two media. The details are given by Nkoma *et al* (1974) and some results are illustrated in figure 6.4. It is seen in particular that the net power flow is negative in the surface-active (lower) medium, although the flow in the upper medium is larger in magnitude so that the total flow is positive. One point that is not clear from that figure is that the field intensity is strongly concentrated near the interface. It is seen from (6.1) that since q_{1z} and q_{2z} are pure imaginary (when damping is neglected) the fields die away exponentially with distance from the surface. The field components can be found from (6.1)–(6.5), and it is seen that for the Ag/air interface, for example, for which ε_2 is negative and large in magnitude, the field enhancement in the SPP is very strong. This has significant implications for applications. For example, the SPP can be used to investigate molecules adsorbed at the interface, as will be seen in sections 6.3 and 6.4. The large field at the interface can also be exploited in Raman spectroscopy (section 6.5) and in nonlinear optical effects (section 11.3).

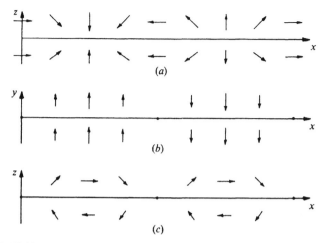

Figure 6.4. Field amplitudes and Poynting vector for an interface with $\varepsilon_1 > 0$ (upper medium) and $\varepsilon_2 < 0$ (lower medium). (a) Electric field pattern in the xz plane. The field magnitudes are oscillatory in the x direction but decrease exponentially away from the boundary. (b) An analogous diagram of the magnetic field pattern in the xy plane. (c) Poynting vector pattern.

Before proceeding further it is worthwhile to mention a special limiting case. The frequency ω_S for which $q_x \to \infty$ in dispersion curves like figures 6.2 and 6.3 is of particular interest. For example, in electron energy-loss experiments on metals the momentum transfer is relatively large, and a loss peak is observed at ω_S. The condition $q_x \gg \omega/c$ means that the wavelength $2\pi/q_x$ of the wave along the surface is much smaller than the free-space wavelength $2\pi c/\omega$. Retardation effects, arising from the nonzero propagation time of an electromagnetic signal, are therefore unimportant, and the asymptotic wave with $\omega \approx \omega_S$ and $q_x \gg \omega_S/c$ is called an *electrostatic surface wave*. It is seen from (6.7) that its (fixed) frequency is given by

$$\varepsilon_1 + \varepsilon_2 = 0 \qquad (6.16)$$

with explicit expressions for the vacuum-bounded polar dielectric and simple plasma given by (6.11) and (6.14) respectively. In view of the importance of the result, we shall now derive (6.16) again by means of an explicit electrostatic calculation, which is the analogue of that presented for the magnetostatic surface wave in section 5.1.

In the electrostatic approximation, media 1 and 2 are described by potentials V_1 and V_2 satisfying

$$\nabla^2 V_i = 0, \qquad i = 1, 2. \qquad (6.17)$$

A plane wave localized at the interface and travelling in the x direction is described by

$$V_1 = V_{10} \exp(-\kappa_1 z) \exp[i(q_x x - \omega t)] \tag{6.18}$$

with a similar expression for V_2. Substitution into (6.17), which involves only spatial derivatives, gives $\kappa_i^2 = q_x^2$ for $i = 1, 2$. Thus the potentials are

$$V_i = V_{i0} \exp(\mp q_x z) \exp[i(q_x x - i\omega t)] \tag{6.19}$$

with the upper (negative) sign for medium 1 and the lower sign for medium 2. The corresponding fields are

$$\mathbf{E}_i = (-iq_x, 0, \pm q_x) V_i \tag{6.20}$$

with the upper sign again corresponding to medium 1. Equations (6.19) and (6.20) have the property, which they share with magnetostatic waves in the Voigt configuration, that the decay constant normal to the interface is equal to the wavevector along the interface. This follows from the fact that the potentials satisfy Laplace's equation (6.17). It is seen further from (6.20) that the field components E_x and E_z in either medium are equal in magnitude but 90° out of phase. This is a special case of the general surface-polariton field patterns illustrated in figure 6.4.

In order to derive the frequency of the surface wave, we apply standard boundary conditions to the fields given by (6.20). Continuity of E_x gives

$$V_{10} = V_{20} \tag{6.21}$$

and then continuity of D_z gives

$$\varepsilon_1 q_x V_{10} = -\varepsilon_2 q_x V_{20}. \tag{6.22}$$

Taken together, these two equations lead to the previously derived (6.16).

6.1.2   Anisotropic media

An important extension of the results so far derived is to the case when one of the media is anisotropic. A brief history with references is given by Mirlin (1982); reviews are given by Borstel *et al* (1974) and by Borstel and Falge (1977, 1978). Here we simply quote one of the main results which will be useful later on.

We continue to use the notation of figure 6.1 so that (x, y, z) are 'surface-related' axes and we consider a mode propagating in the x direction, $\mathbf{q}_{\parallel} = (q_x, 0)$. We suppose that medium 1 is isotropic and medium 2 anisotropic, with optical properties characterized by a dielectric tensor $\bar{\varepsilon}_2(\omega)$. This tensor has principal axes, say (x', y', z'), and considerable algebraic complexities arise because in general these axes are in an arbitrary orientation with respect to (x, y, z). However, it is sufficient for a simple illustration to consider the special case when the y axis is a principal axis of $\bar{\varepsilon}_2(\omega)$, $y = y'$

say. In this case the boundary conditions do not mix s- and p-polarizations and we choose to deal with the latter, taking $\mathbf{E} = (E_x, 0, E_z)$. With these simplifications, it can be shown that the dispersion relation is

$$q_x^2 = \varepsilon_1 \frac{\omega^2}{c^2} \frac{\varepsilon'_x \varepsilon'_z - \varepsilon_1 \varepsilon_{zz}}{\varepsilon'_x \varepsilon'_z - \varepsilon_1^2}. \tag{6.23}$$

Here ε'_x and ε'_z are the principal values of $\bar{\varepsilon}_2$ in the $x'z'$ plane, while ε_{zz} is the component of $\bar{\varepsilon}_2$ in the (x, y, z) axes. It is a relatively straightforward exercise to derive (6.23) for the special case when the (x, y, z) and (x', y', z') axes coincide.

6.1.3 Two-dimensional electron gas and charge-sheet polaritons

In some circumstances it is possible to have an electron gas in the form of a charge sheet that is sufficiently thin that it can be treated as a 2D electron gas. Two important examples are a charge sheet at the surface of liquid helium and a charge sheet at a semiconductor surface or interface. Liquid helium has a dielectric constant of about 1.06, which is sufficient for electrons to be trapped in the weak image potential at the surface. The areal density of the 2D electron gas trapped in this way can be varied by a static electric field applied normal to the surface. A review of the topic is given by Cole (1974). In semiconductors, a static electric field applied normal to a surface, as in a field-effect transistor, can attract a charge sheet of electrons or holes to the surface, and again the areal density can be varied by changing the field. In addition, electrons can be trapped at a heterojunction between semiconductors, e.g. between GaAs and $Ga_{1-x}Al_xAs$.

We mentioned at the beginning of section 2.4.2 that calculation of the response of the 3D electron gas is fundamentally a problem in quantum-mechanical many-body theory, although as we saw the simpler hydrodynamic method yields useful results. It was first pointed out by Stern (1967) that the quantum-mechanical analysis applies, with necessary modifications, to the 2D system. Later, Fetter (1973) showed how the hydrodynamic approach of section 2.4.2 can be applied to the 2D electron gas. Subsequently Karsono and Tilley (1977) extended the analysis to the case where the embedding media are different on either side of the layer and they discussed a line of mobile charge, i.e. the 1D electron gas.

Both Stern (1967) and Fetter (1973) also discussed electromagnetically-coupled modes propagating parallel to the charge sheet and localized on the sheet, although Stern exclusively and Fetter mainly use the electrostatic approximation. Here we derive the dispersion relation, including retardation, by extending the analysis of section 6.1.1 to include a charge sheet whose dynamics are described by the hydrodynamic equations. We use the axes of figure 6.1 and the notation of section 6.1.1, but we now suppose that a 2D charge sheet of particle density $\nu\,m^{-2}$ is localized at the interface $z = 0$. Equations (6.1)–(6.4) still apply, but (6.5) must be replaced, as it is equivalent to the continuity of the field component H_y. There is now a sheet of mobile charge at the interface so the correct form is

$$H_{2y} - H_{1y} = j_x \qquad (6.24)$$

where j_x is the current density in the charge sheet. As in the 3D plasma, j_x is driven by the alternating electric field at the interface, E_{1x}, which from (6.4) is equal to E_{2x}. Ignoring collisions in the electron gas, we can write from a 2D equivalent of (2.128)

$$j_x = i\nu e^2 E_{1x}/m\omega. \qquad (6.25)$$

Substituting this into (6.24) leads to the relation that replaces (6.6), namely

$$\frac{\varepsilon_1}{\kappa_1} + \frac{\varepsilon_2}{\kappa_2} = \frac{\Omega_S c}{\omega^2} \qquad (6.26)$$

where κ_1 and κ_2 are still defined by $q_{1z} = i\kappa_1$ and $q_{2z} = -i\kappa_2$. The frequency Ω_S is

$$\Omega_S = \nu e^2/\varepsilon_0 mc \qquad (6.27)$$

where m is the effective mass of the electrons. In the electrostatic limit, κ_1 and κ_2 are both replaced by q_x. Equation (6.26) then simplifies to

$$\omega = [\Omega_S c q_x/(\varepsilon_1 + \varepsilon_2)]^{1/2}. \qquad (6.28)$$

The general dispersion relation (6.26) was first derived by Nakayama (1974) by the method described here. Equation (6.28) was derived earlier by Stern (1967), who as mentioned made use of many-body theory to describe the electronic response.

An important property of both (6.26) and (6.28) is that a localized mode can propagate even when ε_1 and ε_2 are both positive. This is in striking contrast to the ordinary polariton, for which as we saw ε_1 and ε_2 must have opposite signs. The dispersion curve for the case of $\varepsilon_1 = \varepsilon_2$ is shown in figure 6.5. An analogous derivation can be carried out when there is a polarizable, rather than conducting, layer at the interface; (6.25) is replaced by $P_x = \varepsilon_0 \chi(\omega) E_{1x}$, where $\chi(\omega)$ is a frequency-dependent susceptibility. The resulting dispersion relation has the same form as (6.26) but with the right-hand side replaced by $-\chi(\omega)$.

The first experimental observation of a charge-sheet plasmon mode was by means of grating coupling (Batke and Heitmann 1984); an account of this technique will be given in section 6.4.

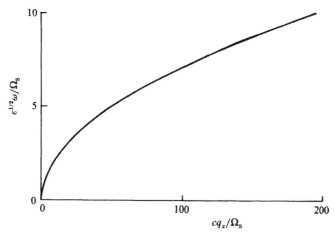

Figure 6.5. Charge sheet dispersion curve corresponding to (6.26) for $\varepsilon_1 = \varepsilon_2 \equiv \varepsilon$ (>0).

6.1.4 Surface exciton polaritons

In section 2.5.3 it was seen that the centre-of-mass momentum of the exciton leads to significant spatial dispersion in the dielectric function. We now discuss qualitatively how the calculation of surface-polariton properties, as given in section 6.1.1, is modified by spatial dispersion.

In the absence of spatial dispersion, the surface polariton is found in some frequency range determined by (6.8) and (6.9) and lying within the bulk-polariton stop band. As shown in figure 2.14, however, the bulk exciton–polariton has no stop band; the branch labelled 2 there has the large-q behaviour $\omega \approx Dq^2$ because of the spatial dispersion. The situation is analogous to that described in section 5.4 for dipole-exchange modes in ferromagnetic films. In the absence of exchange effects, a pure magnetostatic surface mode (the DE mode) propagates with frequencies greater than those of the bulk modes, as seen in section 5.1.1. The inclusion of exchange, which is required for slightly larger q values, means that the DE branch in the dispersion relation 'crosses' the discrete bulk modes which are now at higher frequencies (e.g. see figure 5.10). Hence the surface magnetostatic mode becomes 'leaky' in the sense that it radiates into the bulk mode with which it is degenerate in frequency. Similarly in the present case there is no uncoupled surface exciton–polariton; instead we have a leaky surface mode.

A treatment of the surface exciton–polariton analogous to that of section 6.1.1 was given for the first time by Maradudin and Mills (1973). The medium is described by the isotropic dielectric function of (2.158). It is assumed that the mode is p-polarized, so that in the non-surface-active medium the electric field is given by (6.1) with wavevector component q_{1z}

determined by (6.2). In the excitonic medium, however, q_{2z} is found as a function of q_x and ω from (2.159). In the absence of damping, (2.159) has one real root for q as a function of ω, namely the one called branch 2 in figure 2.15, for $\omega < \omega_L$, and three real roots for $\omega > \omega_L$. There is no frequency range in which only imaginary roots are found. Likewise, when (2.159) is regarded as giving q_{2z} as a function of q_x and ω, the root corresponding to branch 2 is always real. With damping present, this root is off the real axis, but for a realistic value of the damping it is still predominantly real. The other two roots for q_{2z} are predominantly imaginary for small ω and real for larger ω.

As in the calculation of reflection and transmission of p-polarized light, mentioned in section 2.5.3, the electric field in the excitonic medium has three components, so that the analogue of (6.1) is

$$\mathbf{E} = \sum_{\alpha=1}^{3} (E_{2x}^{\alpha}, 0, E_{2z}^{\alpha}) \exp(iq_x x - i\omega t) \exp(iq_{2z}^{\alpha} z). \tag{6.29}$$

Here α labels the three solutions of (2.159) for q_{2z}. The amplitude E_{2z}^{α} is related to E_{2x}^{α} by $\nabla \cdot \mathbf{D} = 0$ for the transverse modes, and by $\nabla \times \mathbf{E} = 0$ for the longitudinal mode. In all, then, four field amplitudes have been defined, namely E_{1x} and E_{2x}^{α} for $\alpha = 1, 2, 3$. Four linear homogeneous equations in these amplitudes are found from the two electromagnetic boundary conditions and two additional boundary conditions (ABCs) as discussed in section 2.5.3. The condition for these equations to have a solution gives the dispersion relation $q_x(\omega)$. As anticipated, it is found that even in the absence of damping q_x is complex, so that the surface wave is attenuated due to energy leakage into the bulk branch 2. With the damping parameter in (2.158) included, the attenuation of the surface mode increases. As was seen in section 2.5.1, the definition of a dispersion curve becomes ambiguous in the presence of attenuation, since the dispersion relation is then between two complex variables ω and q_x. Nevertheless, for small attenuation a dispersion curve can be a useful guide to the eye, and an example of a calculated curve is shown in figure 6.6. The theory outlined here assumed an isotropic medium. However, many excitonic media are anisotropic, and the calculation must then be appropriately generalized.

Experimental work on surface exciton–polaritons began with studies of the ZnO surface by Lagois and Fischer (1976) and by DeMartini *et al* (1977); a detailed account of this and later work on ZnO is given by Lagois and Fischer (1982). The former group used the method of attenuated total reflection (ATR) and an account of their results will be given in section 6.3.3. The latter group used the nonlinear-optics technique of frequency-doubling.

The phrase 'surface exciton polariton' has been used in a somewhat different sense by Yang *et al* (1990a) to describe a long-range mode supported

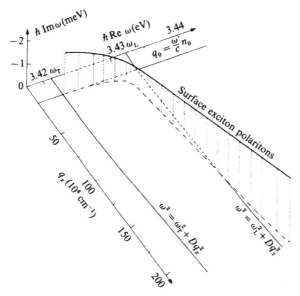

Figure 6.6. Dispersion curve of a surface exciton–polariton for a complex frequency ω and real wavenumber q_x. The calculation is for a ZnO exciton using the additional boundary condition $\mathbf{P} = 0$. After Lagois and Fischer (1978).

by a thin metal film with a predominantly imaginary dielectric function. Their results will be discussed in section 6.2.1.

6.1.6 Surface magnon–polaritons

In section 2.6.1 the gyromagnetic susceptibilities of ferromagnets and anti-ferromagnets were derived; in many cases the susceptibility takes the form of (2.165) where the tensor components χ_a and χ_b have poles at the relevant resonance frequency. In section 2.6.2 we discussed the corresponding bulk polariton dispersion curves, which may, as in figures 2.16 and 2.17, have a reststrahl-like stop band. It is here that a surface polariton mode can appear, considering now the interface between the magnetic and a non-magnetic medium. We will derive the main properties of these modes, starting with the dispersion relation, as first derived by Hartstein *et al* (1973). Reviews are given by Sarmento and Tilley (1982) and Abraha and Tilley (1996), the main emphasis in the latter being on antiferromagnets.

We give explicit results for the Voigt geometry, where the static field $\mathbf{B}_0$ and the magnetization are in the plane of the surface and propagation is perpendicular to $\mathbf{B}_0$. As in section 5.1.1, we take the z axis along $\mathbf{B}_0$ and the x axis as the normal to the surface, so that here the propagation vector $\mathbf{q}_\parallel$ is in the $\pm y$ direction. The magnetic medium is in the half-space $x < 0$.

Following Hartstein *et al*, we work in terms of a permeability tensor

$$\bar{\mu} = \begin{pmatrix} \mu_{xx} & \mu_{xy} & 0 \\ -\mu_{xy} & \mu_{xx} & 0 \\ 0 & 0 & \mu_{zz} \end{pmatrix} \tag{6.30}$$

where

$$\mu_{xx} = \mu(1 + \chi_a), \qquad \mu_{xy} = i\mu\chi_b, \qquad \mu_{zz} = \mu. \tag{6.31}$$

This is based on the susceptibility (2.165) with the addition of μ as the 'background permeability', analogous to the factor ε_∞ first introduced in (2.129) and accounting for contributions of possible higher-frequency resonances.

For the magnetic medium, Maxwell's equations give

$$\nabla^2 \mathbf{h} - \nabla(\nabla \cdot \mathbf{h}) - (\varepsilon/c^2)\bar{\mu}\, \partial^2\mathbf{h}/\partial t^2 = 0 \tag{6.32}$$

and

$$\nabla \cdot (\bar{\mu}\,\mathbf{h}) = 0. \tag{6.33}$$

Some tensor products are implied in (6.32) and in (6.33), while in the vacuum

$$\nabla^2 \mathbf{h} - (1/c^2)\partial^2\mathbf{h}/\partial t^2 = 0. \tag{6.34}$$

In order to find the dispersion relation of the surface mode, we substitute into (6.32), (6.34) and the boundary conditions the forms

$$\mathbf{h} = \mathbf{h}_1 \exp(iq_{\parallel}y - i\omega t)\exp(-\kappa_1 x) \qquad x > 0 \tag{6.35}$$

$$\mathbf{h} = \mathbf{h}_2 \exp(iq_{\parallel}y - i\omega t)\exp(\kappa_2 x) \qquad x < 0. \tag{6.36}$$

Substitution of (6.36) into the z component of (6.32) gives $h_{2z} = 0$, and the boundary condition on tangential $\mathbf{h}$ then gives $h_{1z} = 0$. It follows that the surface polariton is s-polarized with $\mathbf{E}$ in the z direction. (Here we use the convention that s and p denote the direction of the $\mathbf{E}$ field, just as for non-magnetic materials). The x and y components of (6.32), with (6.36), yield

$$\kappa_2^2 - q_{\parallel}^2 + (\varepsilon\omega^2/c^2)\mu_V = 0 \tag{6.37}$$

where ε is the dielectric constant of the medium (taken as a scalar) and μ_V is called the *Voigt permeability*. It is defined by

$$\mu_V = \mu_{xx} + \mu_{xy}^2/\mu_{xx}. \tag{6.38}$$

Equations (6.34) and (6.35) give

$$\kappa_1^2 - q_{\parallel}^2 + \omega^2/c^2 = 0. \tag{6.39}$$

Equation (6.37) is equivalent to the bulk-polariton dispersion equation (2.180) (with imaginary q_x) for this geometry. Equations (6.37) and (6.39) are the analogues of (6.2) for the dielectric problem; they characterize the individual media with no reference to the boundary conditions.

The Voigt permeability is central to this geometry. For the particular case of a ferromagnet, substitution of (2.166) and (2.167) gives

$$\mu_V/\mu = [(\omega_0 + \omega_m)^2 - \omega^2]/(\omega_0^2 + \omega_0\omega_m - \omega^2) \qquad (6.40)$$

from which it is seen that μ_V has a pole at the frequency ω_B of (5.6). As mentioned there, ω_B is the frequency of the bulk magnetostatic mode in the Voigt geometry.

The derivation of the dispersion relation now follows in a standard way. The divergence condition (6.33) gives expressions for the ratio h_x/h_y in the two media. The boundary conditions of continuous h_y and E_z are simultaneously satisfied only if

$$\kappa_2 = -\kappa_1\mu_V + iq_{\parallel}\mu_{xy}/\mu_{xx}. \qquad (6.41)$$

As is seen from (6.31), as long as the resonance modes are undamped μ_{xy} is pure imaginary, and thus (6.41) involves only real quantities.

Equations (6.37), (6.39) and (6.41) are simultaneous equations for the three quantities κ_1, κ_2 and $q_{\parallel}$. Elimination of κ_1 and κ_2 yields an equation, quoted explicitly by Hartstein *et al* (1973), for the dispersion function $q_{\parallel}(\omega)$. The most important property of the dispersion curve can be seen from (6.41). Since $q_{\parallel}$ occurs there linearly, the directions $+q_{\parallel}$ and $-q_{\parallel}$ are not equivalent. This is the property of non-reciprocity that was already noted in the reflectivity curves of figure 2.18 and in the magnetostatic modes discussed in section 5.1.

As mentioned in section 2.6.1, the ferromagnetic resonance frequencies are typically in the microwave region so that retarded modes of the type under discussion are not expected to occur in normal size samples. By contrast, since antiferromagnetic resonance frequencies may be in the infrared region, the corresponding retarded modes can be (and have been) detected. Still, the ferromagnet is the simplest case and although the dispersion curve is only of formal significance we show its form in figure 6.7. The $+q_{\parallel}$ mode is the generalization of the Damon–Eshbach (DE) surface mode to include retardation. For larger $q_{\parallel}$ the curve coincides with the DE one, being asymptotic to the surface frequency ω_S of (5.8), which with μ included becomes

$$\omega_S = [\omega_0 + \mu(\omega_0 + \omega_m)]/(\mu + 1). \qquad (6.42)$$

For small $q_{\parallel}$ the DE curve has the unphysical property that it crosses the vacuum photon line and reaches $q_{\parallel} = 0$ at the frequency ω_B of (5.6). The calculation with retardation corrects this; the mode still terminates at ω_B, but this occurs at a nonzero value q_V given by

$$q_V = (\omega_B/c)[(\omega_0 + \omega_m)/\omega_m]^{1/2}. \qquad (6.43)$$

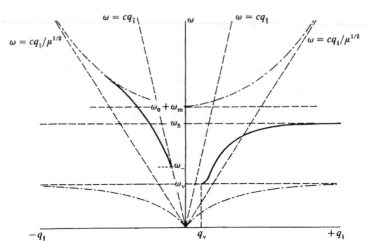

Figure 6.7. Surface polariton dispersion curves (full line) for the $+q_{\parallel}$ and $-q_{\parallel}$ modes on a ferromagnet in the Voigt geometry. Also shown is the bulk dispersion curve (chain line) and $\omega_V = \omega_B$. After Hartstein *et al* (1973).

In addition to the $+q_{\parallel}$ mode, figure 6.7 shows a mode propagating in the $-q_{\parallel}$ direction. This mode only exists for $\mu > 1$; it starts on the vacuum photon line at frequency

$$\omega_- = (\mu - 1)^{-1/2}[\omega_0^2 + \mu\omega_0(\omega_0 + \omega_m)]^{1/2} \qquad (6.44)$$

and merges with the bulk-polariton dispersion curve at a finite frequency.

Hartstein *et al* (1973) give a brief qualitative account of the ferromagnetic surface polariton as wavevector $\mathbf{q}_{\parallel}$ moves away from the Voigt direction. As for the magnetostatic surface mode, the surface polariton exists only for a range of angles ϕ (see figure 5.1) corresponding to $\phi_c < \phi < 180° - \phi_c$, where ϕ_c is given by (5.15).

We now turn to uniaxial antiferromagnets, starting with the simplest case where the c axis is in the surface plane, $\mathbf{B}_0$ is along c, and $\mathbf{q}_{\parallel}$ is transverse to $\mathbf{B}_0$ (the Voigt geometry). The permeability tensor is again given by (6.30) but with the susceptibility elements of (2.173) and (2.174). The surface-polariton dispersion relation is therefore given by (6.37)–(6.39) and (6.41). As remarked in section 2.6.1, the special case $\mathbf{B}_0 = 0$ corresponds to a diagonal permeability. In consequence the magnetic properties become reciprocal and a simple calculation shows that the dispersion equation then takes the form

$$q^2 = \frac{\omega^2}{c^2} \frac{\mu_{xx}(\mu_{xx} - \varepsilon)}{\mu_{xx}^2 - 1}. \qquad (6.45)$$

Calculated dispersion curves for FeF_2 in zero field and with an applied field of 0.3 T are shown subsequently in comparison with experimental data in figure 6.18. It was seen in chapter 2 that the reststrahl regions in anti-ferromagnets are typically very narrow, as may be seen in the reflectivity curves of figure 2.18 for example, and the surface polaritons are confined to these regions. For zero field, figure 6.18(a), the dispersion curves are, as stated, reciprocal, i.e. the same for $+q$ and $-q$. In a nonzero field the reststrahl band is Zeeman split into two, as seen in figure 2.18, and the surface-polariton dispersion becomes markedly non-reciprocal. Further discussion of the dispersion curves is deferred to section 6.3.4.

The Voigt geometry is the easiest case theoretically and the one that has been most studied experimentally. In general, three basic directions are involved: c axis, magnetic field and surface normal. It will be recalled from section 2.6.1 that the form of $\bar{\mu}$ depends on the relative directions of the field and the c axis. Thus the number of special cases that might be discussed is rather large. Abraha and Tilley (1996) give a fairly general account of surface antiferromagnetic polaritons in different geometries.

6.2 Two-interface modes

6.2.1 Nonmagnetic media

We now extend the calculations of section 6.1.1 for nonmagnetic media, i.e. phonon– and plasmon–polaritons, to a geometry with *two* plane interfaces, using the notation of figure 6.8. As before, we take the z axis normal to the interfaces, which are situated at $z = 0$ and $z = -L$, and the x axis is the direction of propagation. The dispersion equation can be derived for general forms of the three dielectric functions but the usual application is to a surface-active film on a substrate, so that ε_2 is taken as frequency-dependent while ε_1 and ε_3 are taken as constants. We focus first on the

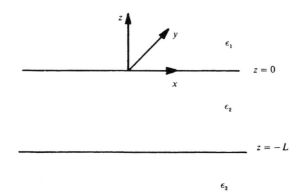

Figure 6.8. Notation for two-interface calculations. A film of thickness L and dielectric function ε_2 is bounded by media with dielectric functions ε_1 and ε_3.

direct extension of the surface polariton analysis introduced in section 6.1.1 so it will be assumed that (neglecting damping) the wavevector q_{2z} in the film is pure imaginary and we take the case of p-polarization ($E_y = 0$) since it is easy to show that no modes of this type occur in s-polarization.

As we discussed in section 3.2.1 for Rayleigh waves, for sufficiently large L independent surface polaritons propagate on each interface. If we imagine L decreasing, the fields of the two surface polaritons propagating for a given q_x start to overlap, and the degeneracy in frequency between them is lifted. Thus we expect to see two modes, essentially a bonding and an antibonding combination. We should mention that, because of non-reciprocity, the behaviour of magnon–polaritons in films is rather different, as will be seen in the next section.

For a semi-infinite medium, q_{2z} has to be imaginary in order for the fields to become small at infinity. This restriction does not apply to a film, so that *guided-wave modes* in which q_{2z} is (predominantly) real can also occur. It will be seen that these occupy a different region of the (q_x, ω) plane from the surface modes and they occur in both s- and p-polarization. The formal derivation, which now follows, holds for any q_{2z} so it applies for both surface-type and guided modes. The calculation is a straightforward extension of that given in section 6.1.1, and some of the equations presented there still apply.

We assume p-polarization and write the **E** fields in the three media as

$$\mathbf{E} = (E_{1x}, 0, -(q_x/q_{1z})E_{1x}) \exp(iq_{1z}z) \exp(iq_x x - i\omega t), \qquad z > 0 \quad (6.46)$$

$$\mathbf{E} = \{a \exp[iq_{2z}(z + L/2)] + b \exp[-iq_{2z}(z + L/2)], 0,$$
$$- (q_x/q_{2z})a \exp[iq_{2z}(z + L/2)] + (q_x/q_{2z})b \exp[-iq_{2z}(z + L/2)]\}$$
$$\times \exp(iq_x x - i\omega t), \qquad 0 > z > -L \quad (6.47)$$

$$\mathbf{E} = (E_{3x}, 0, -(q_x/q_{3z})E_{3x}) \exp[iq_{3z}(z + L)] \exp(iq_x x - i\omega t), \qquad z < -L. \quad (6.48)$$

Here (6.3) has been applied to relate the z and x components in each medium. The localization conditions are $\text{Im}(q_{1z}) > 0$ and $\text{Im}(q_{3z}) < 0$, which require

$$q_x^2 > \varepsilon_i \omega^2/c^2, \qquad i = 1, 3. \quad (6.49)$$

The four amplitudes E_{1x}, a, b and E_{3x} are related by four boundary conditions, two at each interface. They take the form

$$af + bf^{-1} = E_{1x} \quad (6.50)$$

$$\varepsilon_2(af - bf^{-1})/q_{2z} = \varepsilon_1 E_{1x}/q_{1z} \quad (6.51)$$

$$af^{-1} + bf = E_{3x} \quad (6.52)$$

$$\varepsilon_2(af^{-1} - bf)/q_{2z} = \varepsilon_3 E_{3x}/q_{3z} \quad (6.53)$$

where $f = \exp(iq_{2z}L/2)$. Equations (6.50)–(6.53) are four homogeneous equations in the amplitudes, and the solvability condition gives the dispersion equation. It is readily seen that

$$\frac{\varepsilon_1 q_{2z} + \varepsilon_2 q_{1z}}{\varepsilon_2 q_{1z} - \varepsilon_1 q_{2z}} \exp(-iq_{2z}L) = \frac{\varepsilon_3 q_{2z} + \varepsilon_2 q_{3z}}{\varepsilon_2 q_{3z} - \varepsilon_3 q_{2z}} \exp(iq_{2z}L). \tag{6.54}$$

This is somewhat too general for comfort, and we therefore specialize to the symmetric geometry where each bounding medium is the same ($\varepsilon_1 = \varepsilon_3$). Bearing in mind that we are omitting damping, so that ε_1, ε_2 and ε_3 are real, the localization condition becomes

$$q_{1z} = -q_{3z} = i\kappa_1 \qquad \text{with } \kappa_1 > 0 \tag{6.55}$$

and (6.54) is seen to have two solutions

$$\exp(iq_{2z}L) = \pm(\varepsilon_1 q_{2z} + i\varepsilon_2\kappa_1)/(\varepsilon_1 q_{2z} - i\varepsilon_2\kappa_1). \tag{6.56}$$

Equation (6.56) applies equally to surface-polariton modes, with q_{2z} imaginary, say $q_{2z} = i\kappa_2$ and to p-polarized guided-wave modes, with q_{2z} real. For the former, the two solutions of (6.56) can be reorganized into the forms

$$\varepsilon_2/\varepsilon_1 = -(\kappa_2/\kappa_1)\tanh(\kappa_2 L/2) \tag{6.57}$$

$$\varepsilon_2/\varepsilon_1 = -(\kappa_2/\kappa_1)\coth(\kappa_2 L/2). \tag{6.58}$$

As $L \to \infty$, the hyperbolic functions in (6.57) and (6.58) tend to unity, so that both equations are asymptotically the same as the dispersion equation for the single-interface mode given by (6.6). As expected, (6.57) and (6.58) describe the bonding and antibonding combinations of the surface-polariton modes. The general form of the solutions is illustrated for different thicknesses of LiF film in figure 6.9. It is seen that like the single-surface polariton, curve 3, the modes fall within the surface-mode window $\omega_T < \omega < \omega_L$. Figure 6.9 also shows the dispersion curves obtained within the electrostatic approximation. These agree with the retarded curves for $q_x \gg \omega/c$ but they have the unphysical property that for small q_x they cross into the region $q_x < \omega/c$. The dispersion equations in the electrostatic approximation can be studied either by letting $c \to \infty$ in (6.57) and (6.58) or by generalizing the potential analysis given in section 6.1.1.

So far in this discussion of the surface-type modes we have made the approximation of assuming that all three dielectric constants are real, that is, damping has been neglected. Fukui *et al* (1979) and Sarid (1981) carried out numerical studies of the effect of including damping. In both cases, the starting point was the dispersion equation, i.e. (6.54) in general, or (6.57) and (6.58) for a symmetric case. Fukui *et al* took q_x real and calculated $\text{Im}(\omega)$, while Sarid took ω real and calculated $\text{Im}(q_x)$. These different assumptions correspond, ultimately, to different experimental arrangements. For example, real ω corresponds to a wave launched along the film at that frequency and $1/\text{Im}(q_x)$ is then a measure of the decay length of the wave

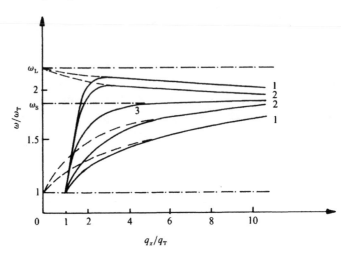

Figure 6.9. Two-interface surface polaritons for LiF film bounded by vacuum. 1, $q_T L = 0.1$ ($L = 0.5\,\mu m$); 2, $q_T L = 0.2$ ($L = 1.0\,\mu m$); 3, $q_T L > 2$ ($L > 10\,\mu m$). ———, The calculation with retardation; – – –, without retardation, and $q_T = \omega_T/c$. After Bryksin *et al* (1974).

along the film. Both sets of authors considered the particular case of metal films and plasmon–polaritons, so that the dielectric function of the film had the form of (2.132). The important result concerns the higher-frequency mode. Sarid's result is that its decay length increases as the film thickness decreases, and correspondingly Fukui *et al* find that the lifetime $1/\text{Im}(\omega)$ increases as the thickness decreases. The mode is therefore called the *long-range surface plasmon* (LRSP). In the LRSP for a film between identical dielectrics the electric-field pattern is antisymmetric across the film, while in the lower-frequency mode the pattern is symmetric. The long decay length of the LRSP may be seen as arising from the fact that as the film thickness decreases a smaller proportion of the mode energy is transported within the film.

Experimental confirmation of these results has been obtained by a number of groups, one of the earliest being that by Craig *et al* (1983). A related ATR curve will be shown in section 6.3.3. In applications in nonlinear optics it is helpful to use long-range modes, so the LRSP is of potential technical importance, as will be discussed in section 11.3. As stated, the earlier work on the LRSP was concerned with surface plasmon–polaritons. The extension to surface phonon–polaritons was carried out by a number of authors; a good account is given by Wendler and Haupt (1986a). They also discussed the surface plasmon–phonon–polariton in a doped semiconductor when ω_p is close to ω_T (Wendler and Haupt 1986b). The starting point for this is the present formalism with ε_2 in the form of (2.150).

The study of long-range modes was carried a stage further by Yang *et al* (1990a,b, 1991a), who considered the LRSP in a film with a dielectric function which is large and mainly imaginary. One example is an excitonic material like ZnO near to resonance, and perhaps for this reason the authors dubbed the mode a surface exciton–polariton. However, the properties of the mode do not involve spatial dispersion so this double-interface mode is quite different from the spatially dispersive single-interface mode that was discussed in section 6.1.4. The dispersion relation of the mode in question is given by (6.57) and the point is that for ε_2 imaginary and large in magnitude the mode has small damping and is therefore long range. The calculations of the long range were supported by measurements of attenuated total reflection (ATR) spectra as will be reviewed in section 6.3.3.

We now turn to discussion of the guided modes, for which q_{2z} is real. For a symmetric film in p-polarization, the two forms of (6.56) reduce to

$$\varepsilon_2/\varepsilon_1 = (q_{2z}/\kappa_1)\tan(q_{2z}L/2) \tag{6.59}$$

$$\varepsilon_2/\varepsilon_1 = -(q_{2z}/\kappa_1)\cot(q_{2z}L/2) \tag{6.60}$$

as can also be seen directly from (6.57) and (6.58). The numerical, or graphical, task of solving these is similar to that of finding the odd and even parity modes in a quantum-mechanical square well, or dispersion curves for the Love waves mentioned in section 3.2.2. For small L there is only a small number of modes; the number increases with L until for large enough L the guided-wave modes form a quasi-continuum, which in the limit becomes the bulk continuum. The effect of introducing a finite thickness L is to quantize the bulk continuum. The dispersion curves for guided phonon–polaritons in a symmetric film, with ε_2 in the form of (2.146), are shown in figure 6.10. Comparison with figure 6.2 shows how the curves occupy the parts of the (q_x, ω) plane that as $L \to \infty$ become the bulk continuum.

The calculation for the s-polarized guided-wave modes is similar to that for the p-polarized modes. Instead of (6.46)–(6.48) the $\mathbf{E}$ fields (all in the y direction) are written

$$E_y = E_1 \exp(iq_{1z}z)\exp(iq_x x - i\omega t), \qquad z > 0 \tag{6.61}$$

$$E_y = \{c\exp[iq_{2z}(z + L/2)] + d\exp[-iq_{2z}(z + L/2)]\}$$
$$\times \exp(iq_x x - i\omega t), \qquad 0 > z > -L \tag{6.62}$$

$$E_y = E_3 \exp[iq_{3z}(z + L)]\exp(iq_x x - i\omega t), \qquad z < -L. \tag{6.63}$$

The boundary conditions involve the tangential magnetic-field component H_x, which is derived from the Maxwell equation for $\nabla \times \mathbf{E}$. It emerges that the equations for p-polarization can be converted to those for s-polarization by substituting $\varepsilon_2 q_{iz}/\varepsilon_{i1} q_{2z}$ with q_{2z}/q_{iz} ($i = 1, 3$). Thus the

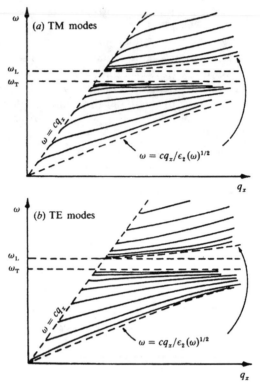

Figure 6.10. Dispersion curves of guided-wave polaritons in a film placed in vacuum. (a) p-polarized, (b) s-polarized. After Ushioda and Loudon (1982).

general result, taken over from (6.54), is

$$\frac{q_{1z} + q_{2z}}{q_{2z} - q_{1z}} \exp(-iq_{2z}L) = \frac{q_{3z} + q_{2z}}{q_{2z} - q_{3z}} \exp(iq_{2z}L) \tag{6.64}$$

while for a symmetric geometry the equations are

$$1 = (q_{2z}/\kappa_1)\tan(q_{2z}L/2) \tag{6.65}$$

$$1 = -(q_{2z}/\kappa_1)\cot(q_{2z}L/2). \tag{6.66}$$

The corresponding dispersion curves are shown in figure 6.10b.

In figure 6.10 we illustrated the guided modes for a highly dispersive film, since dispersive media are one of the main themes of this book. However, guided modes are also of great importance in optical engineering, and in that case one would more likely consider media in which the dielectric function, or equivalently the refractive index, is only a *slowly-varying* function of frequency. For most applications the refractive indices of the film and the bounding media are real and positive. In that case modes of the surface-polariton type do not occur and use is made of the guided waves. The primary condition for guided waves to occur is that q_{2z} should be real while q_{1z} and q_{3z} are imaginary. In view of (6.2), this means that n_2, the refractive index of the film, should be larger than the other two refractive indices:

$$n_2 > \max(n_1, n_3). \tag{6.67}$$

This is the *wave guiding condition* for the geometry in figure 6.9.

When the refractive indices of the film and the bounding media are not very different it is convenient to use transformed variables instead of q_x and ω to represent the dispersion curves of the guided modes (see e.g. Adams 1981, Yariv 1985, Sibley 1995). The phase velocity is used to define an effective refractive index n_{eff}:

$$\omega/q_x = c/n_{\mathrm{eff}}. \tag{6.68}$$

For a symmetric film with refractive index n_F bounded by media with refractive index n_B the transformed variables are

$$V = (\omega L/c)(n_F^2 - n_B^2)^{1/2}, \qquad b = (n_{\mathrm{eff}}^2 - n_B^2)/(n_F^2 - n_B^2). \tag{6.69}$$

Here V is a scaled dimensionless combination of frequency and thickness, while b is used instead of n_{eff}. The guided mode dispersion can then be represented as a set of universal curves of b versus V.

It is not our purpose here to go much further into engineering applications. However, we mention an interesting result concerned with the number of guided modes and introduce the important idea of a *single-mode*, or *monomode, waveguide*. An analysis (e.g. by graphical methods) of our results in (6.65) and (6.66) for s-polarized guided-wave modes shows that (6.65) has just one solution for sufficiently small L, whereas (6.66) has no solutions for small L. Hence there is a value L_C such that for $L < L_C$ the film is single-moded. In terms of the variables defined in (6.69) it may be shown that the monomode region is given by $V < \pi/2$.

6.2.2 Magnetic media

The surface-type polariton modes of a magnetic film of finite thickness L were first investigated by Karsono and Tilley (1978) and the discussion was extended to guided modes by Marchand and Caillé (1980). We give only a brief summary since there has not yet been any experimental work on these modes; details are given by Sarmento and Tilley (1982).

As in section 6.1.5 we consider the Voigt geometry with the static field $\mathbf{B}_0$ in the plane of the film and propagation transverse to $\mathbf{B}_0$; axes are chosen as z along $\mathbf{B}_0$ and x as the film normal so that propagation is along y. The fluctuating field is given by the equivalents of (6.46)–(6.48) for the non-magnetic film, that is,

$$\mathbf{h} = \mathbf{h}_1 \exp(-\kappa_1 x) \exp(iq_{\parallel} y - i\omega t), \qquad x > 0 \qquad (6.70)$$

$$\mathbf{h} = [\mathbf{a} \exp(iq_x x) + \mathbf{b} \exp(-iq_x x)] \exp(iq_{\parallel} y - i\omega t), \qquad 0 > x > -L \quad (6.71)$$

$$\mathbf{h} = \mathbf{h}_3 \exp(\kappa_1 x) \exp(iq_{\parallel} y - i\omega t), \qquad x < -L \qquad (6.72)$$

for a symmetrically-bounded film. The component q_x is real for guided modes and imaginary ($q_x = i\kappa_2$) for surface-type modes; κ_2 and κ_1 are given by (6.37) and (6.39). As is the case for the single-interface modes, it follows from the boundary conditions that the surface-type mode is s-polarized ($\mathbf{E}$ along z). The dispersion equation is found from the boundary conditions in the same way as for nonmagnetic modes and the result, as quoted by Marchand and Caillé (1980), is

$$[\kappa_1^2 \mu_V^2 + \kappa_2^2 + q_{\parallel}^2 (\mu_{xy}^2/\mu_{xx}^2)] \tanh(\kappa_2 L) + 2\kappa_1 \kappa_2 \mu_V = 0, \qquad (6.73)$$

where μ_V is defined in (6.38). This generalizes the Damon–Esbach result (5.5) to include retardation as well as the result of Hartstein *et al* (1973) to finite thickness. Dispersion curves for the surface-type and guided modes are shown by Sarmento and Tilley (1982), for example.

6.3   Attenuated total reflection

6.3.1   Basic principles

A basic property of the surface polariton discussed in section 6.1.1 is that it is confined to the region of the (q_x, ω) plane satisfying $q_x^2 > \varepsilon_1 \omega^2/c^2$ where, as in figure 6.1, ε_1 is the dielectric constant of medium 1, occupying the half space $z > 0$. This means that the surface mode cannot be excited by a light beam incident in the normal way in medium 1, since such a beam satisfies

$$q_x^2 + q_z^2 = \varepsilon_1 \omega^2/c^2 \qquad (6.74)$$

with q_z real. For this reason the surface polariton is called *nonradiative*. The restriction to real q_z is removed in the method of attenuated total reflection (ATR). Here we give an introductory account; more detail is to be found in review articles by Otto (1974, 1976), Borstel *et al* (1974) and Abeles (1986) and a helpful introduction to both ATR and grating coupling is given by Sambles *et al* (1991). Some later developments, particularly for ATR used in conjunction with other techniques, are mentioned in the book by Kawata (2001).

Three ways of coupling an incident beam in medium 1 to nonradiative modes are illustrated in figure 6.11. In the grating method, figure 6.11(a), a diffraction grating is laid down on the surface of interest. The incident

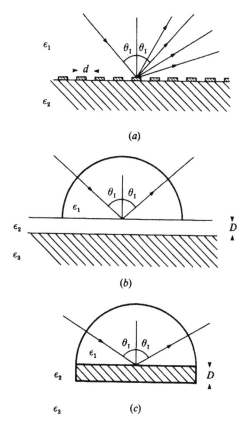

Figure 6.11. Techniques for coupling light to nonradiative modes: (a) grating method; (b) ATR, Otto configuration; (c) ATR, Raether–Kretschmann configuration. In each case, the surface-active medium is shown shaded.

light has surface wavevector $\varepsilon_1^{1/2}(\omega/c)\sin\theta_1$. By Bloch's theorem (see section 1.1), this couples to modes with

$$q_x = \varepsilon_1^{1/2}(\omega/c)\sin\theta_1 + 2m\pi/d \qquad (6.75)$$

where d is the grating periodic distance and m is an integer. Clearly for appropriate values of m and d coupling to nonradiative modes is possible. When ω, θ_1 and the order m are such that a surface mode is excited, the intensity of the diffracted light in that order is reduced. These reduced intensities observed with a diffraction grating are known as *Wood's anomalies* (Wood 1935). Further discussion of grating coupling is deferred to section 6.4.

Two different versions of the ATR method are sketched in figures 6.11(b) and (c). Both methods involve a three-layer geometry and in both the light is incident in a prism of dielectric constant ε_1. In the *Otto configuration*, figure 6.11(b), a spacing medium, typically vacuum, is adjacent to the prism, with the surface-active medium as medium 3. The prism is chosen with a relatively large refractive index, $\varepsilon_1 > \varepsilon_2$, and the angle of incidence θ_1 is larger than the critical angle θ_C for total internal reflection at the 1–2 interface, $\theta_1 > \theta_C = \sin^{-1}(\varepsilon_2^{1/2}/\varepsilon_1^{1/2})$. Thus if medium 3 were absent, the incident light would be totally reflected. However, total reflection involves the presence of an evanescent mode with decreasing amplitude (imaginary q_z) in medium 2. The tail of this evanescent mode can excite a surface mode on the 2–3 interface; this mode removes energy and thus gives a reduction (attenuation) of the total reflection. More concisely, the in-plane wavevector component

$$q_x = \varepsilon_1^{1/2}(\omega/c)\sin\theta_1 \tag{6.76}$$

is the same in all three media, and with sufficiently large ε_1 and θ_1 it can be made to satisfy the condition

$$q_x > \varepsilon_2^{1/2}\omega/c \tag{6.77}$$

necessary for excitation of a non-radiative surface mode at the 2–3 interface. Then when the incident values of ω and q_x lie on the dispersion curve of a surface mode the reflectivity is reduced below unity. The main technical problem is the control and uniformity of the spacer thickness, i.e. the thickness of medium 2. There is a short discussion of the experimental techniques in Otto (1976).

In the *Raether–Kretschmann configuration*, figure 6.11(c), the surface-active medium is deposited direct on the surface of the prism. Thus the method is most readily applicable when the material of medium 2 is easily evaporated, and the configuration has been mostly used to study surface plasmon–polaritons on metals such as Al and Ag. In this case ε_2 is negative, and the electromagnetic field is automatically evanescent in medium 2 since (6.74) must be satisfied (with ε_2 replacing ε_1) so that q_z is necessarily imaginary in medium 2. The surface mode of interest is at the 2–3 interface, and the necessary condition for excitation is

$$q_x > \varepsilon_3^{1/2}\omega/c \tag{6.78}$$

in place of (6.77). As in the Otto configuration, the reflectivity is reduced when ω and q_x lie on the dispersion curve.

Two forms of experimental scan are possible in ATR in either geometry. First, there is a frequency scan with θ_1 fixed. It follows from (6.76) that in this case a straight-line trajectory is scanned in the (q_x, ω) plane. Second, there is a scan over angle at fixed frequency; the trajectory is then a line parallel to the q_x axis. The two possibilities are shown schematically in figure 6.12. In either case, where the scan line crosses the surface-mode dispersion curve a dip should occur in the reflection coefficient, since the surface mode is excited.

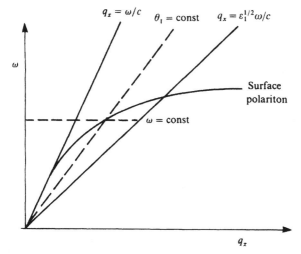

Figure 6.12. Schematic comparison of frequency and angular scan in ATR. After Otto (1974).

Since both the Otto and the Raether–Kretschmann configurations involve a three-layer geometry, the properties of the surface mode at the 2–3 interface are perturbed by the presence of the 1–2 interface. If the thickness D of medium 2 is too small, then the system is 'overcoupled' and the properties of the isolated 2–3 interface will not be observed. If D is too large, on the other hand, the system is 'undercoupled' and the 2–3 interface mode is only weakly excited, so that only a small reflectivity dip is seen. Choice of optimal D is therefore an integral part of experimental design. Fortunately, as we shall now see, the theoretical expression for the reflectivity is readily derived, and numerical simulations can be performed to give a guide to the choice of D value. In simple cases, the rule of thumb $D \approx \lambda_0$, where λ_0 is the free-space wavelength of the incident radiation, may suffice.

6.3.2 Theory

The theory of ATR in either configuration is formally quite similar to the theory of the two-interface polaritons given in section 6.2. Since surface polaritons are p-polarized modes with the **H** field in the y direction and the **E** field in the xz plane, ATR coupling to the surface polaritons occurs only in p-polarization. In medium 1, the **E** field takes the form

$$\mathbf{E} = (E_0, 0, q_x E_0/q_{1z}) \exp(-iq_{1z}z) \exp(iq_x x - i\omega t)$$
$$+ r(E_0, 0, -q_x E_0/q_{1z}) \exp(iq_{1z}z) \exp(iq_x x - i\omega t). \tag{6.79}$$

Here the first term describes the incident wave, with E_0 determined by the incident intensity. The second term is the reflected wave and the object of the calculation is to find the complex amplitude reflection coefficient r. In media 2 and 3, the **E** field can be written in the form of (6.47) and (6.48). The next step is to apply the two electromagnetic boundary conditions at each interface; this gives four equations for the four coefficients r, a, b and E_{3x} in terms of E_0. The calculation is straightforward leading to

$$r = \frac{(\varepsilon_3 q_{2z} + \varepsilon_2 q_{3z})(\varepsilon_2 q_{1z} - \varepsilon_1 q_{2z}) + g(\varepsilon_3 q_{2z} - \varepsilon_2 q_{3z})(\varepsilon_2 q_{1z} + \varepsilon_1 q_{2z})}{(\varepsilon_3 q_{2z} + \varepsilon_2 q_{3z})(\varepsilon_2 q_{1z} + \varepsilon_1 q_{2z}) + g(\varepsilon_3 q_{2z} - \varepsilon_2 q_{3z})(\varepsilon_2 q_{1z} - \varepsilon_1 q_{2z})} \quad (6.80)$$

where $g = \exp(-2iq_{2z}L)$.

Equation (6.80) is a kind of response function, and like other response functions it is most useful when damping is included. However, one or two comments can be made about its significance in the absence of damping. The condition for the numerator to vanish is formally identical to (6.54), while the condition for the denominator to vanish is (6.54) with the sign of q_{1z} reversed. This is because (6.54) is the condition for Maxwell's equations to have a solution with only one field term in medium 1, namely (6.46), whereas (6.80) is found from (6.79), in which two field terms are present, except in the limits $r \to 0$ and $r \to \infty$.

Experimental work to date has mainly concentrated on the intensity reflection coefficient $R = |r|^2$ although, as pointed out by Abeles (1986), by using the techniques of ellipsometry it should be possible to measure both the amplitude and the phase of r. Our discussion will largely be concerned with R.

6.3.3   Experimental results: nonmagnetic media

ATR curves for an Ag sample in the Otto configuration are shown in figure 6.13. The importance of the choice of gap width L is quite clear. For large L (curve a) the system is undercoupled, and the reflectivity dip is small. For optimum L (curve b) the reflectivity drops almost to zero. For small L (curve c) the system is overcoupled. The ATR dip is then broadened and its minimum is shifted from the point found in optimum coupling. We should note here the comment by Faulkner *et al* (1993) that Ag is not a simple free-electron metal and the observed surface mode is really a coupled mode between the free-electron plasmon and electronic transitions in the band arising from the atomic d levels.

The Otto configuration has more often been applied in the infrared, usually to study surface phonon–polaritons. Early experimental results for GaP are shown in figure 6.14; some similar studies of semiconductor superlattices will be reviewed in section 8.3.4. It may be noted that a Si prism was used for these measurements. Undoped Si, which is readily available, is transparent in the far infrared, because the TO phonon does not carry a dipole moment, and it has a high refractive index of 3.4 which gives for the critical angle for total internal reflection $\theta_C = \sin^{-1}(1/3.4) = 17.1°$. Thus Si

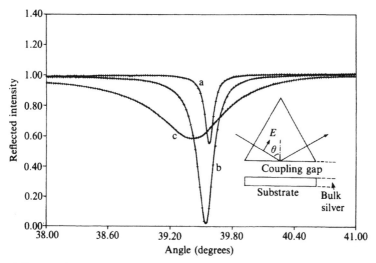

Figure 6.13. Experimental and theoretical spectra for Otto configuration with glass prism, vacuum spacer and Ag sample; detailed geometry in inset. ×, experimental points; ——, theoretical curves obtained by least-squares fit of (6.80). Curves are for angle scan at the wavelength $\lambda = 633$ nm. Gap widths (a) 951 nm, (b) 918 nm, (c) 581 nm. Courtesy of J R Sambles and K Welford.

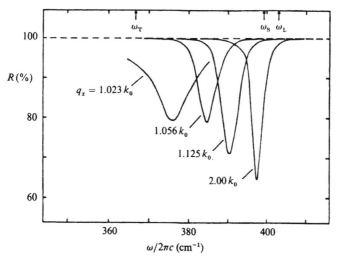

Figure 6.14. ATR spectra (frequency scan) for GaP/air interface with Si prism. The curves are labelled with q_x from (6.76) where $k_0 = \omega/c$. Gap values were $L = 40$, 25, 12, 2.5 μm, increasing as q_x increases. After Marschall and Fischer (1972).

is an ideal prism material for ATR studies in this region. Analysis of the results shown in figure 6.14 gives good agreement with the dispersion curve for the surface phonon–polaritons on GaP.

In section 6.1.4 it was mentioned that ATR was one of the principal techniques used to study surface exciton–polaritons. Comparing sections 6.3.2 and 6.1.4 shows us how to adapt the theory for this case. For the Otto geometry, the surface-active medium is 3, and as in section 6.1.4 it is described by three field amplitudes (two transverse, one longitudinal) rather than the one used in the derivation of (6.80). Corresponding to the two additional amplitudes there are two extra boundary conditions, derived for example from (2.160). Thus it is still possible to find an expression for the reflected amplitude r, although the expression is more complicated than (6.80). The first such calculation was reported by Maradudin and Mills (1973) and experiments on ZnSe were reported by Tokura *et al* (1981).

We have described the principle of the ATR method and presented examples of spectra obtained in the Otto configuration on three-medium geometries as illustrated in figure 6.11(b). Raether–Kretschmann spectra show similar features. In fact the ATR method has been applied widely to study more complicated samples consisting of a number of layers. As an example, figure 6.15 shows an angle-scan ATR spectrum for a prism/ MgF_2/Ag/air geometry. Here Ag is the surface-active medium and the modes of interest are the surface-type modes in the Ag film. Although the Ag film is bounded asymmetrically, with MgF_2 on one side and air on

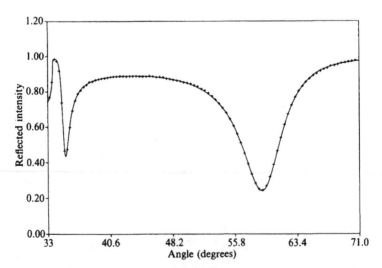

Figure 6.15. ATR spectrum (angle scan) for a system comprising prism/MgF_2/Ag/air at wavelength 632.8 nm. ×, experimental points; ——, least-squares fit of theory, giving thicknesses 203 and 28.2 nm for MgF_2 and Ag films respectively. Courtesy of J R Sambles and M D Tillin.

the other, nevertheless the higher-frequency mode behaves like the long-range mode (LRSP) that was described for a symmetric geometry in section 6.2.1. This is clearly seen in figure 6.15. The higher-frequency mode corresponds to the lower angle, as can be seen from figure 6.12, and it is observed that this mode gives a much sharper resonance (smaller damping) than the lower-frequency mode.

In section 6.1.4 we mentioned the use of the phrase 'surface exciton–polariton' by Yang *et al* (1990a) to describe the long-range mode supported by a film with a dielectric constant that is imaginary and large in magnitude. Figure 6.16(a) illustrates this property by means of an angle-scan ATR spectrum taken in p-polarization in the near infrared. The low damping is apparent as a narrow linewidth in the spectrum. The physical reason for the low damping is that the electric-field intensity is low in the absorptive film, and this is nicely illustrated by the Poynting vector shown in figure 6.16(b). Later papers extended this work to ATR spectra in the visible on very thin Ag films in the form of islands on a quartz substrate (Yang *et al* 1991b) and in the near infrared (Bryan-Brown *et al* 1991) on Cr films. The latter measurements used grating coupling, to be discussed in section 6.4, as well as ATR.

For a four-layer geometry, as in figures 6.15 and 6.16, it is possible to derive an explicit formula like (6.80) for the reflection amplitude r. In practice, even for four layers, and certainly for any larger number, it is preferable to use a matrix formulation to express the boundary conditions relating the field amplitudes in layers n and $n + 1$. Calculation of r then involves multiplying together the matrices for all the successive interfaces (Heavens 1965). A version of this formalism will be employed for superlattices in sections 8.1.1 and 8.3.1.

6.3.4 Experimental results: magnetic media

It was mentioned in section 6.1.5 that because ferromagnetic resonance frequencies are in the microwave region the wavelengths are on the scale of centimetres so that experiments on ferromagnetic surface polaritons would require a sample of too large a size to be practical. Many antiferromagnets, on the other hand, have resonance frequencies in the far infrared so that measurements are possible in principle. The main problem, as discussed in section 2.6.2, is that the magnetic reststrahl bands are typically very narrow and resolution on the scale between 0.1 and 0.01 cm^{-1} is necessary. The standard instrument for this region is the Michelson interferometer, applied as a Fourier-transform spectrometer, and the resolution is of the order of M^{-1}, where M is the traverse of the moving mirror. Thus a value of M in the range between 10 cm and 1 m is required, compared with typically 1 cm in a standard spectrometer. A number of instrumental difficulties arise because of this large value of M, but these have been overcome; a detailed

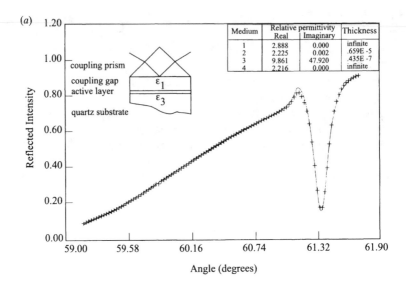

(a)

Medium	Relative permittivity Real	Relative permittivity Imaginary	Thickness
1	2.888	0.000	infinite
2	2.225	0.002	.659E -5
3	9.861	47.920	.435E -7
4	2.216	0.000	infinite

coupling prism

coupling gap
active layer

quartz substrate

ε_1

ε_3

Reflected Intensity

Angle (degrees)

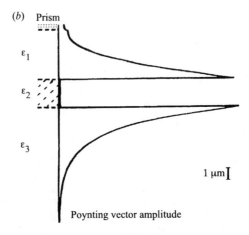

(b)   Prism

ε_1

ε_2

ε_3

1 μm

Poynting vector amplitude

Figure 6.16. (a) Angle-resolved ATR spectrum in p-polarization at fixed wavelength 3.391 μm from a thin vanadium film. ×, experimental; ——, calculated. The experimental geometry and values used in fitting are in the insets. (b) Poynting vector versus distance at 61.38°, corresponding to the reflection dip in (a). After Yang *et al* (1990).

review is given by Brown *et al* (1999). Figure 2.18 shows high-resolution reflectivity spectra on the uniaxial antiferromagnet FeF_2 and later ATR measurements on the same sample produced the first observation of the surface magnetic polariton (Jensen *et al* 1995) at a resolution of $0.06 \, cm^{-1}$. Here we review later results (Jensen *et al* 1997) obtained in higher resolution, $0.02 \, cm^{-1}$ with several prisms of different angles.

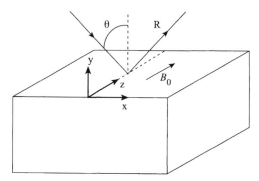

Figure 6.17. Voigt geometry used for reflectivity and ATR measurements on FeF$_2$. After Jensen *et al* (1997).

Like the reflectivity spectra of figure 2.18, the measurements were made in the Voigt geometry, depicted in figure 6.17. The FeF$_2$ sample was oriented with the uniaxis along the applied field, so that the magnetic susceptibility is that of (2.173) and (2.174). The plane of incidence was normal to the field, as shown, and the measurements were taken in s-polarization, i.e. with the r.f. **E** field along the B_0 direction so that the r.f. **h** field coupled to the resonant magnetic susceptibility. As in the GaP spectra of figure 6.14, the ATR studies used a Si prism.

Surface-polariton dispersion curves and the associated bulk continuum regions for these measurements are shown in figure 6.18 for $B_0 = 0$ and $B_0 = \pm 0.3$ T. The surface-polariton curves are calculated by simultaneous solution of (6.37), (6.39) and (6.41), as mentioned in section 6.1.5. For comparison with reflectivity spectra, the continuation of these curves as '*surface resonances*' (Stamps and Camley 1989) into the bulk continua is shown. The idea of these is as follows. With damping included in the susceptibility, (6.37), (6.39) and (6.41) can be solved for frequencies just inside a bulk continuum to give a relation between ω and q_x, both taken as complex variables. Where the values are not too far off the real axis, i.e. the imaginary parts are small compared with the real parts, then the curve of Re(ω) versus Re(q_x) defines the surface resonance.

For $B_0 = 0$ the surface-polariton dispersion curve, as expected, occupies the gap between the two bulk-continuum regions and the whole dispersion curve is contained within a frequency interval of order 0.1 cm^{-1}; obviously, the curves are the same for positive and negative q_x. For $B_0 = \pm 0.3$ T the reststrahl splits into two parts separated by a narrow bulk continuum region. The surface-polariton dispersion curves are strikingly non-reciprocal. In the lower reststrahl region the + mode rises in frequency compared with that for $B_0 = 0$, while the − mode decreases in frequency to run along the top of the lowest bulk continuum. In the upper reststrahl region, a mode is seen on the − side but there is none on the + side.

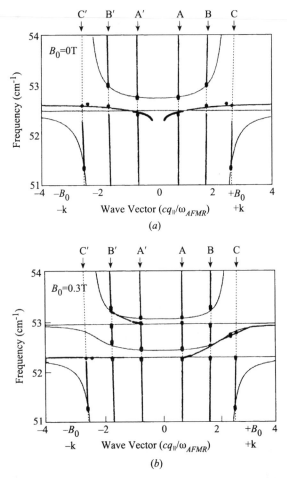

Figure 6.18. Surface-polariton dispersion curves and bulk continuum regions for FeF_2 in Voigt geometry. (a) $B_0 = 0$ and (b) $B_0 = \pm 0.3$ T. The surface-polariton curves are shown as thick lines and their extension into the continua as surface resonances (see text) as thicker lines. Experimental points, marked as ●, are determined as explained in the text. Lines A and A′ are scan lines for 45° reflectivity while B, B′ and C, C′ are scan lines for ATR with 30° and 50° Si prisms. After Jensen *et al* (1997).

The scan lines marked A and A′ in figure 6.18 correspond to 45° ordinary reflectivity measurements. These are a lower-field, higher-resolution version of those shown previously in figure 2.18, and they are discussed in detail in Jensen *et al* (1977).

ATR spectra with a 30° Si prism are shown in figure 6.19. The theoretical curves are obtained by a generalization of the theory outlined in section 6.3.2 to include the gyromagnetic permeability. The spectra follow scan lines B and B′ in figure 6.18. In particular, note the appearance of the surface mode

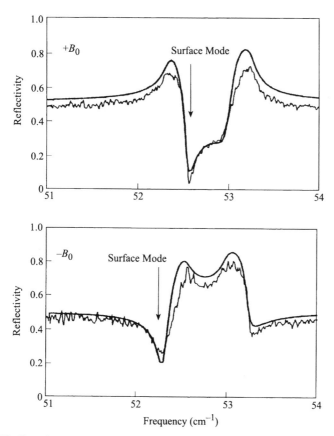

Figure 6.19. Experimental (thin line) and theoretical (thick line) ATR spectra of FeF$_2$ at 1.7 K in the geometry of figure 6.17 with a 30° Si prism. Resolution 0.02 cm^{-1}. Fields $B_0 = \pm 0.3$ T as marked. After Jensen *et al* (1997).

just at the top of the lowest bulk region for $B_0 = -0.3$ T but at a markedly higher frequency for $B_0 = +0.3$ T. Likewise, the higher-frequency surface mode seen in figure 6.18 for $B_0 = -0.3$ T appears as a dip just above the upper reststrahl in figure 6.19 but no such dip is seen for $B_0 = +0.3$ T. Experimental points determined from figure 6.19 are marked on figure 6.18. That figure also marks points from spectra obtained with 45° and 50° prisms. The original paper contains, in addition, detailed results for $B_0 = \pm 0.6$ T and a discussion of using the results to deduce the magnetic parameters of antiferromagnets.

The detailed study reviewed here, which used a bulk crystal of FeF$_2$, opens the way for much further work on bulk, thin-film and superlattice samples. The necessary theory for a wide range of experimental geometries using bulk samples, for example different relative orientations of field,

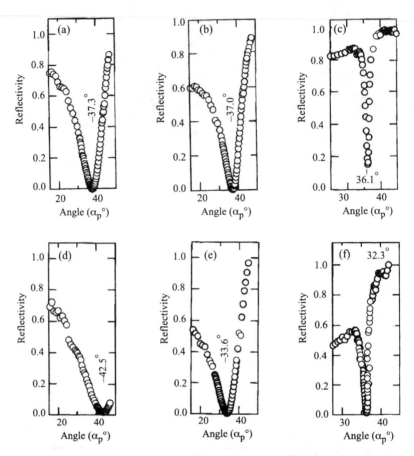

Figure 6.20. Angle resolved p-polarized ATR spectra on sapphire/air gap/ copper at wavelengths (a) 457.9 nm, (b) 514.5 nm, (c) 632.8 nm. Dip positions are marked. Corresponding spectra on sapphire/air gap/CuPc(5 nm)/copper are shown in (d), (e) and (f). After Futamata (1997).

uniaxis, surface normal and plane of incidence, has been given by Abraha and Tilley (1996). Of particular note is the proposal by Abraha (see Abraha and Tilley 1994) that the use of ATR would lead to detection of magnetic-resonance features in spectra of metallic materials. The proposal, based on computed spectra in which the magnetic material was given a high conductivity, has received some experimental support (T J Parker, private communication) and if further substantiated it resolves a long-standing problem in spectroscopy. Of particular interest, perhaps, are the rare-earth metals, in which the f electron spins, which carry the magnetism, interact via a long-range and oscillatory exchange force. Partly because of this oscillatory nature, a wide range of equilibrium configurations occurs in the bulk materials (Legvold 1980) and of even more interest many rare-earth/metal superlattices display magnetic orderings that are characteristic of the superlattices themselves (Majkrzak *et al* 1991). It is to be hoped that further experimental studies of a variety of magnetic materials will appear in due course.

6.3.5   Applications of ATR

As has been mentioned, the surface polariton is characterized by a strong field maximum at the surface, and similarly all total reflection is accompanied by such a field maximum. This property has been used to develop sensors and spectroscopic techniques for layers deposited on surfaces and for adsorbed molecules. The developments in the application of both ATR and grating coupling have been very extensive, and one may say that they are now the basis of surface-specific optical spectroscopy. Here we briefly discuss ATR and some comments about applications of grating coupling will be made in section 6.4.3.

Two ways of applying ATR to surface effects may be distinguished. First, the presence of an overlayer on Ag, for example, leads to a shift and often a broadening of the surface plasmon–polariton (SPP) dip in an ATR spectrum. Second, what would otherwise be total reflection can be reduced at the optical absorption frequencies of adsorbed molecules. Both effects can be sensitive to monolayer-thickness overlayers. We should mention a third effect: Raman scattering intensities can be enhanced when the incident and scattered radiation propagate as surface polaritons. This will be discussed later, in section 6.5.2.

As an example of the first technique, figure 6.20 shows angle-resolved ATR spectra for three incident wavelengths from a sapphire/gap/Cu sample (the Otto geometry). The first three spectra are for a pure Cu surface and show the usual SPP dip. In the second set of spectra the Cu is coated with a 5 nm layer of copper phthalocyanine (CuPc). Two effects are evident. First, the position of the SPP dip shifts by several degrees and second the lineshape of the dip changes significantly. This technique is of fairly wide application for various overlayer systems; for example Vukusic *et al* (1992a) use Raether–Kretschmann ATR from coated gold films to characterize spin-coated phthalocyanine films while Geddes *et al* (1994) report results for immunoglobin on gold.

The second technique mentioned was the observation via ATR of the absorption lines of adsorbed molecules. The example of CO_2 adsorbed on a NaCl single-crystal surface is shown in figure 6.21. It is seen that sharp spectral lines occur in both polarizations. The authors attribute the enhancement in the sensitivity in ATR compared with ordinary transmission measurements to two factors. The first is the large field strength at the interface in ATR, which has been mentioned already, and the second is the

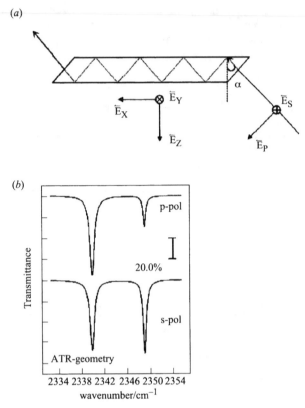

Figure 6.21. (a) Experimental arrangement for multiple-reflection ATR from a single-crystal NaCl prism, (b) Polarized infrared ATR absorption spectra of CO_2 adsorbed on NaCl (001) surface. After Heidberg *et al* (1999).

use of the multiple-reflection geometry shown in figure 6.21(a). They mention as a subsidiary advantage that there are two components of electric field in p-polarization, x and z in the notation of figure 6.21(a), and the y component is the field direction in s-polarization. Thus all three directions are probed by the use of both polarizations. The main purpose of the paper by Heidberg *et al* (1999) was a study of adsorbed H_2 and D_2; they show the CO_2 spectra only as an example of sensitivity. However, it would take us too far afield to discuss their main results. A theoretical analysis of the multiple reflection geometry shown in figure 6.21(a) is given by Loring and Land (1998). Among the large number of ATR studies of adsorbates we may cite further the work by Tsuchida *et al* (1999) on hydrogen adsorbed on SiC and the investigation by Rudkevich *et al* (1998) of the Ge-covered Si surface which made use of optical absorption by surface Ge–H and Si–H bonds.

The applications of ATR are now very extensive, partly because it can be used for a wide variety of surfaces. It is not the aim of this brief section to give a full review but the examples that have been shown together with the references in the papers cited should serve as an introduction to the field.

6.4   Grating coupling

The basic idea of grating coupling to non-radiative modes was illustrated in figure 6.11, and (6.75) gives the relation between grating period d, angle of incidence θ_1 and in-plane wavevector q_x. It can be seen from (6.75) that the grating coupling has an advantage over ATR in that access to a wide range of q_x values is possible by choice of d. Clearly, however, (6.75) is only a starting point. First, it describes only the kinematics and does not account for reflected intensities. Second, the simplest use involves assuming that the properties of the surface mode are not affected by the presence of the grating, which is obviously an approximation. The main correction is that the presence of the grating introduces Brillouin-zone boundaries at positions $q_x = n\pi/d$ and frequency gaps in the dispersion curve must appear at these points. We take these two questions in turn. In section 6.4.1 we describe the basic results for grating coupling to surface modes, then in section 6.4.2 we discuss the modification of the surface-mode properties due to the presence of a grating. In both these sections, the emphasis is on experimental results and we give a description rather than a full account of the theory. Like ATR, grating coupling has found a range of applications and these are discussed in section 6.4.3.

Clearly the grating coupling is important in its own right. In addition, the study of gratings is a step towards understanding optically rough surfaces, since such a surface can be represented as a Fourier series of gratings with different periods and directions. In the same way as a grating, a rough surface makes possible the direct coupling of surface modes to the electromagnetic field in the adjoining medium. One example is the light-emitting tunnel junction to be discussed in section 6.5.3.

6.4.1   Coupling to surface modes

It will be helpful to start with some key experimental results obtained by Zaidi *et al* (1991a,b) in a detailed study of coupling to surface plasmon–polaritons (SPP) on a silver grating. The gratings were fabricated by electron-beam evaporation of a 100 nm thick Ag film on to a holographic grating prepared on a Si substrate. The measurements were made on a range of gratings of different periods and depths, the shallow gratings having a sinusoidal profile with the deeper ones more rectangular. The measurements were made as angle scans with a He–Ne (633 nm) laser in

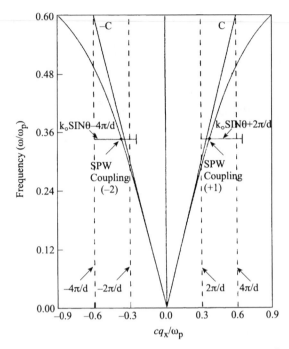

Figure 6.22. SPP dispersion curve for Ag/air interface together with grating coupling lines from (6.75). The axes are scaled as cq_x/ω_p and ω/ω_p, where ω_p is the Ag plasma frequency. After Zaidi *et al* (1991b).

p-polarization because this is the polarization of the SPP. As in figure 6.11, the plane of incidence was normal to the grating lines. Figure 6.22 shows the coupling condition (6.75) for a grating with period 870 nm together with the flat-surface SPP dispersion curve (6.7) for the plasma dielectric function $\varepsilon(\omega) = 1 - \omega_p^2/\omega^2$. The Bragg vectors $2n\pi/d$ are marked. The horizontal lines at the laser frequency show the range of q_x values accessible when θ_1 is varied from 0 to $\pi/2$ (normal to grazing incidence). It is seen that there are two coupling points, $n = +1$ and $n = -2$ with the coupling angle smaller for the former.

Experimental reflectance curves in zero order, i.e. with angle of reflection equal to angle of incidence, for a range of gratings of the period used for figure 6.22 and different depths h are shown in figure 6.23. The $n = +1$ coupling is seen as a prominent sharp dip at about $19°$ and the $n = -2$ coupling is the smaller dip at about $25°$. These dips become deeper, with little change of shape, as h increases from 55 to 85 nm, so that at 55 nm the incident light is undercoupled to the SPP and 85 nm represents optimum coupling. As the grating depth h increases further, the dips broaden and eventually merge.

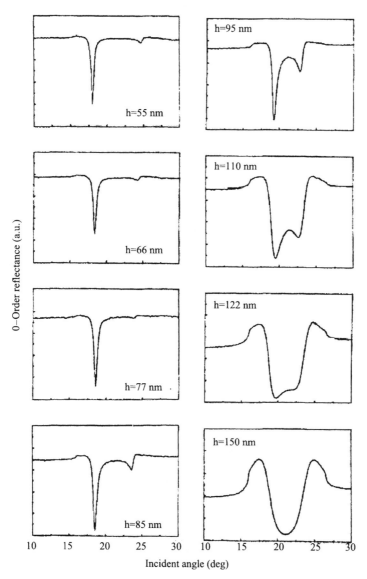

Figure 6.23. Zero-order reflectance (specular) from Ag coated gratings of differing depths *h* (marked). After Zaidi *et al* (1991b).

The theory of reflection from a grating has been developed by a number of groups and we refer mainly to the method of Toigo *et al* (1975) as applied by Weber and Mills (1983) and Weber (1986). The starting point is the use of the classical Green function for the electromagnetic field to express the fields above and below the surface as integrals over the value of the field at the surface. The periodicity of the grating means that the surface field can be written as a Fourier sum so that the reflected field can also be written as a sum. Calculations involve truncating these infinite sums and convergence problems can arise for deep enough gratings. Weber and Mills make numerical calculations for sawtooth (triangular) gratings and these can explain the main features of experimental data like those in figure 6.23.

At first sight, the increase in linewidth for larger h values in figure 6.23 is surprising, since the dielectric function of Ag has only a very small imaginary (dissipative) part and therefore the intrinsic linewidth of the Ag SPP is very small, as seen, e.g. in the ATR spectra of figure 6.13. Weber and Mills point out that the open structure of the grating allows direct coupling of the SPP to free-space radiation, so that the linewidth is due to radiative damping.

The 'classical' grating geometry, as described above, typically uses a single corrugated metal/dielectric interface, and the SPP propagates along this *corrugated* boundary. An alternative grating geometry, e.g. used by Hibbins *et al* (2000), employs a corrugated dielectric overlayer deposited on a metallic substrate with planar surface, along which the SPP propagates.

Among the other theoretical approaches that have been applied to this problem we may mention the work of Chandezon *et al* (1982) who use a transformation of coordinates to a new frame in which a sinusoidal surface corresponds to a constant value of one of the new coordinates. The development of the theory of reflection by bigratings, i.e. gratings that are periodic in both directions on the surface, for example by Glass *et al* (1983) and Glass (1987), is of considerable importance.

6.4.2 Modified surface modes

As mentioned, the presence of a grating on the surface leads to a modification of the properties of the SPP and in particular to the appearance of energy gaps at the zone boundaries. One of the first theoretical studies of this was due to Laks *et al* (1981). The radiative damping that has just been discussed means that an account of the modified properties in terms of simple dispersion curves requires caution. Nevertheless, we start with figure 6.24 (Glass *et al* 1984), which shows dispersion curves for a triangular grating on Ag calculated by an extension of the method of Weber and Mills (1983). As mentioned before, in the presence of damping both ω and q_x are in principle complex variables but in a particular experiment one or other may be real. For example, in an angle scan experiment, ω, which is fixed, is real. It is seen in figure 6.24 that gaps appear at the $n = 2$ and $n = 3$ zone boundaries for the real-q_x plot but not for the real-ω plot. This was taken further by Weber and Mills (1986) who pointed out that a more logical representation is of a single surface representing the measured reflectivity versus the parameters ω and angle of incidence θ_I, in effect ω

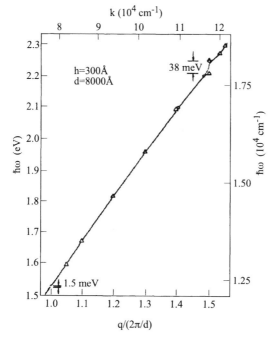

Figure 6.24. Calculated dispersion relations for SPP on a sawtooth grating (height 30 nm, period 800 nm) on Ag. ——, $\text{Re}(\omega)$ versus q_x (assumed real); – – –, ω (assumed real) versus $\text{Re}(k_x)$. After Glass *et al* (1984).

and q_x. An experimental scan is then a section parallel to one or other of these axes. If a point on a dispersion curve is defined as the position of a reflectivity dip then quite different curves are deduced from constant-frequency and constant-angle scans. A direct experimental confirmation of these ideas was given by Heitmann *et al* (1987). They applied a version of the light-emitting tunnel junction; their results will be deferred to the section on that topic.

The distinction between constant-ω and constant-q_x scans appears also in the optical study of SPP gaps by Nash *et al* (1995), who carried out angle scans on a gold grating of period 800 nm prepared by a similar technique to that used by Zaidi *et al* (1991a,b).

It is not necessary that the plane of incidence should contain the normal to the grating lines. As an example of theoretical and experimental work in which a more general relation is obtained we cite the study of Watts *et al* (1997) on theory and experiment for reflection from a silver grating. We mentioned bigratings (2D periodic structures) in section 6.4.1 and some properties of SPPs on gold bigratings are explored by Watts *et al* (1996).

6.4.3   Applications of grating coupling

Some applications of ATR were discussed in section 6.3.5 and the general comments made there apply also to applications of grating coupling. The key property is the enhancement of the field at the interface in a SPP that makes the mode sensitive to thin interface layers. Here we review a small number of examples rather than attempt a comprehensive survey.

Much of the work parallels similar results obtained by ATR. For example, Lyndin *et al* (1999) carried out theoretical and experimental studies of thin Cu films deposited on corrugated glass substrates. They demonstrated a very narrow reflection peak arising from coupling to the long-range surface plasmon. Calculations show that the angular position of the peak depends strongly on overlayers even of monolayer thickness, so that this structure is potentially useful as a sensor for adsorbates. Earlier calculations by Syguchov *et al* (1997) had already suggested this.

Sambles and coworkers make the point that the interface between a grating and a metal is protected from contamination so measurements of SPP reflectivity give a highly-accurate determination of the optical dielectric function of the metal. They have applied this technique to Cu, Ag, Zn and Al (Nash and Sambles 1995, 1996, 1998, Hallam *et al* 1999). The first three metals involved angle scan measurements for a range of wavelengths between 400 and 900 nm with grating pitch of 800 nm while the Al determination was at the single ultraviolet wavelength of 325 nm and grating pitch 800 nm. These studies were extended to narrow-ridged short-pitched gratings by Hooper and Sambles (2002).

The variation of the angle between the plane of incidence and the grating lines leads to a polarization conversion on reflection. This is illustrated in figure 6.25 which shows angle scans of R_{ps}, the reflected s intensity for p-polarized incident light of 633 nm wavelength incident on a silver grating of period 843 nm. ϕ is the angle between the normal to the plane of incidence and the direction of the grating lines, so that $\phi = 0$ corresponds to the usual arrangement in which all the wavevector transfer is perpendicular to the grating lines. As expected from symmetry, there is no conversion for $\phi = 0$ or for $\phi = 90°$ and not surprisingly the maximum conversion is at $\phi = 45°$. The peak is resonant corresponding to excitation of the SPP; its position moves to larger angles of incidence as ϕ increases. Vukusic *et al* (1992b) have exploited the sharpness of the line to devise a detector for adsorbed gases. This is possible because for a given ϕ the peak of R_{ps} in figure 6.25 shifts as molecules are adsorbed.

We mentioned in section 6.1.4 that the charge-sheet plasmon mode was first demonstrated experimentally by Batke and Heitmann (1984) using grating coupling. Studies of this kind have been pursued in detail by Hughes and collaborators. As an example, figure 6.26 shows transmission spectra obtained by grating coupling to this mode in a 2D electron gas (2DEG) in

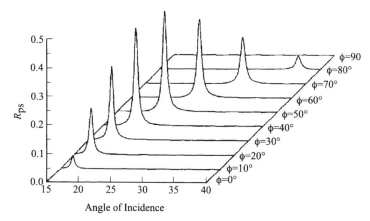

Figure 6.25. Angle scans of reflection of p- into s-polarized light for an angled grating. ϕ is the angle between the normal to the plane of incidence and the grating lines. After Bryan-Brown *et al* (1990).

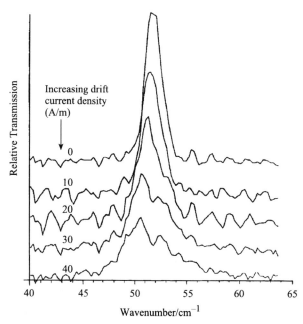

Figure 6.26. Far infrared transmission spectra for grating coupling to a drifting 2DEG in a AlGaAs/As structure. Grating period 0.75 mm, doping density 5×10^{12} cm^{-2}. After Tyson *et al* (1993).

an AlGaAs/GaAs heterostructure. The spectra were measured with the 2DEG at rest relative to the grating (top curve) and also for various drift velocities of the 2DEG that were induced by passing a current along it. The transmission peaks result from coupling to the plasmon and a Doppler shift resulting from the drift velocity can be seen. Analysis shows that the shift is nonlinear in the drift velocity; this is discussed by Tyson *et al* (1993).

6.5   Other experimental methods

In this section we group together several related forms of spectroscopy. First, we deal with the application of Raman scattering to observe surface polaritons (SPs). As mentioned at the end of section 2.5.1, much of the experimental investigation of bulk polaritons was carried out by means of Raman spectroscopy. By contrast, compared with ATR spectroscopy, Raman scattering by SPs is a technically difficult experiment and only a relatively small number of groups have published measurements. This work is reviewed in section 6.5.1.

It was pointed out in section 6.1.1 that the field strength is large at the surface in the SP. This means that *surface-enhanced Raman scattering* measurements can be performed in which the incident and scattered beams propagate as SPs. An introduction to this topic is given in section 6.5.2.

As was discussed in section 6.4, the presence of a grating, or more generally surface roughness, on a surface means that the SP becomes radiative, that is, it can couple directly to the electromagnetic field in the bounding medium. The way in which this leads to radiative damping of the SP on a grating was described in section 6.4.2. Another implication is that if the SP is excited in some way it can decay radiatively with emission of radiation. This is achieved in the *light-emitting tunnel junction*, in which the SP is generated by the tunnelling current in a metal/oxide/metal junction; this forms the subject of section 6.5.3.

In addition, we should mention that the properties of surface plasmon–polaritons (SPPs) are related to interesting advances in near-field optics, particularly in scanning near-field optical microscopy (SNOM). These developments are the focus of the book edited by Kawata (2001).

6.5.1   Raman scattering by surface polaritons

It was thought at one time that SPs might be seen by backscattering from the surface of a semi-infinite medium. This is true in principle, but in most cases the cross section is expected to be smaller by a factor of 10^2 or more than that for forward scattering from a film sample. The reason was first explained in detail by Chen *et al* (1975). The theory is similar to that given for Brillouin scattering in section 3.3.2 and in particular the backscattered field contains a factor $(Q_z + k_{\bar{1}2}^{\bar{z}} + k_{\bar{S}2}^{\bar{z}})^{-1}$ like that seen in (3.49). Here $k_{\bar{1}2}^{\bar{z}}$ and $k_{\bar{S}2}^{\bar{z}}$ are the z components of the wavevectors of the incident and scattered light. It is seen that they *add* to generate a large denominator for back scattering. By contrast, for near-forward scattering the corresponding factor is $(Q_z + k_{\bar{1}2}^{\bar{z}} - k_{\bar{S}2}^{\bar{z}})^{-1}$. The z components of the optical wavevectors now *subtract* to produce a small denominator and consequently a larger cross section. The combination of a small scattering volume with the need to employ near-forward scattering angles makes the experiments technically demanding.

The first experimental results for the single-interface polariton were reported by Valdez and Ushioda (1977) using a GaP sample of thickness $L = 20\,\mu\text{m}$. A correct choice of L in such work is crucial. The frequency of the argon-ion laser line used by Valdez and Ushioda is close to the GaP band gap, so that resonant enhancement of the cross section occurs (Hayes and Loudon 1978) and the sample is fairly opaque; L was chosen large enough so that the upper- and lower-branch polariton dispersion curves effectively coincide on the single-interface curve, and small enough for some light to be transmitted. Rather than their work we show in figure 6.27(a) spectra from a later study by Denisov *et al* (1987) of a GaP film of thickness 11 µm. The exciting radiation was the second harmonic of a pulsed Nd^{3+} laser (wavelength 532 nm, photon energy 2.33 eV, compared with the 2.34 eV indirect gap of GaP). The spectra are dominated by the large peak due to the LO phonon at $\omega_{\text{LO}}/2\pi c = 403\,\text{cm}^{-1}$ which is identified from the comparison spectrum at $\theta = 135°$; the surface-polariton scattering is the feature marked by arrows. It is seen that its frequency changes with scattering angle θ, corresponding to the dispersion of the surface-polariton curve. The peak position is plotted and compared with the theoretical dispersion curve and with earlier experimental results in figure 6.27(b). Although the spectra in figure 6.27(a) are for unpolarized light, Denisov *et al* (1987) go on to show and discuss polarized spectra. The published work on single-surface modes includes the experimental and theoretical study by Watanabe *et al* (1989) of the SP on a GaP surface on which a photolithographic grating had been prepared.

The Raman work in the literature includes detailed investigations of all the polariton modes of a film, as discussed in section 6.2.1. Evans *et al* (1973) studied the lower surface-type mode of a GaAs film on sapphire. Subsequently Prieur and Ushioda (1975) detected the upper mode as well while Davydov *et al* (1997) observed the upper mode on a free-standing GaN film. Exceptionally, backscattering was employed in this paper; the high optical reflectivity of the lower surface of the GaN film leads to a breakdown in the general argument of Chen *et al* (1975) (or one might say that the spectrum is detected in forward scattering from the reflected beam).

The possibility of observing guided-wave polaritons emerged from theoretical work by Mills *et al* (1976) and by Subbaswamy and Mills

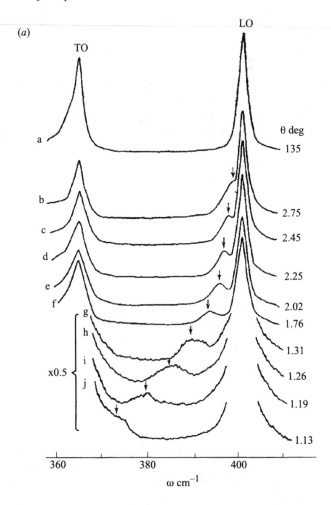

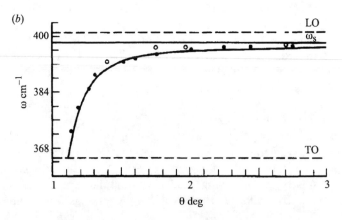

(1978) and detailed spectra were than obtained on a GaP film and compared with the dispersion curves (Ushioda and Loudon 1982). Similar work was reported by Denisov *et al* (1988). Full calculations, essentially along the lines indicated in section 3.3.2, were published by Nkoma and Loudon (1975) and Nkoma (1975) as well as by Mills *et al* (1976). These and other papers are reviewed by Cottam and Maradudin (1984).

6.5.2 Raman scattering via surface plasmon–polaritons

We commented in section 6.1.1 that the electromagnetic field intensity in a surface polariton is strongly enhanced near the interface. In addition, Weber and Mills (1983) showed that there is a substantial field enhancement near the ridge point in a triangular grating; this is an example of the enhancement that is found at any sharp metallic ridge or point. Both of these effects have been applied to produce a very large increase in the sensitivity of Raman spectroscopy, particularly in the study of surface adsorbates. Typically, the interface studied is between a metal and the adsorbate, so the surface mode that is involved is a surface plasmon–polariton (SPP).

The first effect to be discovered, by Fleischmann *et al* (1974), was *surface-enhanced Raman scattering* (SERS). This refers to an enormous increase, by a factor of up to 10^6 relative to the unenhanced value, in the Raman cross section for scattering by molecules adsorbed on a rough metal surface. As mentioned in section 6.4, a rough surface may be described as a random Fourier sum of gratings. The roughness therefore works like a grating in producing coupling of the incident and scattered beams to SPPs in the surface and in giving additional field enhancement at points with a small radius of curvature. It would take us too far afield to discuss this substantial field, and we refer to the comprehensive review by Moskovits (1985).

A more precise technique than SERS involves an arrangement whereby the incident and scattered radiation propagate as SPPs. In the pioneering work by Ushioda and Sasaki (1983) prism coupling in the Raether–Kretschmann geometry was used, as in figure 6.28. The incident light was the 514.5 nm line of an argon ion laser and the Raman spectrum studied was that due to the 2835 cm^{-1} vibrational line of CH_3OH (methyl alcohol). In a similar way to the Otto geometry, figure 6.13, the incident beam produces an intense SPP at the lower Ag interface in a narrow linewidth centred around an angle θ_{I0}, calculated as 73° in this case. The corresponding resonance angle for the scattered beam is $\theta_{S0} = 68°$; the two angles are not

---

Figure 6.27. (a) Unpolarized Raman spectra of a single-crystal GaP film for near-forward scattering angles θ marked. (b) Marked peak positions versus scattering angle. ● from (a); ○ data of Ushioda *et al* (1979); ——, theoretical. LO phonon and electrostatic surface mode (ω_S) frequencies marked. After Denisov *et al* (1987).

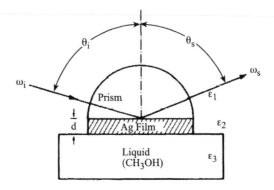

Figure 6.28. ATR/Raman sample used by Ushioda and Sasaki (1983).

the same because of the frequency shift in the scattering. It was found that the Raman intensity is a resonant function of both the angle of incidence θ_I and the angle of the scattered beam θ_S, peaking at $\theta_I = \theta_{I0}$ and $\theta_S = \theta_{S0}$. The maximum intensity is a factor of about 4×10^4 larger than that observed in a subsidiary conventional Raman measurement on methyl alcohol in an optical cell. This measured factor agrees well with a prior theoretical estimate (Sakoda *et al* 1982). The increase in intensity is lower than can be observed in SERS because the experiment takes advantage of the flat-surface field enhancement but not of the additional enhancement found at rough points. The technique can be applied to adsorbed layers, as for instance in the study by Bruckbauer and Otto (1998) of pyridine adsorbed on single-crystal Cu surfaces.

As is clear from earlier sections, an alternative to a prism-coupling scheme like that in figure 6.28 is the use of grating coupling. From our earlier discussion, it will be clear that the grating should not be too deep since otherwise the radiative broadening of the coupling lines would be unacceptable. The estimated increase in sensitivity for grating coupling is of order 10^4 (Nemetz and Knoll 1992), comparable with that in ATR coupling and lower than in SERS. This sensitivity means that, as with ATR coupling, the technique is sensitive to monolayer coverage. Together with the use of CCD detectors, grating coupling has been developed into a technique for imaging adsorbed layers by means of their Raman spectra (Knobloch and Knoll 1991).

6.5.3 Direct radiation via surface plasmon polaritons

The direct coupling of surface polaritons to external radiation by means of a grating or a rough surface has been the basis of the effects discussed in a number of previous sections. With the exception of the scattered beam in

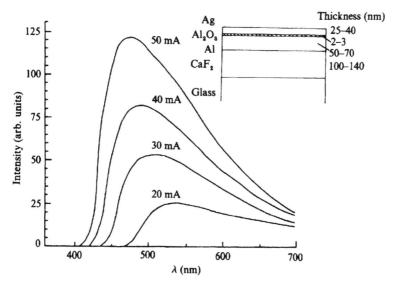

Figure 6.29. Spectra of light emitted in normal direction by rough $Al/Al_2O_3/Ag$ tunnel junctions. Curves are labelled by tunnelling current I_0. Inset shows sample geometry. After Dawson *et al* (1984).

Raman scattering, the effects have made use of the generation of SPPs by incident radiation. Here we describe experiments involving the inverse effect: an already generated SPP radiates light away from the surface.

The first method of excitation that was employed is the use of the tunnel current flowing between two metals. If the interfaces are rough or if the substrate is a diffraction grating, the junction emits light, typically in the visible part of the spectrum. An example of such light emission is shown in figure 6.29, with the sample configuration shown schematically in the insert. The first layer deposited on the glass substrate is CaF_2; this has a rough top surface, and consequently all subsequent interfaces are rough. The tunnel junction itself is $Al/Al_2O_3/Ag$, the barrier being formed by oxidation of the Al film. In the experiment a d.c. bias voltage V_0 is applied between the Al and the Ag and a tunnel current I_0 flows through the Al_2O_3 barrier. The spectrum of light emitted by the junction is seen to change from red to blue as V_0 increases. The actual spectrum emitted in the direction of the sample normal is shown in figure 6.29. It should be noted that the spectrum is unpolarized, and the maximum photon energy $\hbar\omega_0$, corresponding to the short-wavelength cut-off of the spectrum, is given to a good approximation by $\hbar\omega_0 = eV_0$. Thus at the short-wavelength end virtually all of the energy of a tunnelling electron is converted into photon energy.

Two other important forms of light-emitting tunnel junction are, first, a junction deposited on a diffraction grating and, second, a junction in which the top electrode consists of discrete metal particles. In the first case the spectrum is p-polarized, since this is the polarization of the SPP, and has a sharp peak whose wavelength varies with angle of emission relative to the sample normal. In the second case the spectrum is predominantly p-polarized but is broadband.

As implied at the beginning of this section, in all three types of light-emitting junction the tunnel current excites a surface plasmon that because of the surface roughness is able to decay radiatively to produce the observed spectrum. The nature of the plasmon excited depends on the form of the junction; for a detailed discussion and further references see Dawson *et al* (1984).

As is clear, the function of the tunnel junction in these experiments is simply to generate the SPP and the junction can be replaced by other excitation mechanisms. An ingenious scheme was proposed by Gruhlke *et al* (1986). 488 nm radiation is incident on the silica substrate and excites a broadband fluorescence in the photoresist which in turn generates SPPs of a range of frequencies at the lower Ag interface. Detailed accounts of these experiments are given in the original paper and in Dawson *et al* (1994).

References

Abeles F 1986 in *Electromagnetic Surface Excitations* ed R F Wallis and G I Stegeman (Berlin: Springer)

Abraha K and Tilley D R 1994 *Infrared Phys. Tech.* **35** 681

Abraha K and Tilley D R 1996 *Surf. Sci. Rep.* **24** 125

Adams M J 1981 *An Introduction to Optical Waveguides* (New York: Wiley)

Agranovich V M and Mills D L (eds) 1982 *Surface Polaritons* (Amsterdam: North-Holland)

Batke E and Heitmann D 1984 *Infrared Phys. Tech.* **24** 189

Borstel G, Falge H J and Otto A 1974 *Springer Tracts in Modern Physics* **74** 107

Borstel G and Falge H J 1977 *Phys. Stat. Solidi (b)* **83** 11

Borstel G and Falge H J 1978 *Appl. Phys.* **16** 211

Brown D E, Dumelow T, Jensen M R F and Parker T J 1999 *Infrared Phys. Tech.* **40** 219

Bruckbauer A and Otto A 1998 *J. Raman Spect.* **29** 665

Bryan-Brown G P, Sambles J R and Hutley M C 1990 *J. Modern Optics* **37** 1227

Bryan-Brown G P, Yang F, Bradberry G W and Sambles J R 1991 *J. Opt. Soc. Am. B* **8** 765

Bryksin V V, Mirlin D N and Firsov Yu A 1974 *Usp. Fiz. Nauk* **113** 29; *Sov. Phys. Usp.* **17** 305

Burstein E and DeMartini F (eds) 1974 *Polaritons* (New York: Pergamon)

Chandezon J, Dupuis M T, Cornet G and Maystre D 1982 *J. Opt. Soc. Am.* **72** 839

Chen Y J, Burstein E and Mills D L 1975 *Phys. Rev. Lett.* **34** 1516

Cole M W 1974 *Rev. Mod. Phys.* **46** 451

Cottam M G and Maradudin A A 1984 in *Surface Excitations* ed V M Agranovich and R Loudon (Amsterdam: North-Holland)

Craig A E, Olsen G A and Sarid D 1983 *Opt. Lett.* **8** 380

Davydov V Yu, Subashiev A V, Cheng T S, Foxon V T, Goncharuk I N, Smirnov A N and Zolotareva R V 1997 *Solid State Commun.* **104** 397

Dawson P, Walmsley D G, Quinn H A and Ferguson A J L 1984 *Phys. Rev. B* **30** 3164

Dawson P, Bryan-Brown G and Sambles J R 1994 *J. Modern Optics* **41** 1279

DeMartini F, Colocci M, Kohn S E and Shen Y R 1977 *Phys. Rev. Lett.* **38** 1223

Denisov V N, Mavrin B N and Pobobedov V B 1987 *Sov. Phys. JETP* **65** 1042

Denisov V N, Leskova T A, Mavrin B N and Pobobedov V B 1988 *Sov. Phys. JETP* **67** 1013

Evans D J, Ushioda S and McMullen J D 1973 *Phys. Rev. Lett.* **31** 372

Faulkner J, Flavell W R, Davies J, Sunderland R F and Nunnerley C S 1993 *J. Electron Spect.* **64/65** 441

Fetter A L 1973 *Ann. Phys.* **81** 367

Fleischmann M, Hendra P J and McQuillan A J 1974 *Chem. Phys. Lett.* **26** 163

Fukui M, So V C Y and Normandin R 1979 *Phys. Stat. Sol. (b)* **91** K61

Futamata M 1997 *Applied Optics* **36** 364

Geddes N J, Martin A S, Caruso F, Urquhart R S, Furlong D N, Sambles J R, Than K A and Edgar J A 1994 *J. Immunological Meth.* **175** 149

Glass N E 1987 *Phys. Rev. B* **35** 2647

Glass N E, Maradudin A A and Celli V 1983 *J. Opt. Soc. Am.* **73** 1240

Glass N E, Weber M and Mills D L 1984 *Phys. Rev. B* **29** 6584

Gruhlke R W, Holland W R and Hall D G 1986 *Phys. Rev. Lett.* **56** 2838

Hallam B T, Sambles J R and Kitson S C 1999 *J. Mod. Optics* **46** 1099

Hartstein A, Burstein E, Maradudin A A, Brewster R and Wallis R F 1973 *J. Phys. C* **6** 1266

Hayes W and Loudon R 1978 *Scattering of Light by Crystals* (New York: Wiley)

Heavens O S 1965 *Optical Properties of Thin Solid Films* (New York: Dover)

Heidberg J, Vossberg A, Hustedt M, Thomas M, Briquez S, Picaud S and Girardet C 1999 *J. Chem. Phys.* **110** 2566

Heitmann D, Kroo N, Schultz C and Szentirmay Z 1987 *Phys. Rev. B* **35** 2660

Hibbins A P, Sambles J R and Lawrence C R 2000 *J. Appl. Phys.* **87** 2677

Hooper I R and Sambles J R 2002 *Phys. Rev. B* **66** 205408

Jensen M R F, Parker T J, Abraha K and Tilley D R 1995 *Phys. Rev. Lett.* **75** 3756

Jensen M R F, Feiven S A, Parker T J and Camley R E 1997 *J. Phys.: Condens. Matter* **9** 7233

Karsono A D and Tilley D R 1977 *J. Phys. C* **10** 2123

Karsono A D and Tilley D R 1978 *J. Phys. C* **11** 3487

Kawata S (ed) 2001 *Near-Field Optics and Surface Plasmon Polaritons* (Berlin: Springer)

Knobloch H and Knoll W 1991 *J. Chem. Phys.* **94** 835

Kim O K and Spitzer W G 1979 *J. Appl. Phys.* **50** 4362

Lagois J and Fischer B 1976 *Phys. Rev. Lett.* **36** 380

Lagois J and Fischer B 1978 *Phys. Rev. B* **17** 3814

Lagois J and Fischer B 1982 in *Surface Polaritons* ed V M Agranovich and D L Mills (Amsterdam: North-Holland)

Laks B, Mills D L and Maradudin A A 1981 *Phys. Rev. B* **23** 4965

Landau L D and Lifshitz E M 1971 *Electrodynamics of Continuous Media* (Oxford: Pergamon)

Legvold S 1980 in *Ferromagnetic Materials* vol 1, ed E P Wohlfarth (Amsterdam: North-Holland) ch 3

Loring J S and Land D P 1998 *Appl. Optics* **37** 3515

Lyndin N M, Salakhutdinov I F, Sychugov V A, Usievich B A, Pudonin F A and Parriaux O 1999 *Sensors and Actuators B* **54** 37

Majkrzak C F, Kwo J, Hong M, Yafet Y, Gibbs D, Chien C L and Bohr J 1991 *Adv. Phys.* **40** 99

Maradudin A A and Mills D L 1973 *Phys. Rev. B* **7** 2787

Marchand M and Caillé A 1980 *Sol. State Commun.* **34** 827

Marschall N and Fischer B 1972 *Phys. Rev. Lett.* **28** 811

Mills D L, Chen Y J and Burstein E 1976 *Phys. Rev. B* **13** 4419

Mirlin D N 1982 in *Surface Polaritons* ed V M Agranovich and D L Mills (Amsterdam: North-Holland)

Moskovits M 1985 *Rev. Mod. Phys.* **57** 783

Nakayama M 1974 *J. Phys. Soc. Japan* **36** 393

Nash D J and Sambles J R 1995 *J. Mod. Optics* **42** 1639

Nash D J and Sambles J R 1996 *J. Mod. Optics* **43** 81

Nash D J and Sambles J R 1998 *J. Mod. Optics* **45** 2585

Nash D J, Cotter N P K, Wood E L, Bradberry G W and Sambles J R 1995 *J. Mod. Optics* **42** 243

Nemetz A and Knoll W 1992 *J. Chem. Phys.* **97** 7835

Nkoma J S 1975 *J. Phys. C* **8** 3919

Nkoma J S and Loudon R 1975 *J. Phys. C* **8** 1950

Nkoma J S, Loudon R and Tilley D R 1974 *J. Phys. C* **7** 3547

Otto A 1974 *Festkörperprobleme* **XIV** 1

Otto A 1976 in *Optical Properties of Solids: New Developments* ed B O Seraphim (Amsterdam: North-Holland) p 678

Palik E D, Kaplan R, Gammon R W, Kaplan H, Wallis R F and Quinn J J 1976 *Phys. Rev. B* **13** 2497

Prieur J-Y and Ushioda S 1975 *Phys. Rev. Lett.* **34** 1012

Rudkevich E, Feng Liu, Savage D E, Kuech T F, McCaughan L and Lagally M G 1998 *Phys. Rev. Lett.* **81** 3467

Sakoda K, Ohtaka K and Hanamura E 1982 *Solid State Commun.* **41** 393

Sambles J R, Bradberry G W and Yang F 1991 *Contemp. Phys.* **32** 173

Sarid D 1981 *Phys. Rev. Lett.* **47** 1927

Sarmento E F and Tilley D R 1982 in *Electromagnetic Surface Modes* ed A D Boardman (Chichester: Wiley) ch 16

Sibley M J N 1995 *Optical Communications* (Basingstoke: Macmillan)

Stamps R L and Camley R E 1989 *Phys. Rev. B* **40** 596

Stern F 1967 *Phys. Rev. Lett.* **18** 546

Subbaswamy K R and Mills D L 1978 *Solid State Commun.* **27** 1085

Sychugov V A, Tiscchenko A V, Lyndin N M and Parriaux O 1997 *Sensors and Actuators B* **38/39** 360

Toigo F, Marvin A, Celli V and Hill N R 1975 *Phys. Rev. B* **15** 5618

Tokura Y, Koda T, Hirabayashi I and Nakada S 1981 *J. Phys. Soc. Japan* **50** 145

Tsuchida H, Kamata I and Izumi K 1999 *J. Appl. Phys.* **85** 3569

Tyson R E, Stuart R J, Hughes H P, Frost J E F, Ritchie D A, Jones G A C and Shearwood C 1993 *Int. J. Infrared and Millimeter Waves* **14** 1237

Ushioda S and Loudon R 1982 in *Surface Polaritons* ed V M Agranovich and D L Mills (Amsterdam: North-Holland)

Ushioda S and Sasaki Y 1983 *Phys. Rev. B* **27** 1401

Ushioda S, Aziza A, Valdez J B and Mattei G 1979 *Phys. Rev. B* **19** 4012

Valdez J B and Ushioda S 1977 *Phys. Rev. Lett.* **38** 1088

Vukusic P S, Sambles J R and Wright J D 1992a *J. Mater. Chem.* **2** 1105

Vukusic P S, Bryan-Brown G P and Sambles J R 1992b *Sensors and Actuators B* **8** 155

Watanabe J, Uchinokura K and Sekine T 1989 *Phys. Rev. B* **40** 7860

Watts R A, Harris J B, Hibbins A P, Preist T W and Sambles J R 1996 *J. Mod. Optics* **43** 1351

Watts R A, Preist T W and Sambles J R 1997 *Phys. Rev. Lett.* **79** 3978

Weber M G 1986 *Phys. Rev. B* **33** 909

Weber M G and Mills D L 1983 *Phys. Rev. B* **27** 2698

Weber M G and Mills D L 1986 *Phys. Rev. B* **34** 2893

Wendler L and Haupt R 1986a *Phys. Stat. Sol. (b)* **137** 269

Wendler L and Haupt R 1986b *J. Phys. C* **19** 1871

Wood R W 1935 *Phys. Rev.* **48** 928

Yang F, Sambles J R and Bradberry G W 1990a *Phys. Rev. Lett.* **64** 559

Yang F, Sambles J R and Bradberry G W 1990b *J. Mod. Optics* **37** 1545

Yang F, Sambles J R and Bradberry G W 1991a *Phys. Rev. B* **44** 5855

Yang F, Sambles J R and Bradberry G W 1991b *Phys. Rev. Lett.* **66** 2030

Yariv A 1985 *Optical Electronics* (New York: Holt, Rinehart and Winston)

Zaidi S H, Yousak M and Brueck S R J 1991a *J. Opt. Soc. Am. B* **8** 770

Zaidi S H, Yousak M and Brueck S R J 1991b *J. Opt. Soc. Am. B* **8** 1348

Chapter 7

Other surface excitations

In the preceding chapters we have presented a survey of the dynamical properties of materials with planar surfaces and interfaces, assuming geometries for either semi-infinite media or thin films. The selection of surface excitations, with which we illustrated the general principles, was fairly extensive but inevitably incomplete. Our choice was guided by the existence of appropriate theoretical models, by the availability of data to compare theory and experiment, and in particular by recent advances in fabricating low-dimensional solid structures.

Several topics relevant to the study of surface excitations in single-interface and film geometries have been omitted, or mentioned only in passing, in the previous chapters. It is useful now to provide some brief comments on these topics, together with references to just a few illustrative examples in each case, rather than giving a detailed account.

The topics fall broadly into two categories. The first concerns other kinds of excitations in film geometries, including single-electron surface states (as in section 7.1) and 'mixed' excitations involving coupled waves of two different kinds (as in section 7.2). The second category deals with excitations in other geometries. Thus in section 7.3 we discuss excitations in systems where there are planar surfaces that are non-parallel, e.g. giving edges and wedges in the simplest case and structures such as long wires (or rods) with a rectangular cross section if more surfaces are involved. Then in sections 7.4 and 7.5 we describe surfaces and interfaces that are non-planar, either through being curved (e.g. as in cylindrical and spherical samples), rough (e.g. in a random fashion or with regular corrugations) or patterned (e.g. with steps or arrays of on-grown features). As we will discuss, there is a strong theoretical and experimental interest nowadays in such systems as exemplified by various kinds of nanowire structures that can be fabricated either singly or in arrays, or deposited on different substrates.

7.1 Single-electron surface states

In earlier chapters of this book we have seen how in some circumstances a surface mode can arise at a frequency within a forbidden gap for the corresponding bulk mode. For example, the calculation of section 3.1 for a semi-infinite diatomic lattice shows such a mode appearing in the frequency gap between the acoustic and optic branches, and we have given several analogous examples for other excitations.

We now turn our attention to *single-electron* surface states, by contrast with the collective electron properties already discussed in chapter 6 in terms of surface plasmon–polaritons. Interest in this particular topic has a long history, since Tamm (1932) first pointed out that electronic surface states could exist. He considered solutions of Schrödinger's equation for a model 1D system comprising a semi-infinite lattice of square potential wells, i.e. the semi-infinite analogue of the familiar Kronig–Penney model in solid-state physics (see e.g. Kittel 1986). For about 20 years after his paper was published, various other calculations for model 1D systems were performed corresponding to different choices for the potential wells. Attention then shifted to the question of performing realistic calculations for actual surfaces, the theoretical methods being developed from those used for calculation of bulk electronic band structure. A full treatment of this later work would involve an extensive account of band-structure theory, and is beyond the scope of this book. Instead, we restrict our attention to a brief account of the model calculations, along with references to more recent theories and a description of some relevant experimental results. Some good general references for further details are the books by Ehrenreich *et al* (1980), Lannoo and Friedel (1991) and Lüth (1995).

7.1.1 Theory

As mentioned, the first simple calculation was that given for a 1D semi-infinite lattice by Tamm (1932). Soon afterwards, a number of related calculations appeared, notably those by Goodwin (1939) and Shockley (1939), who both dealt with finite rather than semi-infinite 1D arrays of potentials and made other choices for the potential wells. We summarize here the complete and illuminating discussion given by Shockley.

The model considered is sketched in figure 7.1(a). It consists of N repeat units of a general potential $V(x)$ with cells of size d. The potential outside the specimen has a fixed value V_0, related to the work function of the material in question. Shockley draws attention to two differences between his model (essentially a nearly-free electron theory) and that used by Goodwin, which is based on the tight-binding approximation. First, the potential within a cell is more realistic in Goodwin's model, represented in figure 7.1(b), since it corresponds more closely to an atomic potential. Second,

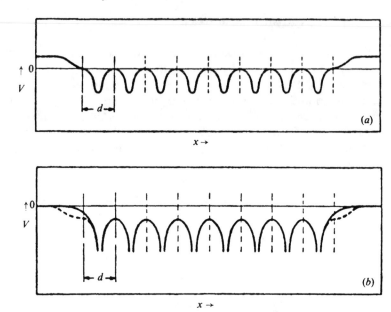

Figure 7.1. Potentials used by Shockley (a) and Goodwin (b) in surface-state calculations. After Shockley (1939).

the potential in the end cells within the specimen differs from that in all other cells.

Within any one interior cell of figure 7.1(a), centred at $x = 0$ say, the general solution of Schrödinger's equation may be written in terms of the two complex amplitudes a and b as

$$\psi(x) = ag(x) + ibu(x) \tag{7.1}$$

where $g(x)$ and $u(x)$ are respectively even and odd functions with respect to the mid-point of the cell. The phase factor i is inserted in the second term for later algebraic convenience. Use of the 1D Bloch's theorem shows that in the neighbouring cell, centred at $x = d$, the wave function is

$$\psi(x) = \exp(iqd)[ag(x - d) + ibu(x - d)]. \tag{7.2}$$

Continuity of ψ and its derivative $d\psi/dx$ at the point $x = d/2$ yields

$$ag + ibu = \exp(iqd)(ag - ibu) \tag{7.3}$$

$$ag' + ibu' = \exp(iqd)(-ag' + ibu') \tag{7.4}$$

where the functions g and u and their derivatives (denoted by g' and u') are all evaluated at $x = d/2$, and use has been made of the symmetry of g and u. Solvability of (7.3) and (7.4) for the amplitudes a and b requires

$$\tan^2(qd/2) = -g'u/gu'. \tag{7.5}$$

In the limit of an infinite 1D crystal (7.5) determines the band structure. For a given energy E the functions $g(x)$ and $u(x)$ appearing in (7.1) are determined. Thus the right-hand side of (7.5) is also determined. If it is positive then (7.5) has a solution in which q is real, so E lies within a pass band. Alternatively, if the right-hand side is negative, then (7.5) has no solution for real q, and E lies within a stop band.

For determination of the energy-level spectrum and wavefunctions of the finite-length system sketched in figure 7.1(a) it is necessary to match the Bloch functions to wavefunctions for the outside regions where the potential is V_0. In each of these regions the wavefunction has the form of a simple exponential that decays with distance away from the specimen. It is proportional to $\exp(\lambda x)$ on the left ($x \rightarrow -\infty$) and $\exp(-\lambda x)$ on the right ($x \rightarrow \infty$), where λ is real and positive:

$$\lambda = \sqrt{2m(V_0 - E)/\hbar^2}. \tag{7.6}$$

The implications for bulk and surface states (as the matched solutions with E falling within a pass band or a stop band, respectively) are discussed in detail by Shockley. The main results may be described with reference to the example in figure 7.2, which shows (in the spirit of the tight-binding approximation) the energy bands derived from three of the energy levels of the isolated atom. The curves are drawn for a finite specimen comprising eight equally spaced atoms, so each 'band' is represented by eight relatively close curves. As $d \rightarrow \infty$, the energy-level diagram simply shows three

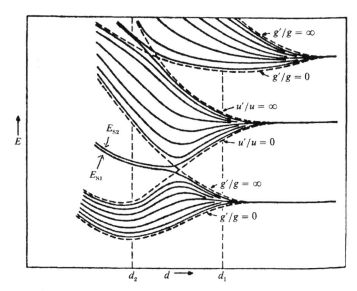

Figure 7.2. Energy E versus lattice parameter d for a 1D specimen with eight atoms. After Shockley (1939).

eight-fold degenerate eigenvalues at the positions of the atomic energy levels. For finite but relatively large d, for example the value d_1 marked, the usual broadening into tight-binding bands is seen. Then, when d is sufficiently small, for example d_2 as marked, two eigenvalues labelled E_{S1} and E_{S2} appear between the two lower bands. These correspond to symmetric and antisymmetric combinations of states with amplitudes that are maxima at one or other surface. As would be expected by analogy with the Rayleigh waves in a finite elastic slab (see section 3.2.1) or magnons in a Heisenberg ferromagnetic film (see section 4.2.2), the separation in energy $E_{S2} - E_{S1}$ decreases exponentially with increase in the total thickness $L = Nd$ of the specimen.

An example of an analytic calculation, based on the above Shockley method, is presented in the book by Lüth (1995) for the *semi-infinite case* ($N \to \infty$) and thus for a single surface (taken at $x = 0$). On the right side ($x > 0$) a constant potential V_0 is assumed, so the wavefunction is proportional to $\exp(-\lambda x)$ with λ given by (7.6). On the left side ($x < 0$) a sinusoidal potential function is chosen:

$$V(x) = V_g \cos(2\pi x/d), \qquad x < 0 \tag{7.7}$$

where $V_g < V_0$. This implies a discontinuity in the potential at $x = 0$ so it is not a realistic choice; nevertheless it serves as a model calculation. If the solutions for the energy bands corresponding to (7.7) are examined it is found that there is a gap in the bulk-mode spectrum at the Brillouin-zone edge π/d that extends from $E = (\hbar^2 \pi^2 / 2md^2) - \frac{1}{2}V_g$ to $(\hbar^2 \pi^2 / 2md^2) + \frac{1}{2}V_g$. This gap of magnitude V_g is just the result from the familiar nearly-free-electron model described in solid-state physics textbooks. Lüth (1995) shows explicitly how to construct the wavefunction for the half-space region $x < 0$. Then, proceeding as we outlined before, it can be verified that there is indeed a single surface-state solution for E in the gap region. Similar calculations, but with slightly different assumptions, are provided in the books by Zangwill (1988) and Lannoo and Friedel (1991). Lannoo and Friedel also describe tight-binding calculations and iterative Green function methods for the electronic states of semi-infinite media.

The above results are presented for a 1D model. A difference in the 3D case, as we have seen before, is that the modes are described in terms of the wavevector $\mathbf{q}_\parallel$ in the plane of the surface. Thus, as for phonons (e.g. figures 3.3 and 3.4) or magnons (e.g. figure 4.4), in the 3D case the surface electronic states form a band characterized by the dispersion relation $E = E(\mathbf{q}_\parallel)$. On an E versus $\mathbf{q}_\parallel$ plot, the bulk bands appear as a quasi-continuum in the usual way (see e.g. Lüth 1995).

In terms of the usual description of energy bands, the surface states of figure 7.2 arise within zone-boundary gaps, since they correspond to extreme values of g'/g or u'/u as indicated in the 1D case. In generalized calculations of 3D band structures, gaps between extrema can also arise

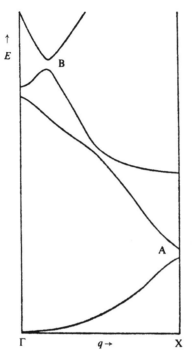

Figure 7.3. Zone-edge (A) and non-zone-edge (B) gaps in a hypothetical band structure. After Forstmann (1970).

inside the Brillouin zone. That is, as sketched in figure 7.3, gaps of type B can arise as well as gaps of type A. The question then arises whether surface states can be associated with type B gaps. This was considered by Fortsmann (1970) and by Pendry and Gutman (1975), who concluded that surface modes do indeed appear in these gaps.

As previously indicated, we have restricted attention to relatively simple model calculations. Later work was concerned with realistic calculations for surface states within the general framework of electron-band theory. The techniques used were generally more advanced than the wave-function matching described here, often requiring intensive numerical calculations. For example, the surface Green function matching (SGFM) technique mentioned in section 3.2.3 has been widely applied in electron problems, along with other techniques from the many-body theory. A good introductory account is given by Heine (1980), while other articles in the same volume, edited by Ehrenreich *et al* (1980), provide more detail. A particularly clear and comprehensive account, covering the basics as well as more recent work, is given in the books by Lannoo and Friedel (1991) and Lüth (1995). These latter authors describe several variations and elaborations of the 1D calculations and deal with generalizations to 3D (with surface reconstruction effects included where needed). They make a distinction in the 3D case between two kinds of surface states (gap solutions):

(i) The state is in a true gap (with forbidden energies in all wavevector space directions). The surface state is then a true *surface state* and may not connect to the region of bulk modes.
(ii) There exists at least one direction in wavevector space where the energies corresponding to the gap are degenerate with allowed bulk states. This is a *surface-resonant state* that may connect to a bulk state. It has an increased amplitude and density of states over a finite range near the surface compared with the bulk region.

Lannoo and Friedel (1991) and Lüth (1995) provide illustrative examples and calculations of both types of surface states, covering applications to transition metals, covalent semiconductors (such as Si) and compound semi-conductors (such as GaAs). Localized electronic states associated with interfaces between different materials are also discussed. Further background to some of the computational methods involved is to be found in the book by Harrison (2000).

7.1.2 Some experimental results

The electronic surface states described in section 7.1.1 are often referred to as *intrinsic surface states,* since they are a property of microscopically clean surfaces, albeit surfaces that may have undergone surface reconstruction, as in Si for various crystallographic orientations (see e.g. Lannoo and Friedel 1991). The experimental study of these states involves working in ultrahigh vacuum, of order 10^{-10} torr, since at higher pressures contamination of the surface takes place very rapidly. A great deal of attention, both experimental and theoretical, has also been given to the *extrinsic surface states* associated with adsorbed or absorbed molecules. An example is provided by the Si(111) surface which undergoes various adsorbate-induced surface reconstructions when covered by a metal, as discussed by Lannoo and Friedel (1991). Some interesting cases are the coverages corresponding to $(\sqrt{3} \times \sqrt{3})$ Al, Ga or In (we employ the surface notation introduced in section 1.2.2). From a comparison between some first-principle calculations (Northrup 1984) and photoemission data (Hansson *et al* 1981, Nicholls *et al* 1987), it has been possible to decide between two alternative reconstruction schemes; details are given by Lannoo and Friedel. Other experimental work, on both extrinsic and intrinsic states, is reviewed by Eastman and Nathan (1975), Park (1975) and Plummer *et al* (1975). Here we do no more than briefly review some results obtained by photoelectron spectroscopy that give clear evidence of the presence of intrinsic surface states.

As mentioned in section 1.2.3, in the technique of photoelectron spectroscopy the surface under study is irradiated with monochromatic

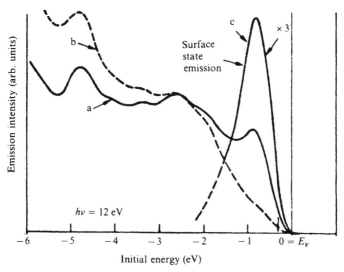

Emission intensity (arb. units)

Surface
state
emission

$h\nu = 12$ eV

$\times 3$

| | | | | | | |
|-6|-5|-4|-3|-2|-1|0 = E_F|

Initial energy (eV)

Figure 7.4. Intensity of photoemitted electrons from Si(111) surface at incident photon energy of 12 eV. Curve a, freshly cleaved surface; curve b, after 7 h exposure at 3×10^{-10} torr; curve c, difference between a and b. After Eastman and Grobman (1972).

(fixed-energy) ultraviolet or x-ray photons. The spectrum of emitted electrons gives information about transitions within the specimen. Only the electrons generated within an escape depth l of the surface are emitted. Since l is restricted by strong electron–electron scattering to a value of typically 1 to 2 nm, photoelectron spectroscopy is very sensitive to surface states and is extensively used as a tool for surface characterization.

Depending on its energy relative to the Fermi level, a surface band in a semiconductor can be either occupied or unoccupied. Results indicating the presence of a filled band on an Si surface are shown in figure 7.4. Curve b is taken after sufficient time for oxygen to be adsorbed to the surface; it is known in this case that the chemical binding of the oxygen to the surface Si atoms removes the intrinsic surface states. The difference between curves a and b, namely curve c, therefore corresponds to photoemission from a filled surface band, situated at an energy just below the top of the bulk valence band.

The technique used for figure 7.4 cannot be applied for an unoccupied surface band. Eastman and Freeouf (1974) overcame this difficulty by making measurements at a fixed energy E of the emitted electrons, but with the incident photon energy $\hbar\omega$ variable. The primary absorption of the photon involves excitation of an electron from an occupied 3d core state, either $3d_{5/2}$ or $3d_{3/2}$, into an unoccupied state. The core state is then repopulated by nonradiative (or Auger) transitions, and the Auger electrons generate the photoemitted electrons by electron–electron scattering. When the energy of the incident photon is resonant with the energy difference

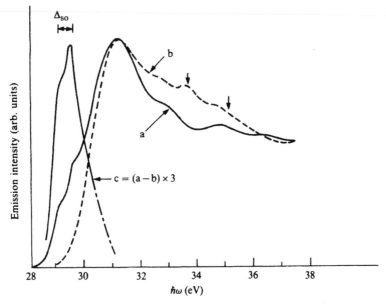

Figure 7.5. Photoemission from Ge(111) surface. Intensity is plotted versus photon energy $\hbar\omega$ at fixed emission energy $E = 4\,\text{eV}$. Curve a, clean surface; curve b, surface coated with a monolayer of Sb; curve c, difference between a and b (surface state contribution). $E_B(d_{5/2}) = 29.1\,\text{eV}$; $\Delta_{SO} = 0.55 \pm 0.05\,\text{eV}$. After Eastman and Freeouf (1974).

between an occupied and an unoccupied state, the photoemitted electron intensity is relatively large. Thus the emitted intensity reflects features of the photoabsorption spectrum. An example of such a spectrum is shown in figure 7.5. As in the previous example, the adlayer of Sb suppresses the intrinsic surface states. Curve a shows two shoulders at around $\hbar\omega = 29\,\text{eV}$; the separation between them corresponds to the spin–orbit splitting Δ_{SO} between the $d_{5/2}$ and $d_{3/2}$ states, which also shows up on the difference curve c. Further analysis shows that this unoccupied surface band falls in the gap between the bulk valence and conduction bands.

Finally, in figure 7.6 we show an example of some experimental and theoretical dispersion curves. This is again for an Si(111) surface, which when produced by cleavage from a bulk crystal reconstructs with a 2×1 pattern. The experimental points from angle-resolved photoemission (Uhrberg *et al* 1982) show a good agreement with the theory curve (Northrup and Cohen 1982; also Himpsel 1985), thereby providing strong evidence for a particular model for the reconstructed surface, the so-called π-bonded chain model due to Pandey (1981).

Experimental work on the electronic states at magnetic surfaces is reviewed by Donath (1993). He gives emphasis to studies on Ni and Fe samples with various crystallographic orientations of surfaces using the

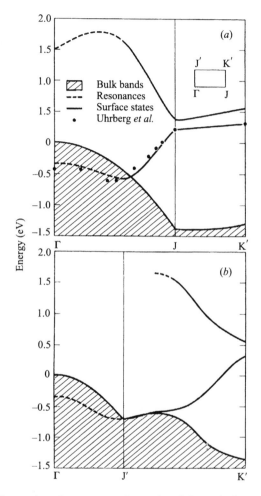

Figure 7.6. (a) Comparison between experimental and theoretical results for the surface state dispersion on $Si(111) - 2 \times 1$ for $\Gamma - J - K'$. Theory is in full lines (π-bonded chain model); data points are discrete symbols. Inset shows labelling of 2D Brillouin zone. (b) Theory for $\Gamma - J' - K'$. After Northrup and Cohen (1982).

technique of spin-polarized inverse photoemission (SPIPE). By contrast with photoemission (PE), which provides information about states either below the Fermi level E_F or above the vacuum level E_V, the inverse photoemission (IPE) probes the *empty* states above E_F, thus making the intermediate range between E_F and E_V accessible. In the spin-polarized version of the inverse technique, information is gained about the magnetic states at the surface since the incident spin-polarized beam selectively probes either the spin minority or majority bands.

7.2   Mixed excitations

Apart from polaritons, which are the excitations formed when a fundamental crystal excitation couples to a photon and which were extensively discussed already in chapters 2 and 6, we have not considered 'mixed' modes of two excitations. These may be important for appropriate ranges of the frequency and wavevector, and as some examples we will briefly discuss here *magneto-elastic surface waves* and *piezoelectric surface waves*. Before doing that, it is relevant to comment that a third example, which in fact we touched upon earlier but did not develop, could be *surface magnetoplasmon–polaritons*. These are just the surface polaritons resulting when the magnetic and plasmon contributions become coupled, e.g. when the dielectric function has a gyrotropic form with plasmon and magnetic-field terms as in (2.133)–(2.135). Their properties for a single-interface or two-interface geometry can be studied by a direct extension of the analysis given in sections 6.1 and 6.2. For a good review covering theory and experiment we refer to Wallis (1982); later work has been directed towards superlattices.

7.2.1   Magnetoelastic waves

We first take the case of coupled magnetic and vibrational modes. In the long-wavelength limit, where the solid can be treated as an elastic continuum, these modes are known as magnetoelastic waves. The form of coupling to produce these waves was first considered by Turov and Irkhin (1956) and Kittel (1958) in bulk systems. For a ferromagnetic solid with simple-cubic lattice structure and static magnetization in the z direction, the energy of the magnetoelastic coupling for the magnon regime, where only terms linear in transverse components of the magnetization $\mathbf{M}$ are retained, takes the form (Kittel 1958)

$$H_{\text{me}} = (b/M_0) \left[M_x(\bar{u}_{xz} + \bar{u}_{zx}) + M_y(\bar{u}_{yz} + \bar{u}_{zy}) \right]. \tag{7.8}$$

Here b is a phenomenological coupling constant, M_0 is the static magnetization, and the $\bar{u}_{\mu\nu}$ terms are elements of the strain tensor (see section 2.2.1). A more complete expression is quoted by Schlömann (1960) retaining terms up to second order in longitudinal and transverse magnetization components, namely

$$H_{\text{me}} = \frac{B_1}{M_0^2} \sum_{\mu} M_\mu^2 \bar{u}_{\mu\mu} + \frac{B_2}{M_0^2} \sum_{\mu,\nu(\neq\mu)} M_\mu M_\nu \bar{u}_{\mu\nu}$$

$$+ \frac{A_1}{M_0^2} \sum_{\mu,\nu,\eta(\neq\nu)} \frac{\partial M_\mu}{\partial \mu} \frac{\partial M_\mu}{\partial \eta} \bar{u}_{\nu\eta} + \frac{A_2}{M_0^2} \sum_{\mu,\nu} \left(\frac{\partial M_\mu}{\partial \mu} \right)^2 \bar{u}_{\nu\nu}. \tag{7.9}$$

The summations are over Cartesian x, y and z coordinates. The first two terms in (7.9) represent the dependence of the magnetic dipole–dipole interaction and the spin–orbit interaction, respectively, on elastic strain. In an isotropic ferromagnet we would have $B_1 = B_2$; however, in the ferrimagnet YIG, which has been widely studied, we have $B_1 \simeq 0.5B_2 \simeq 3.5 \times 10^6$ for the dimensionless coefficients. Also, we remark that the result in (7.8) is contained within the second term of (7.9), i.e. where one of the subscripts is a z and so $b \equiv B_2$. The next two terms in (7.9) come from modulation of the exchange coupling by the strain. All of the terms can provide contributions to the effective bulk and surface anisotropy coefficients, which may then show up in the magnetic resonance and light-scattering data (see e.g. Bland and Heinrich 1994). The A_1 and A_2 terms are of interest mainly from the point of view of nonlinear processes (involving the interaction of two magnons with a phonon) and will not be considered further here.

In the long-wavelength (magnetostatic) regime (see figure 2.5), where exchange is unimportant, we may therefore focus on (7.8) as providing the magnetoelastic coupling of interest. First, in the absence of such coupling ($b = 0$) we recall that the excitations in a film will consist of:

(i)   the bulk and surface (Rayleigh) elastic waves obtained from solving the equation of motion (2.24) or (2.25), together with appropriate boundary conditions as discussed in section 3.2; and

(ii) the bulk and surface magnetostatic modes obtained from the torque equation (2.163) and the susceptibility relations, together with boundary conditions as discussed in section 5.1.

When $b \neq 0$ the equations of motion must be supplemented by additional terms arising due to the coupling (7.8). This has been discussed, for example, by Scott and Mills (1977). These authors concentrated on a theoretical study of surface acoustic-type waves propagating in a semi-infinite ferromagnetic medium with magnetization parallel to the plane (i.e. the geometry of figure 5.1). In particular, they presented calculations for two types of surface acoustic wave, one being in essence a Rayleigh wave (see section 3.2.1) modified by the magnetoelastic coupling and the other a shear-like magnetoelastic surface mode that exists only because of the presence of the magnetoelastic coupling. Both of these waves have non-reciprocal propagation characteristics, a consequence of their being formed by coupling to the Damon–Eshbach surface magnetostatic mode, which is itself non-reciprocal as we discussed in section 5.1. The behaviour is illustrated schematically for the Voigt geometry in figure 7.7, where there are two surface branches. Scott and Mills conclude that these magnetoelastic surface waves provide a useful and sensitive probe of the bulk and surface magnetostatic modes, especially at larger values of the in-plane wavevector $\mathbf{q}_{\parallel}$ in the range of interest.

Other references to magnetoelastic surface waves in ferromagnets are to be found in the review articles by Maradudin (1981) and Cottam and Maradudin (1984). There is a good discussion of the basics in the books

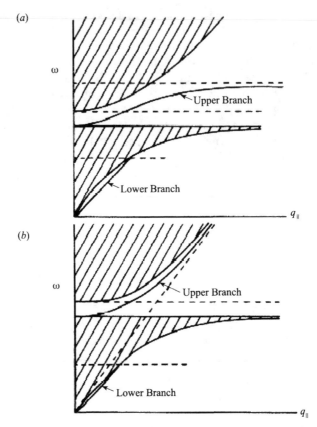

Figure 7.7. Dispersion relation for the shear polarized magnetoelastic surface wave in the Voigt geometry for nonzero external magnetic field, showing the upper and lower branches (full lines) and the bulk continuum (shaded): (a) $\phi = 90°$, (b) $\phi = -90°$. After Scott and

by Lvov (1991) and Gurevich and Melkov (1996), along with extensive references to experimental and theoretical studies of related nonlinear processes in the ferrimagnet YIG and other materials (e.g. by techniques such as parallel pumping with a microwave field). We will discuss nonlinear magnetic processes later in chapter 12. Some thorough experimental studies of magnetoelastic waves in thin films of YIG have been reported by Temiryazev *et al* (1996) under conditions where exchange effects had to be taken into account.

Calculations have also been performed for Rayleigh waves in paramagnetic rare-earth systems (Camley and Fulde 1981, Lingner and Lüthi 1981). In this case the magnetoelastic interactions describe the coupling of the strain and the rotational part of the deformation tensor to the unfilled $4f$ electron shell. Paramagnetic crystals are of interest because their elastic moduli typically exhibit a strong dependence on temperature and applied magnetic field, leading in turn to strong temperature and magnetic field dependences of the Rayleigh wave propagation. Experiments to measure the Rayleigh wave velocity have been reported, for example, for paramagnets $CeAl_2$ and SmSb (Lingner and Lüthi 1981), and are in good agreement with the theoretical predictions.

7.2.2 Piezoelectric waves

As another example of a mixed excitation we may mention piezoelectric materials. These are materials that, when deformed, produce an electric field; or conversely, applying an electric field to a piezoelectric material gives a deformation. Thus when an elastic wave propagates, the elastic deformation is accompanied by an electric field. In the case of a surface acoustic wave on a piezoelectric medium, this electric field may in principle be coupled to external electrical circuits, to other surface waves that have an electromagnetic component, or to charge carriers in semiconductors. Hence there is much potential for applications. The starting point in developing a theory of surface piezoelectric waves is to generalize the elastic-wave theory developed in sections 2.2 and 3.2 to include the extra coupling. Thus the Hooke's Law relationship in (2.20) becomes

$$\sigma_{ij} = \lambda_{ijkl}\bar{u}_{kl} - e_{kij}E_k \qquad (7.10)$$

where the notation is as before and we employ the summation convention for repeated indices. In addition, **E** is the macroscopic electric field and e_{kij} denotes components of the piezoelectric tensor. Conversely, we also now require the result that the electric displacement **D** is given by

$$D_i = \varepsilon_0\varepsilon_{ij}E_j + e_{ijk}\bar{u}_{jk} \qquad (7.11)$$

by extension of (2.112), where ε_{ij} denotes elements of the dielectric tensor. Equations (7.10) and (7.11) are the constitutive equations of piezoelectricity (Cady 1946). They were employed by Albuquerque (1979, 1981) to calculate the dispersion relations and scattering properties for surface modes in a piezoelectric semi-infinite medium. He paid particular attention to the polarization requirements for production and scattering of a mixed wave.

Sections on piezoelectric surface waves are included in the review articles by Maradudin (1981) and Cottam and Maradudin (1984), the latter including some Green-function calculations. More recent developments are described in detail in the two-volume work by Royer and Dieulesaint (2000). They make extensive use of the concept of a *piezoelectric surface permittivity* to give a simplified account of waveguide geometries involving piezoelectric/dielectric interfaces and they have a good discussion of applications, e.g. to interdigital-electrode transducers.

7.3 Excitations at edges, wedges and wires

7.3.1 Edges and wedges

It is an interesting, but rather complicated, topic to generalize the parallel-sided slab geometry, which we have discussed in many applications, to the case where two planar surfaces meet at an angle θ to form a wedge-shaped sample. The surfaces are now such that they remove the translational symmetry in two different directions, with obvious consequences for the symmetry of the excitations. In the geometry of figure 7.8 the apex of the wedge corresponds to the line $x = y = 0$. There is a translational invariance along the z direction (assuming the wedge dimensions to be effectively infinite in this direction) but not along the x or y directions. This means that, under appropriate conditions, there may be localized excitations known generally as *wedge modes* that propagate in a wave-like fashion along the apex of the wedge, but decay with increasing distance into the wedge from its apex and faces. The case of a right-angled wedge ($\theta = 90°$) often leads to simplifications, particularly in microscopic calculations involving cubic materials, and we refer to the localized excitations in this situation as *edge modes*. However, the terms 'wedge modes' and 'edge modes' are often used interchangeably in the literature.

The first calculations were for acoustic waves in wedges and edges formed by isotropic, elastic media (see Lagasse 1972, Maradudin *et al* 1972). These wedge modes have certain convenient properties, such as being essentially nondispersive, which make them of interest for waveguide applications. The number of wedge modes, and their speed of propagation, can be controlled by varying the wedge angle θ. In general, as θ is decreased the number of modes increases and their propagation speeds become lower than that of Rayleigh waves. This is illustrated by a numerical example in figure 7.9. Apart from the ideal infinite wedge, calculations have been

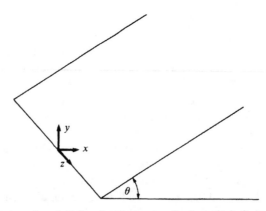

Figure 7.8. A wedge of angle θ showing the assumed orientation of coordinate axes.

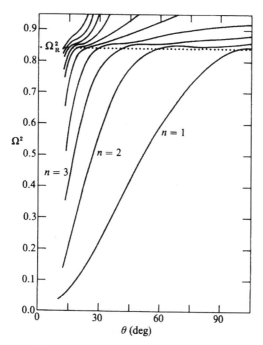

Figure 7.9. The square of the speed of propagation of wedge modes (in units of the speed of bulk transverse acoustic waves) of T_2 symmetry as a function of wedge angle θ. The speed corresponding to Rayleigh waves is denoted by Ω_R. After Moss *et al* (1973).

applied to more realistic waveguide structures, e.g. where the wedge-tip is rounded rather than sharp, so we no longer have planar surfaces. A thorough review of acoustic wedge modes has been given by Maradudin (1981).

Electromagnetic wedge modes have also attracted some theoretical attention (e.g. see Maradudin 1981). Most calculations have been worked out in the electrostatic approximation (see section 6.1.1) since this simplifies the analysis. In an infinite dielectric wedge surrounded by vacuum, one may attempt to solve Laplace's equation by the method of separation of variables to obtain a scalar potential proportional to $\exp(iqz)$ and localized about the apex of the wedge. Details of such a calculation are given by Dobrzynski and Maradudin (1972). When retardation effects are included, one is faced with the much harder task of solving the full set of Maxwell's equations in the wedge geometry. In an analytic theory mathematical difficulties arise at the apex of the wedge with regard to matching the solutions inside and outside the material, through application of the appropriate boundary conditions. From another approach, an elegant calculation of retarded wedge modes in a parabolic wedge has been given by Boardman *et al* (1981).

We turn now to a microscopic, rather than macroscopic, example. The magnons localized at a $90°$ edge of a Heisenberg ferromagnet with simple-cubic structure have been studied by Sharon and Maradudin (1973) for the long-wavelength region, and extended to shorter wavelengths by Maradudin *et al* (1977). They used a model in which all exchange interactions (including those involving spins at the surface) have their bulk values, denoted by J for nearest neighbours and J' for next-nearest neighbours. At long wavelengths and in the absence of surface anisotropy fields, the magnon amplitudes of the edge modes have the approximate dependence

$$\exp[-\kappa(x+y)]\exp[i(qz-\omega t)] \tag{7.12}$$

where the attenuation constant κ is related to the propagation wave vector q by

$$\kappa = aq^2\left(\frac{J'}{J+4J'}\right) \tag{7.13}$$

provided $\kappa > 0$. This is an acoustic-type branch occurring split off below the surface magnon, which is itself split off below the bulk continuum. This behaviour is illustrated in figure 7.10. The simple functional dependence

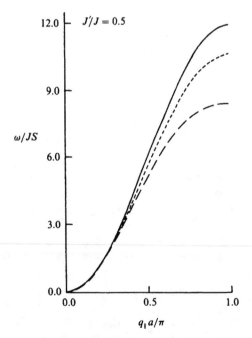

Figure 7.10. Calculation of dispersion relations for bulk magnons (full line), surface magnons (short dash) and edge magnons (long dash) in a simple-cubic Heisenberg ferromagnet with a 90° edge. The bulk dispersion curve refers to the lower edge of the continuum. After Maradudin *et al* (1977).

on x and y in (7.12) does not hold near the apex ($x = y = 0$) or for general values of q. Instead a complicated expansion has to be made in terms of a set of suitable orthogonal functions (Gottlieb functions in this case), and this is typical of other edge and wedge calculations.

In recent years there has been rather more attention given to the generalization of simple edge or wedge geometry to the case where additional planar surfaces are involved. This is the situation for long wires (or rods) with rectangular cross section, which have four edges, each of 90°, as we now discuss.

7.3.2 Rectangular wires

For a long wire we still have translational symmetry in one direction (taken as the z axis), but the system is finite in the xy plane, contrasting with the geometry in figure 7.8. This introduces a new physical effect, namely the spatial quantization of modes in two different directions (x and y), while the dependence along z for any wave is in accordance with Bloch's theorem. Advances in fabrication techniques, involving lithography and differential etching of samples, have made it feasible to produce high-quality wires (either singly or in arrays) with lateral dimensions of the order of a few nm up to about 1 μm. In what follows, we will outline some theoretical and experimental work on ferromagnetic wires, since this is a particularly active field due to potential applications in the magnetic recording and information storage industries (see e.g. Chou 1997).

First, for Heisenberg magnetic materials, we mention that the discrete-lattice calculation for a 90° edge (see section 7.3.1) has been generalized to the case of a rectangular nanowire (Sharon and Maradudin 1977a). More recently Nozières *et al* (1998) used a Heisenberg model with effective anisotropy terms to interpret their measurements for the temperature dependence of the magnetization, due to thermal excitation of magnons, in YCo_2 nanowires. They assumed a square cross section of N atoms by N atoms and found satisfactory qualitative agreement. A more detailed theory was developed by Ferchmin and Puszkarski (2001). This was for an $N \times N$ structure in the exchange-dominated limit, but they included additional effects of anisotropy and obtained conditions for surface modes to exist.

There have been several calculations for long ferromagnetic wires in the magnetostatic and/or dipole-exchange regimes, allowing comparisons to be made with Brillouin light scattering (BLS) experiments. Typically the samples are prepared by starting with a thin metallic film of thickness L on a nonmagnetic substrate (e.g. FeNi on Si); then, by lithography and etching, long wires of width w can be produced in arrays with chosen separation (so that the wires are either well-separated and independent, or closer together and coupled). Here we focus on individual wires, leaving arrays of coupled wires until section 7.4.3 on patterned surfaces. In most

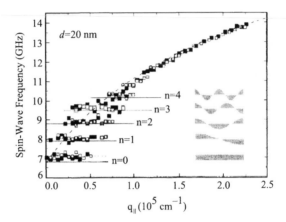

Figure 7.11. Spin–wave dispersion curves for wires with thickness $L = 20$ nm and width $w = 1.8\,\mu m$ in an array with lateral separation $0.7\,\mu m$ (open symbols for experimental points) and $2.2\,\mu m$ (closed symbols). The solid horizontal lines are theory (see text), and for comparison the dashed line shows the calculated DE surface mode for a continuous film with $d = 20$ nm. The mode profiles are illustrated on the right. After Demokritov *et al* (2001).

of the experiments L is a few tens of nm and w is of the order of 1 or 2 μm, so it is a good approximation to take $w \gg L$, which greatly simplifies the analysis of the results. In most BLS studies the magnetization direction is in the plane of the original film and along the axis (symmetry direction) of the wires; the in-plane wavevector $\mathbf{q}_{\parallel}$ is usually perpendicular to the wires as in the Voigt configuration. One of the early experiments was by Gurney *et al* (1991), who reported a splitting of the spin–wave spectrum of the film into discrete components. In later experiments a more complete and quantitative characterization of the modes was obtained, and we mention in particular the papers by Mathieu *et al* (1998) and Jorzick *et al* (1999a). Two excellent reviews of this topic have been given by Demokritov *et al* (2001) and Demokritov and Hillebrands (2002).

In figure 7.11 we show a comparison between theory and experiment for the spin–wave dispersion. The data points, which are from BLS, indicate a splitting at small $q_{\parallel}$ into discrete modes that exist over a characteristic range of $q_{\parallel}$, but show little dispersion. The theoretical interpretation was provided by Jorzick *et al* in terms of modifying the macroscopic dipole-exchange theory for a film (see section 5.4.1). They argued that the in-plane wavevector of the Damon–Eshbach (DE) surface mode would become quantized as

$$q_{\parallel,n} = n\pi/w, \qquad n = 0, 1, 2, \ldots \tag{7.14}$$

due to the finite wire width. If this replacement of $q_{\parallel}$ is made in (5.7), which still holds to a good approximation for the DE mode frequency in the dipole-exchange region, we obtain the horizontal lines indicated in figure 7.11. The agreement with the experimental frequencies is good, and Jorzick *et al* also calculated BLS intensities to estimate the range of $q_{\parallel}$ values for each mode. With the condition $w \gg L$, they showed that

$$I_n \quad \propto \quad |m_n(q_{\parallel})|^2, \qquad n = 0, 1, 2, \ldots \tag{7.15}$$

for the BLS intensity of the quantized modes, where $m_n(q_{\parallel})$ is the wavevector Fourier transform of a function $m_n(y)$ describing the spatial variations of the spin–wave amplitude across the width of the wire. Since $m_n(y)$ is *not* a periodic function due to the truncation, $m_n(q_{\parallel})$ and hence I_n are nonzero over a finite continuous range. Jorzick *et al* were also able to study hybridization between the modified bulk modes and the modified DE modes of the wire by using samples with other values of L.

It should be noted that the experiments described above showed little effect of interactions between wires, despite the relatively small lateral separation in some cases, so the theory was for a single wire with large pinning appropriate to the assumed quantization.

7.4 Excitations at spherical and cylindrical surfaces

So far throughout this book we have idealized the surfaces and interfaces as being planar and smooth. We now look at some other situations, starting in this section with some generalizations to curved samples, in particular to cylinders and spheres where there are applications on the nanometre and micrometre length scales. Then in the following section, the topic of surface roughness, which may be either random or periodic, is treated in section 7.5.1. This leads into section 7.5.2 which is concerned with the recent developments for the properties of patterned surfaces.

7.4.1 Magnetic excitations

We start by considering magnetostatic modes in ferromagnetic samples with curved surfaces. The case of spherical and spheroidal samples was worked out by Mercereau and Feynman (1956) and Walker (1957), followed a few years later by the case of a cylinder (Fletcher and Kittel 1960, Joseph and Schlömann 1961). These results are reviewed by Wolfram and Dewames (1972). We briefly consider the calculation for a long circular cylinder magnetized along its axis of symmetry (taken as the z direction), which is also the direction of the applied magnetic field B_0. A scalar potential ψ may be introduced as in section 2.6.3, and this must satisfy (2.183) and (5.2) for the regions inside and outside the sample, respectively. After transforming from Cartesian to cylindrical coordinates (r, θ, z) in the usual way, one seeks separable solutions of the form

$$\psi(r, \theta, z) = \rho(r) \, \exp(im\theta) \, \exp(iq_z z). \tag{7.16}$$

Here m has to be a positive or negative integer (in order for ψ to be single-valued), and q_z is the propagation wavevector in the z direction (in accordance with the 1D Bloch's theorem). The radial part $\rho(r)$ of the potential has to satisfy

$$(1 + \chi_a)\left(\frac{d^2\rho}{dr^2} + \frac{1}{r}\frac{d\rho}{dr} - \frac{m^2}{r^2}\rho\right) - q_z^2\rho = 0 \tag{7.17}$$

in the interior, where χ_a is given by (2.166). The acceptable solutions which are regular at $r = 0$ correspond to

$$\rho(r) = a J_m(kr) \tag{7.18}$$

where a is a constant, $J_m(kr)$ is a Bessel function of the first kind, and k is a complex quantity that satisfies

$$[\omega_0(\omega_0 + \omega_m) - \omega^2](k^2 + q_z^2) - \omega_0\omega_m q_z^2 = 0. \tag{7.19}$$

When k is real the solutions of (7.19) for ω just describe the continuum of magnetostatic bulk modes extending from ω_0 to $[\omega_0(\omega_0 + \omega_m)]^{1/2}$ (as in section 2.6). However, k is imaginary when $\omega > [\omega_0(\omega_0 + \omega_m)]^{1/2}$ and so there is the possibility of surface modes in this region. This can be investigated (see Sharon and Maradudin 1977b) by writing down the solution in the region outside the cylinder, where ψ satisfies a similar equation to (7.17) but with χ_a set equal to zero. The appropriate form that vanishes as $r \to \infty$ is $\rho(r) = bK_m(q_z r)$, where b is a constant and $K_m(q_z r)$ is a modified Bessel function of the second kind. It then remains to apply the boundary conditions at $r = R$, where R is the radius of the cylinder; these take the form of continuity of ψ and continuity of the radial component of the magnetic flux density $\mathbf{b}$ at $r = R$. The resulting surface-mode dispersion relation consists of a sequence of solutions, characterized by $m = 1, 2, \ldots$. At long wavelengths (with $q_z R \ll 1$) the frequencies simplify:

$$\omega_S(q_z) = \begin{cases} \omega_0 + \tfrac{1}{2}\omega_m - \tfrac{1}{4}\omega_m(q_z R)^2[0.366 - \ln(q_z R)], & m = 1 \\ \omega_0 + \tfrac{1}{2}\omega_m - \tfrac{1}{4}\omega_m(q_z R)^2(m^2 - 1)^{-1}, & m > 1. \end{cases} \tag{7.20}$$

These surface modes all have frequencies above those of the bulk manifold. In general, the localization at the surface of the cylinder increases with increasing q_z; details are to be found in the original references.

As mentioned, the magnetostatic theory has also been applied to spherical ferromagnets. The calculation proceeds in an analogous fashion to the cylindrical case, except that spherical polar coordinates are used. The solutions for the potential ψ involve Legendre functions instead of the Bessel functions of the cylindrical case.

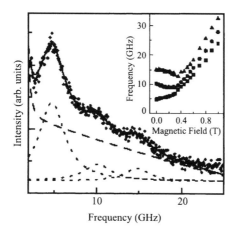

Figure 7.12. Brillouin spectrum of a 30 nm diameter Ni nanowire sample in zero applied magnetic field, showing three spin–wave peaks. Experimental data are denoted by dots. The spectrum is fitted with Lorentzian functions (dotted curves) and a background (dashed curve); the full fitted spectrum is shown as a solid curve. The inset shows peak frequencies versus applied magnetic field. After Wang *et al* (2002).

On the experimental side, spin waves in cylindrical nanowire arrays of Ni have been investigated by ferromagnetic resonance (Encinas-Oropesa *et al* 2001) and Brillouin scattering (Wang *et al* 2002), the latter technique allowing a more thorough study. An example of a Brillouin spectrum obtained in a zero applied field is shown in figure 7.12. When the instrumental background is subtracted from the spectrum, the remaining signal can be resolved into three peaks that diminish in intensity with increasing frequency. For comparison the bulk spin wave in a bulk sample of Ni occurs at a lower frequency, namely $\sim$3 GHz (Sandercock and Wettling 1979). An explanation requires dipole-exchange, rather than magnetostatic, theory, and such a generalization for a ferromagnetic cylinder was reported by Arias and Mills (2001) using a macroscopic theory analogous to that described in section 5.4.1. Wang *et al* (2002) employed a modified version of this theory to explain their zero-field data; they predicted that the higher-frequency part of the spectrum should consist of standing bulk modes of the nanowire. Assuming the pinning at the wire surface to be small, Wang *et al* deduced that the frequencies are given approximately by

$$\omega = \left\{ D\left(\frac{a_m}{R}\right)^2 \left[D\left(\frac{a_m}{R}\right)^2 + \omega_m \right] \right\}^{1/2}, \qquad m = 1, 2, 3, \ldots \qquad (7.21)$$

where D is the exchange 'stiffness' constant (see section 5.4.1) and R is the nanowire radius. The radial term in the magnetostatic potential is expressible in terms of Bessel functions of order m as in (7.18), and the numerical

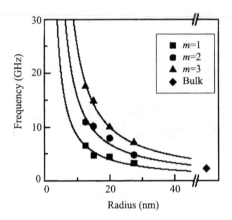

Figure 7.13. Bulk spin–wave frequencies versus Ni nanowire radius in zero applied magnetic field. The experimental points for three values of m are indicated. The solid lines are best fits from theory obtained using equation (7.21). After Wang *et al* (2002).

constants a_m in the above equation correspond to the values of argument x at which $\mathrm{d}J_m(x)/\mathrm{d}x = 0$; the three lowest values are $a_1 \simeq 1.84$, $a_2 \simeq 3.05$ and $a_3 \simeq 4.20$. We notice that the expression in (7.21) is independent of wavevector q_z along the longitudinal nanowire axis. This approximation holds provided $q_z^2 \ll (a_m/R)^2$, which was well satisfied in the Brillouin scattering experiments. Figure 7.13 shows a comparison between theory and experiment for spin–wave frequency versus radius for a series of measurements; the agreement is excellent. The observed dependence of the spin–wave frequencies on magnetic field, applied *perpendicular* to the nanowire axis, is shown in the inset to figure 7.12. This causes a canting of the magnetization direction, which becomes aligned in the perpendicular direction for fields greater than about 0.3 T.

7.4.2 Nonmagnetic excitations

Turning now to nonmagnetic excitations, the theory of elastic surface waves in homogeneous isotropic materials, as discussed in section 3.2 for a parallel-sided slab geometry, can be extended relatively straightforwardly to cylindrical and spherical geometries. The propagation of Rayleigh-type surface modes and/or various kinds of standing modes can then be studied in appropriate cases. The first calculation was due to Hudson (1943) for elastic waves travelling along the axis of a long circular cylinder; subsequently several authors (e.g. Viktorov 1958, Rulf 1969) calculated the Rayleigh waves propagating circumferentially. Other related work is reviewed by Maradudin (1981). The theory of acoustic modes in a homogeneous elastic sphere with a free surface was formulated by Lamb (1882), who predicted two types of modes: the spheroidal and the torsional modes.

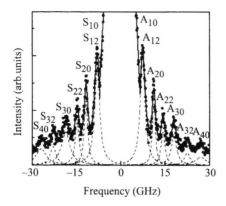

Figure 7.14. Brillouin spectrum in p–p polarization of a 340 nm-diameter SiO_2 nanosphere sample. The spectrum is fitted with Lorentzian functions (dashed curves) and the resultant fitted spectrum is shown as a solid curve. The Stokes and anti-Stokes peaks arising from the confined acoustic modes are labelled by S_{nl} and A_{nl} respectively. After Kuok *et al* (2003).

On the experimental side, the studies are much more recent and now include investigating excitations in nanoparticles by inelastic light scattering. Initially low-frequency Raman scattering was employed by several groups (see e.g. Tanaka *et al* 1993), but it was not until the work of Kuok *et al* (2003) using higher-resolution Brillouin scattering on improved samples (SiO_2 nanospheres in a 3D ordered array) that a thorough analysis was possible. The elastic modes in a sphere can be labelled by the angular momentum quantum number l, where $l = 0, 1, 2, \ldots$ for spheroidal modes. The selection rules for light scattering from these samples (see Duval 1992) preclude scattering from the torsional modes, while allowing only the spheroidal modes with $l = 0$ or 2 to contribute. For a given l the sequence of modes, in increasing energy, is indexed by n ($= 1, 2, 3, \ldots$), where $n = 1$ corresponds to surface modes and higher n values refer to discrete bulk modes. An example of a Brillouin spectrum measured by Kuok *et al* is shown in figure 7.14, where the Stokes and anti-Stokes peaks are labelled by their (n, l) values. Their mode frequencies were found to be in good agreement with previous theories (Nishiguchi and Sakuma 1981, Tamura *et al* 1982), as can be seen from figure 7.15. In particular the predictions that the frequencies are inversely proportional to the diameter are well satisfied.

Electromagnetic modes connected with spheres have been extensively discussed. In a series of papers, Englman and Ruppin discussed both the retarded (electromagnetic) and non-retarded (electrostatic) modes of spheres; this work includes discussions of the equivalents of phonon–polaritons, using a dielectric function of the form of (2.146), and plasmon–polaritons, using (2.132). The work is fully reviewed by Ruppin in Boardman

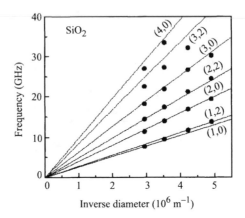

Figure 7.15. Dependence of the Brillouin peak frequency on the inverse nanosphere diameter. Experimental data are denoted by dots and the lines represent theory (see text). After Kuok *et al* (2003).

(1982). Further references to the early literature are contained in the review articles by Economou and Ngai (1974) and Kliewer and Fuchs (1974). More recently, following the development of new techniques for producing micro-metre-size materials and the need for improved understanding and design of various near-field optical devices (see Kawata 2001), there has been renewed attention on such electromagnetic excitations.

Electromagnetic modes of cylinders have received a great deal of attention, with most of the emphasis on modes propagating along a single straight cylinder with a field pattern decaying with distance from the cylinder. These are the analogue of the surface and guided modes of a film that were discussed in section 6.2.1, and in optical engineering they are the modes that carry the signals in optical fibres. A simple fibre comprises a glass core inside a cladding of a different material, and in the engineering literature both core and cladding are usually characterized by a refractive index. From the point of view of this book, it is more natural to use dielectric functions, which may be negative in regions of strong dispersion. It is convenient here to draw on the unified account given by Khosravi *et al* (1991). We use cylindrical polar coordinates (r, θ, z). The region $r < a$ is occupied by a cylinder, the core, with dielectric function $\varepsilon_2(\omega)$, while $r > a$ is occupied by the cladding with dielectric function $\varepsilon_1(\omega)$. The modes in question are given by the solution of Maxwell's equations, together with the electromagnetic boundary conditions at $r = a$, and the modes propagating along the cylinder have an electric field of the form $\mathbf{F}(r) \exp(iqz + im\theta - i\omega t)$. Here q is the propagation wavenumber and m is the azimuthal integer. We have written the radial function as a vector $\mathbf{F}(r)$ because, unlike the modes of a film, there is no separation of the modes into pure transverse E and pure transverse H, except for $m = 0$. On substitution into the wave equation it is found that $\mathbf{F}(r)$ can be expressed in terms of Bessel functions and substitution into the boundary conditions gives the dispersion equation. This can be written in terms of the auxiliary quantities

$$W_i^2 = -U_i^2 = q^2 - \varepsilon_i \omega^2 / c^2 \qquad (7.22)$$

and takes the form

$$\frac{\omega^2}{c^2}(\eta + \gamma)(\varepsilon_2 \eta + \varepsilon_1 \gamma) = \frac{m^2 q^2}{a^2}\left(\frac{1}{U_2^2} + \frac{1}{W_1^2}\right)^2 \qquad (7.23)$$

for $U_2^2 > 0$ (guided modes) and

$$\frac{\omega^2}{c^2}(\gamma - \eta_2)(\varepsilon_1 \gamma - \varepsilon_2 \eta_2) = \frac{m^2 q^2}{a^2}\left(\frac{1}{W_1^2} - \frac{1}{W_2^2}\right)^2 \qquad (7.24)$$

for $U_2^2 < 0$ (surface-type modes). In (7.23) and (7.24)

$$\gamma = K_m'(W_1 a)/W_1 K_m(W_1 a) \qquad (7.25)$$

$$\eta = J_m'(U_2 a)/U_2 J_m(U_2 a) \qquad (7.26)$$

and

$$\eta_2 = I_m'(W_2 a)/W_2 I_m(W_2 a) \qquad (7.27)$$

where the notation for Bessel functions is that of Abramowitz and Stegun (1965).

The guided-mode equation (7.23) characterizes the propagation modes of a core-plus-cladding optical fibre, otherwise described as a step-index fibre. In the engineering literature it is written in terms of refractive indices $n_i = \varepsilon_i^{1/2}$ and usually in terms of the dimensionless variables V and b defined in (6.69). For the guided mode to occur, the radial dependence of the field must be oscillatory (J_m) in the core and decaying (K_m) in the cladding. For this to be possible, the refractive indices must satisfy the wave-guiding condition $n_2 > n_1$ analogous to (6.67). One of the most important properties of (7.23) is that the fibre is monomode, with only the lowest $m = 1$ mode propagating, for $V < 2.405$; this is analogous to the condition $V < \pi/2$ mentioned for a slab waveguide in section 6.2.1. Mono-mode fibres are of great practical importance since they are used for long-haul optical communications. The condition $V < 2.405$ translates for typical glass-fibre parameters into a requirement for a very small core diameter $a \approx 10 \ \mu m$. The monomode fibres used in practice are step-index, but generally consist of a number of layers, rather than being simply core plus cladding. For short-distance communications a multimode fibre is often used, since it can have larger core radius a, so that the manufacturing requirement is less stringent. These fibres are often graded-index, i.e. the refractive index varies continuously with radius r rather than changing discontinuously. In general the dispersion equation of a graded-index fibre has to be found numerically, although an analytic solution exists for the parabolic dependence $n^2(r) = n_0^2(1 - \eta r^2)$ (see Yariv 1989). Much more detailed discussion of propagation in optical fibres is to be found in the engineering literature, for example, Kapany and Burke (1972), Marcuse (1982), Snyder and Love (1983), Gowar (1993) or Syms and Cozens (1992).

Modes analogous to the surface polaritons of chapter 6 are found when the core dielectric function $\varepsilon_2(\omega)$ is negative. A detailed discussion is given by Khosravi *et al* (1991) for both the phonon form (2.146) and the plasmon form (2.132); we mention some of the results for the former. The frequency intervals in which guided and surface type modes occur are the same as those for a film with dielectric function $\varepsilon_2(\omega)$ embedded in a medium with dielectric function ε_1, assumed constant for simplicity. There are two guided-wave windows $0 < \omega < \omega_T$ and $\omega_2 < \omega$, where $\varepsilon_2(\omega_2) = \varepsilon_1$; the former only extends down to zero frequency provided $\varepsilon_1 < \varepsilon_2(0)$ and the latter exists only if $\varepsilon_1 < \varepsilon_\infty$. Both these conditions are satisfied if the cladding is vacuum. The surface-mode range is $\omega_T < \omega < \omega_S$, where $\varepsilon_2(\omega_S) = -\varepsilon_1$. The dispersion curves for these cylinder polaritons are more complicated than for the film discussed in section 6.2.1 because there is a set of curves for each value of the azimuthal integer m. For both the upper and lower guided-wave windows there is a monomode frequency region; only the lowest $m = 1$ mode propagates in some frequency interval at the bottom of the window. A surface mode is found for each value of m. Khosravi *et al* explore only $m = 0$ and $m = 1$; for some ranges of radius a and propagation wavenumber q the dispersion curve for $m = 1$ has the surprising property of negative group velocity, $d\omega/dq < 0$.

More recently there have been calculations for polaritons in solid and hollow cylinders of anisotropic materials (e.g. Nobre *et al* 1998, 2000, Farias *et al* 2002). Finally we mention that Kottmann and Martin (2001) have studied the plasmon–polariton resonant coupling in a pair of metallic nanowires. Using two 50 nm diameter Ag nanowires in configurations that could either be non-touching (but close together) or intersecting, they demonstrated dramatic field enhancements between the wires, reaching a hundredfold of the illumination. Some implications for surface-enhanced Raman scattering, a topic which we mentioned in section 6.5, are discussed in their paper.

7.5   Excitations at rough and patterned surfaces

To some extent the two topics in this section are overlapping, but we shall tend to use the term *rough* to refer to cases where there is either a random deviation from an otherwise planar surface or where there is a periodic corrugation (e.g. a ruled grating). The term *patterned* will be preferred in cases where there is some other feature (which might include steps, ridges, platelets, adsorbed atoms or clusters etc.) on an otherwise planar surface, but this is not a clear-cut distinction. It is the latter situation that has received a lot of attention in recent years.

For the special case of surface polaritons, some discussion has already been given in chapter 6. Section 6.4 deals with grating coupling, and we saw that in the simplest approximation a shallow grating couples external radiation to the surface polariton, whose properties may be taken as unmodified by the grating. For deeper gratings, a more careful analysis is required, because the grating modifies the properties of the surface mode itself. Coupling to external radiation is also found at a rough surface, two examples being surface-enhanced Raman scattering (section 6.5.2) and direct radiation (section 6.5.3). We now consider these matters more generally, first for rough surfaces in section 7.5.1, then for patterned surfaces in section 7.5.2.

7.5.1 Rough surfaces

All real solid surfaces are rough to some degree. This may be because of the way in which the sample has been prepared, e.g. by vapour deposition on to a substrate material, or because of subsequent treatment of the surface, e.g. polishing of the surface by an abrasive substance of a particular particle size. It is therefore relevant to determine how this roughness affects the properties of the excitations and how these properties differ from those in samples in an 'ideal' plane surface.

In many cases it can be assumed that the presence of surface roughness acts simply as a small perturbation on effects that already occur with plane surfaces, e.g. the frequency and/or damping of an excitation may be perturbed due to the roughness. Many calculations have concentrated on this regime. For example, the modifications of Rayleigh surface waves propagating along a randomly rough surface have been studied by Urazakov and Falkovsii (1972) and by Maradudin and coworkers (see e.g. Maradudin 1981). These latter calculations were carried out by a perturbation method correct up to order $(\delta/b)^2$, where δ is the root-mean-square departure from flatness and b is the mean distance between consecutive peaks and valleys on the surface. More generally, a randomly-roughened surface perpendicular to the z axis may be defined by $z = \xi(\mathbf{r}_{\parallel})$, where $\mathbf{r}_{\parallel} = (x, y)$ and the surface profile function $\xi(\mathbf{r}_{\parallel})$ is assumed to be a stationary stochastic process characterized by the following two statistical results:

$$\langle \xi(\mathbf{r}_{\parallel}) \rangle = 0 \tag{7.28}$$

$$\langle \xi(\mathbf{r}_{\parallel})\xi(\mathbf{r}'_{\parallel}) \rangle = \delta^2 \, W(|\mathbf{r}_{\parallel} - \mathbf{r}'_{\parallel}|) \tag{7.29}$$

where the angular brackets denote an average over the ensemble of realizations of the surface profile and $\delta^2 = \langle \xi^2(\mathbf{r}_\parallel) \rangle$, as defined above. The correlation function W can be chosen to model different types of roughness; for example, a Gaussian form is often convenient:

$$W(|\mathbf{r}_\parallel - \mathbf{r}'_\parallel|) = \exp(-|\mathbf{r}_\parallel - \mathbf{r}'_\parallel|^2/b^2). \qquad (7.30)$$

As we mentioned earlier, the region between the minimum and maximum values of $\xi(\mathbf{r}_\parallel)$ is called the selvedge region. Several perturbative-type Green function calculations have been carried out for excitations at randomly rough surfaces and interfaces described according to the description provided by (7.28)–(7.30). This generally involves matching the Green functions within the selvedge region, treated statistically, to those corresponding to the 'unperturbed' media on either side. Some such applications to dielectric and elastic media are reviewed by Cottam and Maradudin (1984).

In other cases the random roughness can produce qualitative changes; an example is the roughness-induced splitting of the surface-plasmon dispersion curve observed experimentally (e.g. Palmer and Schnatterly 1971) and studied theoretically (e.g. Kretschmann *et al* 1979). It can also produce large quantitative effects, such as the giant enhancement (by several orders of magnitude) of the Raman intensity for scattering from molecules adsorbed on certain metal surfaces, and we have already discussed surface enhanced Raman scattering (SERS) in section 6.5. Although there are various mechanisms proposed for this enhancement, it is claimed that surface roughness can give rise to a significant part of it (Burstein *et al* 1982). For example, this is supported by the observation by Ushioda *et al* (1979) that the Raman intensity for scattering from surface polaritons in GaP increased by a factor of four to five when the surface of the sample was roughened.

Likewise, for magnetic materials, the giant magnetoresistance (GMR) in layered structures (see the discussion in section 4.5) is influenced by roughness at the interfaces (see e.g. Heinrich and Bland 1994, Hartmann 2000). Also we referred in section 4.6 to ferromagnetic resonance studies by Suran *et al* (1998) on ultrathin Ni films that were either grown epitaxially on a W substrate or were sandwiched between thick W layers. Localized surface magnetic modes were clearly observed when one or more of the interfaces were deliberately roughened. There is some further discussion of interface roughness in the case of nonmagnetic materials in chapter 8.

Apart from the random roughness, which we have discussed in the preceding paragraphs, there can be periodically corrugated surfaces, e.g. as in the case of a diffraction grating ruled on the surface. As mentioned, the use of a grating to couple external light to a surface polariton was examined in section 6.4.1. Indeed, grating surfaces are of considerable technological interest because of their role in optical surface acoustic wave devices, e.g. as surface mode to bulk wave transducers (see Royer and Dieulesaint 2000). Furthermore, from the theoretical point of view, they allow a treatment of surface roughness in situations where a perturbative approach would fail, because the Bloch's theorem associated with the grating periodicity at the surface can be utilized. For example, calculations for surface excitations propagating across the grooves of large-amplitude gratings have been given by Laks *et al* (1981) for surface polaritons, Glass and Maradudin (1981) for surface plasmons, and Glass *et al* (1981) for Rayleigh waves. Another theoretical approach has developed from calculations by Pereira *et al* (1998, 2000) for surface electromagnetic waves. They extended earlier work by Maradudin and Visscher (1985) on surface shape resonances of a single 1D surface protuberance (e.g. a ridge) to apply to an *array* of such ridges, thus simulating a grating. They considered various ridge shapes (Lorentzian, Gaussian and sinusoidal) on semi-infinite dielectrics and films. Experimentally it is now possible to create ridges or grooves on a surface by standard photolithographic techniques (see Lopez-Rios *et al* 1998) or direct ablation (see Smolyaninov *et al* 1997) with good control of sizes and shapes.

Finally, we have focused here on the dynamical properties of rough surfaces, in keeping with the overall theme of this book. However, there is also considerable literature on the static properties and morphology of rough surfaces. Some general references are Gonis and Stocks (1992), Lüth (1995), Pimpinelli and Villain (1998) and Holy *et al* (1999).

7.5.2 Patterned surfaces

We continue the discussion of rough surfaces with some topics that might be grouped under the heading of patterned surface. One possibility involves *steps* on surfaces. A good introduction to this subject is given in the review article by Wagner (1979), who describes the morphology and characterization of stepped surfaces (e.g. by the low-energy electron diffraction, or LEED, technique described in section 1.2). Another important experimental tool for stepped surfaces is He atom scattering (see e.g. Hulpke 1992).

In particular, steps can occur if a clean surface is produced by cleaving a crystal under ultrahigh vacuum (UHV) conditions along an orientation that is slightly misaligned from a high-symmetry direction. The so-called *vicinal surface* that is formed consists of an array of atomic steps with a regular separation by terraces. This is illustrated in figure 7.16, where there is a side-view sketch (in the xz plane) of a surface that is vicinal to the (001) orientation; it has a unit atomic step height and has five atoms along each terrace. In the standard notation this would be designated as [5(001) × 1(100)]. The general form is

$$[n(hkl) \times n'(h'k'l')] \qquad (7.31)$$

where n gives the number of rows on the terrace of (hkl) orientation, n' is the step height and $(h'k'l')$ is the orientation of the step face. Ordered step structures on metal surfaces usually exhibit monoatomic step height, but this is not necessarily the case for semiconductors, e.g. some (111) and

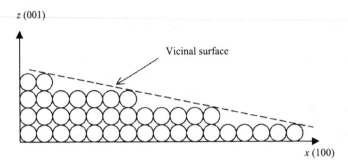

Figure 7.16. Sketch (in side view) of a stepped surface vicinal to the (001) plane of a cubic crystal structure. The step height is 1 and the terrace length is 5 in units of the atomic spacing.

(100) vicinals of Ge are found to have double step heights. In some other cases the step height is effectively non-integer because the atoms near the surface may undergo surface reconstruction. Steps with greater heights than those mentioned above, or steps with edge angles other than 90°, can be produced by lithographic and etching methods.

We shall mention just a few examples where excitations at stepped surfaces have been studied. Following Knipp (1991), who developed a thorough theoretical analysis for phonons, the surface excitations may be broadly divided into two classes. One class (which we henceforth refer to as Type I) is characteristic of the terrace and consists of surface modes reflected and transmitted at the step edges. The other class (Type II) is localized at the edges of the steps, along which they propagate freely. The features of Type I surface excitations are rather well approximated by a 'folding' of the 2D Brillouin zone of the terrace (in an analogous way to the Brillouin zone folding for periodic superlattices described later in sections 8.1 and 8.2). On the other hand, the Type II surface modes can be regarded as a generalization of the (single) edge-mode calculations described in section 7.3. An example of Type I surface modes is provided by Riste and Stamps (2002) who calculated the spin–wave properties of a monolayer Fe film grown on a stepped W surface. Their assumption of a monolayer on a non-magnetic substrate made it feasible to calculate the dispersion relations with both exchange and dipolar effects included. Type II surface spin waves at steps have been studied in the exchange regime by Jiang and Cottam (1999) and Abou Ghantous and Khater (1999) for ferromagnets and by Tamine (2003) for antiferromagnets.

While there is currently a lack of experimental data regarding spin–wave excitations at steps, this is not the case for nonmagnetic excitations. For example, in a seminal paper Niu *et al* (1995) used inelastic He atom scattering to measure the phonons propagating at a stepped surface of Ni. Their thorough analysis of the dispersion relations, allowing comparison with the theory due to Knipp (1991), allowed them to identify different types of edge and terrace phonon modes. They also concluded that some of the force constants near the step edge were reduced compared with the bulk values. Examples of other experimental studies of excitations on stepped surfaces involve plasmon–polaritons (Puygranier *et al* 2001) using photon scanning tunnelling microscopy and electronic states (Namba *et al* 1996, Ogawa *et al* 2003) using angle-resolved photoelectron spectroscopy.

Patterning of surfaces can be produced in various other ways than by means of steps and gratings, both mentioned earlier. In addition, in section 7.1.2 we gave some examples related to the extrinsic surface states and the surface reconstruction that can take place when certain atoms or molecules are adsorbed on an otherwise clean surface. Further discussion is given by Zangwill (1988).

However, the field of patterned magnetic nanostructures has been particularly active, largely because of the drive towards producing nanoscale devices for data storage and sensors. An excellent review is given by Chou (1997), who discusses the growth of patterned arrays (both in 1D and 2D) of features on substrate surfaces. The materials involved are mostly metals and alloys, and the surface features include flat platelets, dots, vertical pillars etc. The magnetic excitations, namely the dipole-exchange spin waves, at surfaces patterned by arrays of rectangular metallic platelets and circular discs (or 'dots') have been extensively studied using Brillouin scattering (Jorzick *et al* 1999a,b); see also the reviews by Demokritov *et al* (2001) and Demokritov and Hillebrands (2002). This is an extension of the work done on rectangular wires by the same group of authors described in section 7.3.2. Their motivation was to study the spatial confinement

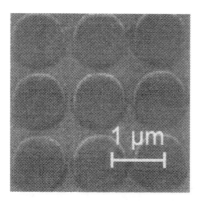

Figure 7.17. Scanning electron micrographs of a patterned surface made up of permalloy dots with a diameter of 1 μm and a closest separation of 0.1 μm, as used for Brillouin scattering. After Demokritov *et al* (2001).

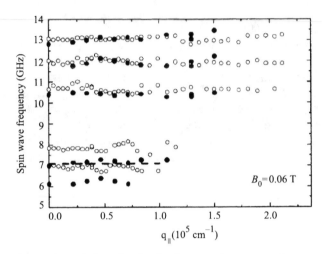

Figure 7.18. Dispersion relations for the five lowest spin–wave modes, as measured by Brillouin scattering on two patterned surfaces with permalloy dots of the same diameter (1 μm) but two values of the separation, namely 0.1 μm (full symbols) and 1 μm (open symbols). The horizontal dashed line represents the uniform precession mode of a single dot. After Demokritov *et al* (2001).

introduced by the patterning. In the case of arrays of rectangular platelets, the results are fairly well understood as a direct generalization of the situation for wires. Specifically, the effective quantization of the in-plane wavevector in 1D for wires, as represented by (7.14) in section 7.3.2, is replaced by a similar quantization for both components of the 2D in-plane wavevector $\mathbf{q}_{\parallel}$. This comes about because there are two length scales, say w_1 and w_2, corresponding to the sides of the rectangles. For the case of dots, we show in figure 7.17 a scanning electron micrograph for a surface patterned with permalloy dots. Then in figure 7.18 we show some data for the dispersion relations of the quantized spin waves; two cases are illustrated corresponding to the same dot diameter (1 μm) but two different separations (0.1 μm and 1 μm). By comparison the film thickness is 40 nm. The effect of inter-dot interactions is evident from the frequency shifts of the two lowest branches in each data set (Demokritov *et al* 2001).

References

Abou Ghantous M and Khater A 1999 *Eur. Phys. J. B* **12** 335
Abramowitz M and Stegun I A 1965 *Handbook of Mathematical Functions* (New York: Dover)
Albuquerque E L 1979 *Phys. Stat. Solidi (b)* **96** 475
Albuquerque E L 1981 *Phys. Stat. Solidi (b)* **104** 667
Arias R and Mills D L 2001 *Phys. Rev. B* **63** 134439

Bland J A C and Heinrich B (eds) 1994 *Ultrathin Magnetic Structures I* (Berlin: Springer)

Boardman A D, Aers G C and Teshima R 1981 *Phys. Rev. B* **24** 5703

Boardman A D (ed) 1982 *Electromagnetic Surface Modes* (New York: Wiley)

Burstein E, Lundquist S and Mills D L 1982 in *Surface Enhanced Raman Scattering* ed R K Chang and T E Furtak (New York: Plenum) p 67

Cady W G 1946 *Piezoelectricity* (New York: McGraw-Hill)

Camley R E and Fulde P 1981 *Phys. Rev. B* **23** 2614

Chou S Y 1997 *Proc. IEEE* **85** 652

Cottam M G and Maradudin A A 1984 in *Surface Excitations* ed V M Agranovich and R Loudon (Amsterdam: North-Holland) p 1

Demokritov S O and Hillebrands B 2002 in *Spin Dynamics in Confined Magnetic Structures I* ed B Hillebrands and K Ounadjela (Berlin: Springer) p 65

Demokritov S O, Hillebrands B and Slavin A N 2001 *Phys. Rep.* **348** 441

Dobrzynski L and Maradudin A A 1972 *Phys. Rev. B* **6** 3810

Donath M 1993 in *Magnetism and Structure in Systems of Reduced Dimension* ed R C F Farrow *et al* (New York: Plenum) p 243

Duval E 1992 *Phys. Rev. B* **46** 5795

Eastman D E and Freeouf J L 1974 *Phys. Rev. Lett.* **33** 1601

Eastman D E and Grobman W D 1972 *Phys. Rev. Lett.* **28** 1378

Eastman D E and Nathan M 1 1975 *Physics Today* **28** April, p 44

Economou E N and Ngai K L 1974 *Adv. Chem. Phys.* **27** 265

Ehrenreich H, Seitz F and Turnbull D 1980 *Solid State Phys.* vol 35 (New York: Academic)

Encinas-Oropesa A, Demand M, Piraux L, Huynen I and Ebels U 2001 *Phys. Rev. B* **63** 104415

Farias G A, Nobre E F, Moretzsohn R, Almeida N S and Cottam M G 2002 *J. Opt. Soc. Am. A* **19** 2449

Ferchmin A R and Puszkarski H 2001 *J. Appl. Phys.* **90** 5335

Fletcher P C and Kittel C 1960 *Phys. Rev.* **120** 2004

Forstmann F 1970 *Z. Physik* **235** 69

Glass N E, Loudon R and Maradudin A A 1981 *Phys. Rev. B* **24** 6843

Glass N E and Maradudin A A 1981 *Phys. Rev. B* **24** 595

Gonis A and Stocks G M (eds) 1992 *Equilibrium Structure and Properties of Surfaces and Interfaces* (New York: Plenum)

Gowar J 1993 *Optical Communication Systems* (New York: Prentice-Hall)

Goodwin E T 1939 *Proc. Camb. Phil. Soc.* **35** 205 221 232

Gurevich A G and Melkov G A 1996 *Magnetization Oscillations and Waves* (Boca Raton, FL: CRC Press)

Gurney B A, Baumgart P, Speriosu V, Fontana R, Patlac A, Logan T and Humbert P 1991 *Proc. Int. Conf. Magn. Films Surf. Glasgow* (London: IOP) p 474

Hansson G V, Bachrach R Z, Bauer R S and Chiaradia P 1981 *Phys. Rev. Lett.* **46** 1033

Harrison P 2000 *Quantum Wells, Wires and Dots* (New York: Wiley)

Hartmann U (ed) 2000 *Magnetic Multilayers and Giant Magnetoresistance* (Berlin: Springer)

Heine V 1980 *Solid State Physics* vol 35, ed H Ehrenreich, F Seitz and D Turnbull (New York: Academic) p 1

Heinrich B and Bland J A C (eds) 1994 *Ultrathin Magnetic Structures II* (Berlin: Springer)

Himpsel F 1985 *Appl. Phys. A* **38** 205

Holy V, Pietsch U and Baumbach T 1999 *High-Resolution X-Ray Scattering from Thin Films and Multilayers* (Berlin: Springer)

Hudson G E 1943 *Phys. Rev.* **63** 46

Hulpke E (ed) 1992 *Helium Atom Scattering from Surfaces* (Berlin: Springer)

Jiang L J and Cottam M G 1999 *J. Appl. Phys.* **85** 5495

Jorzick J, Demokritov S O, Mathieu C, Hillebrands B, Bartenlian B, Chappert C, Rousseaux F and Slavin A N 1999a *Phys. Rev. B* **60** 15194

Jorzick J, Demokritov S O, Hillebrands B, Bartenlian B, Chappert C, Decanini D, Rousseaux F and Cambril E 1999b *Appl. Phys. Lett.* **75** 3859

Joseph R I and Schlömann E 1961 *J. Appl. Phys.* **32** 1001

Kapany N S and Burke J J 1972 *Optical Waveguides* (New York: Academic)

Kawata S (ed) 2001 *Near-Field Optics and Surface Plasmon Polaritons* (Berlin: Springer)

Khosravi H, Tilley D R and Loudon R 1991 *J. Opt. Soc. Am. A* **8** 112

Kittel C 1958 *Phys. Rev.* **110** 836

Kittel C 1986 *Introduction to Solid State Physics* 6th edition (New York: Wiley)

Kliewer K L and Fuchs R 1974 *Adv. Chem. Phys.* **27** 355

Knipp P 1991 *Phys. Rev. B* **43** 6908

Kottmann J P and Martin O J F 2001 *Optics Express* **8** 655

Kretschmann E, Ferrell T L and Ashley J C 1979 *Phys. Rev. Lett.* **42** 1312

Kuok M H, Lim H S, Ng S C, Liu N N and Wang Z K 2003 *Phys. Rev. Lett.* **90** 255502

Lagasse P E 1972 *Electron. Lett.* **8** 372

Laks B, Mills D L and Maradudin A A 1981 *Phys. Rev. B* **23** 4965

Lamb H 1882 *Proc. London Math. Soc.* **13** 189

Lannoo M and Friedel P 1991 *Atomic and Electronic Structure of Surfaces* (Berlin: Springer)

Lingner C and Lüthi B 1981 *Phys. Rev. B* **23** 256

Lopez-Rios T, Mendoza D, Garcia-Vidal S J, Sanchez-Dehesa J and Pannetier D 1998 *Phys. Rev. Lett.* **81** 665

Lüth H 1995 *Surfaces and Interfaces of Solid Materials* (Berlin: Springer)

Lvov V S 1991 *Wave Turbulence under Parametric Excitation* (Berlin: Springer)

Maradudin A A 1981 in *Festkörperprobleme Adv. in Solid State Physics* **21** 25

Maradudin A A, Moss S L and Cunningham S L 1977 *Phys. Rev. B* **15** 4490

Maradudin A A and Visscher W M 1985 *Z. Phys. B* **60** 215

Maradudin A A, Wallis R F, Mills D L and Ballard R L 1972 *Phys. Rev. B* **6** 1106

Marcuse D 1982 *Light Transmission Optics* (New York: Van Nostrand Reinhold)

Mathieu C, Jorzick J, Frank A, Demokritov S A, Slavin A N, Hillebrands B, Bartenlian B, Chappert C, Decanini D, Rosseaux F and Cambril E 1998 *Phys. Rev. Lett.* **81** 3968

Mercereau J and Feynman R P 1956 *Phys. Rev.* **104** 63

Moss S L, Maradudin A A and Cunningham S L 1973 *Phys. Rev. B* **8** 2999

Namba H, Nakanishi N, Yamaguchi T, Ohta T and Kuroda H 1996 *Surf. Sci.* **357** 238

Nicholls J M, Reihl B and Northrup J E 1987 *Phys. Rev. B* **35** 4137

Nishiguchi N and Sakuma T 1981 *Solid State Commun.* **38** 1073

Niu L, Gaspar D J and Sibener S J 1995 *Science* **268** 847

Nobre E F, Costa Filho R N, Farias G A and Almeida N S 1998 *Phys. Rev. B* **57** 10583

Nobre E F, Farias G A and Almeida N S 2000 *J. Opt. Soc. Am. A* **17** 173

Northrup J E 1984 *Phys. Rev. Lett.* **53** 683

Northrup J E and Cohen M L 1982 *Phys. Rev. Lett.* **49** 1349

Nozières J P, Ghidini M, Pinettes C, Lacroix C, Gervais B, Suran G and Dufeu D 1998 *Phys. Rev. B* **57** R14040

Ogawa K, Nakanishi K and Namba H 2003 *Solid. State Commun.* **125** 517
Palmer R E and Schnatterly S E 1971 *Phys. Rev. B* **4** 2329
Pandey K C 1981 *Phys. Rev. Lett.* **47** 1913
Park R L 1975 *Physics Today* **28** April p 52
Pendry J B and Gutman S J 1975 *Surf. Sci.* **49** 87
Pereira J M, Costa Filho R N, Freire V N and Farias G A 1998 *Eur. Phys. J. B* **3** 119
Pereira J M, Costa Filho R N, Freire V N and Farias G A 2000 *Eur. Phys. J. B* **13** 589
Pimpinelli A and Villain J 1998 *Physics of Crystal Growth* (Cambridge: Cambridge University Press)
Plummer E W, Gadzuk J W and Penn D R 1975 *Physics Today* **28** April p 63
Puygranier B A F, Dawson P, Lacroute Y and Goudonnet J P 2001 *Surf. Sci.* **490** 85
Riste N S and Stamps R L 2002 *J. Mag. Mag. Mat.* **242** 1041
Royer D and Dieulesaint E 2000 *Elastic Waves in Solids* vols I and II (Berlin: Springer)
Rulf B 1969 *J. Acoust. Soc. Am.* **45** 493
Sandercock J R and Wettling W 1979 *J. Appl. Phys.* **50** 7784
Schlömann E 1960 *J. Appl. Phys.* **31** 1647
Schrieffer J R and Soven P 1975 *Surf. Sci.* **49** 63
Scott R Q and Mills D L 1977 *Phys. Rev. B* **15** 3545
Sharon T M and Maradudin A A 1973 *Solid State Commun.* **13** 187
Sharon T M and Maradudin A A 1977a *J. Phys. Chem. Solids* **38** 971
Sharon T M and Maradudin A A 1977b *J. Phys. Chem. Solids* **38** 977
Shockley W 1939 *Phys. Rev.* **56** 317
Smolyaninov I, Mazzoni D L, Mait J and Davis C C 1997 *Phys. Rev. B* **56** 1601
Suran G, Rothman J and Meyer C 1998 *IEEE Trans. Magnetics* **34** 849
Syms R and Cozens J 1992 *Optical Guided Waves and Devices* (London: McGraw-Hill)
Tamine M 2003 *J. Mag. Mag. Mat.* **260** 305
Tamm I 1932 *Physik. Z. Sowjetunion* **1** 733
Tamura A, Higeta K and Ichinokawa T 1982 *J. Phys. C* **15** 4975
Tanaka A, Onari S and Arai T 1993 *Phys. Rev. B* **47** 1237
Temiryazev A G, Tikhomirova M P and Zilberman P E 1996 in *Nonlinear Microwave Signal Processing: Towards a New Range of Devices* ed M Marcelli and S A Nikitov (Dordrecht: Kluwer) p 165
Turov E A and Irkhin Yu P 1956 *Fiz. Met. Metalloved.* **3** 15
Uhrberg R I G, Hansson G V, Nicholls J M and Flodstrom S A 1982 *Phys. Rev. Lett.* **48** 1032
Urazokov E I and Falkovskii L A 1972 *Zh. Eksp. i Teor. Fiz.* **63** 2297 [1973 *Sov. Phys. JETP* **36** 1214]
Ushioda S, Aziza A, Valdez J B and Mattei G 1979 *Phys. Rev. B* **19** 4012
Viktorov I A 1958 *Akust. Zh.* **4** 131 [*Sov. Phys.—Acoustics* **4** 131]
Wagner H 1979 in *Solid Surface Science* (Berlin: Springer) p 151
Walker L R 1957 *Phys. Rev.* **105** 390
Wallis R F 1982 in *Electromagnetic Surface Modes* ed A D Boardman (New York: Wiley) p 575
Wang Z, Kuok M H, Ng S C, Lockwood D J, Cottam M G, Nielsch K, Wehrspohn R B and Gosele U 2002 *Phys. Rev. Lett.* **89** 027201
Wolfram T and Dewames R E 1972 *Prog. Surf. Sci.* **2** 233
Yariv A 1989 *Quantum Electronics* 3rd edition (New York: Wiley)
Zangwill A 1988 *Physics at Surfaces* (Cambridge: Cambridge University Press)

PART THREE

EXCITATIONS IN MULTILAYER STRUCTURES

Chapter 8

Nonmagnetic layered structures and superlattices

In section 1.2 a number of growth techniques for the production of epitaxial-layer systems were described. These techniques can be used to prepare multilayer samples consisting of alternating layers of thickness d_1 of constituent 1 and thickness d_2 of constituent 2. Samples can be prepared so that d_1 and d_2 have any values from two or three atomic spacings up to the order of 100 nm or more. We refer to them as *periodically layered structures* or *superlattices* (see also section 1.1.3); the latter has a specific meaning for electronic states in semiconductor structures, but in other contexts it is often convenient to use the two terms as synonyms. Many of the physical properties are greatly modified by the existence of the long (compared with the lattice parameter) spatial period $D = d_1 + d_2$. The most important general consequence is that as a result of Bloch's theorem a new Brillouin-zone edge appears at wavevector component π/D perpendicular to the interfaces. This can be much smaller than the zone-edge wavevector π/a related to the lattice constant a. Dispersion curves, such as those for acoustic phonons for example, develop band gaps at these new zone edges.

It is natural to define what might be called 'bulk' modes of an infinitely extended superlattice and 'surface' modes that may occur in a multilayer sample with one or two 'free' (or end) surfaces. That is, the former propagate with a real Bloch wavevector component Q perpendicular to the interfaces, while the latter have a complex Q, i.e. $\mathrm{Im}(Q)$ is nonzero. Since a 'bulk' superlattice consists of an extended array of interfaces, the bulk modes are constructed as a superposition of interface modes of the types that were discussed in earlier chapters. For this reason, we give a detailed account of the bulk modes.

As in the earlier chapters, the emphasis here is on acoustic and optical properties. In section 8.1 we introduce the main ideas with a detailed study of the continuum-acoustic modes of a bulk superlattice. Using a formalism similar to that of Albuquerque *et al* (1986a,b), we show that the system can be described by means of a *transfer matrix* T which relates the field amplitude at coordinate $z + D$ to that at z, where z is measured perpendicular to the interfaces. The dispersion relation is simply expressed in terms of T. Subsequent sections are also largely based on the transfer-matrix formalism. In section 8.2 we turn to the lattice dynamics of a simple 1D model of a diatomic superlattice and an introduction to optical properties is given in section 8.3. Then the collective-electron, i.e. plasma, properties are discussed in section 8.4. Finally we discuss quasiperiodic multilayer structures in section 8.5; these have a different growth rule for the layers, as mentioned in section 1.1.3.

As indicated, this chapter is mainly concerned with epitaxial-crystal solids in which the long period is produced by variation of crystal-growth conditions. Long periods also exist in some other types of material. *Intercalated compounds*, for example intercalated graphite, are basically layered materials in which long organic molecules are interposed to space out the layers of the starting materials (Dresselhaus 1986). The length of the organic molecules can be chosen to give a stipulated layer-to-layer spacing. Liquid crystals (see e.g. Chandrasekhar 1992, Collings 1990, De Gennes and Prost 1993) containing chiral, that is screw-like, organic molecules, can undergo a helical distortion in which the mean orientation of the molecules spirals around an axis in space; the two phases of most importance are the *cholesteric* and the *smectic-C**. In these the pitch of the helix is a function of temperature and concentration of chiral molecules and can be varied in addition by applied electric and magnetic fields; it can have any value from a few tens to many hundreds of nm. While the properties of these long-period materials have much in common with those of superlattices, it would take us too far afield to discuss them.

8.1 Continuum acoustics: folded acoustic modes

8.1.1 Propagation in one dimension

We first discuss the propagation of an acoustic wave, either longitudinal or transverse, in a direction normal to the interfaces of an infinite superlattice. The notation is shown in figure 8.1. Within the layers the displacement u satisfies the 1D wave equation

$$\rho_i \frac{\partial^2 u}{\partial t^2} = C_i \frac{\partial^2 u}{\partial z^2}, \qquad i = 1, 2, \ldots . \tag{8.1}$$

We aim to solve these equations with the boundary conditions (see section 2.2.2) that u and $C \, \partial u / \partial z$ are continuous at each interface, and we consider a wave of angular frequency ω so that $\partial^2 u / \partial t^2 = -\omega^2 u$. Since the system has

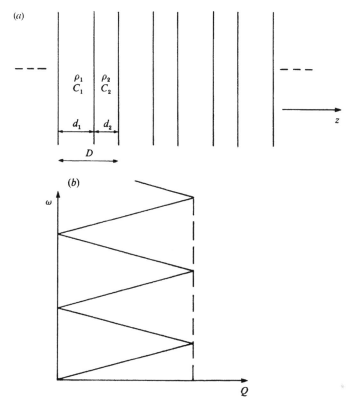

Figure 8.1. (a) Infinite superlattice. The layers are characterized by thickness d_i, density ρ_i and elastic modulus C_i (appropriate to longitudinal or transverse waves). The z axis is taken normal to the interfaces. (b) Acoustic-phonon dispersion curve for infinite superlattice when layers are impedance matched. Slopes of lines are v, given by $v^{-1} = (d_1 v_1^{-1} + d_2 v_2^{-1})/D$.

the translational period $D = d_1 + d_2$ in the z direction we can introduce the wavevector component Q by means of Bloch's theorem:

$$u(z + D) = \exp(iQD)\, u(z). \qquad (8.2)$$

The solution of the wave equation in either medium is a superposition of a forward- and a backward-travelling wave. It is convenient to represent the solution within any layer in two alternative forms, one with the phases referred to the left-hand end of the layer and the other with them referred to the right-hand end. Thus we write

$$u = a_l^L \exp[iq_1(z - lD)] + b_l^L \exp[-iq_1(z - lD)]$$
$$= a_l^R \exp[iq_1(z - lD - d_1)] + b_l^R \exp[-iq_1(z - lD - d_1)] \qquad (8.3)$$

where $lD \leq z \leq lD + d_1$ and $q_1 = \omega/v_1$. The amplitudes are related by

$$|u_l^R\rangle = F_1|u_l^L\rangle \tag{8.4}$$

where we denote

$$|u_l^{L,R}\rangle = \begin{pmatrix} a_l^{L,R} \\ b_l^{L,R} \end{pmatrix} \tag{8.5}$$

and the 2×2 matrix F_1 is

$$F_1 = \begin{pmatrix} f_1 & 0 \\ 0 & f_1^{-1} \end{pmatrix} \tag{8.6}$$

with $f_1 = \exp(iq_1 d_1)$. In a layer of the second medium, the displacement is

$$\begin{aligned} u &= d_l^L \exp[iq_2(z - lD - d_1)] + e_l^L \exp[-iq_2(z - lD - d_1)] \\ &= d_l^R \exp[iq_2(z - (l+1)D)] + e_l^R \exp[-iq_2(z - (l+1)D)] \end{aligned} \tag{8.7}$$

where $lD + d_1 \leq z \leq (l+1)D$. The amplitudes are related by

$$|w_l^R\rangle = F_2|w_l^L\rangle \tag{8.8}$$

with similar notations and definitions to (8.5) and (8.6).

The boundary conditions at $z = lD + d_1$ relate $|u_l^R\rangle$ to $|w_l^L\rangle$:

$$Q_1|u_l^R\rangle = Q_2|w_l^L\rangle \tag{8.9}$$

where

$$Q_1 = \begin{pmatrix} 1 & 1 \\ Z & -Z \end{pmatrix}, \qquad Q_2 = \begin{pmatrix} 1 & 1 \\ 1 & -1 \end{pmatrix} \tag{8.10}$$

and $Z = C_1 q_1/C_2 q_2$ is the ratio of the acoustic impedances of the media. Similarly the boundary conditions at $z = (l+1)D$ give

$$Q_2|w_l^R\rangle = Q_1|u_{l+1}^L\rangle. \tag{8.11}$$

The same matrices occur in (8.9) and (8.11) because the formalism is symmetric between the $+z$ and $-z$ directions. Now combining (8.4), (8.8), (8.9) and (8.11) we find

$$|u_{l+1}^L\rangle = T|u_l^L\rangle \tag{8.12}$$

where the *transfer matrix* T is given by

$$T = Q_1^{-1}Q_2 F_2 Q_2^{-1} Q_1 F_1 \tag{8.13}$$

and explicit evaluation gives

$$T = \begin{pmatrix} f_1 c_2 + \frac{1}{2}i(Z + Z^{-1})f_1 s_2 & -\frac{1}{2}i(Z - Z^{-1})f_1^{-1} s_2 \\ \frac{1}{2}i(Z - Z^{-1})f_1 s_2 & f_1^{-1} c_2 - \frac{1}{2}i(Z + Z^{-1})f_1^{-1} s_2 \end{pmatrix} \tag{8.14}$$

with notation

$$c_2 = \cos(q_2 d_2), \qquad s_2 = \sin(q_2 d_2). \tag{8.15}$$

It can be verified from (8.14) that the transfer matrix is unimodular, $\det T = 1$, as is required for conservation of energy. The same result follows more easily from (8.13) because products of matrices have the property (see e.g. Kreyszig 1988) that $\det(M_1 M_2 \ldots M_n) = \det(M_1)\det(M_2)\ldots\det(M_n)$, from which it follows also that $\det(M^{-1}) = [\det(M)]^{-1}$. The fact that $\det T = 1$ then follows from (8.13) since F_1 and F_2 are both unimodular.

We next apply Bloch's theorem (see section 1.1). It follows from (8.2) that

$$|u_{l+1}^L\rangle = \exp(iQD)|u_l^L\rangle \tag{8.16}$$

and so

$$[T - \exp(iQD)I]|u_l^L\rangle = 0 \tag{8.17}$$

where I is the 2×2 unit matrix. Equally from the equation relating $|u_{l-1}^L\rangle$ to $|u_l^L\rangle$,

$$[T^{-1} - \exp(-iQD)I]|u_l^L\rangle = 0 \tag{8.18}$$

so that combining (8.17) and (8.18)

$$[\tfrac{1}{2}(T + T^{-1}) - I\cos(QD)]|u_l^L\rangle = 0. \tag{8.19}$$

This holds for the amplitude $|u_l^L\rangle$ in any cell l and so the matrix must vanish. Hence we have

$$I\cos QD = \tfrac{1}{2}(T + T^{-1}) = \tfrac{1}{2}I \operatorname{tr} T = \tfrac{1}{2}I (\lambda_1 + \lambda_2) \tag{8.20}$$

where I is the 2×2 unit matrix and $\operatorname{tr} T$ denotes the trace. We have introduced the eigenvalues λ_1 and λ_2 of T. The two alternative expressions in (8.20) follow from $\det T = \lambda_1 \lambda_2 = 1$ since T is a 2×2 matrix. With (8.14) for T, (8.20) gives the explicit form of the dispersion equation:

$$\cos QD = \cos q_1 d_1 \cos q_2 d_2 - [(1 + Z^2)/2Z] \sin q_1 d_1 \sin q_2 d_2. \tag{8.21}$$

This equation is long established and was first derived by Rytov (1956).

Equation (8.21) has some elementary properties of interest. First, if $Z = 1$ the layers are impedance matched and the medium is effectively continuous acoustically. The equation then reduces to

$$\cos QD = \cos(q_1 d_1 + q_2 d_2) \qquad (8.22)$$

which has solutions

$$\omega\left(\frac{d_1}{v_1} + \frac{d_2}{v_2}\right) = \pm QD + n\pi. \qquad (8.23)$$

This is the equivalent of the 'empty-lattice' approximation of electron-band theory, since we have applied Bloch's theorem but without any reflections at the interfaces. As illustrated in figure 8.1(b), equation (8.23) corresponds to the folding of the acoustic-phonon dispersion curve into the reduced zone. Note, however, that the slope of the curve is an average of the velocities in the two media.

For $Z \neq 1$ we have the inequality $1 + Z^2 > 2Z$. If (8.21) is reorganized into a sum of two terms in $\cos(q_1 d_1 \pm q_2 d_2)$ it is readily seen that for some values of ω the absolute value of the right-hand side is greater than 1. For these values of ω there is no solution with real Q. Thus for $Z \neq 1$ there are some stop bands in the dispersion curve. It can also be seen that if $q_1 d_1$ and $q_2 d_2$ are not commensurate, the dispersion curve is not, strictly speaking, periodic in ω. That is, the magnitudes of the stop bands vary irregularly with ω. These features are illustrated in figure 8.2.

It is helpful to look at the stop bands in terms of the eigenvalues λ_1 and λ_2. The condition $\lambda_1 \lambda_2 = 1$ can be satisfied if

(i) λ_1 and λ_2 lie on the unit circle and are complex conjugates, or
(ii) λ_1 and λ_2 both lie on the real axis.

In the former case, $(\lambda_1 + \lambda_2)/2 < 1$ so that (8.20) is solved by real Q. Thus case (i) corresponds to a pass band. Conversely, when λ_1 and λ_2 are both real, $(\lambda_1 + \lambda_2)/2 > 1$ since $\lambda_2 = 1/\lambda_1$. In case (ii), therefore, Q is not real so case (ii) corresponds to a stop band. We may make more precise statements in terms of the trajectories followed by λ_1, λ_2 and Q as the frequency increases from zero in figure 8.2. At first, Q is real and λ_1 and λ_2 move in opposite directions around the unit circle starting from $\lambda_1 = \lambda_2 = 1$. At the top of the first pass band, Q has reached π/D while $\lambda_1 = \lambda_2 = -1$. Through the first stop band, Q makes an excursion of the form $Q = \pi/D + iy$, returning to $Q = \pi/D$ at the top of the stop band. Meanwhile λ_1 and λ_2 make excursions in opposite directions along the real axis, returning to $\lambda_1 = \lambda_2 = -1$. Through the second pass band, Q decreases through real values to zero while λ_1 and λ_2 return around the unit circle to $\lambda_1 = \lambda_2 = 1$. In the second stop band, $Q = iy$ while λ_1 and λ_2 make excursions along the real axis from $\lambda_1 = \lambda_2 = 1$. It will be seen in section 8.1.2 how particular discrete non-real values of Q correspond to surface modes of a semi-infinite superlattice.

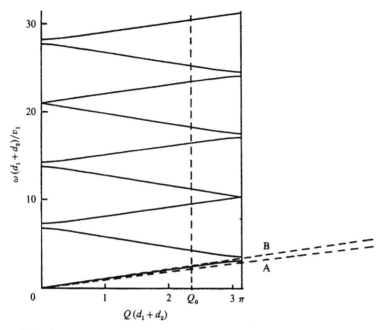

Figure 8.2. The folded-phonon dispersion curve of (8.21). The parameters are $d_1/d_2 = 3/7$, $v_1/v_2 = 0.856$ and $Z = 0.779$. These are chosen arbitrarily to illustrate the stop bands and the nonperiodicity of the dispersion curve. The scan line $q = Q_0$ for a typical light-scattering experiment is shown. After Babiker *et al* (1985a).

The calculation that has been presented here, like most of the rest of this chapter, is concerned with superlattices in which the unit cell consists of two layers. The method can be extended to the general case where the unit cell consists of N different layers. A range of theoretical results for this case has been presented by Djafari-Rouhani and Dobrzynski (1987).

8.1.2 Surface modes

We now discuss, in outline, how some values of the Bloch vector Q away from the real axis may correspond to surface modes on a semi-infinite superlattice. For definiteness, consider $Q = iy$ which as we have just seen corresponds to one of the zone-centre stop bands in figure 8.2. Bloch's theorem (8.16) now takes the form

$$|u_{l+1}^L\rangle = \exp(-yD)|u_l^L\rangle. \tag{8.24}$$

In a semi-infinite superlattice comprising cells $l = 1, 2, 3, \ldots$ this describes exponential decrease of the acoustic amplitude with distance from the surface. Thus this imaginary value of Q is a candidate for a surface mode similar to that discussed for a diatomic lattice in section 3.1. However, there is a boundary condition to be satisfied where the superlattice terminates at one end of layer 1; for a free surface, for example, this will be a condition of zero stress. Thus we may expect that at most certain discrete values of Q correspond to surface modes. Examples to illustrate this general discussion will be given in some of the subsequent sections.

8.1.3   Light scattering

Inelastic light scattering has been used to investigate folded acoustic-phonon dispersion curves like those of figure 8.2. The first such observation was reported by Feldman *et al* (1968) on long-period SiC polytypes. We concentrate here, however, on results obtained on semiconductor superlattices. As always, inelastic light scattering essentially probes the dispersion curve along a vertical line in the (q, ω) plane, say $q = Q_0$ as illustrated in figure 8.2. By appropriate choice of sample period D and incident-light frequency, the value of Q_0 can be chosen to be on a scale comparable with π/D so that much of the superlattice Brillouin zone $0 < q < \pi/D$ is accessible. Furthermore, substitution of typical values for D and the acoustic velocities v_1 and v_2 show that the excitation frequencies in dispersion curves like figure 8.2 can correspond to tens of wavenumber units. This means that Raman scattering can be used, at least for the higher branches, rather than the more difficult technique of Brillouin scattering that is normally required for light scattering off acoustic modes.

The first experimental results on the folded acoustic spectrum were obtained by Colvard *et al* (1980) and a more recent Raman spectrum is shown in figure 8.3. The experiment was carried out in backscattering and the polarization selection rules are the same as those for backscattering off the surface of an ordinary sample, as discussed in section 3.3.2. Thus in (x, x) polarization the LA phonons are seen, and there is no scattering by phonons in (x, y) polarization. The (x, x) spectrum shows three prominent doublets as marked on the horizontal axis and the positions fall on the folded dispersion curve as indicated in the inset. As anticipated, the (x, y) spectrum is featureless.

Comprehensive studies along these lines were carried out by Jusserand *et al* (1983, 1984a,b) as well as by Colvard *et al* (1985), and a review of experimental work, primarily inelastic light scattering, on both folded acoustic phonons and confined optic phonons, to be discussed in section 8.2.1, is given by Klein (1986).

8.1.4   Green functions, oblique propagation and surface modes

Some developments of the Rytov-type theory presented in section 8.1.1 should be mentioned. By analogy with the detailed treatment of Brillouin

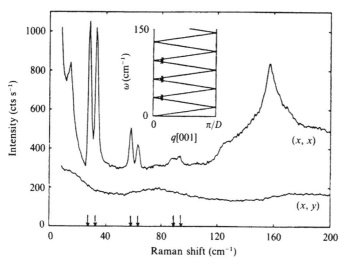

Figure 8.3. Raman backscattering from a sample nominally consisting of a repeat unit of 4.2 nm of GaAs and 0.8 nm of $Al_{0.3}Ga_{0.7}As$. Arrows indicate peak frequencies corresponding to the superlattice period of 5.22 nm determined by x-ray diffraction. Inset shows a dispersion curve corresponding to (8.21) for LA phonons, with crosses at peak positions in (x, x) spectrum. After Colvard *et al* (1985).

scattering off isotropic samples (section 3.3.2), a full theory of lineshapes and strengths in light-scattering experiments should start with a derivation of the acoustic Green function. Babiker *et al* (1985a) found the Green function for longitudinal or transverse phonons propagating normally to the interface in an infinitely extended superlattice, and the result was used for an approximate theory of the light-scattering cross section (Babiker and Tilley 1984, Babiker *et al* 1985b).

The theory of acoustic propagation at a general angle to the interface planes was given by Camley *et al* (1983) for s-transverse waves. The derivation of the dispersion relation of the bulk modes is not much more complicated than was given above for normal incidence. Camley *et al* go on to discuss surface and interface modes of a semi-infinite superlattice in contact with either vacuum or a bulk elastic medium. It was seen in section 8.1.2 that surface modes may occur at frequencies within the stop bands of dispersion curves like figure 8.2 and it is found that for the present case they do indeed occur. Similar results are presented by Bulgakov (1985). It should be emphasized that in contrast to the Rayleigh and Stoneley waves discussed in section 3.2 these surface modes occur in pure s-polarization, and are a consequence of the alternation of elastic properties depicted in figure 8.1.

The propagation of longitudinal plus p-transverse waves is discussed by Djafari-Rouhani *et al* (1983). The elastic displacement in each layer is

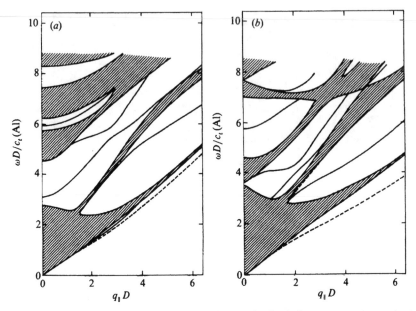

Figure 8.4. Bulk continua and surface bands for longitudinal plus p-transverse modes in the Al/W superlattice (full line, Al on surface; broken line, W on surface). (a) $d_{Al} = d_W$; (b) $d_{Al} = 5d_W$. The surface-layer thickness is taken the same as that of a corresponding bulk layer. The variables are dimensionless with $c_t(Al)$ denoting the velocity of transverse sound in Al and $D = d_{Al} + d_W$. After Djafari-Rouhani *et al* (1983).

described by four amplitudes rather than the two of (8.3) or (8.7), since there are two amplitudes for the longitudinal and two for the transverse component of the displacement. Consequently the transfer matrix is 4×4; the explicit form is given by Djafari-Rouhani *et al*. The authors present numerical illustrations for the Al/W superlattice, showing both the modes of an infinite superlattice as a bulk continuum and the surface modes as dispersion curves in a graph of ω versus $q_\parallel$. Examples of their results are shown in figure 8.4.

8.2 Lattice dynamics: folded and confined modes

In the discussion of acoustic modes of bulk materials in chapter 2 we started with lattice dynamics in section 2.1, then moved on to the continuum acoustics in section 2.2. As was seen there, the latter can be derived either as the continuum limit of the lattice-dynamic formalism or independently by the methods of macroscopic elasticity theory. By contrast, in this chapter, we have started with continuum acoustics, because it is the simplest example, and we now turn to lattice dynamics. If we were to follow section 2.1, we should start with a superlattice composed of monatomic materials.

However, the samples of most interest are based on III–V semiconductors such as GaAs and these must be modelled as diatomic materials. We therefore discuss in detail only this case.

The previous section was concerned with the folding of the acoustic-phonon dispersion curve into the mini-Brillouin zone resulting from the long period D. We are now dealing with diatomic materials, which support optic as well as acoustic phonons. The former are also folded in a super-lattice, although in some cases it is more convenient to refer to the modes as confined rather than folded. We begin this section with a detailed discussion of the simplest 1D model, then discuss more realistic models and Raman scattering results, which are reviewed in more detail by Klein (1986). Detailed information on optic-phonon modes has also been obtained by far-infrared spectroscopy, but discussion of this is deferred to section 8.3 on optical properties.

8.2.1 One-dimensional model

The model to be discussed is summarized in figure 8.5. It is assumed that the interatomic spacing a and the elastic constant C have the same values in the two components of the superlattice, the difference between the two being the change from mass m_1 to mass m_2. The model may be regarded as a simple description of a $GaAs/Al_xGa_{1-x}As$ superlattice, m_0 being the As mass.

The dynamical behaviour of an infinite 1D diatomic lattice was summarized in section 2.1.2. The dispersion equation is given in (2.16) and illustrated in figure 2.2. For the superlattice we take solutions in cell l of the form of (8.3) and (8.7). Now, of course, q_1 and q_2 are related to ω by (2.16) with appropriate mass values substituted. There is no need to restrict attention to frequency intervals in which q_1 and q_2 are both real; it will be seen in fact that there are phonon modes of the superlattice in which only one is real. The algebra takes the same form in any case.

The amplitudes at the left and right ends of a layer are still related by matrices F_1 and F_2 as in (8.6). In section 8.1.1 the amplitudes in the two

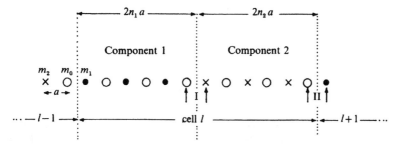

Figure 8.5. Microscopic model for a diatomic superlattice in the region of unit cell l. The masses are coupled by nearest-neighbour forces with spring constant C.

constituents of cell l were related by the two boundary conditions at the interface to yield (8.9) and (8.10). As may be seen from section 2.1.1, the lattice-dynamical equivalent is provided by the equations of motion of the atoms on either side of the interface. These are the masses indicated by arrows in figure 8.5 and their equations of motion lead to equations of the form of (8.9) and (8.11) with

$$Q_1 = \begin{pmatrix} 1 & 1 \\ \alpha_1 s_1^{-1} & \alpha_1 s_1 \end{pmatrix}, \qquad Q_2 = \begin{pmatrix} 1 & 1 \\ \alpha_2 s_2^{-1} & \alpha_2 s_2 \end{pmatrix} \qquad (8.25)$$

where

$$s_i = \exp(iq_i a), \qquad i = 1, 2. \qquad (8.26)$$

Also α_1 and α_2 are the ratios of the displacements of the two masses in the relevant component, as found from (2.14) and (2.15):

$$\alpha_i = \frac{2C - m_0 \omega^2}{2C \cos(q_i a)} = \frac{2C \cos(q_i a)}{2C - m_i \omega^2}, \qquad i = 1, 2. \qquad (8.27)$$

Equations (8.12) and (8.13) continue to define the transfer matrix T and the detailed algebra gives for the matrix elements:

$$T_{11} = T_{22}^* = \frac{1}{4}\left(\frac{\tan(q_1 a)}{\tan(q_2 a)} + \frac{\tan(q_2 a)}{\tan(q_1 a)}\right) f_1(f_2 - f_2^{-1}) + \tfrac{1}{2}f_1(f_2 + f_2^{-1}) \quad (8.28)$$

$$T_{21} = T_{12}^* = -\frac{1}{2}\left(\frac{\cot(2q_1 a)}{\sin(2q_2 a)} - \frac{\cot(2q_2 a)}{\sin(2q_1 a)}\right) f_1(f_2 - f_2^{-1}) \qquad (8.29)$$

where $f_n = \exp(iq_n d_n)$ for $n = 1, 2$. It can be confirmed from (8.28) and (8.29) that $\det T = 1$, as also follows from (8.13). The dispersion relation continues to be given by (8.20) and can be brought into the explicit form

$$\cos(QD) = \cos(q_1 d_1)\cos(q_2 d_2) - \frac{1}{2}\left(\frac{\tan(q_1 a)}{\tan(q_2 a)} + \frac{\tan(q_2 a)}{\tan(q_1 a)}\right)$$

$$\times \sin(q_1 d_1)\sin(q_2 d_2). \qquad (8.30)$$

This result was first given by Jusserand *et al* (1984a) and the explicit derivation by a Green function technique was due to Djafari-Rouhani *et al* (1985).

Equation (8.30) contains a considerable amount of information, which we now explore. In the continuum limit $q_1 a \ll 1$ and $q_2 a \ll 1$ it reduces to (8.21), as it should. The result of numerical solution of (8.30) for a specific choice of parameters is shown in figure 8.6(b). In order to interpret the result, we include in figure 8.6(a) the dispersion curves for the two bulk media. This enables us to divide the frequency axis into six regions, as indicated on the figure, and Table 8.1 shows the character of the bulk wave numbers q_1 and q_2 in each of these regions. Comparison with figure 8.6(b) shows that where q_1 and q_2 are both real, in regions I and IV, the superlattice

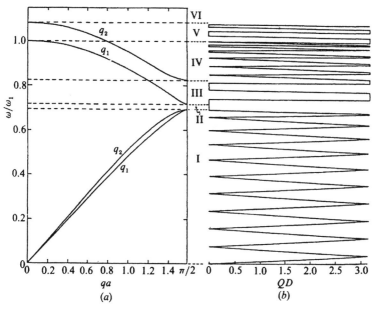

Figure 8.6. (a) Bulk dispersion curves, (2.16), for two constituents of a diatomic superlattice with $m_1/m_0 = 0.93$ and $m_2/m_0 = 0.70$. Frequency axis is ω/ω_1, where $\omega_1 = [2C(m_0^{-1} + m_1^{-1})]^{1/2}$ is the zone-centre optic-phonon frequency in medium 1. (b) Dispersion curve for a superlattice of these constituents with $n_1 = n_2 = 10$. After Albuquerque *et al* (1988).

dispersion curves have broad pass bands and narrow stop bands. This is similar to what was found in section 8.1.1 for the continuum description. In regions III and V, however, where only one of q_1 and q_2 is real, the pass bands are narrow and the stop bands broad. A qualitative reason for this can easily be seen from (8.30). In region V, for example, $q_1 = iy$, where y is real. The trigonometric functions then become hyperbolic functions: $\cos(q_1 d_1) = \cosh(y d_1)$, $\sin(q_1 d_1) = i\sinh(y d_1)$, and $\tan(q_1 a) = i\tanh(ya)$.

Table 8.1. Frequency intervals corresponding to figure 8.6 (R = real; I = imaginary; C = complex).

Frequency interval	q_1	q_2	Superlattice dispersion curve
I	R	R	Broad pass bands; narrow stop bands
II	C	C	No solutions
III	R	C	Narrow pass bands; broad stop bands
IV	R	R	Broad pass bands; narrow stop bands
V	I	R	Narrow pass bands; broad stop bands
VI	I	I	No solutions

Thus the right-hand side of (8.30) is essentially of the form $A_1 \cos(q_2 d_2) + A_2 \sin(q_2 d_2)$ with $|A_1| \gg 1$ and $|A_2| \gg 1$. This can be rewritten as $B \cos(q_2 d_2 + \theta)$ with $|B| \gg 1$. Thus as ω moves up through region V, equation (8.30) has solutions for real Q only in the very narrow frequency bands where $|\cos(q_2 d_2 + \theta)| < 1/|B|$. Region III is similar. The displacements in regions III and V have an oscillatory spatial dependence in the medium with real q and an exponentially decaying spatial dependence in the other medium. They are usually called *confined phonon modes*. The frequency varies only slightly with Q, and is the same as would be found in a single slab of the real-q medium embedded in the other medium.

The lattice-dynamical model discussed here has been extended to include next-nearest-neighbour interactions by Hadizad *et al* (1991a,b) for monatomic and diatomic materials, respectively. The main technical change is that the transfer matrix T becomes 4×4 rather than 2×2 and in consequence the discussion of the mode spectrum becomes considerably more complicated. The form of the superlattice dispersion curve is best understood in terms of the complex-plane trajectories mapped out by the eigenvalues of T as the frequency changes.

Schematic 1D models of the kind discussed so far give a simple picture of folded and confined optic modes and it will be seen that they are in qualitative agreement with Raman-scattering results. However, a full account must include the long-range Coulomb interaction and of course must also include 3D lattice dynamics. One way of expressing the former point is to say that it is necessary to find the contribution to the dielectric function of the confined optic modes. The first theory of this kind was given by Yip and Chang (1984); related work by Samson *et al* (1992) will be discussed in section 8.3.5 on optical properties of short-period superlattices. As reviewed briefly by Dumelow *et al* (1993), full 3D lattice-dynamic calculations, particularly for $(\text{GaAs})_{n_1}/(\text{AlAs})_{n_2}$ superlattices, have now been carried out by a number of authors. In particular, Molinari *et al* (1992) reported full *ab initio* calculations.

8.2.2   Light scattering

The experimental technique that was first applied in the study of the folded-optic mode spectrum is Raman scattering. The calculation presented above is concerned with phonons propagating normally to the interfaces, so the relevant experimental results are from 180° backscattering at normal incidence. The selection rule for this geometry is that only longitudinal modes contribute to the cross section, so the data involve folded LO phonons. Figure 8.7 shows spectra obtained by Jusserand *et al* (1984b, 1985) from a number of $\text{GaAs}/\text{Al}_{0.3}\text{Ga}_{0.7}\text{As}$ superlattice samples. The peaks labelled 1 to 4 are the highest four confined states; peak A is assigned to an AlGaAs LO mode. The peaks correspond to LO modes confined in

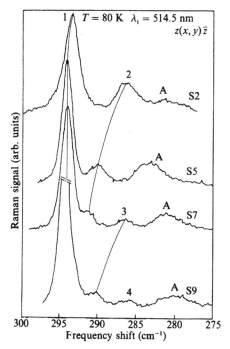

Figure 8.7. Raman spectra in the LO phonon region from four samples consisting of repeat units having n monolayers of GaAs and n' monolayers of $Al_{0.3}Ga_{0.7}As$. The (n, n') values from top to bottom are: (6, 4), (9, 9), (12, 7), and (17, 12). After Jusserand *et al* (1984b).

GaAs by AlGaAs 'phonon barriers', so that their positions should depend on the GaAs monolayer number n but not significantly on the AlGaAs monolayer number n'; this was indeed found to be the case. It may be noted that the line intensity decreases with increasing folding number, as was seen for folded acoustic phonons in figure 8.3. This agrees with the theoretical predictions (Samson *et al* 1992).

8.2.3 Surface modes

Equation (8.30) and the subsequent discussion were concerned with the 'bulk' modes of an infinite superlattice. As was discussed in section 8.1.2 in the context of continuum acoustics, surface modes in a semi-infinite super-lattice may occur in the stop bands, such as those of figure 8.6(b). A brief discussion of these modes for the 1D model is given by Djafari-Rouhani *et al* (1985). Their calculation is analogous to that of Wallis (1957) for the surface mode of a semi-infinite 1D diatomic crystal, which was outlined in section 3.1.

8.3   Optical properties

The basic theme of section 2.5 on bulk polaritons and chapter 6 on surface polaritons is that optical propagation is governed by the dielectric function. As was seen in section 2.4.1, this in turn is determined by the spectrum of elementary excitations in the material. Two basic examples were given: the plasma dielectric function of (2.132), which is the free-carrier response, and the 'reststrahl' form of (2.146), which is determined by the long-wavelength ($q \approx 0$) optic-phonon response.

How do these principles apply for superlattices? Let us consider the specific example of a superlattice constructed from ionic materials without free carriers, for example NaCl/KCl (alkali halides), $PbTiO_3/BaTiO_3$ (ferroelectrics) or undoped GaAs/AlAs (ionic semiconductors), so that the dielectric function is determined by the optic phonons. We have just seen in section 8.2.1 that the optic-phonon spectrum is modified by the superlattice structure and as shown in figure 8.6 the TO phonon spectrum becomes a series of confined modes. This means that the single bulk $q = 0$ TO phonon at frequency ω_T is replaced by a series of $Q = 0$ modes at a number of different frequencies. In general, therefore, the problem is to calculate the dipole strengths of these modes and hence the dielectric response of the superlattice. Having done so the next step is to investigate the optical propagation. This is a formidable task, but fortunately it can be simplified for *long-period superlattices*, those satisfying

$$d_1 \gg a_1, \qquad d_2 \gg a_2 \qquad (8.31)$$

where a_1 and a_2 are the lattice constants of the component media. The mini Brillouin zone then becomes narrow since $\pi/D \ll \pi/a_i$ and the TO dipole strength becomes concentrated in the mode at frequency ω_T. In other words, from the optical point of view the superlattice behaves as a series of layers each with its bulk dielectric function ε_1 or ε_2. This description may be called the *bulk-slab model* of the superlattice. Within this model one is left with the task of applying classical electrodynamics to find the optical propagation. It is worth commenting that when the component materials of the superlattice are cubic, which means optically isotropic, as in the examples listed above, then the optical symmetry of the superlattice is uniaxial. This is obvious for the bulk-slab model, and holds generally.

In this section, we first give the basic formalism for the bulk-slab model and discuss some implications for non-dispersive media. Then we introduce the *effective-medium approximation* that holds for long-wavelength radiation satisfying $\lambda \gg D$ and discuss its application to dispersive media, particularly semiconductor superlattices. Doped semiconductors of course exhibit plasma as well as phonon response in the dielectric function, but the topic of superlattice plasma modes is sufficiently important to be left to a separate section. In section 8.3.5, we turn to short-period superlattices, in which the effects of phonon confinement are important, and discuss the problem of constructing the expression for the dielectric function. The existence of structure-induced band gaps is of potential importance in optical engineering, and an introduction to the general topic of *photonic band gaps* is given in section 8.3.6.

8.3.1 Bulk-slab model: formulation and non-dispersive media

The complete formal theory of optical propagation in an infinite periodic layered medium, in which the constituents are isotropic or cubic, was given by Yeh *et al* (1977) and Yariv and Yeh (1977). An account is also given by Yariv and Yeh (1984). Raj and Tilley (1989) give a detailed review with some emphasis on the interaction between frequency-dispersion of the dielectric functions and the effects of the superlattice periodicity.

The theory of optical propagation is very similar to the continuum acoustics of section 8.1. There is, however, the simplification that in optics only transverse modes occur even in oblique incidence. Thus the theory for waves travelling in a general direction is not much harder than that for waves travelling normal to the interfaces and we proceed straight away to the general case.

As indicated in figure 8.8, the notation is similar to that used for the acoustics calculation. Each medium is characterized by an isotropic dielectric function (ε_1 or ε_2) or equivalently a refractive index ($\eta_1 = \varepsilon_1^{1/2}$ or $\eta_2 = \varepsilon_2^{1/2}$). The z axis is taken normal to the layers, as before, and the x axis is chosen so that the propagation vector lies in the xz plane. As usual, therefore, we can distinguish s-polarization, with **E** in the y direction, from p-polarization, with **E** in the xz plane. We start with the former, for which the formalism

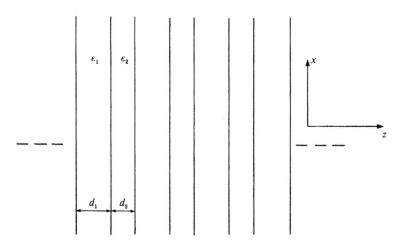

Figure 8.8. Notation for optical calculation.

of section 8.1 applies with minor modifications. All fields are taken to have x-and t-dependences proportional to $\exp(iq_x x - i\omega t)$. The **E** fields in a layer of medium 1 can be written in either of the forms of (8.3), where now q_1 is given by

$$q_1^2 + q_x^2 = \varepsilon_1 \omega^2 / c^2. \tag{8.32}$$

Equation (8.4) shows how the phase matrix F_1 relates the amplitudes at the two ends of the layer. The **E** field in medium 2 is written as (8.7), with (8.8) relating the two forms.

The boundary conditions at an interface are that E_y and H_x are continuous, i.e. E_y and $\partial E_y / \partial z$ are continuous. They may be written in the form of (8.9) and (8.11), with the Q matrices given by

$$Q_i = \begin{pmatrix} 1 & 1 \\ q_i & -q_i \end{pmatrix}, \qquad i = 1, 2. \tag{8.33}$$

The transfer matrix T is still given formally by (8.13) and evaluation gives for the diagonal elements

$$T_{11} = f_1 [f_2 (q_1 + q_2)^2 - f_2^{-1} (q_1 - q_2)^2] / 4 q_1 q_2 \tag{8.34}$$

$$T_{22} = f_1^{-1} [f_2^{-1} (q_1 + q_2)^2 - f_2 (q_1 - q_2)^2] / 4 q_1 q_2 \tag{8.35}$$

where, as before, $f_n = \exp(iq_n d_n)$. As in the earlier examples T is unimodular, $\det T = 1$, because of (8.13) for example. The application of Bloch's theorem as given in section 8.1.1 now shows after some straightforward algebra that the dispersion relation is

$$\cos(QD) = \cos(q_1 d_1) \cos(q_2 d_2) - g_s \sin(q_1 d_1) \sin(q_2 d_2) \tag{8.36}$$

where

$$g_s = \tfrac{1}{2}(q_2/q_1 + q_1/q_2). \tag{8.37}$$

The calculation for p-polarization is similar. As noted in section 6.2.1, the boundary conditions for s-polarization can be converted to those for p-polarization by the replacement $q_2/q_1 \to \varepsilon_2 q_1 / \varepsilon_1 q_2$. Thus the dispersion relation is given by (8.36), with g_s replaced by

$$g_p = \tfrac{1}{2}(\varepsilon_2 q_1 / \varepsilon_1 q_2 + \varepsilon_1 q_2 / \varepsilon_2 q_1). \tag{8.38}$$

Equation (8.36) gives the dispersion equation $\omega = \omega(q_x, Q)$; q_x and ω enter via (8.32) for q_1 and q_2. A very full discussion of the implications when ε_1 and ε_2 are positive and independent of frequency is given by Yeh *et al* (1977) and Yariv and Yeh (1977). As for the acoustic case, band gaps occur at the points $QD = n\pi$. It is convenient to start with the extended-zone representation; a typical curve of ω versus Q for either polarization is sketched in figure 8.9(a). As mentioned, the macroscopic symmetry of the superlattice is uniaxial with z as the uniaxis, and in order to make contact

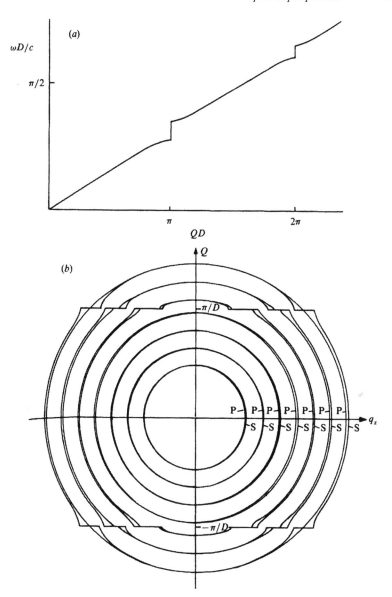

Figure 8.9. (a) Dispersion curves ω versus Q for fixed q_x for propagation through an optical superlattice in either polarization. In the extended-zone scheme, band gaps appear at $Q = n\pi/D$. (b) Constant-frequency contours in the (q_x, Q) plane.

with the usual discussion of the optics of uniaxial media (Landau and Lifshitz 1971, Born and Wolf 1980, Lipson *et al* 1995) it is helpful to draw the contours of constant frequency in the (q_x, Q) plane, as shown in figure 8.9(b). The contours for s- and p-polarization are different, except that they touch in the direction $q_x = 0$. This corresponds to propagation normal to the interfaces, for which the two polarizations are indistinguishable. As in electron-band theory, the band gaps at $Q = \pi/D$ in figure 8.9(a) have the consequence that the contours in the first zone are pulled out towards the zone edge. The contours may be compared with those drawn for a conventional uniaxial medium, as shown in most optics texts (e.g. Lipson *et al* 1995). In that case, the distinction is drawn between the ordinary wave, for which the contours are circles and the phase velocity (or effective refractive index) does not depend on the direction of propagation, and the extraordinary wave, for which the contours are ellipses and the phase velocity does depend on the direction of propagation. For the superlattice neither of the sets of contours in figure 8.9(b) is circular, so in a sense propagation in both polarizations is extraordinary. Further, for a frequency and direction corresponding to a gap region, there can be no or only one propagating mode rather than the two that are always found in a conventional medium with positive dielectric constants.

The stop bands in figure 8.9(a) have come to be known as *photonic band gaps* and it has been seen that they arise naturally as a result of the 1D superlattice period. A natural question to ask is how the theory generalizes to structures with a 2D or 3D long period, and most attention has focused on the possibility of 3D structures in which there are frequency ranges in which propagation cannot occur in any direction at all. This question will be discussed in more detail in section 8.3.6.

8.3.2 Bulk-slab model: effective-medium approximation and dispersive media

Figure 8.9 is for a superlattice in which the dielectric constants of both media are positive and non-dispersive, or at most weakly dispersive. However, the superlattices that have attracted most attention are composed of semi-conductors or magnetic media, for which frequency dispersion cannot be neglected. A completely general discussion would be very demanding and has perhaps never been attempted. Here we mainly review treatments of the properties of semiconductor superlattices in the frequency ranges where dispersion is due to plasmons and/or phonons.

A useful starting point is the approximation known as the *effective-medium theory*, which applies to superlattices with a long period D. The exact requirement is that the inequalities

$$q_x D \ll 1, \qquad QD \ll 1, \qquad q_1 D \ll 1, \qquad q_2 D \ll 1 \qquad (8.39)$$

should hold; in terms of figure 8.9(b) we are concerned with contours well inside the first Brillouin zone. The inequalities (8.31) for the bulk-slab model continue to apply. We may summarize (8.31) and (8.39), somewhat loosely, by saying that the superlattice period D is long compared with the atomic scale a and short compared with the optical wavelength λ

($a \ll D \ll \lambda$). Since for visible light, or longer wavelengths, a and λ differ by at least three orders of magnitude, (8.31) and (8.39) are satisfied in many cases of practical interest.

When (8.39) holds, all the trigonometric functions in (8.36) can be expanded. Retention of the first two terms in the cosines and the first terms in the sines leads to

$$\varepsilon_{xx}^{-1}(Q^2 + q_x^2) = \omega^2/c^2 \qquad \text{s-polarization} \qquad (8.40)$$

$$\varepsilon_{xx}^{-1}Q^2 + \varepsilon_{zz}^{-1}q_x^2 = \omega^2/c^2 \qquad \text{p-polarization} \qquad (8.41)$$

where

$$\varepsilon_{xx} = (\varepsilon_1 d_1 + \varepsilon_2 d_2)/D \qquad (8.42)$$

$$\varepsilon_{zz}^{-1} = (\varepsilon_1^{-1}d_1 + \varepsilon_2^{-1}d_2)/D. \qquad (8.43)$$

The derivation in this form is due to Raj and Tilley (1985); the same results, with alternative derivations and different physical discussion, were obtained at about the same time by Agranovich and Kravtsov (1985) and by Liu *et al* (1985). The argument presented by the former authors is particularly illuminating. At long wavelengths the boundary conditions that E_x and D_z are continuous means that these field components have the same values over many periods of the superlattice. Thus we write for the fields in layer i:

$$E_{xi} = \bar{E}_x, \qquad D_{zi} = \bar{D}_z, \qquad i = 1, 2 \qquad (8.44)$$

where $\bar{E}_x$ and $\bar{D}_z$ are the constant values of the fields. The other components are

$$D_{xi} = \varepsilon_i \varepsilon_0 E_{xi}, \qquad E_{zi} = \varepsilon_i^{-1}\varepsilon_0^{-1}D_{zi}, \qquad i = 1, 2. \qquad (8.45)$$

The spatial averages $\bar{D}_x = (d_1 D_{x1} + d_2 D_{x2})/D$ and $\bar{E}_z = (d_1 E_{z1} + d_2 E_{z2})/D$ are then expressed as

$$\bar{D}_x = \varepsilon_{xx}\varepsilon_0 \bar{E}_x, \qquad \bar{E}_z = \varepsilon_{zz}^{-1}\varepsilon_0^{-1}\bar{D}_z \qquad (8.46)$$

where the dielectric-tensor components are given by (8.42) and (8.43).

The superlattice may be regarded as an example of a *composite medium*, in which layers of medium 2 are dispersed in the matrix of medium 1. Other examples include dispersions of spheres or cylindrical rods in a matrix. Effective-medium expressions can be derived for the dielectric and other tensor properties of composite media (Bergman and Stroud 1992) and (8.42) and (8.43) are the simplest examples of these expressions.

Equations (8.40) and (8.41) show that in the long-wavelength limit the superlattice behaves as a uniaxial medium with the dielectric tensor given by (8.42) and (8.43). These relations have the basic property that ε_{xx} has *poles* at the frequencies of the poles of ε_1 and ε_2, whereas ε_{zz} has *zeros* at the frequencies of the zeros of ε_1 and ε_2. As a particular example Raj

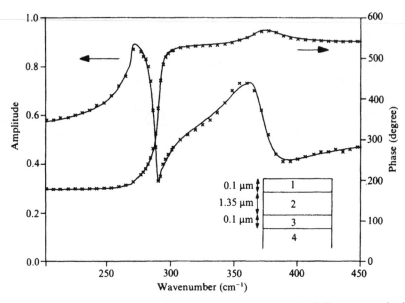

Figure 8.10. Far-infrared reflectance spectrum for a sample including a superlattice comprising 60 periods, each composed of 5.5 nm of GaAs and 17 nm of $Al_{0.35}Ga_{0.65}As$. The insert shows the sample structure with thicknesses given in µm. Regions 1 and 3, $Al_{0.35}Ga_{0.65}As$; region 2, superlattice; region 4, GaAs substrate. In the main part of the figure, the crosses are experimental points and the curves are theoretical. After Maslin *et al* (1986).

and Tilley (1985) discuss the reststrahl regions of $GaAs/Al_xGa_{1-x}As$. The frequency dependence of ε_{xx} and ε_{zz} is quite complicated since $Al_xGa_{1-x}As$ exhibits a two-mode behaviour (see section 6.1.1), while GaAs has a single TO resonance. For this system, (8.42) was used by Maslin *et al* (1986) to give an account of normal-incidence far-infrared reflectivity measurements. The experiments were carried out by dispersive Fourier-transform spectroscopy (DFTS), which as explained by Parker (1990) is a technique for making simultaneous measurements of the amplitude and phase of the complex reflectivity amplitude. Experimental results and theory are compared in figure 8.10, in which the theoretical curves are a least-squares fit to the data. The theoretical expressions are given by the usual classical optics of a multilayer system, with the superlattice described by (8.42). The dielectric functions appearing there are taken in the form of (2.146) for GaAs and (6.15) for $Al_xGa_{1-x}As$, and the parameters in those expressions are found from the fit to experiment. The amplitude curve shows two reststrahl-type features, one from about 260 to 290 cm^{-1} arising from the GaAs TO phonon, and a broader feature around 360 cm^{-1} that is related to the AlAs-like TO phonon and is produced by the $Al_{0.35}Ga_{0.65}As$

regions. The phase curve rises rapidly at the onset of the lower reststrahl region, as happens for a simple driven harmonic oscillator, and the response is so strong that the phase locks on to $540°$ rather than returning to the reference level of $180°$. This is possible for a layered sample although not for reflection off a simple semi-infinite medium (Chambers 1975). The phase response at the upper resonance is much weaker.

Normal-incidence measurements like those of figure 8.10 are analysed purely by means of the expression (8.42) for ε_{xx}; the other component ε_{zz} requires polarized oblique-incidence measurements, as were first performed by Lou *et al* (1988). To be precise, spectra in s-polarization are analysed in terms of ε_{xx} alone, while p-polarized spectra involve both ε_{xx} and ε_{zz}. Examples of such a spectrum will be shown later, in figure 8.16, for the more complicated case of a short-period superlattice. Some discussion of p-polarized spectra for long-period samples, which are related to (8.43) for ε_{zz}, is given by Dumelow *et al* (1993).

In section 2.5.2 it was seen that reflectivity curves like that of figure 8.10 mirror the propagation characteristics of the bulk polaritons. Equations (8.42) and (8.43) can be used together with standard results for optical propagation in a uniaxial medium to give an account of bulk-polariton propagation through the superlattice in the effective-medium approximation. This was done by Raj *et al* (1987). A very full discussion of the classical optics of this unusual kind of uniaxial medium has been given by Dumelow and Tilley (1993).

8.3.3   Bulk-slab model: Green functions

Linear-response theory has been used by Babiker *et al* (1987a) to evaluate the electric-field Green functions for an infinitely-extended two-component bulk-slab dielectric superlattice, as depicted in figure 8.8. The calculation involves using Maxwell's equations to write down expressions for the electric field vectors in each constituent layer of the superlattice in the presence of an externally applied polarization field $\mathbf{P}_{\text{ext}}$. Standard electromagnetic boundary conditions are applied at each interface, just as in the corresponding homogeneous calculation in section 8.3.1, and the periodicity of the superlattice is utilized through Bloch's theorem.

The required Green functions, which are of the form $\langle\langle E_\mu(Q); E_\nu(Q')\rangle\rangle_\omega$, are obtained from the linear response to $\mathbf{P}_{\text{ext}}$ in accordance with general results in the Appendix. Here μ and ν denote Cartesian components, and Q and Q' are components of wavevector in the z direction. For an infinitely extended superlattice the Green functions are all proportional to $\delta(Q - Q')$. It is found that the Green functions with μ and ν both equal to y describe s-polarized modes; they have poles corresponding to the dispersion relation (8.36). Likewise the Green functions with μ and ν equal to either x or z describe p-polarized modes. The general results are rather complicated and will not be given here. As shown by Babiker *et al* (1986c, 1987a), the results simplify considerably in the non-retarded limit and some applications to light scattering from plasma modes will be quoted later.

8.3.4   Bulk-slab model: surface modes

As was discussed in general terms in section 8.1.2, surface modes can appear on a semi-infinite superlattice in the stop bands of the folded dispersion curve. A striking illustration occurs with s-transverse acoustic modes, where as mentioned in section 8.1.3 surface modes are predicted on super-lattices even though they do not occur on the corresponding bulk samples. Similarly, surface optical modes can occur simply as a result of the periodic alternation of purely real refractive indices, and modes of this kind were already discussed by Yeh *et al* (1977). In addition, by analogy with the discussion of chapter 6, it is to be expected that in optics surface modes can also occur because one or both component media has a negative di-electric constant and thus is surface-active. The range of possibilities is rather large.

Surface polaritons of the second kind in reststrahl frequency regions have been discussed within the effective-medium approximation of section 8.3.2 by Raj and Tilley (1985) and by Raj *et al* (1987). Within that approx-imation the superlattice is described as a uniaxial medium with the uniaxis in the z direction and the dielectric tensor components given by (8.42) and (8.43). The surface-polariton dispersion equation is therefore given by (6.23), with the simplification that the x and z directions are principal direc-tions of the dielectric tensor. The calculated dispersion curves for a semi-infinite $GaAs/Al_xGa_{1-x}As$ superlattice bounded by vacuum are shown in figure 8.11. Surface polaritons of this kind must satisfy the localization condition $\varepsilon_{xx} < 0$. In figure 8.11 a distinction is made, following Hartstein *et al* (1973), between modes with $\varepsilon_{zz} < 0$, called *real-excitation* modes, and those with $\varepsilon_{zz} > 0$, called *virtual-excitation* modes. In general, modes of the former kind continue to exist as $q_x \to \infty$, while the latter terminate at a finite value of q_x. Thus real-excitation modes have an electrostatic limit, but virtual-excitation modes do not. Figure 8.11 shows three modes in the GaAs-like reststrahl region between 260 and 290 cm^{-1}; these arise because of the complicated frequency dependence of ε_{xx} and ε_{zz} mentioned earlier. In addition, a mode occurs in the $Al_xGa_{1-x}As$ reststrahl around 360 cm^{-1}. An experimental ATR spectrum corresponding to these curves is shown in figure 8.12. The dips marked S1 to S4 correspond closely to the crossing of the ATR scan line with the branches of the dispersion curve in figure 8.11 and are therefore identified with the surface-polariton branches. The dip marked G1 is a guided mode of the kind discussed in section 6.2.1 in which the field amplitudes are oscillatory within the superlattice but decaying in the substrate and vacuum. The assignments as surface and guided-wave

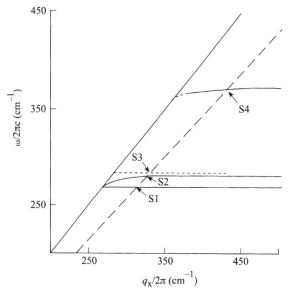

Figure 8.11. Calculated dispersion curves for surface polaritons at a $GaAs/Al_{0.35}Ga_{0.65}As$ superlattice-to-vacuum interface. Virtual modes are indicated by broken lines and real modes by full lines. The vacuum light line (full) and the ATR scan line (broken) for 20° incidence in Si are also shown. The crossing points whose frequencies are identified as those of the surface-mode dips S1 to S4 in the ATR data of figure 8.12 are identified. After Raj *et al* (1987), El-Gohary *et al* (1989).

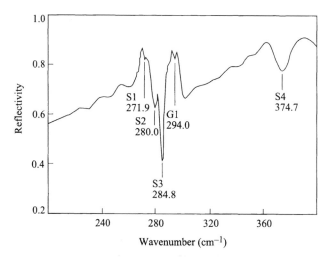

Figure 8.12. Measured ATR spectrum for p-polarized radiation incident at 20° in an Si prism on the sample of figure 8.10. After El-Gohary *et al* (1989).

modes are confirmed by calculation of the field profiles for frequencies at the bottom of the dips.

A range of general results for the p-polarized polaritons of a finite super-lattice were given by Haupt and Wendler (1987). In addition to the dispersion curves, they present a perturbation calculation in which the damping of the surface modes is found to first order in the damping constants Γ appearing in the expressions for the dielectric functions of the constituent media. Their most important conclusion is that, just as for the homogeneous film discussed in section 6.2.1, the surface mode of higher frequency has a long propagation distance. That is, the superlattice supports a long-range surface polariton.

8.3.5   Short-period superlattices: confined modes

At the beginning of this section the general problem of the contribution of confined optical modes to the dielectric function was raised. For long-period superlattices, satisfying (8.31), there is no difficulty in using the bulk-slab model. However, molecular beam epitaxy (MBE) or metallo-organic chemical vapour deposition (MOCVD) growth is capable of producing superlattices of as little as four monolayers per period, and for these short-period superlattices a fuller analysis is needed. Here we describe the main ideas, following Dumelow *et al* (1993), to which reference may be made for a detailed account.

Confined modes occur in frequency intervals like that denoted III in figure 8.6 and table 8.1, where q_1 is real and q_2 is not. The confined modes then have very nearly the character of a series of standing waves in medium 1, since the displacement is unable to penetrate any great distance into medium 2. This is illustrated in figure 8.13(a) from which it is seen that the wavenumber components of the confined modes are given by

$$q_1 = m\pi/d_1, \qquad m = 1, 2, \ldots, n_1 \tag{8.47}$$

where $n_1 = d_1/a$ is the number of lattice units in layer 1. The frequencies of the modes are then found approximately from the bulk TO phonon dispersion curve as shown in figure 8.13(b), the values being in good agreement with those found from the exact dispersion equation (8.30). In practice d_1 in (8.47) is often replaced by $(n_1 + \delta)a$ with $\delta \approx 0.8$ in order to allow for a small amount of penetration into medium 2. It can be seen from figure 8.13(a) that modes with odd m carry an overall dipole moment so these are the ones that contribute to the dielectric function.

Various models have been applied to calculate the dielectric tensor including the contributions of the higher-order confined modes. They all lead to expressions of the kind

$$\varepsilon_{xx} = \varepsilon_{xx}^{\infty}\left(1 - \sum_{\mu} \frac{S_{\mathrm{T}\mu}}{\omega^2 - \omega_{\mathrm{T}\mu}^2 + i\omega\Gamma_{\mathrm{T}\mu}}\right) \tag{8.48}$$

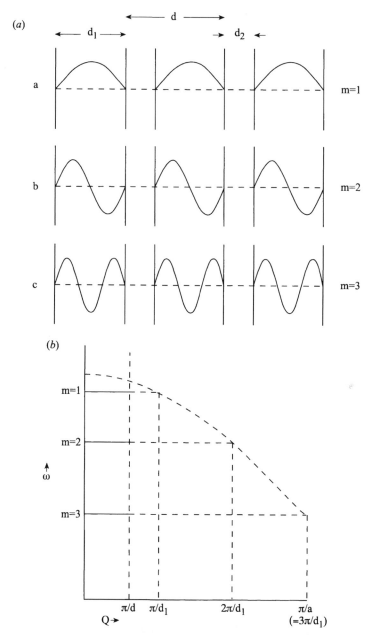

Figure 8.13. (a) Superlattice ionic displacements for $Q = 0$ in the case for which a phonon propagates through medium 1 but not through medium 2; m represents the number of half wavelengths within layer 1. (b) Confined optic-phonon frequencies deduced from the bulk dispersion curve (dashed) of component 1 for $n_1 = 3$, i.e. $d_1 = 3a$. After Dumelow *et al* (1993).

$$\frac{1}{\varepsilon_{zz}} = \frac{1}{\varepsilon_{zz}^{\infty}} \left(1 + \sum_{\mu} \frac{S_{L\mu}}{\omega^2 - \omega_{L\mu}^2 + i\omega\Gamma_{L\mu}} \right) \qquad (8.49)$$

where $\omega_{T\mu}$ and $\omega_{L\mu}$ are the frequencies of the confined TO and LO phonons, $S_{T\mu}$ and $S_{L\mu}$ their oscillator strengths and $\Gamma_{T\mu}$ and $\Gamma_{L\mu}$ their damping parameters. The summation over μ covers all the modes in the phonon bands of both constituents. In the simple phonon-confinement description, $S_{T\mu}$ and $S_{L\mu}$ are zero for even m and for odd m they are proportional to $1/m^2$.

Equations (8.48) and (8.49) have some similarities to the simple effective-medium expressions (8.42) and (8.43). Like those, they conform to the macroscopic uniaxial symmetry, and in addition for small damping the *poles* of ε_{xx} are at the TO frequencies $\omega_{T\mu}$ while the *zeros* of ε_{zz} are at the LO frequencies $\omega_{L\mu}$.

The values for the mode frequencies, derived from (8.47) by means of figure 8.13(b), have been tested by Raman scattering, as in figure 8.7, and indirectly by far-infrared spectroscopy. Measured Raman mode frequencies for confined LO modes in a series of short-period superlattices are shown in figure 8.14; the experiments used a backscattering geometry with polarizations set to select LO modes. The frequencies of the experimental points are the Raman peaks and the values of m and n are converted to an effective value of q determined from (8.47), with the $\delta \approx 0.8$ correction mentioned. If the simple theory outlined above held, then the experimental points would all fall on the bulk dispersion curve. However, it can be seen that they do not; in the examples shown the experimental points are low, but in some cases, for

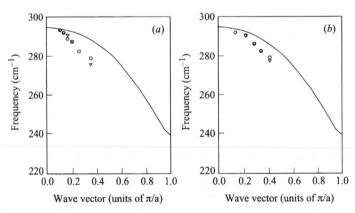

Figure 8.14. LO confined-mode frequencies in the GaAs optic-mode frequency range in a series of short-period $(GaAs)_n/(AlAs)_n$ superlattices as measured by Raman scattering (O) and as calculated for roughness parameter $W = 1.4$ (∇). The data pairs for modes of index m are shown at effective wavevector $q = m/(n + 0.8)$. The solid curves show the bulk GaAs LO dispersion curves. (a) $m = 1$, (b) $m = 2$. The data, from left to right, correspond to $n = 8, 6, 5, 4, 3, 2$ in (a) and $n = 6, 6, 5, 4$ in (b). After Samson *et al* (1992).

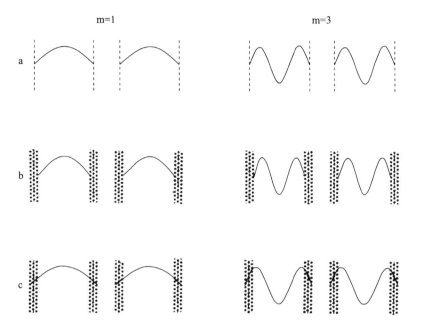

Figure 8.15. Sketch of the effect of interface roughness on phonon confinement. Ionic displacements are shown (a) for no roughness, (b) when the confined-phonon frequency falls outside the range of the 50% alloy dispersion curve and (c) when the confined-phonon frequency falls within the alloy range. After Dumelow *et al* (1993).

example the confined TO modes in the AlAs-like reststrahl region, they are high. The difference is attributed to the effect of interface broadening. In practice, it is impossible to grow a superlattice with ideally sharp interfaces, as sketched in figure 8.13(a), and within a monolayer or so on either side of an interface the material is an alloy. This is often called *interface roughness*. In a simple approach, the interfaces in GaAs/AlAs might be modelled by a layer of the alloy $Al_{0.5}Ga_{0.5}As$, as sketched in figure 8.15. Now if the LO phonon occurs in a stop band of $Al_{0.5}Ga_{0.5}As$ then the effective q value of a confined mode is increased by the roughness, as seen in the comparison of a and b in figure 8.15, and in the converse case q is decreased, as in c. Because of the form of the bulk dispersion curve, the frequency goes down in the former case and up in the latter. The theoretical points in figure 8.14 are calculated from a more refined model of this kind, in which the probability of occurrence of a Ga atom at distance z into the GaAs side from the nominal interface is written

$$P(z) = [1 + \mathrm{erf}(z/Wa)]/2 \qquad (8.50)$$

in terms of the error function (Kreyszig 1988) which increases from 0 to 1 as the argument goes from 0 to ∞. The best fit in figure 8.14 is obtained with $W = 1.4$, corresponding to an interface broadening of around one monolayer.

The detailed dielectric-function expressions (8.48) and (8.49) are required for analysis of polarized oblique-incidence far-infrared reflection spectra. As was mentioned in section 8.3.2 in connection with long-period superlattices, s-polarization involves only ε_{xx} while both ε_{xx} and ε_{zz} are required for the analysis of p-polarized reflection. Spectra of this kind on two samples from the series used for figure 8.14 are shown in figure 8.16. In the s-spectra features related, via (8.48), to the $m = 1$ and $m = 3$ confined TO phonons are seen, while the p-spectra also includes structure due to the lowest LO mode, whose frequency enters (8.49). It is notable that the fit to experiment is improved by the inclusion of the interface-roughness correction of (8.50). Thus despite the very long wavelength of the probe, far infrared spectroscopy is sensitive to material structure on the atomic scale. ATR measurements have also been made on short-period superlattices and are discussed in Dumelow *et al* (1993).

8.3.6   Photonic band gaps

As illustrated in figure 8.9, the primary consequence of the 1D periodicity of an optical superlattice is the appearance of stop bands at the zone boundaries. A question that arises, which was first posed by Yablonovitch (1987), is whether it is possible to devise 3D arrays in which some frequency intervals are stop bands for all directions of propagation and for both polarizations. A structure of this kind is described as having a *photonic band gap* (PBG). The physical motivation for Yablonovitch's paper was that the band gap would prohibit spontaneous emission of radiation in optical transitions and this can be turned to advantage in the design of semiconductor laser structures. However, the possibility of devising PBG structures has many other implications and the subject has therefore attracted a great amount of attention. Other possible applications are in antennas, millimetre wave devices, efficient solar cells and photocatalytic processes. PBGs also give rise to interesting new physics, such as cavity electrodynamics, localization, disorder and photon-number-state squeezing. In fact, at around the same time as Yablonovitch's original paper, related localization problems were studied by John (1987). Excellent accounts of this field are to be found in the review by Berger (1999) and the books by Sakoda (2001) and Johnson and Joannopoulos (2002). The earlier developments are well covered in the books by Burstein and Weisbuch (1995), Soukoulis (1995), Joannopoulos *et al* (1995) and Rarity and Weisbuch (1996).

As was already made clear by Yablonovitch, the most promising periodic structures are those for which the first Brillouin zone does not depart too greatly from spherical or circular. This is because, in terms of

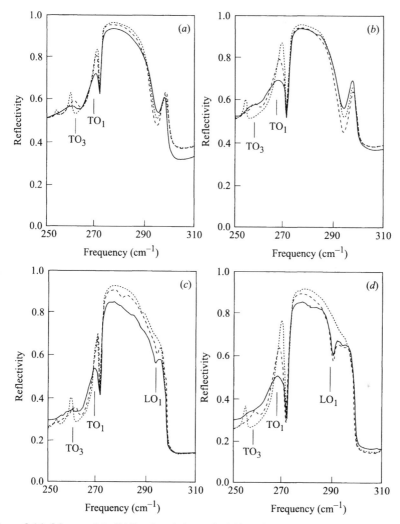

Figure 8.16. Measured (solid lines) and theoretical (dotted curves, $W = 0$; dashed curves $W = 1.4$) oblique-incidence (angle of incidence 45°) FIR reflectivity spectra of $(GaAs)_n/(AlAs)_n$ superlattices. (a) $n = 6$, s-polarization, (b) $n = 4$, s-polarization, (c) $n = 6$, p-polarization, (d) $n = 4$, p-polarization. After Samson *et al* (1992).

the analogy with the nearly-free-electron approximation for electronic bands, the underlying 'free-photon' energy does not vary too much around the surface of the zone. For this reason, most efforts have been devoted to hexagonal structures in 2D and close-packed structures in 3D. Another requirement, obviously, is a substantial difference between the refractive indices of the component materials. PBGs potentially have applications

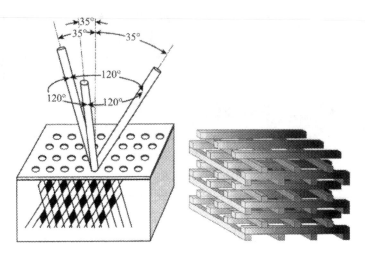

Figure 8.17. Two photonic band gap structures. (a) 'Yablonovite' (Yablonovitch *et al* 1991), (b) 'Woodpile' (Özbay *et al* 1994a). After Berger (1999).

over a very wide region of the spectrum, from microwaves to visible and beyond. However, the period has to be of the same order as the electromagnetic wavelength, so that fabrication problems are very much easier to solve at the longer-wavelength end.

The first demonstration of a full PBG, i.e. a stop band for all propagation directions and polarizations, was given for 15 GHz radiation by Yablonovitch *et al* (1991). The structure, depicted in figure 8.17(a), was made by drilling holes in a dielectric block. This kind of structure is obviously difficult to scale to short wavelengths, and a more versatile structure is the 'woodpile' of dielectric rods (Özbay *et al* 1994a), shown in figure 8.17(b). This has been constructed to show PBGs up to 25 THz (Özbay *et al* 1994a,b, 1996, Lin *et al* 1998). It can be seen that each layer is strongly polarizing and the importance of this to the existence of the PBGs is discussed by Reynolds and Arnold (1998).

Various techniques used for theoretical modelling of the optical characteristics of PBG structures are discussed in the references already cited, particularly Berger (1999) and Johnson and Joannopoulos (2002). Band-structure calculations are performed by a number of the methods that have been developed for electronic band structures, for example the application of the tight-binding method by Lidorikis *et al* (1998). Calculations for transmission and reflection by bounded structures are usually carried out by the transfer-matrix methods that have been a central theme of this chapter. An example is the experimental and theoretical study of transmission through a Ag/MgF_2 superlattice by Bloemer and Scalora

(1998). Attention should also be drawn to the general study by Cornelius and Dowling (1999) of black-body radiation in PBG structures.

An important step forward in the realization of a practical 3D PBG was reported by Blanco *et al* (2000). Rather than using the structures depicted in figure 8.17, which are difficult to synthesize on a large scale, they produced crystals of an Si 'inverse opal' and demonstrated a complete 3D photonic bandgap centred on 1.46 μm. These are produced by growing Si inside the voids of an opal template of close-packed fcc silica (SiO_2) spheres that are connected by small 'necks' formed during sintering, followed by removal of the silica template. The synthesis method is relatively simple and inexpensive, yielding PBGs of pure Si that are easily integrated with existing Si-based microelectronics.

Although most attention has been devoted to 3D PBG materials, some effort has also gone into the question of devising 2D PBG structures for surface waves, particularly surface-plasmon–polaritons (SPPs). The problem is, perhaps, rather easier than in 3D, first because of the lower dimension and second because SPPs exist only in p-polarization. The holographic method of producing 1D gratings on metal films that was mentioned in section 6.4 can be developed to make gratings that are periodic in both surface directions. Kitson *et al* (1996) fabricated a hexagonal grating with a period of 300 nm on a silver film and were able to demonstrate a full 2D PBG in the visible part of the spectrum. Further discussion of this structure is given by Barnes *et al* (1996, 1997).

At the time of writing, new applications and fabrication methods are providing a sustained interest in PBG materials.

8.4 Plasma modes

Both semiconductor and metal superlattices contain free charges so that plasma-type modes may be found. In the former, electrons or holes can be found within one layer of a superlattice so that in a bulk-slab model the layers have a dielectric function of the form of (2.150), which includes both optic-phonon and plasma terms. It was mentioned in section 6.1.3 that a thin layer of carriers can be trapped at a heterojunction interface, like that between GaAs and $Al_xGa_{1-x}As$. In this case a superlattice may be described quite well as a series of 2D charge sheets. As discussed in section 6.1.3, the presence of a charge sheet modifies the **H**-field boundary condition at the interface into the form (6.24) with the consequence that a localized polariton-type mode propagates along a single interface, even when the dielectric constants of the bounding media are both positive. In this section, therefore, we need to discuss the plasma modes of both a charge-sheet and a bulk-slab superlattice, starting with the former. A full review of superlattice plasmon–polaritons has been given by Albuquerque and Cottam (1993).

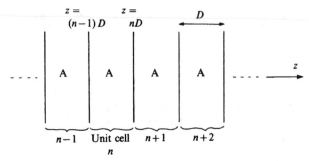

Figure 8.18. Model for a charge-sheet superlattice. Charge sheets of density $\nu\,\mathrm{m}^{-2}$ are equally spaced in a matrix of medium A of dielectric constant ε.

8.4.1   Charge-sheet superlattices

We discuss here the simplest model as depicted in figure 8.18. We choose the x axis so that propagation is in the xz plane and the modes of interest are p-polarized, with **H** field in the y direction. Electromagnetic propagation may be analysed as in section 8.3.1; the only changes are first that the boundary condition of H_y to be continuous is replaced by (6.24) together with (6.25), and second that since the period D contains only a single layer of medium A the transfer matrix is given simply by the product of a phase matrix and an interface matrix rather than (8.13). The outcome is that the dispersion relation is given by

$$\cos QD = \cos q_z D + \frac{\Omega^2 q_z D}{2\omega^2}\sin q_z D \tag{8.51}$$

where

$$q_z = (\varepsilon\omega^2/c^2 - q_x^2)^{1/2} \tag{8.52}$$

and Ω is a characteristic frequency given by

$$\Omega = (\nu e^2/\varepsilon_0\varepsilon m^* D)^{1/2}. \tag{8.53}$$

Here, as in section 6.1.4, ν is the areal density of electrons in the charge sheet and m^* is the effective mass.

  Partly because of the relevance to inelastic-electron scattering, the electrostatic limit is particularly important for plasma modes. For this limit, $q_x^2 \gg \varepsilon\omega^2/c^2$ and consequently $q_z = iq_x$. Equation (8.51) then reduces to the explicit form

$$\omega = \Omega\left\{\frac{q_x D\sinh(q_x D)}{2[\cosh(q_x D) - \cos(QD)]}\right\}^{1/2}. \tag{8.54}$$

This was found previously by Fetter (1974), Das Sarma and Quinn (1982) and Bloss and Brody (1982) by other methods. It is easily verified from

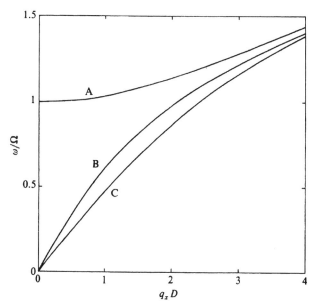

Figure 8.19. Numerical example of the bulk plasmon dispersion relation (8.54). The plots are for three different values of QD: A, 0; B, $\pi/2$; C, π.

(8.54) that, except when $Q = 0$ and $\cos(QD) = 1$, ω is proportional to q_x at long wavelengths such that $q_x D \ll 1$. The behaviour of ω as a function of q_x is illustrated in figure 8.19 for three different values of QD. These are the unretarded modes but the solutions of (8.51) for the retarded modes include additional solutions at higher frequencies such that $\varepsilon \omega^2/c^2 \geq q_x^2$. These have the characteristic of the perturbed photon line folded back in a reduced-zone scheme (Constantinou and Cottam 1986).

Figure 8.18 is, of course, the simplest charge-sheet superlattice and some important generalizations have been discussed (see Albuquerque and Cottam 1993). One extension is to the case where the charge sheets are separated by alternating layers A and B (Constantinou and Cottam 1986) and another is to superlattices with alternating electron and hole layers (Qin *et al* 1983). Also important is the n–i–p–i superlattice (Farias *et al* 1988) with four layers in the unit cell: A (n-type), B (insulating), C (p-type), D (insulating), and with electron and hole layers at the interfaces of the p- and n-regions respectively.

We now turn to the plasmon modes of a semi-infinite superlattice, retaining for simplicity the structure of figure 8.18 but assuming it to occupy the half-space $z \leq 0$. The region $z > 0$ is assumed to be filled with a medium of dielectric constant ε_s and we define $r_s = \varepsilon_s/\varepsilon$. The surface $z = 0$ between the outermost cell of the superlattice and the external medium is different from all the other interfaces, so the electron density ν_0 of the associated charge layer may be different from ν (or even zero). We therefore define a second surface-related parameter $\mu_s = \nu_0/\nu$.

As was seen in general terms in section 8.1.2, a surface mode corresponds to a value of the Bloch wavenumber $Q = i\lambda$ which is off the real axis. It is assumed that the electromagnetic fields in the bounding medium decay as $\exp(-\alpha z)$. The electromagnetic boundary conditions at $z = 0$ then determine λ, or equivalently the frequency ω of the surface mode. For simplicity we give the result for the non-retarded mode, for which as usual $\alpha = q_x$. The equations for λ and ω are then

$$\mu_s(\mu_s - 1)\sinh(q_xD)(q_xD\omega)^2 + \Omega^2(r_s^2 - 1)\sinh(q_xD)$$
$$+ \Omega[r_s(1 - 2\mu_s)\sinh(q_xD) + \cosh(q_xD)] = 0 \tag{8.55}$$

and

$$\exp(-\lambda D) = [r_s - (\mu_s q_x D\omega/\Omega)]\sinh(q_xD) + \cosh(q_xD). \tag{8.56}$$

Solutions of these equations represent surface plasmons only if ω is real and $|\exp(-\lambda D)| < 1$. For the special case $\mu_s = 1$ the surface branch is above the bulk continuum if $r_s < 1$ and below it if $r_s > 1$. The condition on $|\exp(-\lambda D)|$ implies that there is a cut-off wavevector given by $q_xD > \ln|(1 + r_s)/(1 - r_s)|$. Some numerical solutions including cases with $\mu_s \neq 1$ are shown in figure 8.20. The bulk continuum region in this figure corresponds to the region between curves A and C in figure 8.19.

The calculations we have described here are concerned with what are known as *intrasubband plasmons*. The boundary conditions (6.24) and (6.25) imply that the electrons (or holes) form a mobile charge sheet in the xy plane. In fact, the carriers at a semiconductor interface are trapped by a potential well that extends a small distance in the z direction and there may be several bound states in this well. The wave function consists of a z-dependent part describing localization close to the interface multiplied by an (x, y) part corresponding to delocalized motion. This complete model for the motion associated with one of the bound states is called a *subband*. The charge sheet that has been assumed corresponds to a population of carriers in the lowest subband: the carriers are taken as free to move in the xy plane and the extent of the wave function in the z direction is ignored. The account we have given describes the electrodynamic interaction among these carriers. It is possible also to have such interactions together with transitions between subbands and the resulting excitations are known as *intersubband plasmons*. The formalism becomes much more complicated, and we refer to the work of Tsellis and Quinn (1984) and Eliasson *et al* (1987) for details.

The account given here has been concerned with the dispersion equations which result from solution of the homogeneous equations of motion. As has been discussed in earlier parts of this book, solution of inhomogeneous equations gives the Green functions that are particularly

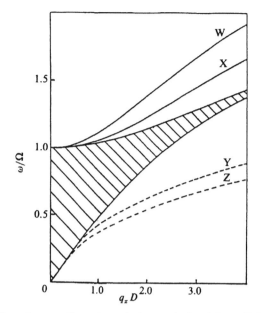

Figure 8.20. Surface plasmon dispersion relations, calculated from (8.55) and (8.56), for several values of the surface parameters. W, $r_s = 0.08$, $\mu_s = 1$; X, $r_s = 0.08$, $\mu_s = 0.75$; Y, $r_s = 4$, $\mu_s = 1$; Z, $r_s = 4$, $\mu_s = 0.75$. The shaded region is the bulk-plasmon continuum. The other parameters are $D = 40$ nm and $\varepsilon = 12.9$. After Constantinou and Cottam (1986).

important in the discussion of inelastic light scattering. Various groups have evaluated Green functions for the charge-sheet model (e.g. Babiker *et al* 1986a,b, Eliasson *et al* 1987). The detailed results for the Green functions in both infinite and semi-infinite charge-sheet superlattices with both a two-layer unit cell and a four-layer unit cell (e.g. n–i–p–i structures) are given by Albuquerque and Cottam (1993).

8.4.2 Bulk-slab model

We now turn to the bulk-slab model, which describes a superlattice in which one or more of the constituent layers contains mobile charge, so that the dielectric function takes the form (2.150). For a two-component superlattice the analysis of section 8.3.1 applies and the dispersion relation is given by (8.36), with factor g_s or g_p according to the polarization. Here we specialize in p-polarization in the non-retarded limit since this corresponds to the light-scattering experiments to be discussed later in this section. In this case (8.32) simplifies to $q_1 = q_2 = iq_x$ so that (8.36), together with (8.38) for g_p, becomes

$$\cos(QD) = \cosh(q_x d_1)\cosh(q_x d_2)$$
$$+ \tfrac{1}{2}(\varepsilon_1/\varepsilon_2 + \varepsilon_2/\varepsilon_1)\sinh(q_x d_1)\sinh(q_x d_2). \tag{8.57}$$

This can be expressed alternatively as (Camley and Mills 1984)

$$\varepsilon_1/\varepsilon_2 = -\beta \pm (\beta^2 - 1)^{1/2} \tag{8.58}$$

where

$$\beta = \frac{\cosh(q_x d_1)\cosh(q_x d_2) - \cos(QD)}{\sinh(q_x d_1)\sinh(q_x d_2)}. \tag{8.59}$$

When ε_1 and ε_2 have the bulk-plasma form (2.132), there are two solutions $\omega^{\pm}$ for the frequencies of the coupled plasmon modes:

$$\omega^{\pm} = \omega_p^{(2)}\left(1 - \frac{s(1 - r^2)}{s + \beta \mp (\beta^2 - 1)^{1/2}}\right)^{1/2} \tag{8.60}$$

where $r = \omega_p^{(1)}/\omega_p^{(2)}$ and $s = \varepsilon_\infty^{(1)}/\varepsilon_\infty^{(2)}$. The lower, or 'acoustic', branch with frequency ω^- in (8.60) is just the analogue of the superlattice plasmon obtained in the 2D-sheet limit. The occurrence of the upper, or 'optic', branch can be associated with the finite charge-layer thickness. The dependence of ω^+ and ω^- on q_x is illustrated in figure 8.21 for GaAs/Al$_x$Ga$_{1-x}$As, taking medium 2 (GaAs) to contain the charges and medium 1 (Al$_x$Ga$_{1-x}$As) to have no charges, corresponding to $r = 0$ in (8.60). The figure is for a fixed superlattice period D and various charge-layer thicknesses d_2. Other numerical values have been assigned in accordance with the Raman measurements of Olego *et al* (1982) for their Sample 2, i.e. $D = 89$ nm, $m_2^* = 6.37 \times 10^{-32}$ kg, $\varepsilon_b^{(1)} = \varepsilon_b^{(2)} = 13.0$, and $n_0 = n_2 d_2$ with $n_0 = 7.3 \times 10^{15}$ m^{-2}. The experimental sample corresponded to $d_2/D = 0.29$ and the data (crosses) are the scattering peaks in measurements like those shown below in figure 8.24.

The above calculations, which were for the dispersion relations of the plasmon–polaritons in finite-thickness charge layers, have been generalized to the evaluation of electric-field Green functions by Babiker *et al* (1986c, 1987a,b), using the method discussed in section 8.3.3.

A limitation of these calculations for the dispersion relations and the Green functions is that they employ a scalar dielectric function for the charge layers with the bulk ω_p value. This is an unsatisfactory approximation for thin enough layers, and some calculations by King-Smith and Inkson (1986, 1987) for this case using a microscopic theory imply that the scalar ε is effectively replaced by a tensor with ε_{zz} different from ε_{xx} and ε_{yy}. Another microscopic theory, also based on the random-phase approximation, has been put forward by Wasserman and Lee (1985).

8.4.3 Magnetoplasma modes

The generalization of the bulk-slab model to the case where a static magnetic field $\mathbf{B}_0$ is applied, so that the scalar ε in the form of (2.132) is replaced by the

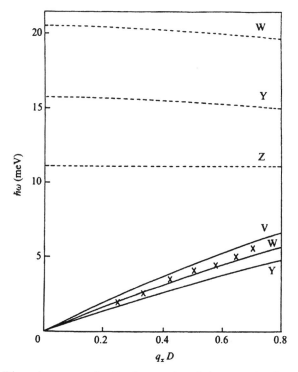

Figure 8.21. Dispersion curves for the frequencies of the acoustic plasmons (ω^-, full curves) and the optic plasmons (ω^+, broken curves) as a function of $q_x D$ (for fixed QD). The corresponding experimental results (Olego *et al* 1982) for $d_2/D = 0.29$ are shown as crosses. The theory curves are given for the following values of d_2/D: V, 0.0; W, 0.29; Y, 0.6; Z, 1.0. After Babiker *et al* (1986c).

gyrotropic tensor of (2.133) to (2.135), has been discussed by a number of authors, e.g. Wallis *et al* (1987), Albuquerque *et al* (1991); the latter discuss surface modes as well as bulk modes, and n–i–p–i structures as well as two-component superlattices. The detailed form of the transfer matrix for a two-component superlattice in the Voigt geometry ($\mathbf{B}_0$ parallel to the interfaces), together with the bulk and surface-mode dispersion relations, is given by Albuquerque and Cottam (1993). An example of the calculated surface and bulk magnetoplasma modes in a two-component semi-infinite superlattice is shown in figure 8.22. The axes are chosen so that z is the normal to the layers and $\mathbf{B}_0$ is applied in the y direction. The surface-mode curves correspond to propagation along x, and the bulk continua are drawn for given q_x and variable q_z. Thus both bulk and surface modes are in the Voigt geometry. It can be shown that in the absence of $\mathbf{B}_0$ there are two surface modes, doubly degenerate, corresponding to propagation along $\pm x$, for a given magnitude of q_x. In the presence of the external

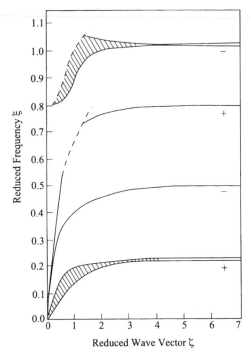

Figure 8.22. Calculated bulk and surface magnetoplasmon dispersion curves for the Voigt geometry in a two-component semi-infinite superlattice. The components are A and B with plasma frequencies and layer thicknesses ω_{pA}, ω_{pB} and d_A, d_B. Dimensionless frequency and wavevector are $\xi = \omega/\omega_{pA}$, $\zeta = cq_x/\omega_{pA}$ and ratios $\omega_{pB}/\omega_{pA} = 0.5$, $d_B/d_A = 2.0$ are chosen. The magnetic field is taken so that $\omega_c/\omega_{pA} = 0.5$, where ω_c is the cyclotron frequency. Surface modes shown as solid curves, bulk-continuum regions shaded. After Wallis *et al* (1987).

magnetic field, this degeneracy is lifted, leading to the four surface modes labelled by the plus and minus signs in figure 8.22. Three of them lie in the gap between the bulk bands, while the other is in the region below the lower bulk band, merging into it when $\zeta \approx 3.0$. The upper plus mode has the curious property that it ceases to exist in an intermediate frequency interval.

8.4.4 Light scattering

The early experimental work on Raman scattering from the plasma oscillations in semiconductor superlattices has been reviewed by Pinczuk (1984). Attention was mostly focused on the GaAs/Al$_x$Ga$_{1-x}$As system with the experiments being carried out under resonance conditions in order to enhance the scattered intensity.

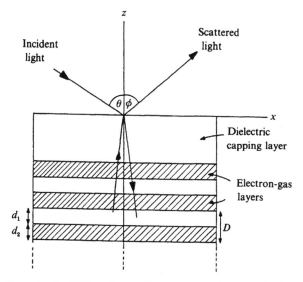

Figure 8.23. Geometry for light scattering from plasmons in a superlattice with electron-gas layers.

The first experiments to demonstrate clearly Raman scattering from plasmons in a GaAs/Al$_x$Ga$_{1-x}$As MQW structure were due to Olego *et al* (1982). Their geometry is sketched in figure 8.23. The incident and scattered light are in the same vertical plane, labelled *xz*, and a right-angled configuration was employed, so that $\theta + \phi = 90°$ for the angles outside the sample. However, because the refractive index η of the semiconductor is fairly large ($\eta \approx 3.6$) the incident and scattered beams inside the sample are near to being antiparallel and parallel, respectively, to the *z* axis. In this configuration the light-scattering wavevector has components

$$q_x = (2\pi/\lambda)\,(\sin\theta - \cos\theta) \tag{8.61}$$

$$q_z = (2\pi/\lambda)\,[(\eta^2 - \sin^2\theta)^{1/2} + (\eta^2 - \cos^2\theta)^{1/2}]$$

$$\cong (2\pi/\lambda)\,2\eta(1 - 1/4\eta^2) \tag{8.62}$$

where λ is the vacuum wavelength and the last line of (8.62) follows from $\eta \gg 1$. It follows from (8.61) and (8.62) that, by varying the angle θ, q_x may be varied while q_z is kept effectively constant. Raman spectra at three different angles θ and with $\lambda \approx 780$ nm are shown in figure 8.24. Similar measurements have been reported by Fasol *et al* (1985) and Sooryakumar *et al* (1985).

Apart from the peak frequencies a complete theory of Raman scattering from the plasmons should predict the spectral width and lineshape and the

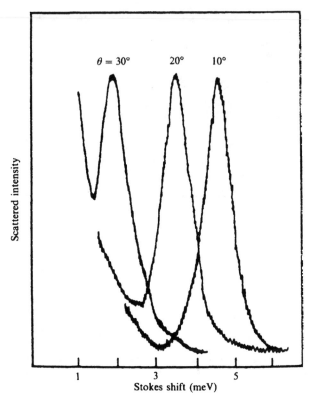

Figure 8.24. Raman spectra for scattering from plasmons in a GaAs/Al$_x$Ga$_{1-x}$As MQW structure for three different angles θ in a right-angle scattering geometry. After Olego *et al* (1982).

integrated intensity, and also the dependence of these quantities on the scattering geometry and the polarizations of the light beams. Light-scattering calculations have been carried out by several authors (e.g. Hawrylak *et al* 1985, Jain and Allen 1985a,b,c, Katayama and Ando 1985) using an approach based on density–density correlations or Green functions. In particular, they dealt either with effects in semi-infinite or finite superlattices, where localized surface plasmons are predicted, or with effects of interactions between plasmons and phonons. They found results for the spectral line-width and shape of the bulk-plasmon and surface-plasmon resonances. A different approach, based on general methods for calculating light scattering from polaritons in superlattices, was adopted by Babiker *et al* (1986a,b,c, 1987b) for both the 2D-sheet and the bulk-slab model. The formalism incorporated the dependence on the scattering geometry and the polarization directions.

8.5 Quasiperiodic structures

We referred briefly to quasiperiodic multilayer structures in section 1.1.3. These can be considered as being generated in terms of a particular mathematical sequence. The usual example is the *Fibonacci* sequence of integer numbers defined in (1.35): following initial conditions, each generation number is defined recursively as the sum of the two previous numbers. There are several more quasiperiodic sequences, but the only other one that we shall describe here is the *Thue–Morse* sequence. In the context of multilayer structures they have the common feature that there is no longer a periodicity length, but there is nevertheless a growth rule (unlike in a random layered structure). As a consequence quasiperiodic superlattices are of interest regarding the localization and multifractal aspects of the excitations. An extensive review has been given by Albuquerque and Cottam (2003). Accounts of the earlier work, mainly for excitations in semiconductor structures, have been given by MacDonald (1988) and Merlin (1989) for theory and experiment (e.g. Raman scattering), respectively. The book edited by DiVincenzo and Steinhardt (1999) provides an account of the static properties and selected characterization techniques.

We focus our attention on two-component quasiperiodic superlattices, where there are two fundamental building blocks, labelled A and B, corresponding to layers of thickness d_A and d_B respectively. Each block may be composed of one or more layers of different materials of arbitrary thickness. Instead of forming an $ABABAB\ldots$ periodic structure as in the previous sections, the building blocks in the quasiperiodic structure are juxtaposed according to a growth rule. This specifies the nth stage of the process S_n, usually in a recursive manner. For a Fibonacci structure the growth rule, analogous to (1.35) for the Fibonacci numbers, is $S_n = S_{n-1}S_{n-2}$ for $n \geq 2$, starting with $S_0 = B$ and $S_1 = A$. It has the property of being invariant under the transformations $A \rightarrow AB$ and $B \rightarrow A$, which provides an inflation (or growth) rule. The first few Fibonacci generations are (see also figure 8.25(a))

$$S_0 = B, \quad S_1 = A, \quad S_2 = AB, \quad S_3 = ABA, \quad S_4 = ABAAB, \quad \ldots$$
$$(8.63)$$

where the number of building blocks in generation n is just the Fibonacci number F_n (= 1, 1, 2, 3, 5, 8, 13 etc. for increasing n) obtained from (1.35). For each n the number of A building blocks is in general different from the number of B blocks, as can be seen from (8.63). It is easily shown that the ratio tends to the golden mean number τ as $n \rightarrow \infty$, where

$$\tau = (1 + \sqrt{5})/2 \approx 1.618. \quad (8.64)$$

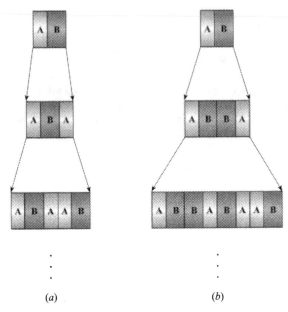

Figure 8.25. Schematic illustration of different quasiperiodic structures: (a) Fibonacci; (b) Thue–Morse. After Albuquerque and Cottam (2003).

A Thue–Morse quasiperiodic structure can also be defined recursively. In this case it is defined by the relations $S_n = S_{n-1}S_{n-1}^+$ and $S_n^+ = S_{n-1}^+ S_{n-1}$ for $n \geq 1$, with $S_0 = A$ and $S_0^+ = B$. Alternatively, it is easier to visualize by application of its inflation rule $A \rightarrow AB$ and $B \rightarrow BA$. The first few Thue–Morse generations are (see also figure 8.25(b))

$$S_0 = A, \quad S_1 = AB, \quad S_2 = ABBA, \quad S_3 = ABBABAAB, \quad \ldots \quad (8.65)$$

The Thue–Morse superlattice structure has the property that the total number of layers (A and B) increases with the generation number n as 2^n, which is faster than for the Fibonacci case. Also there are equal numbers of A and B layers for the Thue–Morse case (for $n \geq 1$), but not for the Fibonacci case. These differences have consequences for the excitation spectra, as we shall discuss.

In the following two sections, which deal with experimental and theoretical aspects respectively, we outline some results for nonmagnetic excitations. Details and further examples are given in the review article by Albuquerque and Cottam (2003).

8.5.1   Some experimental results

Experimental work on quasiperiodic superlattices was pioneered by Merlin *et al* (1985). Using the MBE technique described in section 1.2, they reported

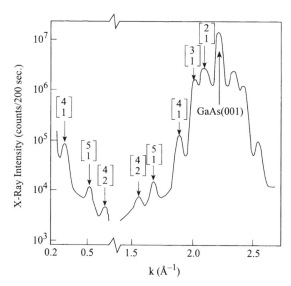

Figure 8.26. X-ray diffraction intensity versus wavevector k perpendicular to the layers for a Fibonacci superlattice comprising AlAs and GaAs layers (see text). The arrows with labels $[n, p]$ indicate the positions of k_{np} (see text). After Merlin *et al* (1985).

the growth of semiconductor heterostructures with a layer structure following a Fibonacci sequence. Their A and B building blocks were each composed of AlAs and GaAs layers, but with different thickness: $A = (1.7 \text{ nm of AlAs})/(4.2 \text{ nm of GaAs})$ and $B = (1.7 \text{ nm of AlAs})/(2 \text{ nm of GaAs})$. By design this gave a ratio d_A/d_B very close in value to the golden mean number τ (a property which is not necessary in general but was done to simplify analysis of the data). The sample consisted of 13 Fibonacci generations. Some initial x-ray diffraction and Raman scattering measurements were carried out; higher-resolution x-ray studies were reported soon afterwards by the same group (see Todd *et al* 1986). An x-ray diffraction spectrum, showing intensity versus the wavevector component k along the growth direction, is seen in figure 8.26. There is a series of sharp peaks at particular values of k characteristic of the Fibonacci structure. Remarkably, all but the weakest peaks occur in a geometric progression with τ as the common ratio. The theory, as we shall discuss later, predicts that the main peaks occur at values of k equal to $2\pi n \tau^p / d_F$ for the case of Fibonacci superlattices with $d_A/d_B = \tau$. Here n and p are positive integers, and

$$d_F = \tau d_A + d_B \tag{8.66}$$

plays the role of an 'average' periodicity length for a Fibonacci superlattice. The predicted k values (see section 8.5.2) are indicated by arrows in figure 8.26 and give excellent agreement with the experimental data.

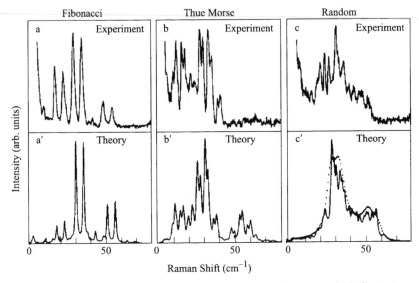

Figure 8.27. Comparisons between experiment (a, b, c) and theory (a′, b′, c′) for Raman scattering from acoustic phonons in GaAs/AlAs superlattices: (a, a′) Fibonacci; (b, b′) Thue–Morse; (c, c′) Random. After Merlin *et al* (1987).

An example of Raman scattering from LA phonons in different types of superlattices is given in figure 8.27 (Merlin *et al* 1987). In this case the building blocks correspond to $A = (2\,nm$ of GaAs) and $B = (2\,nm$ of AlAs). The Fibonacci and Thue–Morse cases are compared with each other and with a random superlattice (i.e. where the sequence of each A and B block is decided on a random basis, such as the toss of a coin). The comparison with theory will be discussed later, but there are obvious differences between the three cases.

Similar experimental work on quasiperiodic superlattices at around the same time was reported by Dharma-wardana *et al* (1987) and Macrander *et al* (1988). They studied Fibonacci superlattices with building blocks of Si and Ge_xSi_{1-x}, and the experimental tools were also x-ray diffraction and Raman scattering from acoustic phonons. Again it was possible to account theoretically for the main peaks in the spectra, and a particularly detailed theoretical analysis for these materials was carried out by Aers *et al* (1989).

Raman scattering was also employed to study the plasmons in Fibonacci superlattices (see Merlin *et al* 1992). Building block A consisted of 27 nm of GaAs and 80 nm of $Al_{0.3}Ga_{0.7}As$, while block B differed only in that the $Al_{0.3}Ga_{0.7}As$ thickness was 43 nm. In the combined structure the $Al_{0.3}Ga_{0.7}As$ barriers were doped with Si donors leading to trapping of carriers by the GaAs wells and, thus, to the formation of a set of quasi-2D

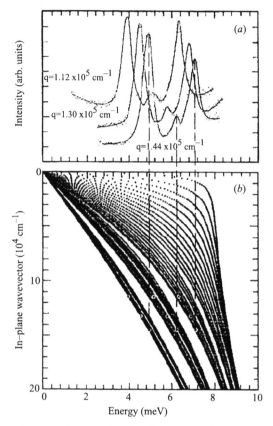

Figure 8.28. Experiment and theory for the plasmons in GaAs/Al$_{0.3}$Ga$_{0.7}$As Fibonacci superlattices. (a) Raman spectra for three value of the in-plane wavevector as marked. (b) Calculated dispersion of the plasmon modes, where the large dots represent the measured maxima. After Merlin *et al* (1992).

electron gas layers. The end result is a quasiperiodic analogue of the charge-sheet periodic superlattices described in section 8.4.1. Figure 8.28(a) shows Raman spectra for scattering from the three lowest plasmon bands at three different values of the in-plane wavevector. The calculated dispersion of the plasmons for a Fibonacci superlattice is shown in figure 8.28(b). As indicated, the agreement for the peak positions is excellent.

There have also been a number of experimental measurements of the optical reflectance and transmission spectra in quasiperiodic structures, mostly with a view to studying light localization and the fractal structure within the spectra. Among earlier work we mention optical transmission data obtained by Gellermann *et al* (1994) on Fibonacci superlattices formed with SiO_2 and TiO_2 layers as building blocks, and reflectance data by Munzar *et al* (1994) on Fibonacci superlattices with GaAs and $Al_xGa_{1-x}As$ as building blocks. In both of these papers a transfer-matrix formalism, analogous to that introduced earlier in this chapter, was employed in the analysis.

8.5.2 Theoretical discussion

We begin by outlining some theoretical background to the x-ray and Raman scattering work discussed in the previous section. Good accounts of the theory, based on defining a structure-dependent form factor, are contained in the review articles by MacDonald (1988) and Merlin (1989). Following them, we may define a structure factor $S(k)$ corresponding to any specified set of points $\{z_j\}$ along the growth axis of a 1D system of length L by

$$S(k) = \sum_j \exp(-ikz_j) \tag{8.67}$$

where k is a component of wavevector in that direction. For a macroscopically large ($L \to \infty$) periodic system with spacing D the set $\{z_j\}$ consists of all integer multiples of D, and so the structure factor consists of a sequence of delta-function peaks, as discussed in most solid-state physics textbooks (see e.g. Kittel 1986). These peaks occur at a sequence of values $k = k_n$, where $k_n = 2\pi n/D$ and n is any integer. Thus, for the corresponding scattering intensity $I(k) \propto |S(k)|^2$, we conclude simply that $I(k)$ is proportional to L^2 for $k = k_n$ and zero otherwise.

If we have a quasiperiodic structure the above result will no longer hold, because the members of the set $\{z_j\}$ are unequally spaced (e.g. for a Fibonacci structure the spacings are related to the sequence of Fibonacci numbers defined previously). This makes the calculation of $S(k)$ extremely complex in general. However, it has been shown that many quasiperiodic structures may be generated mathematically by a projection technique that involves two or more *periodic* structures (see MacDonald 1988). The Fibonacci structure falls in this category (involving just two basic periods whose ratio is the golden mean τ), but the Thue–Morse structure does not. It would take us too far astray to go into details, but the projection method (where applicable) provides a convenient way to calculate $S(k)$ and hence the intensity spectrum. We should comment that some authors use the term quasiperiodic more restrictively to apply only to the former category above. The result for a Fibonacci superlattice is (see e.g. Merlin 1989)

$$S(k) = \frac{L\tau^2}{d_F} \sum_{m,n} \exp(-i\phi_{mn}/\tau^3) \frac{\sin \phi_{mn}}{\phi_{mn}} \delta(k - k_{mn}) \tag{8.68}$$

where d_F is defined in (8.66), and m and n are integers. The other quantities in the above expression are $k_{mn} = 2\pi(m + n\tau)/d_F$ and $\phi_{mn} = \pi\tau^2(md_A - nd_B)/d_F$. The largest peaks in the spectrum are when ϕ_{mn} is close to zero, which therefore emphasizes certain integer pairs m and n. This result is general, but in the special case where the structure is grown so that $d_A/d_B = \tau$, this eventually leads to the prediction quoted in section 8.5.1 for our discussion of figure 8.26.

We note that (8.68) implies $I(k) \propto L^2$ for a Fibonacci superlattice, which is the same dependence found for a periodic superlattice. By contrast, for a Thue–Morse superlattice where the mentioned projection technique is inapplicable, it is found that $I(k) \propto L^{\gamma(k)}$, where the index γ depends on k but is less than 2 (see Merlin 1989). The random superlattice gives $I(k) \propto L$. These differences, among other properties, motivated the studies in figure 8.27.

Of course, a full analysis of x-ray or light-scattering spectra involves far more than just the structure factor, since the properties of the particular excitation are also important. Apart from the references already quoted, there have been Green-function calculations of acoustic and optic phonons by, for example, Fernandez-Alvarez and Velasco (1998) and Perez-Alvarez *et al* (2001) in Fibonacci superlattices; extensions to other quasiperiodic superlattices were made by Velasco and Zarate (2001). The propagation of electronic wave packets in Fibonacci semiconductor superlattices was studied by Diez *et al* (1996).

Other Green-function calculations relating to plasmons and electronic excitations are detailed by Albuquerque and Cottam (2003) for superlattices based on several different quasiperiodic sequences. They include an extensive treatment of the optical transmission spectra, as well as the spectra of the excitations, emphasizing the behaviour as the generation number n becomes large. In particular, it is shown how the transfer matrices corresponding to the Fibonacci, Thue–Morse and other quasiperiodic generations may be calculated by a generalization of the method that we described in section 8.1. Usually each generation S_n is considered to be repeated periodically (in the so-called periodic boundary condition, or PBC, model); the excitation spectrum can then be found from the trace of the transfer matrix, e.g. as in (8.20), and the properties of the spectrum for large n can be examined. In the limit of large n the number of individual layers (and interfaces) increases rapidly; for example, for $n = 10$ the PBC Fibonacci superlattice has 89 layers in each periodic length. A consequence is that for the bulk excitation spectrum of the superlattice the number of energy bands increases with n and rapidly becomes very fragmented.

There are two properties of special interest associated with this, as discussed by Albuquerque and Cottam (2003). One is that the spectra, in some cases, show signs of self-similarity such as might be associated with

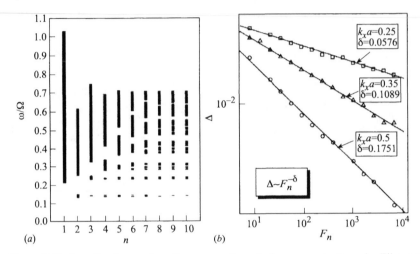

Figure 8.29. Localization and scaling properties of plasmon–polaritons in Fibonacci structures (see text for the parameters): (a) The distribution of band widths versus generation number n (here $\Omega \simeq 23\,\text{Thz}$); (b) Log–log plot of the total width Δ of the allowed regions versus the Fibonacci number for three values of the in-plane wavevector. After Albuquerque and Cottam (2003).

fractals. Roughly speaking, fractals are complex geometric shapes with fine structure at arbitrary small scales. They typically exhibit some degree of self-similarity, meaning that if we magnify a tiny part of the fractal we will see features reminiscent of the whole (see e.g. Strogatz 1994, Nicolis 1995). In simple fractals the self-similarity aspects are related to the concept of a fractal dimension, which describes a property of the above magnifying process; it is usually a non-integer and is often an irrational number. It would take us too far astray into the mathematics to elaborate these ideas here in the context of the quasiperiodic superlattices and it is a topic of ongoing research. We comment that the quasiperiodic structures of interest actually behave as multifractals, rather than simple fractals. Essentially this means that they have to be characterized by a continuous distribution of fractal dimensions, rather than just a single value. The second property of interest with respect to the large-n behaviour is more empirical and concerns identifying certain scaling exponents characteristic of the type of excitation, its wavevector and the type of quasiperiodic structure. We provide an example in the following paragraph to explain this.

We consider a calculation for quasiperiodic superlattices that gener-alizes the charge-sheet and bulk-slab models of the plasmon–polaritons in periodic superlattices (see sections 8.4.1 and 8.4.2). Specifically for the building blocks we choose A to be GaAs described by a frequency-dependent dielectric function of the form (2.150) and plasma frequency $\omega_{pA} = 4.04$ THz, whereas B is SiO_2 with a frequency-independent dielectric constant $\varepsilon_B = 12.3$. In addition, both A and B can have a 2D charge sheet (with areal charge densities ν_A and ν_B) on one side, say their left side. Albuquerque and Cottam (2003) calculated the transfer matrices corresponding to the different Fibonacci and Thue–Morse generations and hence deduced the spectrum of plasmon–polaritons in p-polarization using the PBC model mentioned earlier. When the properties of the plasmon–polariton spectrum for large n are examined, certain scaling relationships can be deduced. Some results are shown in figure 8.29 for Fibonacci structures, taking parameters $\nu_A = \nu_B = 6 \times 10^5$ m^{-2} and $d_A = \frac{1}{2}d_B = 40$ nm. First, in figure 29(a) we show the frequency of bulk bands (at a fixed value in plane wavevector) versus generation number n, where the frequency is in units of Ω defined as in (8.53) but with d_A as the length scale. Note that, as anticipated from our earlier discussion, the energy bands increase in number (because there are more layers per periodic length); also it is found that the bands become narrower for large n, which is an indication of greater localization of the modes. If the total width Δ of the allowed energy regions (i.e. the Lebesgue measure of the energy spectrum) is calculated and analysed for scaling properties, it is found that it decreases with n as the power law $\Delta \approx (F_n)^{-\delta}$. Here F_n is the Fibonacci number and the empirical scaling exponent δ (known as the *diffusion constant* of the spectrum) depends on the in-plane wavevector. We show in figure 8.29(b) a log–log plot of these power laws for three different values of the wavevector. In this case there is a strong dependence of δ, and hence of the localization (see e.g. Hawrylak and Quinn 1986), on the excitation wavevector. In the corresponding case of Thue–Morse structures the scaling is according to $\Delta \approx (2^n)^{-\delta}$, where δ takes different values and now has a much weaker dependence on wavevector (Albuquerque and Cottam 2003).

References

Aers G C, Dharma-wardana M W C, Schwartz G P and Bevk J 1989 *Phys. Rev. B* **39** 1092

Agranovich V M and Kravtsov V E 1985 *Solid State Commun.* **55** 85

Albuquerque E L and Cottam M G 1993 *Phys. Reports* **233** 67

Albuquerque E L and Cottam M G 2003 *Phys. Reports* **376** 225

Albuquerque E L, Fulco P and Tilley D R 1986a *Rev. Bras. Fis.* **16** 315

Albuquerque E L, Fulco P, Sarmento E F and Tilley D R 1986b *Sol. State Commun.* **58** 41

Albuquerque E L, Fulco P and Tilley D R 1988 *Phys. Stat. Solidi (b)* **146** 449

Albuquerque E L, Fulco P, Farias G A, Auto M M and Tilley D R 1991 *Phys. Rev. B* **43** 2032

Babiker M and Tilley D R 1984 *J. Phys. C* **17** L829

Babiker M, Tilley D R, Albuquerque E L and Gonçalves da Silva C E T 1985a *J. Phys. C* **18** 1269

Babiker M, Tilley D R and Albuquerque E L 1985b *J. Phys. C* **18** 286

Babiker M, Constantinou N C and Cottam M G 1986a *Solid State Commun.* **57** 877

Babiker M, Constantinou N C and Cottam M G 1986b *J. Phys. C* **19** 5849

Babiker M, Constantinou N C and Cottam M G 1986c *Solid State Commun.* **59** 751

Babiker M, Constantinou N C and Cottam M G 1987a *J. Phys. C* **20** 4581

Babiker M, Constantinou N C and Cottam M G 1987b *J. Phys. C* **20** 4597

Barnes W L, Preist T W, Kitson S C and Sambles J R 1996 *Phys. Rev. B* **54** 6227

Barnes W L, Kitson S C, Preist T W and Sambles J R 1997 *J. Opt. Soc. Am. A* **14** 1654

Berger V 1999 *Optical Materials* **11** 131

Bergman D J and Stroud D 1992 *Solid State Physics* vol 46, ed H Ehrenreich and D Turnbull (New York: Academic) p 147

Blanco A, Chomski E, Grabtchak S, Ibisate M, John S, Leonard S W, Lopez C, Meseguer F, Miguez H, Mondia J P, Ozin G A, Toader O and van Driel H M 2000 *Nature* **405** 437

Bloemer M J and Scalora M 1998 *Appl. Phys. Lett.* **72** 1676

Bloss W L and Brody E M 1982 *Solid State Commun.* **43** 523

Born M and Wolf E 1980 *Principles of Optics* (Oxford: Pergamon)

Bulgakov A A 1985 *Solid State Commun.* **55** 869

Burstein E and Weisbuch C (eds) 1995 *Confined Electrons and Photons: New Physics and Applications* (New York: Plenum)

Camley R E and Mills D L 1984 *Phys. Rev. B* **29** 1695

Camley R E, Djafari-Rouhani B, Dobrzynski L and Maradudin A A 1983 *Phys. Rev. B* **27** 7318

Chambers W G 1975 *Infrared Phys.* **15** 139

Chandrasekhar S 1992 *Liquid Crystals* (Cambridge: Cambridge University Press)

Collings P J 1990 *Liquid Crystals* (Bristol: Adam Hilger)

Colvard C, Merlin R, Klein M V and Gossard A C 1980 *Phys. Rev. Lett.* **45** 298

Colvard C, Gant T A, Klein M V, Merlin R, Fischer R, Morkoc H and Gossard A C 1985 *Phys. Rev. B* **31** 2080

Constantinou N C and Cottam M G 1986 *J. Phys. C* **19** 739

Cornelius C M and Dowling J P 1999 *Phys. Rev. A* **59** 4736

Das Sarma S and Quinn J J 1982 *Phys. Rev. B* **25** 7603

De Gennes P G and Prost J 1993 *The Physics of Liquid Crystals* (Oxford: Oxford University Press)

Dharma-wardana M C W, MacDonald A H, Lockwood D J, Baribeau J M and Houghton D C 1987 *Phys. Rev. Lett.* **58** 1761

Diez E, Dominguez-Adame F, Macia E and Sanchez A 1996 *Phys. Rev. B* **54** 16972

DiVincenzo D P and Steinhardt P J (eds) 1999 *Quasicrystals: The State of the Art*, 2nd edition (New Jersey: World Scientific)

Djafari-Rouhani B and Dobrzynski L 1987 *Solid State Commun.* **62** 609

Djafari-Rouhani B, Dobrzynski L, Hardouin-Duparc O, Camley R E and Maradudin A A 1983 *Phys. Rev. B* **28** 1711

Djafari-Rouhani B, Sapriel J and Bonnouvrier F 1985 *Superlattices and Microstructures* **1** 29

Dresselhaus M S 1986 in *Magnetic Properties of Low-Dimensional Systems* ed L M Falicov and J L Moran-Lopez (Heidelberg: Springer)

Dumelow T and Tilley D R 1993 *J. Opt. Soc. Am. A* **10** 633

Dumelow T, Parker T J, Smith S R P and Tilley D R 1993 *Surf. Sci. Rep.* **17** 151

El-Gohary A R, Parker T J, Raj N, Tilley D R, Dobson P J, Hilton D and Foxon C T B 1989 *Semicond. Sci. Technol.* **4** 388

Eliasson G, Hawrylak P and Quinn J J 1987 *Phys. Rev. B* **35** 5569

Farias G A, Auto M M and Albuquerque E L 1988 *Phys. Rev. B* **38** 12540

Fasol G, Hughes H P and Ploog K 1985 *Proc. Int. Conf. on Electronic Properties of Two-Dimensional Systems* Tokyo p 742

Feldman D W, Parker J H, Choyke W J and Patrick L 1968 *Phys. Rev. B* **15** 698

Fernandez-Alvarez L and Velasco V R 1998 *Phys. Rev. B* **57** 14141

Fetter A L 1974 *Ann. Phys. NY* **88** 1

Gellermann W, Kohmoto M, Sutherland B and Taylor P C 1994 *Phys. Rev. Lett.* **72** 633

Hadizad R, Tilley D R and Tilley J 1991a *J. Phys.: Condens. Matter* **3** 291

Hadizad R, Tilley D R and Tilley J 1991b *J. Phys.: Condens. Matter* **3** 9697

Hartstein A, Burstein E, Brion J J and Wallis R F 1973 *Solid State Commun.* **12** 1083

Haupt R and Wendler L 1987 *Solid State Commun.* **612** 341

Hawrylak P and Quinn J J 1986 *Phys. Rev. Lett.* **57** 380

Hawrylak P, Wu J W and Quinn J J 1985 *Phys. Rev. B* **32** 5169

Jain J K and Allen P B 1985a *Phys. Rev. Lett.* **54** 947

Jain J K and Allen P B 1985b *Phys. Rev. Lett.* **54** 2437

Jain J K and Allen P B 1985c *Phys. Rev. B* **32** 997

Joannopoulos J D, Meade R D and Winn J N 1995 *Photonic Crystals* (Princeton: Princeton University Press)

John S 1987 *Phys. Rev. Lett.* **58** 2486

Johnson S G and Joannopoulos J D 2002 *Photonic Crystals: The Road from Theory to Practice* (Dordrecht: Kluwer)

Jusserand B, Paquet D, Regreny A and Kervarec J 1983 *Solid State Commun.* **48** 499

Jusserand B, Paquet D, Kervarec J and Regreny A 1984a *J. Physique* **45** C5-154

Jusserand B, Paquet D and Regreny A 1984b *Phys. Rev. B* **30** 6245

Jusserand B, Paquet D and Regreny A 1985 *Superlattices and Microstruct.* **1** 61

Katayama S and Ando T 1985 *J. Phys. Soc. Japan* **54** 1615

King-Smith R D and Inkson J C 1986 *Phys. Rev. B* **33** 5489

King-Smith R D and Inkson J C 1987 *Phys. Rev. B* **36** 4796

Kitson S C, Barnes W L and Sambles J R 1996 *Phys. Rev. Lett.* **77** 2670

Kittel C 1986 *Introduction to Solid State Physics* 6th edition (New York: Wiley)

Klein M V 1986 *IEEE J. Quantum Elect.* **QE-22** 1760

Kreyszig E 1988 *Advanced Engineering Mathematics* (New York: Wiley)

Landau L D and Lifshitz E M 1971 *Electrodynamics of Continuous Media* (Oxford: Pergamon)

Lidorikis E, Sigalas M M, Economou E N and Soukoulis C M 1998 *Phys. Rev. Lett.* **81** 1405

Lin S Y, Flemming J G, Hetherington D L, Smith B K, Biswas R, Ho K M, Sigalas M M, Zubrzycki W, Kurtx S R and Bur J 1998 *Nature* **394** 251

Lipson S G, Lipson H and Tannhauser D S 1995 *Optical Physics* 3rd edition (Cambridge: Cambridge University Press)

Liu W M, Eliasson G and Quinn J J 1985 *Solid State Commun.* **55** 533

Lou B, Sudharsanan R and Perkowitz S 1988 *Phys. Rev. B* **38** 2212

MacDonald A H 1988 in *Interfaces, Quantum Wells, and Superlattices* ed C R Leavens and R Taylor (New York: Plenum) p 347

Macrander A T, Schwartz G P and Bevk J 1988 *Phys. Rev. B* **37** 8459

Maslin K A, Parker T J, Raj N, Tilley D R, Dobson P J, Hilton D and Foxon C T B 1986 *Solid State Commun.* **60** 461

Merlin R 1989 in *Light Scattering in Solids V* ed M Cardona and G Güntherodt (Berlin: Springer) p 214

Merlin R, Bajema K, Clarke R, Juang F Y and Bhattacharya P K 1985 *Phys. Rev. Lett.* **55** 1768

Merlin R, Bajema K, Nagle J and Ploog K 1987 *J. de Physique* **48** C5-503

Merlin R, Valladares J P, Pinczuk A, Gossard A C and English J H 1992 *Solid State Commun.* **84** 87

Molinari E, Baroni S, Giannozzi P and de Gironcoli S 1992 *Phys. Rev. B* **45** 4280

Munzar D, Bocaek L, Humlicek J and Ploog K 1994 *J. Phys.: Cond. Mat.* **6** 4107

Nicolis G 1995 *Introduction to Nonlinear Science* (Cambridge: Cambridge University Press)

Olego D, Pinczuk A, Gossard A C and Wiegmann W 1982 *Phys. Rev. B* **25** 7867

Özbay E, Michel E, Tuttle G, Biswas R, Sigalis M and Ho K M 1994a *Appl. Phys. Lett.* **64** 2059

Özbay E, Michel E, Tuttle G, Biswas R, Ho K M, Bostak J and Bloom D M 1994b *Opt. Lett.* **19** 1155

Özbay E, Temelkuran B, Sigalis M, Tuttle G, Soukoulis C M and Ho K M 1996 *Appl. Phys. Lett.* **69** 3797

Parker T J 1990 *Contemp Phys.* **31** 335

Perez-Alvarez R, Garcia-Moliner F, Trallero-Giner C and Velasco V R 2001 *J. Raman Spectrosc.* **31** 421

Pinczuk A 1984 *J. de Physique* **45** C5-477

Qin G, Giuliani G F and Quinn J J 1983 *Phys. Rev. B* **28** 6144

Raj N and Tilley D R 1985 *Solid State Commun.* **55** 373

Raj N and Tilley D R 1989 in *The Dielectric Function of Condensed Systems* ed L V Keldysh, D A Kirzhnitz and A A Maradudin (Amsterdam: Elsevier)

Raj N, Camley R E and Tilley D R 1987 *J. Phys. C* **20** 5203

Rarity J and Weisbuch C 1996 *Microcavities and Photonic Bandgaps: Physics and Applications* (Dordrecht: Kluwer)

Reynolds A L and Arnold J M 1998 *IEE Proc. Optoelectron.* **145** 436

Rytov S M 1956 *Akust. Zh.* **2** 71 [*Sov. Phys.—Acoust.* **2** 68]

Sakoda K 2001 *Optical Properties of Photonic Crystals* (Berlin: Springer)

Samson B, Dumelow T, Hamilton A A, Parker T J, Smith S R P, Tilley D R, Foxon C T B, Hilton D and Moore K J 1992 *Phys. Rev. B* **46** 2375

Sooryakumar R, Pinczuk A, Gossard A C and Wiegeman W 1985 *Phys. Rev. B* **31** 2578

Soukoulis C (ed) 1995 *Photonic Band Gap Materials* (New York: Plenum)

Strogatz S H 1994 *Nonlinear Dynamics and Chaos* (Reading: Addison-Wesley)

Todd J, Merlin R, Clarke R, Mohanty K M and Axe J D 1986 *Phys. Rev. Lett.* **57** 1157

Tsellis A and Quinn J J 1984 *Phys. Rev. B* **29** 3318

Velasco V R and Zarate J E 2001 *Prog. Surf. Sci.* **67** 383

Wallis R F 1957 *Phys. Rev.* **105** 540

Wallis R F, Szenics R, Quinn J J and Giuliani G F 1987 *Phys. Rev. B* **36** 1218

Wasserman A L and Lee Y I 1985 *Solid State Commun.* **54** 855
Yablonovitch E 1987 *Phys. Rev. Lett.* **58** 2059
Yablonovitch E, Gmitter T J and Leung K M 1991 *Phys. Rev. Lett.* **67** 2295
Yariv A and Yeh P 1977 *J. Opt. Soc. Am.* **67** 438
Yariv A and Yeh P 1984 *Optical Waves in Crystals* (New York: Wiley)
Yeh P, Yariv A and Hong C S 1977 *J. Opt. Soc. Am.* **67** 423
Yip S K and Chang Y C 1984 *Phys. Rev. B* **30** 7037

Chapter 9

Magnetic layered structures and superlattices

We now give an introductory account of superlattices in which one or both of the component materials are magnetic. The nature of the magnon excitations depends crucially on whether the long-range dipole–dipole interactions or the short-range exchange interactions dominate, just as discussed in chapters 4 and 5 for single magnetic layers, and this in turn depends largely on the wavevector of the excitation. Furthermore, when the superlattice consists of magnetic layers alternating with nonmagnetic layers, the coupling across the nonmagnetic spacers may be via the oscillatory exchange (including biquadratic contributions), as in section 4.5, or via the dipole–dipole fields, as in section 5.2.

We first treat the case where dipolar terms are dominant, i.e. the magnetostatic and magnetic-polariton regimes, in sections 9.1 and 9.2 respectively. Just as in the film geometry, there is a strong dependence on whether the magnetization is parallel or perpendicular to the surfaces or interfaces. Then exchange effects are taken into account, starting with the exchange-dominated regime in section 9.3 and then the generalization to the dipole-exchange case given in section 9.4. The chapter is concluded by section 9.5 on quasiperiodic superlattices, where different magnetic models are considered.

The basic theoretical techniques, just as in chapter 8 for the non-magnetic superlattices, involve either employing the transfer matrix method or directly using Bloch's theorem over a periodicity length. It is a case of choosing whichever is more convenient (depending on the magnetic model) to facilitate generalizing the results of chapters 4 and 5 for magnetic films to the case of multilayers. Applications to experiments (including, in particular, light scattering, magnetic resonance and microwave absorption) are made throughout in parallel with the theory. Some examples of reviews of excitations in magnetic superlattices are by Grünberg (1989), Mills (1989), Barnás (1994) and Hillebrands (2000).

9.1 Magnetostatic modes

The simplest case is a superlattice in which ferromagnetic and nonmagnetic layers alternate. The magnetostatic modes of a single film were described in sections 5.1 and 5.3 for ferromagnets and antiferromagnets, respectively. It was shown there, by using the magnetostatic scalar potential ψ, that the electromagnetic fields associated with the excitations extend outside the film. In this exterior region they decay with distance from the surface as $1/q_\parallel$, where $q_\parallel$ is the magnitude of the in-plane wavevector $\mathbf{q}_\parallel$. For a magnetic superlattice in the magnetostatic regime, these fields provide the coupling across the nonmagnetic spacer layers (just as in section 5.2 for the double-layer ferromagnets discussed there).

9.1.1 In-plane magnetization

The most interesting situation for a single ferromagnetic film (see section 5.1) is when the magnetization is parallel to the surface, since a surface mode (the Damon–Eshbach or DE mode) may occur for certain directions of $\mathbf{q}_\parallel$. We therefore start with this case. The theory for superlattices was first worked out by Camley *et al* (1983) and Grünberg and Mika (1983). The former authors calculated the magnetic Green functions as well as the dispersion relations in a semi-infinite superlattice, while the latter authors focused on the dispersion relations but included the case of a finite number of layers in the superlattice.

We now outline a calculation of the dispersion relations for the semi-infinite superlattice shown in figure 9.1, which corresponds to the Voigt configuration (i.e. the in-plane wavevector $\mathbf{q}_\parallel$ is chosen to be perpendicular to the magnetization $\mathbf{M}_0$ and the applied field). A transfer-matrix formalism, analogous to that followed in section 8.1 for acoustic modes, can be utilized. We deal here with the magnetostatic scalar potential written in the form $\psi_1(x)\exp(\mathrm{i}\mathbf{q}_\parallel \cdot \mathbf{r}_\parallel)$ as in section 5.1.1. Then for the magnetic film in the lth cell, i.e. $-(l-1)D > x > -(l-1)D - d_2$, we denote

$$\psi_1(x) = a_l \exp\{\mathrm{i}q_x[x + (l-1)D]\} + b_l \exp\{-\mathrm{i}q_x[x + (l-1)D]\} \quad (9.1)$$

by analogy with (8.3), where a_l and b_l are constants. The quantity q_x is defined in general by (5.4), but for the Voigt geometry we have $q_z = 0$ and so we may replace q_x with $\mathrm{i}q_\parallel$ (where $q_\parallel = |q_y|$). A similar expression, but with different amplitude terms and modified exponential factors, holds in the nonmagnetic film of the lth cell. A transfer matrix T relating the amplitudes in cell $l+1$ to those at the corresponding interface in cell l can be defined as in (8.12), except that $|u_l^L\rangle$ is replaced by the two-component column matrix with elements a_l and b_l. The boundary conditions at the interfaces where $x = -(l-1)D - d_2$ and lD are the same as described in section 5.1.1. With this information it is then a straightforward matter to calculate

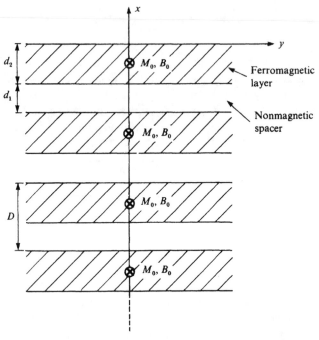

Figure 9.1. The geometry for a semi-infinite superlattice composed of alternating ferromagnetic and nonmagnetic layers, assuming the Voigt configuration.

the elements of the transfer matrix. The results have the explicit form (see e.g. Barnás 1994)

$$T_{11(22)} = \exp(\pm q_{\parallel} d_2) \left[\cosh(q_{\parallel} d_1) \pm Z \sinh(q_{\parallel} d_1) \right] \tag{9.2}$$

$$T_{12(21)} = \exp(\mp q_{\parallel} d_2) Z^{(\pm)} \sinh(q_{\parallel} d_1) \tag{9.3}$$

where the lower signs refer to the indices in parentheses for T, while Z and $Z^{(\pm)}$ are combinations of susceptibility components χ_a and χ_b defined in (2.166) and (2.167):

$$Z = 1 + \chi_a - \frac{\chi_b^2}{2(1 + \chi_a)} \tag{9.4}$$

$$Z^{(\pm)} = -\chi_b \mathrm{sgn}(q_y) \mp \frac{\chi_a(2 + \chi_a) + \chi_b^2}{2(1 + \chi_a)}. \tag{9.5}$$

The spectrum of bulk superlattice modes can be found from the transfer matrix using the property in (8.20). For real values of the Bloch wavevector Q it yields

$$\cos(QD) = \cosh(q_{\parallel} d_1) \cosh(q_{\parallel} d_2) + Z \sinh(q_{\parallel} d_1) \sinh(q_{\parallel} d_2). \tag{9.6}$$

Any surface modes of the semi-infinite superlattice would correspond to solutions for Q that are complex and have the sign for $\text{Im}(Q)$ that describes a decaying envelope function with distance from the surface ($x \to -\infty$). It is also necessary to satisfy the magnetostatic boundary conditions at the surface $x = 0$. This leads to a surface-mode dispersion relation in the implicit form

$$\chi_a + \chi_b \text{sgn}(q_y) + 2 = 0 \tag{9.7}$$

provided $d_2 > d_1$ (i.e. the magnetic films are thicker than the nonmagnetic ones). If $d_2 < d_1$, then no surface mode can exist because the localization condition cannot be satisfied.

The above results are general in the sense that they apply for the magnetic layers being either ferromagnets or antiferromagnets. In either case the appropriate forms of χ_a and χ_b must be substituted, and we now discuss the main features of the predicted excitation spectrum.

For a ferromagnetic/nonmagnetic superlattice, the results for the magnon frequencies may be summarized as follows. The frequencies can be conveniently expressed in terms of a parameter α, as in (5.20) for ferromagnetic double layers, where the modes of the superlattice correspond to:

(i)   a band of bulk modes for which

$$\alpha = 1 - \frac{2 \sinh(q_\| d_1) \sinh(q_\| d_2)}{\cosh(q_\| D) - \cos(QD)} \tag{9.8}$$

where $\mathbf{q} = (Q, \mathbf{q}_\|)$ is the 3D wavevector, and
(ii) a surface-magnon branch that has (for a semi-infinite superlattice)

$$\alpha = 0 \tag{9.9}$$

and exists only if $d_2 > d_1$.

Furthermore, it is a consequence of (9.7) that the surface magnon exists for only one sign of in-plane wavevector component q_y (in fact, $q_y > 0$ in the present example). This non-reciprocal propagation property is analogous to that for the DE surface mode, as discussed in section 5.1.1.

As a numerical example we show in figure 9.2 the values of α for the bulk and surface modes of the superlattice plotted against d_2/d_1 for fixed values of d_2 and $\mathbf{q}_\|$. It can be shown that the bulk-magnon modes of the superlattice are made up of a linear superposition of *surface* waves within each magnetic layer; the net effect is a bulk mode since the amplitudes and phase factors are such that Bloch's theorem is satisfied (see Camley *et al* 1983). The bulk modes always lie in the frequency range from $[\omega_0(\omega_0 + \omega_m)]^{1/2}$ to $(\omega_0 + \frac{1}{2}\omega_m)$, in the notation of chapter 5. The surface mode of the semi-infinite superlattice is also a linear superposition of surface waves within each magnetic layer, but they are combined with an envelope function that decays exponentially with distance into the superlattice. This mode has a frequency equal to

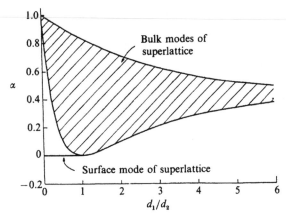

Figure 9.2. Plot of the coefficient α against d_1/d_2 for a semi-infinite ferromagnetic super-lattice, taking $d_2 = 25$ nm and $q_\parallel = 0.017$ nm^{-1}. After Cottam and Lockwood (1986).

$(\omega_0 + \frac{1}{2}\omega_m)$ as can be deduced either from (9.7) or from (5.20) and (9.9). This frequency is the same as that of the DE mode in a semi-infinite ferromagnet (see section 5.1.1); remarkably, it is independent of the layer thicknesses d_1 and d_2 provided $d_2 > d_1$.

Brillouin scattering has been used to test the above prediction for the surface mode of the superlattice, with initial experiments by Grimsditch *et al* (1983) and Kueny *et al* (1984). In figure 9.3(a) and (b) we show examples of spectra obtained for Ni/Mo superlattices with $d_2 > d_1$ and $d_2 < d_1$, respectively. The occurrence of a pair of anti-Stokes lines in figure 9.3(a) when $d_2 > d_1$, and of only one anti-Stokes line in figure 9.3(b) when $d_2 < d_1$, provides confirmation of the prediction concerning the surface magnon of the superlattice. Another experimental study by Hillebrands *et al* (1988) for Fe/Pd superlattices in different applied magnetic fields (see figure 9.4) showed these effects more clearly. In figures 9.4(c) and (d) the Fe layer thickness is larger than the Pd layer thickness and an additional surface-mode peak is seen on one side of the spectra; it switches sides when the applied field is reversed. No such behaviour occurs in figure 9.4(a) and (b) where the Fe thickness is less than that of Pd. However, in figure 9.4(a) there is a weak non-reciprocal band at about 23 GHz. Hillebrands *et al* (1988) attribute this feature as being due to surface-like modes of Fe layers individually, rather than superlattice modes; it is observable because of density-of-states effects when the Fe thickness is very small. Further experimental evidence by Brillouin scattering is reviewed by Grimsditch (1989) and Hillebrands (2000). Some Green-function calculations, together with their application to light scattering, are described, for example, by Camley and Stamps (1994); the approach is analogous to that used in chapter 5 for the film geometry.

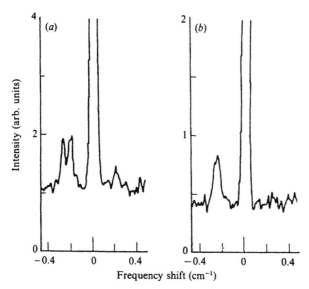

Figure 9.3. Brillouin spectra for Ni/Mo superlattices in a field of 0.093 T: (a) $d_2 = 24.9$ nm, $d_1 = 8.3$ nm; (b) $d_2 = 10$ nm, $d_1 = 30$ nm. Note the additional peak (due to the surface mode) in the anti-Stokes part of (a). After Grimsditch *et al* (1983).

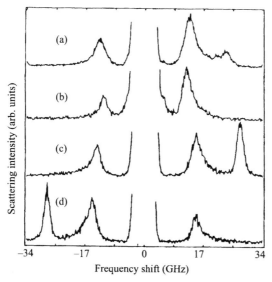

Figure 9.4. Brillouin spectra for Fe/Pd superlattices in a field of 0.1 T: (a) $d_2 = 2.19$ nm, $d_1 = 2.43$ nm; (b) $d_2 = 4.17$ nm, $d_1 = 13.87$ nm; (c), (d) $d_2 = 4.10$ nm, $d_1 = 0.91$ nm. The field direction is reversed in (d) compared with (c), causing the non-reciprocal surface mode to switch to the other side of the spectrum. After Hillebrands *et al* (1988).

As mentioned earlier, the theoretical results of (9.2)–(9.7) also apply in the Voigt geometry when the magnetic layers are antiferromagnetic. A good discussion of this case is provided by Barnás (1994), representing a generalization of the magnetostatic results for an antiferromagnetic film given in section 5.3. In particular, (9.7) gives the surface-mode dispersion relation (provided $d_2 > d_1$) when expressions for χ_a and χ_b are substituted using (2.173) and (2.174). In the case of zero applied magnetic field $\chi_b = 0$, and so the term in $\mathrm{sgn}(q_y)$ vanishes from (9.7). This means that the surface mode is reciprocal when $B_0 = 0$; it makes no difference if $q_y > 0$ or $q_y < 0$. The frequency of the superlattice surface mode is then the same as that for a semi-infinite antiferromagnet, i.e. as in (5.25) with $\omega_0 = 0$. When $B_0 \neq 0$ there are two surface modes (since now $\chi_b \neq 0$); they show non-reciprocal behaviour and their frequencies are found using (9.7). In all cases there are two bulk bands, which move farther apart when B_0 is increased.

Some magnetostatic calculations have also been carried out for superlattice structures in which the magnetic layers are adjacent to one another (such as for Fe/Co superlattices). Examples are included in the review by Barnás (1994). Typically, however, the exchange interactions may play a role in such systems, and so we defer further discussion until section 9.3 on dipole-exchange magnons in superlattices.

9.1.2 Perpendicular magnetization

Magnetostatic theory has also been applied to superlattices made up of alternating ferromagnetic/nonmagnetic and antiferromagnetic/nonmagnetic layers for the case of the magnetization M_0 *perpendicular* to the interfaces (Camley and Cottam 1987). We already discussed in section 5.1.2 that, for a single ferromagnetic film with perpendicular magnetization, no surface magnetostatic modes occur. This is true also for an antiferromagnetic film with sublattice magnetization in the perpendicular orientation. However, Camley and Cottam showed that, when a superlattice of these magnetic layers is formed, the coupling between layers is such as to allow a sequence of surface magnetic modes in appropriate cases. Furthermore, the nature of the bulk modes is quite different from the case of in-plane magnetization.

The calculation is set up with coordinate axes chosen as described in section 5.1.2. For the ferromagnetic superlattice the bulk susceptibility components χ_a and χ_b from (2.166) and (2.167) may again be utilized, except that the applied field B_0 must be replaced by the internal field $B_i = B_0 - N_z M_0$. This reduction is caused by the demagnetizing fields in the film, where the demagnetizing factor $N_z = 1$ in this geometry. A transfer-matrix formalism is followed, as in the previous section. Thus the z-dependent part $\psi_1(z)$ of the magnetostatic potential can be expressed in

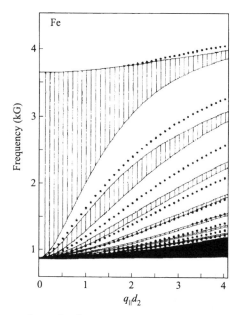

Figure 9.5. Magnetostatic modes for a superlattice with Fe/nonmagnetic structure for applied magnetic field perpendicular to the interfaces. The frequency (expressed in equivalent magnetic field units; $1\,\mathrm{kG} \equiv 0.1\,\mathrm{T}$) is plotted versus $q_{\parallel}d_2$ (with d_2 constant) for bulk bands (shaded) and surface modes (dotted lines). Assumed values are $M_0 = 0.168\,\mathrm{T}$, $B_0 = 0.022\,\mathrm{T}$, $d_1/d_2 = 0.5$ and $d_2^{\mathrm{Surf}}/d_2 = 1.25$. After Camley and Cottam (1987).

terms of two amplitudes just as in (9.1), except that x and q_x are now replaced by z and q_z, where q_z is given by (5.16) for a ferromagnetic layer and reduces to $iq_{\parallel}$ for a nonmagnetic layer. The details are given by Camley and Cottam (1987), who also introduced a surface perturbation by allowing the thickness of the outermost ferromagnetic layer to differ from the bulk value. A calculated dispersion relation for a semi-infinite superlattice with Fe as the magnetic layers is shown in figure 9.5. The bulk bands (shaded regions) occur within the bulk-mode region of a semi-infinite Fe material, but in the superlattice case a series of gaps (stop-bands) appears. In these gaps, and also above the uppermost bulk band, surface modes (dotted curves) can occur. The bulk and surface superlattice modes have a completely different character from those in the previous section. For parallel magnetization we recall that the superlattice modes (bulk and surface) were formed from appropriate combinations of the *surface* modes of a magnetic film. By contrast, in the present case the superlattice modes are formed from the *bulk* modes of a magnetic film. Camley and Cottam gave applications to other magnetic materials, both ferromagnetic and antiferromagnetic.

Although we are not aware of any experiments that are directly relevant to the above theory, it is worthwhile to mention Co/Pd and Co/Pt superlattices. For small Co layer thicknesses (e.g. of several monolayers) they are preferentially magnetized in the perpendicular orientation due to a strong out-of-plane anisotropy. The spin waves have been studied by Brillouin scattering, for example by Krams *et al* (1991) and Hillebrands *et al* (1993) who observed a multiplicity of bulk-like modes. A proper theoretical analysis, however, requires taking into account the large anisotropy and the exchange within the Co layers. We return to this in section 9.4.

9.2   Magnetic polaritons

A few theoretical papers have appeared generalizing the above magnetostatic results (for the case of in-plane magnetization) to include retardation. A clear account of this topic has been given by Tilley (1994). Among the earlier calculations, polariton results for magnetic/nonmagnetic superlattices were obtained by Barnás (1987, 1988) and Raj and Tilley (1987, 1989). All these results are derived for the Voigt configuration.

Following Barnás (1988) and Raj and Tilley (1989), we first consider a transfer-matrix theory that is a direct extension of the magnetostatic calculation in section 9.1.1. We consider s-polarized polaritons in the semi-infinite superlattice depicted in figure 9.1, except that both types of layers may now be magnetic. They are described in terms of the suscepti-bilities $\chi_a^{(j)}$ and $\chi_b^{(j)}$, as before, where the superscript j ($= 1$ or 2) specifies the medium. Attention is restricted to the modes propagating in the xy plane (Voigt geometry), so that the **E** field has a single component E_z. In medium j of cell l its Fourier component at frequency ω may be written as

$$E_{lz}^{(j)} = \exp(i\mathbf{q}_\parallel \cdot \mathbf{r}_\parallel)(a_l^{(j)} \exp\{\kappa_j[x + (l-1)D]\} + b_l^{(j)} \exp\{-\kappa_j[x + (l-1)D]\})$$

$$(9.10)$$

with $\kappa^2 = q_\parallel^2 - \varepsilon\mu_V(\omega^2/c^2)$ for each medium. Here μ_V is the Voigt permeability of the medium; as in (6.38) it is defined in terms of elements of the gyrotropic permeability tensor (and hence χ_a and χ_b) as

$$\mu_V = \mu_{xx} + (\mu_{xy}^2/\mu_{xx}) = 1 + \chi_a + [\chi_b^2/(1 + \chi_a)]. \qquad (9.11)$$

Expressions for the required magnetic-field components $B_{lx}^{(j)}$ and $H_{ly}^{(j)}$ are derived from (9.10) by means of Maxwell's equations; using the fact that these components are continuous across the interfaces, one can relate the a and b coefficients in cell $l + 1$ to the coefficients at the corresponding point in cell l. Hence, as in section 9.1.1, the transfer matrix T can be deduced and the results can be obtained for the superlattice excitations. Generalizing (9.6) the expression for the bulk superlattice spectrum is

$$\cos(QD) = \cosh(\kappa_1 d_1)\cosh(\kappa_2 d_2) + Z' \sinh(\kappa_1 d_1)\sinh(\kappa_2 d_2) \tag{9.12}$$

where $Z' = \Delta^{(1)}\Delta^{(2)} + \frac{1}{2}\Gamma^{(1)}\Lambda^{(2)} + \frac{1}{2}\Lambda^{(1)}\Gamma^{(2)}$ with

$$\Delta^{(j)} = q_\parallel \chi_b^{(j)}/\kappa_j(1 + \chi_a^{(j)}) \tag{9.13}$$

$$\Gamma^{(j)} = [\kappa_j^2(1 + \chi_a^{(j)})^2 - q_\parallel^2(\chi_b^{(j)})^2]/[iq_\parallel \kappa_j \mu_V^{(j)}(1 + \chi_a^{(j)})^2] \tag{9.14}$$

$$\Lambda^{(j)} = -q_\parallel \mu_V^{(j)}/i\kappa_j. \tag{9.15}$$

The above result and the corresponding expression for the surface modes of a semi-infinite superlattice are quite complicated, but they simplify to some extent in various cases. For example, if one of the superlattice materials is nonmagnetic there will be some reduction in (9.13)–(9.15). Also if at least one of the materials is an antiferromagnet and the applied field $B_0 = 0$, then the corresponding $\chi_b^{(j)}$ will vanish. Various numerical examples of spectra are given in the references already quoted. In figure 9.6 we show some calculations for bulk and surface modes due to Barnás (1990) for an antiferromagnetic/nonmagnetic superlattice. The surface mode (broken curve) is seen to exhibit non-reciprocal behaviour when $B_0 \neq 0$, as might be expected by analogy with previous sections. Antiferromagnetic/non-magnetic superlattices have also been considered by Zhu and Cao (1987). Calculations for the polaritons in ferromagnetic/nonmagnetic superlattices by How and Vittoria (1989a,b) have included the case where the directions of magnetization alternate; the magnetostatic limit can also be obtained from their results.

Going beyond the Voigt configuration, Barnás (1991) has used a 4×4 transfer matrix method for both surface and bulk modes in a geometry where the in-plane wavevector $\mathbf{q}_\parallel$ is not necessarily perpendicular to the in-plane magnetization. A 4×4 transfer matrix was also employed for a superlattice inside a metallic waveguide (Zhu and Cao 1987, 1989).

It is not surprising that the full results with retardation are complicated, but fortunately they can be greatly simplified in the long-wavelength limit. As shown by Raj and Tilley (1987), the method due to Agranovich and Kravtsov (1985) can be applied to a magnetic superlattice as follows. Again we consider essentially the same geometry as in figure 9.1, except that both constituents are now taken to be magnetic. For propagation in the xy plane (Voigt geometry) each constituent is characterized by a gyrotropic permeability tensor. We consider an r.f. magnetic field in the xy plane; the wavelength is taken to be much longer than the superlattice period so that the $\mathbf{B}$ and $\mathbf{H}$ fields are the same in successive layers of component 1 and in successive layers of component 2. The boundary conditions are that H_y

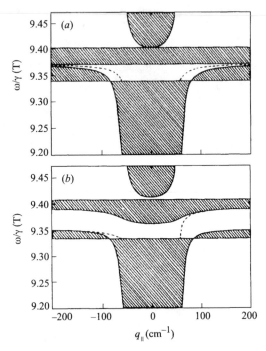

Figure 9.6. Spectrum of bulk polaritons (shaded regions) and surface polaritons (broken curves) for an antiferromagnetic MnF_2/nonmagnetic superlattice with $d_1 = d_2 = 500\,\mu m$: (a) $B_0 = 0$; (b) $B_0 = 0.02\,T$. After Barnás (1990).

and B_x are continuous across interfaces, so these components are everywhere equal to their average values $\bar{H}_y$ and $\bar{B}_x$. In medium 1 the field components are related by

$$B_y/\mu_0 = \mu_{xx}^{(1)}\bar{H}_y + \mu_{xy}^{(1)}H_x \tag{9.16}$$

$$\bar{B}_x/\mu_0 = -\mu_{xy}^{(1)}\bar{H}_y + \mu_{xx}^{(1)}H_x \tag{9.17}$$

so that

$$B_y/\mu_0 = \mu_V^{(1)}\bar{H}_y + (\mu_{xy}^{(1)}/\mu_{xx}^{(1)})H_x \tag{9.18}$$

$$H_x = (\mu_{xy}^{(1)}/\mu_{xx}^{(1)})\bar{H}_y + (1/\mu_{xx}^{(1)})\bar{B}_x/\mu_0. \tag{9.19}$$

Similar equations hold in component 2, and combined with (9.18) and (9.19) lead to expressions for the average fields $\bar{B}_y = (d_1B_y^{(1)} + d_2B_y^{(2)})/D$ and $\bar{H}_x = (d_1H_x^{(1)} + d_2H_x^{(2)})/D$. Reorganization gives relations in terms of the effective-medium permeability tensor:

$$\begin{pmatrix} \bar{B}_x \\ \bar{B}_y \end{pmatrix} = \mu_0 \begin{pmatrix} \bar{\mu}_{xx} & \bar{\mu}_{xy} \\ -\bar{\mu}_{xy} & \bar{\mu}_{yy} \end{pmatrix} \begin{pmatrix} \bar{H}_x \\ \bar{H}_y \end{pmatrix} \tag{9.20}$$

with

$$\bar{\mu}_{xx} = \frac{D\mu_{xx}^{(1)}\mu_{xx}^{(2)}}{d_1\mu_{xx}^{(2)} + d_2\mu_{xx}^{(1)}} \tag{9.21}$$

$$\bar{\mu}_{yy} = \frac{D^2\mu_{xx}^{(1)}\mu_{xx}^{(2)} + d_1 d_2[(\mu_{xx}^{(1)} - \mu_{xx}^{(2)})^2 + (\mu_{xy}^{(1)} - \mu_{xy}^{(2)})^2]}{(d_1 + d_2)(d_1\mu_{xx}^{(2)} + d_2\mu_{xx}^{(1)})} \tag{9.22}$$

$$\bar{\mu}_{xy} = \frac{d_1\mu_{xy}^{(1)}\mu_{xx}^{(2)} + d_2\mu_{xy}^{(2)}\mu_{xx}^{(1)}}{d_1\mu_{xx}^{(2)} + d_2\mu_{xx}^{(1)}}. \tag{9.23}$$

This derivation was given independently by Almeida and Mills (1988).

The form of (9.20) corresponds to a kind of magnetic anisotropy, induced by the superlattice structure itself. It is straightforward to find the dispersion equation for wave propagation in a medium with this kind of magnetic anisotropy. For a TE mode, the relevant component of the dielectric tensor in the effective-medium description is $\varepsilon_{yy} = \bar{\varepsilon} = (\varepsilon_1 d_1 + \varepsilon_2 d_2)/D$. This dispersion equation is

$$\frac{\varepsilon_{yy}\omega^2}{c^2} = \frac{q_x^2}{\bar{\mu}_{yy} + (\bar{\mu}_{xy}^2/\bar{\mu}_{xx})} + \frac{q_y^2}{\bar{\mu}_{xx} + (\bar{\mu}_{xy}^2/\bar{\mu}_{yy})}. \tag{9.24}$$

Raj and Tilley (1987) show that this is the same form as can be obtained by a long-wavelength expansion of the general dispersion equation, derived using transfer matrices.

Further discussion of the above long-wavelength method has been provided by Almeida and Tilley (1990). Focusing on the case of an antiferromagnetic/nonmagnetic superlattice, they showed that the surface polaritons have the dispersion equation

$$q_{\parallel}^2 = \frac{\omega^2}{c^2}\frac{\bar{\mu}_{xx}(\varepsilon_m\bar{\mu}_{yy} - \bar{\varepsilon})}{\bar{\mu}_{xx}\bar{\mu}_{yy} - 1}. \tag{9.25}$$

Here ε_m is the dielectric constant of the bounding medium. Some dispersion curves for the particular case of FeF_2/ZnF_2 are shown in figure 9.7(a). It is seen that the curves occupy a narrow frequency interval near the AFMR frequency, and this interval shrinks as the volume fraction $f = d_2/D$ decreases. For $f > 0.5$ the mode persists to large $q_{\parallel}$, where it becomes the magnetostatic surface mode discussed in section 9.1.1. On the other hand, for $f < 0.5$ the mode terminates at a finite $q_{\parallel}$ and thus has no magnetostatic counterpart. Almeida and Tilley also discussed the application to ATR by the prism coupling method (see section 6.3) and a theoretical spectrum is shown in figure 9.7(b).

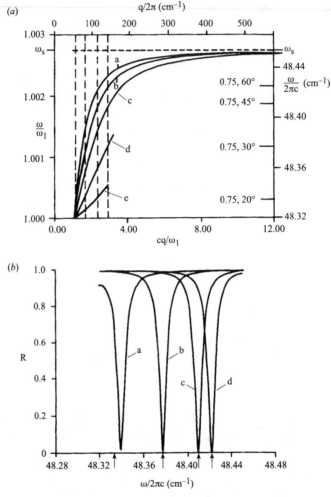

Figure 9.7. (a) Surface-polariton dispersion curves for FeF_2/ZnF_2 superlattices with vacuum as the external medium. The values of the magnetic fraction f are: a, 1.0; b, 0.75; c, 0.5; d, 0.25; e, 0.1. The left scale is frequency relative to the AFMR frequency (denoted by ω_1). (b) Theoretical ATR spectra for $f = 0.75$. These are calculated for the Otto configuration with a Si prism ($n_p = 3.4$) and angles of incidence (degrees) and gaps (cm) equal to (a) 20, 0.04; (b) 30, 0.03; (c) 45, 0.02; (d) 60, 0.015. Corresponding scan lines are shown dashed in (a), with intersections on the dispersion curve marked by arrows in (b). After Almeida and Tilley (1990).

9.3 Exchange-dominated magnons

We now move from the regime where dipolar effects dominate to the oppo-site regime where the exchange effects are dominant. The intermediate and more complicated situation corresponding to dipole-exchange excitations will be deferred until section 9.4.

9.3.1 Magnetic reconstruction

So far we have considered only the dipole–dipole coupling between different magnetic layers that are often separated by a spacer. In the exchange-dominated case the different magnetic materials are either in direct contact (coupled by an interface exchange) or they may be weakly coupled via the long-range exchange through a metal, as discussed in section 4.5. In the latter case we discussed how the exchange may be oscillatory in sign and may have biquadratic, as well as bilinear, contributions. Depending on the materials, the above considerations can lead to complicated schemes for the magnetic ordering near an interface, e.g. there may be phases with canted or twisted spins. We therefore address these matters under the label of magnetic reconstruction.

It is important to recognize at the outset that even the nature of the ground state depends on the specific details of the system under consideration, and that in the presence of external fields and with variation of temperature many forms of spin reconstruction are possible. We illustrate this by the hypothetical ferromagnetic/antiferromagnetic superlattice discussed by Hinchey and Mills (1986a,b).

The system consists of alternating ferromagnetic and antiferromagnetic components, with a ferromagnetic coupling at the interface. The antiferromagnetic structure is one of alternating layers, successively spin-up and spin-down. It can be seen that the nature of the ground state depends crucially on whether the number N_{AF} of layers in the antiferromagnet is even or odd. This is illustrated in figure 9.8. For N_{AF} even, as in figure 9.8(a), successive ferromagnetic layers are aligned oppositely, and the ground state has no net magnetic moment, while for N_{AF} odd (figure 9.8(c)) the ferromagnetic layers are all aligned, and the ground state does have a magnetic moment. Even at zero temperature, the phase diagram of this system in an applied external field is quite complicated. Figure 9.8(b) shows a possible state of the N_{AF}-even superlattice in which all the ferromagnetic spins are aligned with the applied field. The Zeeman energy is lower than in the configuration of figure 9.8(a), but the interface-exchange energy is higher. As the number of spin layers $N_S = N_F + N_{AF}$ in one period increases the relative importance of the interface-exchange energy decreases, so it is to be expected that the critical field B_C for a phase transition out of the state of figure 7.28(a) decreases as N_S increases. Hinchey and Mills find B_C by a 'soft-mode' method, i.e. they deduce B_C as the field for which a spin–wave frequency becomes zero. Typical results are shown in figure 9.9; the field units are such that the spin–flop transition of the bulk antiferromagnet corresponds to $B_0 = 1$. As expected, for even N_{AF} the zero-field

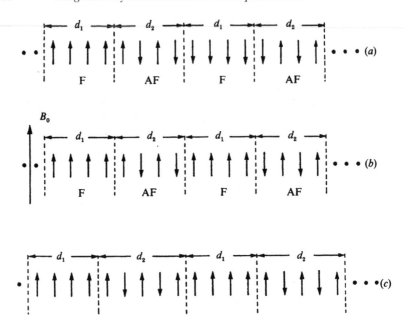

Figure 9.8. Ferromagnetic/antiferromagnetic superlattice. (a) Zero-field ground state when number of antiferromagnetic layers N_{AF} is even; (b) possible high-field state; (c) zero-field ground state when N_{AF} is odd. After Hinchey and Mills (1986).

ground state is rather unstable, with $B_C \to 0$ as $N \to \infty$, while for odd N_{AF} the zero-field ground state is rather stable.

The transition out of the ground state at the critical field B_C in figure 9.9(a) is not straight into the 'Zeeman state' of figure 9.8(b); in fact what occurs is a 'twisted state' in which the spins gain a component such that they remain in the plane of the layers, but are no longer aligned with the magnetic field. There are some analogies with the spin–flop phase of antiferromagnets (see section 4.3). A detailed discussion of the twisted states and phase diagrams is given by Hinchey and Mills. Other related work is reported by Leung (1987).

Spin reconstruction was also discussed by Camley (1987) and Camley and Tilley (1988). They were concerned with the Fe/Gd system, which has the interesting property, as demonstrated by Taborelli *et al* (1986), that although Fe and Gd are ferromagnets, the exchange interaction at an interface is antiferromagnetic. The ground state of an Fe/Gd superlattice is therefore as depicted in figure 9.10(a). This may be expected to be unstable in modest applied fields, and indeed Camley and Tilley find that a phase transition occurs to a twisted state as shown in figure 9.10(b). Numerical exploration of the mean-field expression for the free energy can be used to find the phase diagram in the $B_0 - T$ plane. The detailed results, presented by Camley and Tilley (1988), show that for small values of n_1 and n_2 the phase diagram is sensitive

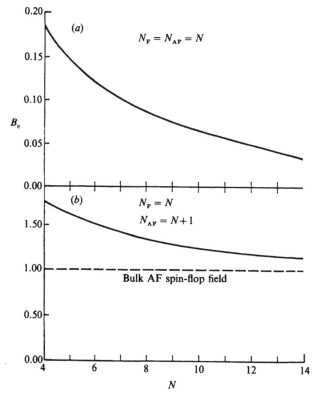

Figure 9.9. Critical field B_c for a transition out of the ground state of the ferromagnetic/antiferromagnetic superlattice as function of $N = N_F$. (a) N_{AF} even, see figure 9.8(a); (b) N_{AF} odd, see figure 9.8(c). The units of field are chosen so that the bulk AF spin–flop field is equal to unity. After Hinchey and Mills (1986).

to the precise number of spins in each layer. Further numerical calculations were developed by Lepage and Camley (1989, 1990).

A formal development that should be of value in discussing reconstruction is the use of Landau-type expansions, as proposed by Fishman *et al* (1987) and by Camley and Tilley (1988). The method applies only for the case when n_1 and n_2 are large, and when the average spin value varies only slowly from site to site. The mean spin $\langle \mathbf{S}_i \rangle$ at site i may then be replaced by a continuous variable $\mathbf{M}(x)$. The basic idea of Landau theory, as expounded in Landau and Lifshitz (1969) for example, is to expand the free energy as a Taylor series in the invariants of the system. For a slab of Fe, extending from $x = -d_1$ to $x = 0$, the expression is

$$F_1 = \int_{-\infty}^{\infty} dy\, dz \int_{-d_1}^{0} dx\{-B_0 M_z + \tfrac{1}{2}a_1(M_y^2 + M_z^2) + \tfrac{1}{2}b_1(M_y^2 + M_z^2)^2$$
$$+ \tfrac{1}{2}c_1[(\nabla M_y)^2 + (\nabla M_z)^2]\}. \qquad (9.26)$$

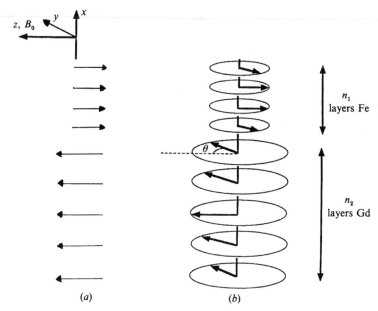

Figure 9.10. Magnetic superlattice of Fe/Gd. (a) Ground state in absence of magnetic field; (b) Twisted state developed in applied field B_0. After Camley and Tilley (1988).

Here a_1, b_1 and c_1 are constants, which may be related to the microscopic parameters for Fe, or may be determined later from comparison with experiment. It is assumed that **M** is confined to the yz plane, corresponding to the twisted state of figure 9.10(b). A similar expression for a term F_2 in terms of a variable **N**, say, holds for the other component of the unit cell, namely a slab of Gd extending from $x = 0$ to $x = d_2$. Finally it is necessary to include a free energy term for the interface. The lowest order invariants are $(M_y^2 + M_z^2)$, $(N_y^2 + N_z^2)$ and $(M_y N_y + M_z N_z)$, each evaluated at $x = 0$, so that the interface free energy is

$$F_1 = \int_{-\infty}^{\infty} dy\,dz\{\tfrac{1}{2}(c_1/\delta_1)(M_y^2 + M_z^2)_{x=0} + \tfrac{1}{2}(c_2/\delta_2)(N_y^2 + N_z^2)_{x=0}$$
$$+ \alpha(M_y N_y + M_z N_z)_{x=0}\}. \tag{9.27}$$

This introduces three more parameters δ_1, δ_2 and α.

Within this model, the equilibrium state is now found as that which minimizes $F_1 + F_2 + F_1$. The minimum is found, in principle, from the Euler–Lagrange equations for **M** and **N**. However, these are quite complicated, and we simplify the problem by use of the *constant-amplitude approximation*, i.e. we put

$$(M_y, M_z) = M(\sin\theta, \cos\theta) \tag{9.28}$$

$$(N_y, N_z) = N(\sin\phi, \cos\phi). \tag{9.29}$$

This approximation should be good at low temperatures, where the reconstruction is described in terms of spins of constant magnitude inclined at angles θ and ϕ to the field direction. Within this approximation, the free energy reduces to

$$
\frac{F}{A} = \int_{-d_1}^{0} dx \left\{ -B_0 M \cos\theta + \tfrac{1}{2}c_1 M^2 \left(\frac{d\theta}{dx}\right)^2 \right\}
$$
$$
+ \int_{0}^{d_2} dx \left\{ -B_0 N \cos\phi + \tfrac{1}{2}c_2 N^2 \left(\frac{d\phi}{dx}\right)^2 \right\}
$$
$$
+ \alpha M N \cos(\theta - \phi)|_{x=0} + \alpha M N \cos(\theta - \phi)|_{x=-d_1}. \quad (9.30)
$$

Here certain constant terms have been omitted, and it is assumed that the equilibrium state has period $d_1 + d_2$, so that (9.30) contains contributions from the two interfaces in the unit cell. A is the specimen area in the yz plane.

The Euler–Lagrange equations for minimum F are

$$
c_1 M \frac{d^2\theta}{dx^2} - B_0 \sin\theta = 0, \qquad c_2 N \frac{d^2\phi}{dx^2} - B_0 \sin\phi = 0 \quad (9.31)
$$

with boundary conditions at interfaces

$$
c_1 M \frac{d\theta}{dx} - \alpha N \sin(\theta - \phi) = 0 \quad (9.32)
$$

$$
c_2 N \frac{d\phi}{dx} - \alpha M \sin(\theta - \phi) = 0. \quad (9.33)
$$

In general the solutions of (9.31)–(9.33) may be given in terms of elliptic functions, as shown by Camley and Tilley (1988). For the particular case of two semi-infinite media in contact, the elliptic functions reduce to hyperbolic functions, and the solutions are

$$
\cos(\tfrac{1}{2}\theta) = \tanh[\lambda_1(y_1 - y)] \quad (9.34)
$$

$$
\cos(\tfrac{1}{2}\phi) = \tanh[\lambda_2(y - y_2)] \quad (9.35)
$$

where $\lambda_1 = (B_0/c_1 M)^{1/2}$ with a similar expression for λ_2. The y_1 and y_2 are constants of integration, found from the boundary conditions, or equivalently from minimization of the free energy. The predicted reconstruction for the case of two semi-infinite media in contact is shown in figure 9.11. It can be seen that, as might be expected, the width of the reconstructed region decreases as the applied field B_0 increases.

The free energy $F_1 + F_2 + F_I$ describes a superlattice of two materials that in their bulk forms have, in general, different critical temperatures. One should therefore ask what is the critical temperature of the superlattice itself. For the expression (9.26) the bulk transitions are second order and it may be expected that the superlattice transition is also second order. In

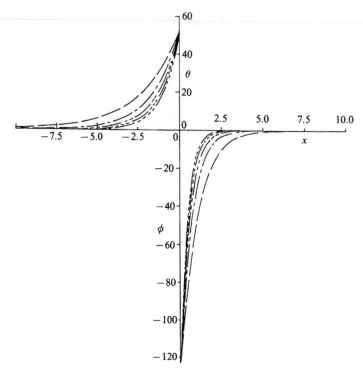

Figure 9.11. Variation of θ and ϕ with position x for a single interface between Fe and Gd. Magnetic field varies from $B_0 = 0.2\,\text{T}$ (broken line) to $1\,\text{T}$ (chain line) in steps of $0.2\,\text{T}$. Horizontal axis scaled by $\lambda_0 = 5.85 \times 10^7\,\text{m}^{-1}$. After Camley and Tilley (1988).

this case the critical temperature is that at which the linearized Euler–Lagrange equations have a solution. A detailed discussion is given in Tilley (1988).

As an aside, the Landau formalism has found its most extensive applications in the theory of ferroelectric and related materials (Tolédano and Tolédano 1987); for a simple ferroelectric the expansion variable is the polarization **P** rather than the magnetization **M**. In a simplified model based on (9.26) and (9.27) with gradient terms omitted, the free energy of a ferroelectric bilayer or superlattice of materials 1 and 2 is

$$F = A_1 P_1^2 L_1 + B_1 P_1^4 L_1 - E P_1 L_1 + A_2 P_2^2 L_2 + B_2 P_2^4 - E P_2 L_2 + N J P_1 P_2 \tag{9.36}$$

where L_1 is the total length of material 1 and N is the number of interfaces. The third and sixth terms describe interaction with an applied electric field E and it is assumed that P_1 and P_2 remain constant and parallel to E in their respective layers. Chew *et al* (2000) show that if the interface coupling is antiferroelectric ($J > 0$) then rather complicated sequences of states occur as E is varied, with the consequence that dielectric hysteresis loops can be 'engineered' by choice of L_1/L_2. Many ferroelectric materials have first- rather than second-order transitions, and the extension of the analysis to these is given by Ong *et al* (2002).

9.3.2 Magnons

In principle the magnon spectrum must be derived for each of the equilibrium states found for a particular superlattice, and in fact Hinchey and Mills (1986a,b) do give an extensive discussion for the ferromagnet/antiferro- magnet superlattices with which they are concerned. The dispersion relation can be found by application of the transfer-matrix method to the equations of motion for the spins. This can be done by either the microscopic or macroscopic methods that were discussed for films in chapter 4. In some cases, however, the equilibrium configuration may be straightforward. Thus we begin with an illustration for what is probably the simplest case, the ferromagnet/ferromagnet superlattice with ferromagnetic exchange at the interfaces, so that reconstruction need not concern us. This has been dis- cussed by Albuquerque *et al* (1986) and by Dobrzynski *et al* (1986).

To be specific, following Albuquerque *et al* (1986), we consider alter- nating simple-cubic ferromagnets in which the interfaces are (001) planes. It is assumed that each component is described by the nearest-neighbour Heisenberg Hamiltonian including an external field, (2.59), and the exchange constants are taken as J_1 in one component, J_2 in the other, and I across the interface. The infinitely-extended superlattice structure is shown in figure 9.12. The problem is essentially 1D, and the transfer matrix may be constructed following closely the methods in chapter 8 and the earlier parts of this chapter. Spin equations of motion for the operators S_i^+ in each atomic layer are written down, and simplified by using the random- phase approximation (RPA) (see section 2.3.2). Within each component

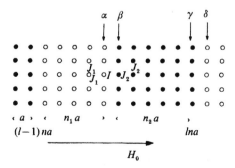

Figure 9.12. Model for simple-cubic magnetic superlattice with nearest-neighbour exchange coupling. After Albuquerque *et al* (1986).

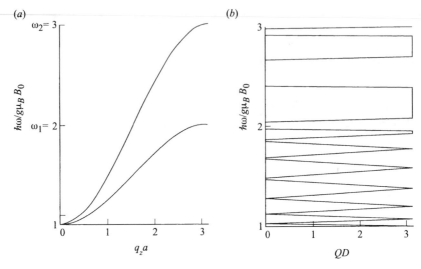

Figure 9.13. Bulk and superlattice magnon dispersion graphs for (001) propagation with parameters $g\mu_B B_0/4SJ_1 = 1$, $J_2/J_1 = 2$, $I/J_1 = 1.5$. Both components have the same values of gyromagnetic ratio g and spin S. Relevant frequencies are $\hbar\omega_1/g\mu_B B_0 = 1 + 4SJ_1/g\mu_B B_0 = 2$, $\hbar\omega_2/g\mu_B B_0 = 1 + 4SJ_2/g\mu_B B_0 = 3$: (a) bulk-magnon dispersion curves for components 1 (lower curve) and 2 (upper curve); (b) bulk superlattice-magnon dispersion curves for $n_1 = n_2 = 10$. After Albuquerque *et al* (1986).

the solution for S_i^+ is written as a sum of forward- and backward-travelling waves, e.g. by analogy with (9.1). The equations of motion of spins adjacent to the interfaces, labelled α, β, γ, δ in figure 9.12, relate amplitudes in adjacent layers, and the 2×2 transfer matrix is built up as in previous examples. Finally, the bulk dispersion equation takes the standard form of (8.20) in terms of the trace of the transfer matrix. The explicit expressions are quite complicated, even for this simple case; for details we refer the reader to Albuquerque *et al* (1986) and also the review article by Barnás (1994).

An example of the dispersion curves for frequency versus Bloch wavevector Q obtained for this model (assuming in-plane wavevector $\mathbf{q}_{||} = 0$) is shown in figure 9.13(b). It has the form of modes that are folded back in the reduced Brillouin zone, and as expected by analogy with some of the non-magnetic examples in chapter 8 (e.g. see section 8.1.1) there are stop bands and pass bands. We take a moment to interpret this behaviour. First, we recall that the magnon dispersion relation for a bulk medium is given by (2.67): when $\mathbf{q}_{||} = 0$ and $T \ll T_C$ there is a single band of frequency from $g\mu_B B_0$ at the zone centre to $g\mu_B B_0 + 4SJ$ at the zone edge. In figure 9.13(a) the corresponding bulk-medium dispersions are drawn for each superlattice component and the Brillouin-zone edge frequencies are denoted by ω_1 and ω_2 (with $\omega_1 < \omega_2$). The resemblance of figure 9.13(b) to figure 8.6 for diatomic-lattice phonons is striking. For $\omega < \omega_2$ the bulk

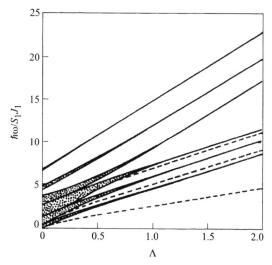

Figure 9.14. Bulk (shaded regions) and surface (broken curves) magnons in a semi-infinite superlattice. The parameters are $n_1 = n_2 = 3$, $J_2/J_1 = 2$, $I/J_1 = 1.4$, $S_2 = S_1$, $B_0 = 0$ and $J_{2S}/J_1 = 0.5$ denoting J_{2S} at the surface. The reduced frequency $\hbar\omega/S_1J_1$ is plotted in terms of $\Lambda = 1 - \frac{1}{2}[\cos(q_x a) + \cos(q_y a)]$ where $\mathbf{q}_{\parallel} = (q_x, q_y)$ is the in-plane wavevector. After Barnás (1994).

wavevectors q_1 and q_2 in the two components are real, and the superlattice dispersion curve consists of broad pass bands with narrow stop bands. For $\omega_1 < \omega < \omega_2$, q_2 is real while q_1 is complex; the superlattice dispersion curve then consists of narrow pass bands and broad stop bands.

Barnás (1994) presents a very thorough treatment based on the transfer-matrix method that generalizes the above formal results for a ferromagnetic/ferromagnetic superlattice in a number of ways. For example, he considers semi-infinite structures, so that surface modes of the superlattice can be studied. A calculation showing the dispersion relation for the superlattice bulk magnons (shaded bands) and surface magnons is given in figure 9.14. Barnás (1994) also includes the case when the repeat unit consists of a finite number N of different ferromagnetic layers; and the effects of uniaxial anisotropy (giving surface and interface pinning) are taken into account.

A further extension was made by Bezerra and Cottam (2002) to include the additional effects of biquadratic coupling across the interfaces in ferromagnetic/ferromagnetic superlattices. As discussed in section 4.4.2 this type of exchange coupling, represented by an extra term in (4.78), can be important at interfaces and may be comparable with the usual bilinear exchange. They showed that as the ratio between the biquadratic J_{BQ} and interfacial bilinear I exchange coupling increases the bulk-magnon bands of the superlattice get narrower. This is because of the competing effects of the two terms, which favour different alignments, and the resulting effect of J_{BQ} can be expressed in terms of a reduction of I to an effective value I_{eff} given by

$$I_{eff} = I - (2S_1 S_2 - S_1 - S_2 + 1)J_{BQ} \qquad (9.37)$$

where S_1 and S_2 are the spin quantum numbers for the two materials. This holds provided $I_{eff} > 0$. For larger biquadratic exchange we have $I_{eff} < 0$ which indicates a spin reorientation to a canted phase. Bezerra and Cottam found that the surface modes in a semi-infinite superlattice were sensitive to the biquadratic exchange; in general both acoustic- and optic-surface modes could occur, but the former were suppressed when J_{BQ}/I exceeded a threshold value.

It is evident that the microscopic approach to calculate the magnon spectrum of superlattices rapidly becomes very complicated when there are more than a few atomic layers in each component of the superlattice. In such cases there are advantages (particularly for the excitations at smaller wavevectors) in using a macroscopic, or continuum, approach, such as that in section 4.2.2 for Heisenberg ferromagnetic films. In fact, we shall predominantly employ the macroscopic approach in section 9.4 to discuss the dipole-exchange modes in magnetic superlattices. Generally the results in the exchange-dominated regime can be deduced by taking the appropriate limit (in which the magnetization, or sublattice-magnetization, terms become small compared with the exchange field in any magnetic layer).

It was mentioned in chapter 1 that, although inelastic neutron scattering is an important technique to study magnetic and nonmagnetic excitations in *bulk* materials, it has not proved to be useful in thin films, mainly because of the large penetration depth of the neutrons and the very weak scattering strength. These technical difficulties have recently been overcome with magnetic superlattices, and the first successful inelastic neutron scattering measurements on a magnetic superlattice was reported by Schreyer *et al* (2000). The key to this important advance lies in achieving a sufficiently large scattering volume and number of interfaces (by having superlattices with many short periods) and with advances in detection and instrumentation. While there are still only preliminary results at the time of writing, it is anticipated that further progress in inelastic neutron scattering will lead to it becoming a valuable tool for probing magnetic excitations in superlattices. It has the great advantage (e.g. compared with inelastic light scattering) that magnetic excitations at larger wavevectors, and thus corresponding to the exchange-dominated regime, can be studied. Schreyer *et al* made their measurements on a Dy(4.3 nm)/Y(2.8 nm) superlattice with 350 periods grown by MBE on a large area substrate (area $\sim$3 cm^2). Although the rare-earth materials have a complicated magnetic structure, they have large moments that make them suitable for these studies. Normally Dy exhibits a ferromagnetic structure below 89 K, but in Dy/Y superlattices a

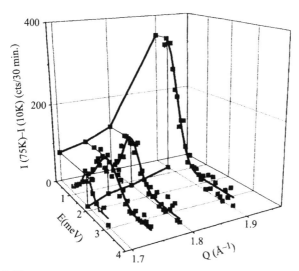

Figure 9.15. Difference spectrum of inelastic neutron scattering intensities taken at 75 K and 10 K with Gaussian fits to the peaks and the resulting dispersion curve plotted in the Q–E plane. Here $E = \hbar\omega$ is the magnon energy and Q denotes the in-plane wavevector. After Schreyer *et al* (2000).

coherent helical phase occurs which extends over many layers, whereas the ferromagnetic phase is suppressed due to magnetoelastic effects (see e.g. Salomon *et al* 1986). Schreyer *et al* performed their inelastic neutron scattering measurements at 75 K to create a sufficiently large magnon scattering cross section (through thermal effects) and subtracted the results from equivalent scans at 10 K (to eliminate background effects). Some results showing essentially the bulk magnon in Dy are represented in figure 9.15. Tentative evidence of zone folding was also reported.

Schreyer *et al* (2000) made further measurements using *elastic* neutron scattering to study the magnetic structure in short-period superlattices of Fe/Cr. The spin orientations in the superlattice can be varied by changing the thickness of the Cr layers, giving rise to various spiral and twisted phases. At low enough temperatures bulk Cr is antiferromagnetic, so the Fe/Cr superlattice has some analogies with the Hinchey and Mills calculation discussed in section 9.3.1. However, the Cr behaviour is rather complex and Schreyer *et al* were able to interpret some of their data on the basis of a theory due to Slonczewski (1995).

Other experimental data on magnetic superlattices where exchange effects play a role arise from studies relating to giant magnetoresistance (GMR) and oscillatory exchange across metallic spacer layers between ferromagnetic layers. These topics were extensively discussed in sections 4.5 and

4.6 for films, but the same papers and review articles cited there are equally applicable for the superlattice studies. For example, the paper by Baibich *et al* (1988), among others, included data on superlattices, since this increased the GMR effect. In the case of inelastic light scattering we have already discussed how the excitation wavevectors are such that a dipole–exchange model is appropriate, which now follows.

9.4   Dipole-exchange modes

Calculations for the spin waves when both the dipole–dipole and exchange effects need to be considered can be made in superlattices by following the same general methods as discussed for films and bilayers in chapter 5. From the experimental side the motivation comes mainly, as before, from inelastic light scattering which typically involves excitation wavevectors in the dipole-exchange regime. A thorough review of this topic, covering both theory and experiment, has been given by Hillebrands (2000).

9.4.1   Macroscopic theories

The theoretical studies for superlattices have developed out of the macroscopic (or continuum) approach described in section 5.4.1 for films. However, apart from the inclusion of both dipolar and exchange terms within any particular magnetic layer, there is the question of how to characterize the coupling between the layers. In the usual case where the superlattice is composed of magnetic layers alternating with nonmagnetic spacer layers, there may be either just the dipolar coupling across the spacer thickness or, in the case of metallic spacers, there may be long-range exchange. In the latter case the exchange may be either ferromagnetic or antiferromagnetic in sign or may include a biquadratic term (see the discussion in section 4.5); all of these considerations affect the relative directions of magnetizations in the layers. Because the range of possibilities is rather large we shall focus on just a few cases of special interest.

The first calculations were reported by Vayhinger and Kronmüller (1986), who considered an infinite stack of ferromagnetic layers, which were all magnetized uniformly in the same direction (either in-plane or perpendicular to the layers) and separated by nonmagnetic spacer layers. Their formalism was the perturbation method based on the tensorial Green-function approach of Vendik and Chartorizhskii (1970), which we mentioned briefly in section 5.4.1. Essentially Vayhinger and Kronmüller gave a generalization of the dipole-exchange theory for a single film due to Kalinikos and Slavin (1986) to the case of a periodic multilayer. In doing so, they simplified the analysis by ignoring exchange coupling between the adjacent magnetic layers. The procedure whereby the interlayer exchange could be included was demonstrated for the special case of a bilayer system in a subsequent paper (Vayhinger and Kronmüller 1988).

At around the same time some rather more extensive calculations were made for the case of ferromagnetic/nonmagnetic superlattices with a finite number of periods by Hillebrands (1988, 1990). He assumed in-plane magnetization for the ferromagnetic layers (by analogy with figure 5.1 for a single film) and allowed for both ferromagnetic exchange and dipole–dipole coupling across the nonmagnetic spacer layers. The effects of interface anisotropies, which can be important for some materials, were also included. The method of calculation can be considered as a direct generalization of the macroscopic dipole-exchange theory for a film, as outlined in equations (5.27)–(5.35). Thus Hillebrands uses the torque equation of motion to deduce the linear response of the fluctuating part of the magnetization $\mathbf{m}(\mathbf{r})$ inside a magnetic layer to a fluctuating magnetic field $\mathbf{h}(\mathbf{r})$, just as in section 5.4.1. Equation (5.27) representing the total effective field can now be supplemented by a term for the interface anisotropy as

$$\mu_0 \mathbf{H} = B_0 \hat{\mathbf{z}} + \mathbf{b}_{\mathrm{Ex}}(\mathbf{r}) + (1/M_0)\nabla_{\alpha} E_{\mathrm{Ani}} + \mu_0 \mathbf{h}(\mathbf{r})\exp(-\mathrm{i}\omega t) \qquad (9.38)$$

where E_{Ani} is the volume magnetic-anisotropy density and ∇_{α} is the gradient operator for which the differentiation variables are the components of the unit vector $\boldsymbol{\alpha}$ pointing in the direction of static magnetization vector $\mathbf{M}_0$. The other notation is as before; in particular the fluctuating part of the exchange field $\mathbf{b}_{\mathrm{Ex}}(\mathbf{r})$ can be written in the form $(D/\gamma)\nabla^2 \mathbf{m}(\mathbf{r})$ as in section 5.4.1. Here D is proportional to the exchange constant (e.g. $D = \langle S^z \rangle J a^2$ in the case of a simple-cubic ferromagnet as discussed in section 2.3.2). It is worthwhile here to comment briefly on other notations in the research literature. Instead of D, another parameter A is employed by some authors; the two quantities are related by

$$D = 2A/M_0. \qquad (9.39)$$

They are both often referred to as the exchange stiffness parameter, but they have different units; D is expressed in V s and A is in J m^{-1}.

The form of E_{Ani} in (9.38) depends on factors such as the symmetry and strain of the magnetic material (see e.g. Bland and Heinrich 1994, Gurevich and Melkov 1996, Hillebrands 2000). Within the bulk of a material the magnetocrystalline anisotropy is often the most important contribution. The free energy of the magnetization, E_{Ani}, is coupled to the crystalline symmetry due to the spin–orbit interaction and becomes a function of the direction of $\mathbf{M}_0$. When this free energy is expanded with respect to functions belonging to the irreducible representations of the crystal symmetry group, various anisotropy constants may arise as coefficients in the expansion. For example, in the case of a cubic crystal the leading-order contribution occurs in fourth order. It involves a volume anisotropy coefficient K_1 and has the form

$$E_{Ani} = K_1(\alpha_x^2\alpha_y^2 + \alpha_y^2\alpha_z^2 + \alpha_z^2\alpha_x^2) \tag{9.40}$$

where the α_i are the direction cosines defined with respect to the crystallographic axes. In a non-cubic material it is often the case that E_{Ani} can be expressed in terms of a uniaxial anisotropy coefficient $K_p^{(2)}$. For example, when y is the uniaxis we have

$$E_{Ani} = K_p^{(2)}\alpha_y^2. \tag{9.41}$$

This favours orientation of the magnetization along the y axis (the 'easy' axis) when $K_p^{(2)} < 0$, otherwise the xz plane is favoured. The physical origin of this contribution can be magnetocrystalline (in second order) or magnetoelastic (see discussion in section 7.2.1).

A similar approach may be adopted for interface anisotropies, which arise as a consequence of the broken translational symmetry. To lowest order, the interface free energy density σ_{Int} is often written in the form (see e.g. Hillebrands 2000)

$$\sigma_{Int} = -k_s\alpha_x^2 + k_p^{(2)}\alpha_y^2 + k_p^{(4)}\alpha_y^2\alpha_z^2. \tag{9.42}$$

Here k_s is the out-of-plane anisotropy constant, while $k_p^{(2)}$ and $k_p^{(4)}$ describe the first two nonvanishing terms (of second and fourth orders, respectively, in the α_i) of the in-plane interface anisotropy. We are following the convention that upper- and lower-case letters are used in denoting the volume (bulk) and surface anisotropy coefficients, respectively. Clearly a large number of cases can arise, depending on the relative magnitudes and signs of the various anisotropy coefficients.

A simplification for the interface anisotropy is often used if the thickness d of any magnetic layer in the superlattice is small enough. Specifically it needs to be smaller than the static exchange correlation length l_{Ex}. This quantity, which provides a measure of the relative importance of exchange and dipolar effects to the spin dynamics, is defined by

$$l_{Ex} = (D/\gamma\mu_0 M_0)^{1/2}. \tag{9.43}$$

It can be regarded as equal to the wavelength at which the dipolar and exchange contributions in the bulk spin–wave dispersion relation are comparable in magnitude, e.g. see (2.95). It typically has the value of some tens of nm for ferromagnetic metals (for example, it is about 30 nm for Fe). If $d < l_{Ex}$ the assumption is often made that the magnetization is relatively homogeneous across the film, and so the interface torques from interface anisotropies can be converted to effective volume torques acting on the entire film magnetization. This is an approximation, but it enables us to replace the interface energy per unit area by a *volume* energy density

$$E_{Int} = 2\sigma_{Int}/d \tag{9.44}$$

with the factor 2 counting the two interfaces of the film; this term can be added to the E_{Ani} volume energy density discussed earlier.

Apart from the interface anisotropy being incorporated into volume terms in the way mentioned above, it also modifies the boundary condition at the interfaces of each magnetic film with the adjacent media. This can be found from the torque equation of motion by writing down the condition that the sum of the interface torques must be zero for each interface. If this is done first for a single magnetic film surrounded by a nonmagnetic medium, the result is of the form

$$\mathbf{M} \times [\nabla_a \sigma_{Surf} - D \, \partial \mathbf{M}/\partial n] = 0. \tag{9.45}$$

Here σ_{Surf} is the surface anisotropy energy, ∇_a is as defined in (9.38) and $\partial/\partial n$ is the partial derivative in the direction of the unit vector $\mathbf{n}$ which is normal to the surface of the film (and pointing inwards at each surface). The boundary condition was obtained in this form from the macroscopic theory by Rado and Weertman (1959). For simple forms of the anisotropy the result becomes equivalent in the linear spin–wave approximation to the boundary conditions employed in chapters 4 and 5 (which we deduced from microscopic theory). In the case of several magnetic layers, the direct generalization of (9.45) leads to the so-called Hoffman boundary condition (Hoffman 1970). Its form for superlattices, as used by Hillebrands (1990, 2000) and others, is as follows. For each magnetic layer i with total magnetization denoted by $\mathbf{M}_i$, the boundary condition at an interface $x = d_i$, including effective exchange coupling to the nearest interface (at $x = d_j$) of the next magnetic layer j with magnetization $\mathbf{M}_j$, is

$$\mathbf{M}_i \times [\nabla_{a_i} \sigma_{Int,i} - D_i \, \partial \mathbf{M}_i/\partial n_i]|_{x=d_i} - \mathbf{M}_i \times D_{ij}[\mathbf{M}_j + a_j \, \partial \mathbf{M}_j/\partial n_j]|_{x=d_j} = 0. \tag{9.46}$$

The first term is just analogous to (9.45) in terms of parameters relating to magnetic layer i, while the second term explicitly takes account of the interlayer exchange through an effective exchange stiffness D_{ij}. The lattice constant in layer j is a_j.

In the dipole-exchange calculations carried out by Hillebrands (1988, 1990) for finite ferromagnetic/nonmagnetic superlattices, the linearized torque equations were deduced using (9.38) with appropriate choices of the anisotropy. From these, expressions were found for components of the fluctuating field $\mathbf{h}$ in each of the N_{Mag} magnetic layers, each of the N_{Sp} spacer layers and the two external media at each end of the superlattice. By analogy with the situation for a single film and a double magnetic layer in chapter 5 (where we employed a discussion in terms of the scalar potential ψ rather than the related $\mathbf{h}$ field), there turns out to be six terms to the solutions within each magnetic layer, two terms within each spacer layer,

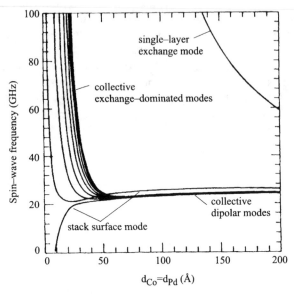

Figure 9.16. Calculated dipole-exchange spin waves in a Co/Pd superlattice with nine periods. The frequencies are plotted versus layer thickness d_{Co} ($= d_{Pd}$) for a fixed in-plane wavevector. The parameter values were deduced from Brillouin scattering data. After Hillebrands *et al* (1990).

and one for each external medium, making a total of

$$N_{Total} = 6N_{Mag} + 2N_{Sp} + 2. \tag{9.47}$$

The solvability condition for the system of N_{Total} linear homogeneous equations can be expressed as the condition for an $N_{Total} \times N_{Total}$ matrix to have its determinant equal to zero. In this way Hillebrands was able to solve numerically for the frequencies of the spin waves in the finite superlattice, provided the system of equations is not too large.

An example of the spin–wave excitations calculated in this way is shown in figure 9.16. The results are for a Co/Pd superlattice with nine periods and bounded by vacuum. Thus $N_{Mag} = N_{Sp} = 9$ and $N_{Total} = 74$. The thicknesses of the magnetic and nonmagnetic layers are taken to be equal ($d_{Co} = d_{Pd}$), and the frequencies are plotted against d_{Co} for a fixed value of in-plane wavevector $q_\parallel$ appropriate to Brillouin light scattering (see Hillebrands *et al* 1990). In this case it is seen that the collective modes are close together and dominated by the dipolar effects for d_{Co} greater than about 5 nm. By contrast, at smaller values of d_{Co} the standing exchange modes split off from the surface mode and shift to higher frequencies. This is broadly analogous to the behaviour in the case of a single film (see section 5.4), but the collective exchange-dominated modes of the superlattice occur at lower frequencies than the single-layer exchange modes, as indicated.

So far we have discussed dipole-exchange spin–wave theories for super-lattices in which the magnetization is parallel to the surfaces and interfaces. There have also been some theories for other orientations. Considering the magnetization to be in the perpendicular direction, Slavin *et al* (1994) and Rojdestvenski *et al* (1996) have developed a comprehensive theory for the mode frequencies and intensities with particular application to Brillouin scattering. Their theory, which applied to superlattices with alternating magnetic and nonmagnetic layers, allows for both dipolar and exchange coupling across the nonmagnetic spacers. It is based on extending the tensorial Green-function perturbation approach mentioned earlier.

In other work, Stamps and Hillebrands (1991) have extended the above-mentioned calculations of Hillebrands (1990) to apply to dipole-exchange modes in finite superlattices with large out-of-plane magnetic anisotropies. Specifically they considered the situation where the out-of-plane anisotropy coefficient k_s in (9.42) for the interface energy σ_{Int} is large and positive. If the thickness d of the magnetic layers is small enough, the magnetization will have a preferred direction perpendicular to the layers. Stamps and Hille-brands studied the situation where there was an applied magnetic field par-allel to the interfaces. Due to competition between the anisotropy and the applied field, the magnetization direction is canted in general, but for the field above a critical value B_C it switches to lie parallel to the interfaces. Some numerical results for spin–wave frequencies versus applied magnetic field are shown in figure 9.17. This is for the case of a superlattice with six periods; the magnetic layers are assumed to be Co with thickness 0.88 nm and the nonmagnetic spacers have thickness 0.76 nm. An abrupt change of behaviour for the frequencies is seen at the critical field. The lowest-frequency branch (both below and above the critical field) is a surface mode. In these calculations it goes 'soft' in a region near the critical field (i.e. its frequency becomes zero). This might otherwise be indicative of a surface phase transition, but in this context the authors regard it as evidence that the magnetization orientation in the Co layers varies across the stack.

9.4.2 Brillouin scattering

It is evident from the previous section that much of the motivation for the theoretical developments has come from light scattering, specifically Brillouin scattering. There is now a wealth of experimental data probing the dipole-exchange spin–wave regime for superlattices, and we mention the two excellent reviews by Grünberg (1989) and Hillebrands (2000). The latter is particularly thorough and comprehensive, so we use it below to summarize the main points.

Although the wavevectors of the ferromagnetic spin waves participating in Brillouin scattering correspond typically to the dipole-exchange region

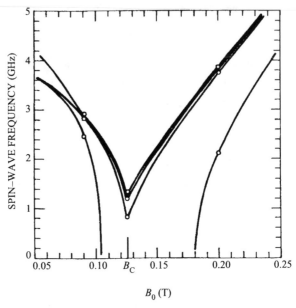

Figure 9.17. Calculated frequencies of the spin waves in a six-period superlattice with large out-of-plane anisotropy versus the in-plane applied magnetic field. Parameters are taken for Co alternating with a nonmagnetic spacer. After Stamps and Hillebrands (1991).

(see figure 2.5) as noted before, it is the case for some materials that approximately the dipolar effects may be dominant. We have given examples of such cases already for ferromagnetic superlattices in section 9.1.1, where studies for Ni/Mo and Fe/Pt with in-plane magnetization were described, and in section 9.1.2, where we referred to systems such as Co/Pd and Co/Pt that have large out-of-plane anisotropies. The more accurate dipole-exchange theory for the latter was described in section 9.4.1. As mentioned, the effective out-of-plane anisotropy becomes more pronounced as the Co layer thickness d_{Co} is reduced, and a magnetization re-orientation may be achieved using an in-plane magnetic field (see figure 9.17 for the dependence of the spin–wave frequencies). In figure 9.18 the Brillouin spectra obtained for Co/Pd superlattices with several different d_{Co} (ranging from 0.8 to 2 nm) and an in-plane applied field of 1 T are shown (Hillebrands *et al* 1993). In the case of $d_{Co} = 0.8$ nm the out-of-plane anisotropy is sufficiently large that the applied field cannot force the magnetization to be in the plane, and a very broad spectrum results. For larger d_{Co} this is no longer true and a well-defined spin–wave band results in accordance with theory. Another system that has been studied because of its out-of-plane anisotropy is Co/Au (Krams *et al* 1991, Hillebrands 2000).

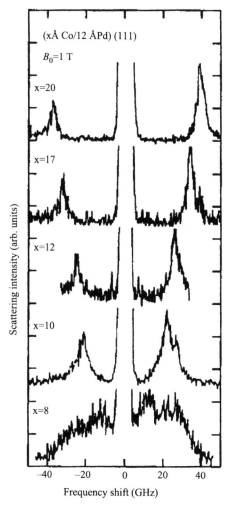

Figure 9.18. Spectra for Brillouin scattering from (111)-oriented Co/Pt superlattices with varying Co layer thicknesses as indicated. After Hillebrands *et al* (1993).

Following the discovery described in section 4.5.1 of oscillatory exchange between the magnetic layers in systems such as Fe/Cr/Fe, there have been Brillouin scattering measurements on finite superlattices for both the ferromagnetic- and antiferromagnetic-exchange cases. In particular, a careful study was made of sputtered Co/Ru superlattices (Fassbender *et al* 1992, 1993). For this system the exchange stiffness parameter D_{ij} in (9.46) oscillates as a function of the Ru layer thickness d_{Ru} with a periodicity of about 1.15 nm. The Brillouin spectra were measured in several samples with a fixed value $d_{Co} = 2.0$ nm for the Co layers and values of d_{Ru} ranging from 0.38 nm to 2.09 nm. Some of the spin–wave modes were observed to shift in frequency depending on whether $D_{ij} > 0$ or $D_{ij} < 0$, in accordance with theoretical estimates made by the authors.

Examples of other magnetic multilayer systems studied by Brillouin scattering are described in the review article by Hillebrands (2000). These include Co/Pt and Co/Au multilayers with spatially-varying anisotropies and Fe/Cu multilayers with growth domains.

9.5   Quasiperiodic structures

The properties of nonmagnetic quasiperiodic multilayers, together with their excitation spectra, were the topic of section 8.5. We now turn to the case of magnetic structures formed from two building blocks A and B (layers of thickness d_A and d_B) as before, where now one block is ferromagnetic or antiferromagnetic, while the other is usually taken to be nonmagnetic. Again we restrict attention here to Fibonacci and Thue–Morse sequences, since these are the most studied; the first few generations are quoted in (8.63) and (8.65) respectively (see also figure 8.25).

9.5.1   Experiment

Although there has been much interest recently in the theory of excitations in magnetic quasiperiodic superlattices, there has at the time of writing been relatively little work on the experimental side. An exception is the heavy rare-earth Gd–Y system, where the Gd layers are responsible for the ferromagnetic properties and there can be long-range coupling across the metallic Y spacer layers. The growth and characterization of various Gd–Y superlattices were reported by Kwo *et al* (1985) and Vettier *et al* (1986); experimental results for Fibonacci structures are described by Majkrzak *et al* (1991). Specifically, in the notation of (8.63) these authors fabricated structures in which the building blocks were chosen as $A = Gd_5Y_{10}$ and $B = Gd_5Y_5$, where the subscripts denote the number of atomic layers of each element grown in the (100) orientation. Thus the magnetic films (Gd) are all of the same thickness and are separated by a nonmagnetic spacer (Y) of thickness equal to either five or ten layers. The long-range exchange across Y is oscillatory, and these spacer thicknesses were chosen so that Y_5 and Y_{10} provide ferromagnetic and antiferromagnetic coupling respectively. A consequence is that the preferred ordering of the magnetizations in the Gd films varies in a complicated way. This is illustrated in figure 9.19 for a simple case of the Fibonacci generation $S_4 = ABAAB$.

Majkrzak *et al* (1991) reported measurements of the overall magnetization versus temperature for high-generation Fibonacci superlattices. They

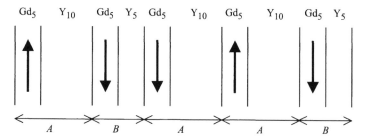

Figure 9.19. Magnetic ordering in a Fibonacci multilayer corresponding to generation $S_4 = ABAAB$ where $A = Gd_5Y_{10}$ and $B = Gd_5Y_5$. It is assumed that the nearest-neighbour interlayer exchange is dominant over any next-nearest neighbour exchange.

found that, due to the lack of periodicity and the presence of both anti-ferromagnetic and ferromagnetic interlayer exchange, the Fibonacci samples were magnetically very hard, particularly at low temperatures (below about 80 K). This is because the antiferromagnetic exchange through a Y_{10} spacer becomes stronger as the temperature is reduced. Neutron diffraction profiles showed main diffraction peaks corresponding to the quasiperiodic Fibonacci chemical structure, as expected on the basis of a structure factor analysed, as described in section 8.5. However, several additional peaks became more intense at low temperatures and provided information about the complex magnetic ordering of these samples, which has yet to be fully resolved.

9.5.2 Theory

A comprehensive account of the theory relating to the spin–wave and other properties of quasiperiodic superlattices is given in the review article by Albuquerque and Cottam (2003). By way of a survey, we cover here three topics: the spin–wave spectra, metallic magnetic multilayers, and spin–wave specific heat.

First, for the spectrum of spin waves in a quasiperiodic superlattice there have been numerous calculations mostly based on extending the transfer-matrix approach described for periodic superlattices in sections 9.1 and 9.3 for the magnetostatic and exchange-dominated regimes respectively. Usually the periodic boundary condition (or PBC) model is employed, just as for nonmagnetic excitations (see section 8.5). This means that each generation S_n of the sequence is considered to be repeated in a periodic fashion, so that the excitation spectrum can then be found from the trace of the transfer matrix, e.g. as in (8.20), and the properties of the spectrum for large n can be examined. In the magnetostatic regime, taking the building blocks A and B to correspond to magnetic and nonmagnetic films respectively, Sy and Feng (1994) and Feng *et al* (1995) considered the case of in-plane magnetization, while Anselmo *et al* (1999, 2000) extended the results to perpendicular magnetization. In the latter case scaling laws were deduced, analogous to those in section 8.5.2 for plasmon–polaritons, and shown to be different for the Fibonacci and Thue–Morse structures. Likewise, for the opposite regime where the exchange effects dominate, Albuquerque and Cottam (2003) take both A and B to be Heisenberg ferromagnets, and they present generalizations for Fibonacci and Thue–Morse superlattices within the PBC approach of the calculation for the periodic case described in section 8.3.2.

The metallic quasiperiodic superlattices, such as the Gd–Y system discussed in the previous section, provide other interesting possibilities because of the long-range exchange across the spacers. This may be variable in sign (for the bilinear exchange part) and may include biquadratic contributions, as in section 4.5. Bezerra *et al* (1999, 2001a) and Mauriz *et al* (2002) made a series of calculations for the magnetoresistance and sample magnetization by using a phenomenological method in which the total magnetic energy per unit area of the multilayer (consisting of m magnetic films separated by spacers) is written as

$$E_{\mathrm{T}} = \sum_{i=1}^{m} [-t_i M_i B_0 \cos(\theta_i - \theta_0) + \tfrac{1}{4} t_i K_a \sin^2(2\theta_i)]$$

$$+ \sum_{i=1}^{m-1} [J_{\mathrm{BL}} \cos(\theta_i - \theta_{i+1}) + J_{\mathrm{BQ}} \cos^2(\theta_i - \theta_{i+1})]. \qquad (9.48)$$

Here the first term is the Zeeman energy due to an external magnetic field B_0 applied in-plane at an orientation θ_0 to the easy axis, taken to be the (100) direction. The second term is a four-fold magneto-crystalline anisotropy energy involving coefficient $K_a > 0$. The third and fourth terms describe effective bilinear and biquadratic exchange energy between adjacent magnetic layers. The thickness, magnitude of magnetization (assumed uniform within each layer) and its orientation in the ith magnetic layer are denoted by t_i, M_i and θ_i respectively. The authors applied their calculations to Fe–Cr structures, for which the parameters are fairly well known from experiments (see section 4.5), and so they chose $A = \mathrm{Fe}$ and $B = \mathrm{Cr}$ for the building blocks. Thus, for example, in the case of the Fibonacci generation $S_5 = ABAABABA$ (as can be deduced from section 8.5) it follows that there are four magnetic (Fe) layers with overall thicknesses corresponding to $t_1 = d_A$, $t_2 = 2d_A$ and $t_3 = t_4 = d_A$. The Cr spacers all have the same thickness d_B in this case (and indeed for any Fibonacci generation).

For any Fibonacci generation and choice of thicknesses d_A and d_B, the equilibrium orientations θ_i of the magnetizations are obtained numerically by minimizing the magnetic energy given by (9.48). The model was applied to calculate normalized values for the magnetoresistance and the sample magnetization in the field direction. Assuming $\theta_0 = 0$ for the field orientation, these quantities are found from

$$\frac{R(B_0)}{R(0)} = \sum_{i=1}^{m-1} \frac{1 - \cos(\theta_i - \theta_{i+1})}{2(m-1)} \qquad (9.49)$$

$$\frac{M(B_0)}{M(0)} = \sum_{i=1}^{m} t_i M_i \cos(\theta_i) \Big/ \sum_{i=1}^{m} t_i M_i. \qquad (9.50)$$

The expression in (9.49) applies when the spin-dependent mechanism (see section 4.5) is responsible for the GMR, as is believed to be the case for Fe–Cr structures. It depends on showing (see Vedyayev *et al* 1994) that the GMR has a term that varies linearly with $\cos(\theta_i - \theta_{i+1})$ if the electrons in the Cr spacers behave as a free-electron gas. The expression in (9.50) simply represents a weighted average of the individual layer magnetizations. Depending on the strength of the biquadratic exchange, Albuquerque and Cottam (2003) present numerical examples using (9.49) and (9.50) to show how self-similar features may appear in the magnetoresistance and magnetization spectra when they are scaled with respect to the Fibonacci generation number n.

As a final example we mention briefly some calculations for the spin–wave contributions to the specific heat of Fibonacci superlattices (Bezerra *et al* 2001b,c). Considering building blocks A and B to be Heisenberg ferromagnets and nonmagnetic metallic spacers, respectively, with both long-range bilinear and biquadratic exchange across the spacers, they predicted oscillations in the specific heat at low temperatures ($T \ll T_C$). These were evident in log–log plots of the specific heat versus temperature and were characteristic of the quasiperiodicity (through the generation number n). Analogous features had been predicted for the polariton specific heat in nonmagnetic quasiperiodic superlattices (Mauriz *et al* 2001), so such oscillations may be a widespread characteristic. However, in the magnetic case there were striking differences between the behaviour for odd and even generation numbers. This is currently a topic of ongoing research.

References

Agranovich V M and Kravtsov V E 1985 *Solid State Commun.* **55** 85
Albuquerque E L and Cottam M G 2003 *Phys. Reports* **376** 225
Albuquerque E L, Fulco P, Sarmento E F and Tilley D R 1986 *Solid State Commun.* **58** 61
Albuquerque E L, Fulco P and Tilley D R 1988 *Phys. Status Solidi* (*b*) **146** 449
Almeida N S and Mills D L 1988 *Phys. Rev. B* **37** 3400
Almeida N S and Tilley D R 1990 *Solid State Commun.* **73** 23
Anselmo D H, Cottam M G and Albuquerque E L 1999 *J. Appl. Phys.* **85** 5774
Anselmo D H, Cottam M G and Albuquerque E L 2000 *J. Phys.: Cond. Mat.* **12** 1041

Barnás J 1987 *Solid State Commun.* **61** 405
Barnás J 1988 *J. Phys. C* **21** 1021
Barnás J 1990 *J. Phys.: Cond. Matt.* **2** 7173
Barnás J 1991 *Phys. Stat. Sol. (b)* **165** 529
Barnás J 1994 in *Linear and Nonlinear Spin Waves in Magnetic Films and Superlattices* ed M G Cottam (Singapore: World Scientific) p 157
Bastard G 1981 *Phys. Rev. B* **24** 5693
Bezerra C G and Cottam M G 2002 *J. Appl. Phys.* **91** 1
Bezerra C G, de Araujo J M, Chesman C and Albuquerque E L 1999 *Phys. Rev. B* **60** 9264
Bezerra C G, de Araujo J M, Chesman C and Albuquerque E L 2001a *J. Appl. Phys.* **89** 2286
Bezerra C G, Albuquerque E L, Mariz A M, da Silva L R and Tsallis C 2001b *Physica A* **294** 415
Bezerra C G, Albuquerque E L and Cottam M G 2001c *Physica A* **301** 372
Bland J A C and Heinrich B (eds) 1994 *Ultrathin Magnetic Structures I* (Berlin: Springer)
Camley R E 1987 *Phys. Rev. B* **35** 3608
Camley R E and Cottam M G 1987 *Phys. Rev. B* **35** 189
Camley R E and Stamps R L 1994 in *Linear and Nonlinear Spin Waves in Magnetic Films and Superlattices* ed M G Cottam (Singapore: World Scientific) p 237
Camley R E and Tilley D R 1988 *Phys. Rev. B* **37** 3413
Camley R E, Rahman T S and Mills D L 1983 *Phys. Rev. B* **27** 261
Chew K-H, Ong L-H, Osman J and Tilley D R 2000 *Appl. Phys. Lett.* **77** 2755
Cottam M G (ed) 1994 *Linear and Nonlinear Spin Waves in Magnetic Films and Super-lattices* (Singapore: World Scientific)
Cottam M G and Lockwood D J 1986 *Light Scattering in Magnetic Solids* (New York: Wiley)
Dobrzynski L, Djafari-Rouhani B and Puszkarski H 1986 *Phys. Rev. B* **33** 3251
Fassbender J, Nörtemann F C, Stamps R L, Camley R E, Hillebrands B, Güntherodt G and Parkin S S P 1992 *Phys. Rev. B* **46** 5810
Fassbender J, Nörtemann F C, Stamps R L, Camley R E, Hillebrands B, Güntherodt G and Parkin S S P 1993 *J. Mag. Mag. Mat.* **121** 270
Feng J W, Jin G J, Hu A, Kang S S, Jiang S S and Feng D 1995 *Phys. Rev. B* **52** 15312
Fishman F, Schwabl F and Schwenk D 1987 *Phys. Lett. A* **121** 192
Grimsditch M 1989 in *Light Scattering in Solids V* ed M Cardona and G Güntherodt (Berlin: Springer) p 285
Grimsditch M, Khan M R, Kueny A and Schuller I K 1983 *Phys. Rev. Lett.* **51** 498
Grünberg P 1989 in *Light Scattering in Solids V* ed M Cardona and G Güntherodt (Berlin: Springer) p 303
Grünberg P and Mika K 1983 *Phys. Rev. B* **27** 2955
Gurevich A G and Melkov G A 1996 *Magnetization Oscillations and Waves* (Boca Raton, FL: CRC Press)
Hillebrands B 1988 *Phys. Rev. B* **37** 9885
Hillebrands B 1990 *Phys. Rev. B* **41** 530
Hillebrands B 2000 in *Light Scattering in Solids VII* ed M Cardona and G Güntherodt (Berlin: Springer) p 174
Hillebrands B, Boufelfel A, Falco C M, Baumgart P, Güntherodt G, Zirngiebl E and Thompson J D 1988 *J. Appl. Phys.* **63** 3880

Hillebrands B, Harzer J V, Güntherodt G, England C D and Falco C M 1990 *Phys. Rev. B* **42** 6839

Hillebrands B, Harzer J V, Güntherodt G and Dutcher J R 1993 *J. Magn. Soc. Jpn.* **17** 17

Hinchey L L and Mills D L 1986a *Phys. Rev. B* **33** 3329

Hinchey L L and Mills D L 1986b *Phys. Rev. B* **34** 1689

Hoffmann F 1970 *Phys. Stat. Sol.* **41** 807

How H and Vittoria C 1989a *Phys. Rev. B* **39** 6823

How H and Vittoria C 1989b *Phys. Rev. B* **39** 6831

Krams P, Hillebrands B, Güntherodt G, Spörl K and Weller D 1991 *J. Appl. Phys.* **69** 5307

Kueny A, Khan M R, Schuller I K and Grimsditch M 1984 *Phys. Rev. B* **29** 2879

Kwo J, Gyorgy E M, McWhan D B, Hong M, DiSalvo F J, Vettier C and Bower J E 1985 *Phys. Rev. Lett.* **55** 1402

Landau L D and Lifshitz E M 1969 *Statistical Physics* (Oxford: Pergamon)

Lepage J G and Camley R E 1989 *Phys. Rev. B* **40** 9113

Lepage J G and Camley R E 1990 *Phys. Rev. Lett.* **65** 1152

Majkrzak C F, Kwo J, Hong M, Yafet Y, Gibbs D, Chien C L and Bohr J 1991 *Adv. Phys.* **40** 99

Mauriz P W, Albuquerque E L and Vasconcelos M S 2001 *Phys. Rev. B* **63** 184203

Mauriz P W, Albuquerque E L and Bezerra C G 2002 *J. Phys.: Cond. Mat.* **14** 1785

Mills D L 1989 in *Light Scattering in Solids V* ed M Cardona and G Güntherodt (Berlin: Springer) p 13

Ong L-H, Osman J and Tilley D R 2002 *Phys. Rev. B* **65** 134108

Rado G T and Weertman J R 1959 *J. Phys. Chem. Solids* **11** 315

Raj N and Tilley D R 1987 *Phys. Rev. B* **36** 7003

Raj N and Tilley D R 1989 in *The Dielectric Function of Condensed Systems* ed L V Keldysh, D A Kirzhnitz and A A Maradudin (Amsterdam: Elsevier)

Rojdestvenski I V, Cottam M G and Slavin A N 1996 *J. Appl. Phys.* **79** 5724

Schreyer A, Schmitte T, Siebrecht R, Bödecker P, Zabel H, Lee S H, Erwin R W, Majkrzak C F, Kwo J and Hong M 2000 *J. Appl. Phys.* **87** 5443

Slavin A N, Rojdestvenski I V and Cottam M G 1994 *J. Appl. Phys.* **76** 6549

Slonczewski J C 1995 *J. Magn. Magn. Mat.* **150** 13

Stamps R L and Hillebrands B 1991 *Phys. Rev. B* **44** 5095

Sy H K and Feng C 1994 *Phys. Rev. B* **50** 3411

Tilley D R 1988 *Solid St. Commun.* **65** 657

Tilley D R 1994 in *Linear and Nonlinear Spin Waves in Magnetic Films and Superlattices* ed M G Cottam (Singapore: World Scientific) p 207

Tolédano J-C and Tolédano P 1987 *The Landau Theory of Phase Transitions* (Singapore: World Scientific)

Vayhinger K and Kronmüller H 1986 *J. Mag. Mag. Mater.* **62** 159

Vayhinger K and Kronmüller H 1988 *J. Mag. Mag. Mater.* **72** 307

Vedyayev A, Dieny B, Ryzhanova N, Genin J B and Cowache C 1994 *Europhys. Lett.* **25** 465

Vettier C, McWhan D B, Gyorgy E M, Kwo J, Buntschuh B M and Batterman B W 1986 *Phys. Rev. Lett.* **56** 757

Zhu N and Cao S 1987 *Phys. Lett. A* **124** 515

Zhu N and Cao S 1989 *J. Phys.: Cond. Matt.* **1** 7977

PART FOUR

NONLINEAR EFFECTS

Chapter 10

Nonlinear optics and nonlinear magnetics of bulk media

The remainder of this book is concerned with some of the nonlinear properties associated with the surfaces and interfaces in films and superlattices. The properties of various linear excitation modes that were described in previous chapters underlie much of the nonlinear behaviour. We shall find in some cases that the nonlinearity, which exists to some extent in all physical systems, modifies the properties of the linear excitations. In other cases it will be shown that new excitations (i.e. those with no counterpart in the linear theory) can emerge. Before covering the nonlinear aspects of films and superlattices, the background material that is required is the formalism for the nonlinear behaviour of bulk samples. This forms the subject matter of the present chapter, while we proceed to the surface-related behaviour of nonmagnetic and magnetic systems in chapters 11 and 12, respectively.

10.1 Nonlinear optics: basic formalism

The development of a theoretical framework for nonlinear optics (NLO) dates from the invention of the laser in the early 1960s, and there are now many excellent textbooks on the subject, for example, by Shen (1984), Yariv (1989), Butcher and Cotter (1990) and Mills (1998). Here we review the relevant parts of the subject and establish the ideas and formalism that will be needed for the subsequent discussion of films and superlattices.

The starting point for *linear* optics is the relationship between the field vectors **P** and **E**, as given in (2.107). This expression introduces the tensor $\gamma_{ij}(\mathbf{r} - \mathbf{r}', t - t')$ and it is convenient to use also the Fourier-transformed version of (2.107):

$$P_i(\mathbf{q}, \omega) = \varepsilon_0 \gamma_{ij}(\mathbf{q}, \omega) E_j(\mathbf{q}, \omega) \tag{10.1}$$

in which $\gamma_{ij}(\mathbf{q}, \omega)$ is the *linear susceptibility tensor*. Equation (10.1) allows for spatial dispersion, that is the **q** dependence, in addition to the frequency dependence of $\gamma_{ij}(\mathbf{q}, \omega)$. As mentioned in section 2.4.1, in most optics spatial dispersion is unimportant and the $\mathbf{q} = 0$ form of (10.1) is adequate.

The standard approach to *nonlinear* optics is to regard (10.1) as the first term in a Taylor expansion, the subsequent terms of which lead to the various nonlinear effects. In making this expansion it is necessary to use the real-variable forms of all the field quantities since, for example, the product of the real parts of two complex numbers is not equal to the real part of the product. In a typical experiment, one or more intense laser beams pass through a nonlinear-optic crystal, and in addition a d.c. field may be applied. Since the laser beams are usually monochromatic, to a good approximation, the electric field is written as

$$\mathbf{E}(t) = \mathbf{E}_0 + \tfrac{1}{2} \sum_{\omega > 0} [\mathbf{E}_\omega \exp(-i\omega t) + \mathbf{E}_{-\omega} \exp(i\omega t)] \tag{10.2}$$

where the condition for $\mathbf{E}(t)$ to be real is $\mathbf{E}_{-\omega} = \mathbf{E}_\omega^*$, $\mathbf{E}_0$ is the d.c. field and the sum is over the frequencies of the laser beams and of any output beams that are produced. The induced polarization is written as a sum of terms proportional to $\mathbf{E}, \mathbf{E}^2, \mathbf{E}^3$ and so on. In the term of nth order a variety of frequencies appears since for each factor $\mathbf{E}$ in the product $\mathbf{E}^n$ any one of the d.c. term $\mathbf{E}_0$, the positive-frequency amplitude $\mathbf{E}_\omega$ and the negative-frequency amplitude $\mathbf{E}_{-\omega}$ can be selected. The full expression for the Taylor expansion is

$$\mathbf{P}(t) = \mathbf{P}_0 + \tfrac{1}{2} \sum_{n, \omega_\sigma} (\mathbf{P}_{\omega_\sigma}^{(n)}) \exp(-i\omega_\sigma t) \tag{10.3}$$

where

$$(\mathbf{P}_{\omega_\sigma}^{(n)})_\mu = \varepsilon_0 K(-\omega_\sigma; \omega_1, \dots, \omega_n) \chi_{\mu\alpha_1\alpha_2\dots\alpha_n}^{(n)}(-\omega_\sigma; \omega_1, \dots, \omega_n)$$
$$\times (\mathbf{E}_{\omega_1})_{\alpha_1} (\mathbf{E}_{\omega_2})_{\alpha_2} \cdots (\mathbf{E}_{\omega_n})_{\alpha_n} \tag{10.4}$$

and ω_σ is defined by

$$\omega_\sigma = \omega_1 + \omega_2 + \cdots + \omega_n. \tag{10.5}$$

Here each of $\omega_1, \omega_2, \dots, \omega_n$ can be positive or negative (or zero) depending on which term in (10.2) is selected, so that ω_σ is the frequency of the output beam. In (10.4) and subsequent equations the summation convention for tensor suffices applies.

The notation in which $-\omega_\sigma$ appears as the first factor in the coefficients in (10.4) is conventional, following Butcher and Cotter (1990), and the sum in (10.3) is over all possible combinations of $\omega_1, \omega_2, \dots, \omega_n$. A given product of the $(\mathbf{E}_{\omega_i})_{\alpha_i}$ field components in (10.4) arises when the fields are selected in any order from the n factors in the form of (10.2). K is a combinatorial factor that gives the number of ways in which this can be done and also allows for a power of 2 arising from the factors $\tfrac{1}{2}$ in (10.2) and (10.3). The explicit form, as given by Butcher and Cotter (1990), is

$$K(-\omega_\sigma; \omega_1, \omega_2, \ldots, \omega_n) = 2^{l+m-n}p \tag{10.6}$$

in which

$$p = n! / (n_1! n_2! \ldots). \tag{10.7}$$

Here n_1 is the number of repetitions of ω_1 in (10.4) and so on. The power of 2 in (10.6) appears because there is no factor of $\frac{1}{2}$ in the d.c. terms in (10.2) and (10.3), and it is given by the rule that m is the number of d.c. fields ($\omega = 0$ terms) appearing in (10.4) and $l = 1$ if $\omega_\sigma \neq 0$ but $l = 0$ if $\omega_\sigma = 0$. The factors K, which are easy to work out, are tabulated for the various nonlinear processes (Butcher and Cotter 1990) and we quote values as needed.

The factor $\chi^{(n)}_{\mu\alpha_1\alpha_2\ldots\alpha_n}$ is a component of an nth order *nonlinear suscept- ibility tensor*, which is seen to be a tensor of rank $n + 1$. Because of the factor of ε_0 that is inserted in (10.4) and because K is dimensionless, it is seen that tensor $\chi^{(n)}$ has dimension $m^{n-1}V^{-(n-1)}$, i.e. the reciprocal of the $n-1$ power of a field. It also follows from the definition in (10.4) that $\chi^{(n)}$ is invariant under all permutations of $(\alpha_1\omega_1), (\alpha_2\omega_2), \ldots, (\alpha_n\omega_n)$. This is called *intrinsic permutation symmetry*.

As is the case for all tensors, the number of non-zero components of $\chi^{(n)}$ is determined by the crystal point-group symmetry, which may also give linear relations between some of the non-vanishing components. A number of books give partial tabulations, and the most comprehensive account is given by Popov *et al* (1995). The most important result is that all $\chi^{(2)}$ tensors vanish in crystals with a centre of inversion symmetry. However, it is very important to recognize that this result applies only for *bulk* crystals. The presence of a surface or interface breaks the inversion symmetry so that *surface-specific* $\chi^{(2)}$ effects arise, and these can be used as probes of surface and interface properties. A further brief discussion of this will be given in chapter 11.

Equations (10.2) to (10.5) contain the general definitions, but in practice nearly all work in nonlinear optics has been concerned with $\chi^{(2)}$ and $\chi^{(3)}$ effects and likewise these will form the subject matter of our account. However, before discussing these specifically there are some general matters to clarify.

Comparison of (10.4) with (10.1) shows that (10.4) is a special case, in that it neglects spatial dispersion; i.e. there is no $\mathbf{q}$ dependence in (10.4). As was seen in earlier chapters, more often than not spatial dispersion can be neglected in linear optics but this is certainly not an invariable rule. The same holds for nonlinear optics and one particularly important case where spatial dispersion must be included is second harmonic generation (SHG) and other effects at the surface of a crystal. This will be discussed in section 11.1. The general formalism for the inclusion of spatial dispersion is an extension of what we have outlined; it is given in the specialist texts (Butcher and Cotter 1990, Popov *et al* 1995) so it is not reviewed here.

Equations (10.2)–(10.5) give the macroscopic definition of the NLO tensors, which is the analogue of parts of the discussion of the dielectric function in section 2.4.1. We also dealt there with the analytic properties of the dielectric function in the complex frequency plane and the present account of the macroscopic properties of the nonlinear-optic tensors can be continued in a similar way (Butcher and Cotter 1990).

Further development of the subject requires first-principles calculation of the NLO tensors from microscopic theory. Most applications of NLO are concerned with signal processing for carrier frequencies in the visible or near infrared. Like the linear dielectric function, the NLO tensors in this spectral region are governed by the electronic response and the $\chi^{(n)}$ can be calculated using quantum-mechanical perturbation theory; here again, the details are given elsewhere (Shen 1984 and Butcher and Cotter 1990). In this approach, E enters as the amplitude of the perturbation so that the result appears naturally in the form of the power series of (10.3) and (10.4).

In addition to the quantum-mechanical method, it is worth considering a classical, or macroscopic, derivation based on the oscillator model used in section 2.4.4 to calculate the dielectric function of ionic crystals. This is easier than the quantum-mechanical derivation and in addition it brings out clearly the origin of NLO effects in nonlinearities in the dynamics of the system. Equation (2.143) gives the relative displacement **u** of the positive- and negative-ion sublattices that is produced by a long-wavelength driving field **E**. The underlying assumption is that the restoring force is linear in **u**, i.e. that it results from a harmonic potential between the sublattices. This should hold for a small magnitude of **E**, but for larger values it may be expected that the magnitude of **u** is large enough that the potential becomes anharmonic. Equation (2.143) is then generalized to

$$(-\omega^2 - i\omega\Gamma)u_i = -\omega_T^2 u_i + \alpha_{ijk}u_j u_k + \beta_{ijkl}u_j u_k u_l + \gamma E_i \qquad (10.8)$$

where terms higher than cubic in **u** are omitted. As a specific example, Osman *et al* (1998) give the values of the nonlinear coefficient tensors $\vec{\alpha}$ and $\vec{\beta}$ for a ferroelectric material. Now the point about (10.8) is that it is linear in **E** but nonlinear in **u**, because the underlying mechanism is the nonlinearity of the interionic potential. However, the definitions in (10.2) to (10.4) have the opposite property: they are linear in **u** (i.e. **P**) but nonlinear in **E**. In order to derive expressions for the NLO tensors, then, it is necessary to invert (10.8) and bring it into the form of (10.4). Examples of this inversion are given by Mills (1998), Osman *et al* (1998) and Murgan *et al* (2002). However, what is involved here is the inversion of an equation giving **E** as a cubic function of **P** to produce an equation where **P** is a cubic function of **E**. A few simple sketches will show that this inversion is not possible unless the nonlinear terms are relatively small so that an assumption of *weak nonlinearity* is implicit in the conventional definition.

An equation of the form of (10.8) is discussed by Popov *et al* (1995) and they give expressions for the $\bar{\alpha}$ and $\bar{\beta}$ tensors for all the crystal point groups. An early discussion was given by Goldstone and Garmire (1984), who stressed that some essential physics may be lost in the passage from (10.8) to the conventional form.

10.2   Second-order nonlinear effects

We review here the standard account of the $\chi^{(2)}$ effects arising from the $n = 2$ term in (10.4), which for convenience we write explicitly as

$$(\mathbf{P}^{(2)}_{\omega_1+\omega_2})_\mu = \varepsilon_0 K(-(\omega_1 + \omega_2); \omega_1, \omega_2)\chi^{(2)}_{\mu\alpha_1\alpha_2}(-(\omega_1 + \omega_2); \omega_1, \omega_2)$$

$$\times (\mathbf{E}_{\omega_1})_{\alpha_1}(\mathbf{E}_{\omega_2})_{\alpha_2}. \tag{10.9}$$

Various physical effects can arise depending on the combination of the frequency factors. For two incident optical beams of different frequencies the possible effects are *sum-frequency generation*, corresponding to the product $(\mathbf{E}_{\omega_1})_{\alpha_1}(\mathbf{E}_{\omega_2})_{\alpha_2}$ with $\omega_\sigma = \omega_1 + \omega_2$ and *difference-frequency generation*, corresponding to the product $(\mathbf{E}_{\omega_1})_{\alpha_1}(\mathbf{E}_{-\omega_2})_{\alpha_2}$ and $\omega_\sigma = \omega_1 - \omega_2$. For a single input optical beam of frequency ω, the corresponding effects are *second-harmonic generation*, $\omega_\sigma = \omega + \omega = 2\omega$ and *optical rectification*, $\omega_\sigma = \omega - \omega = 0$. The second of these describes the appearance of a static polarization $\mathbf{P}^{(2)}_0$. When a d.c. field is included among the inputs, two further effects occur. The first is the *Pockels effect*, corresponding to $\chi^{(2)}_{ijk}(-\omega; 0, \omega)$. This leads to a polarization component at frequency ω which is additional to the polarization resulting from the linear susceptibility, so the Pockels effect appears as a modification of the linear dielectric response. The second is the *quadratic static response*, which is a static polarization $\mathbf{P}^{(2)}_0$ that is quadratic in the static field amplitude E_0 resulting from the susceptibility $\chi^{(2)}_{ijk}(0; 0, 0)$. The combinatorial factors K for these effects may easily be worked out as an example.

10.2.1   Sum-frequency generation

It was stated that the formulation of nonlinear optics by means of a Taylor expansion for the polarization involves an assumption of weak nonlinearity and that this assumption is usually well satisfied. One consequence is that nonlinear effects build up only over distances that are long compared with the optical wavelength. For example, in sum-frequency generation, the input beams must propagate some way through a nonlinear-optic crystal before a significant amplitude is detected in the output beam. This makes it possible to apply a method known as the *slowly-varying-amplitude* (SVA) *approximation*; as the name implies, it is assumed that the amplitude of any of the beams (input or output) varies only over lengths that are long on the scale of the wavelength. Here we give the formalism explicitly for sum-frequency generation; the generalization to difference-frequency generation or second-harmonic generation is obvious.

The wave equation in a nonlinear medium is found from Maxwell's equations in the usual way and is

$$\nabla^2 \mathbf{E} - \frac{1}{c^2}\frac{\partial^2 \mathbf{E}}{\partial t^2} = \mu_0 \frac{\partial^2 \mathbf{P}}{\partial t^2}. \tag{10.10}$$

Here $\mathbf{P}$ is the sum of the linear and nonlinear parts. It is convenient to represent the former by means of the linear dielectric function, $\varepsilon_0 \mathbf{E} + \mathbf{P} = \varepsilon_0 \bar{\varepsilon}(\omega)\mathbf{E}$, so that (10.10) takes the form

$$\nabla^2 \mathbf{E} - \frac{1}{c^2}\bar{\varepsilon}(\omega)\frac{\partial^2 \mathbf{E}}{\partial t^2} = \mu_0 \frac{\partial^2 \mathbf{P}^{\mathrm{NL}}}{\partial t^2} \tag{10.11}$$

where we are emphasizing that the dielectric function is a frequency-dependent tensor and the tensor product is implied in the second term of (10.11). For sum-frequency generation, it is necessary to deal with (10.11) for each of the frequencies ω_1, ω_2 and $\omega_1 + \omega_2$. The third of these is

$$\nabla^2 \mathbf{E}_{\omega_1+\omega_2} + \frac{1}{c^2}(\omega_1+\omega_2)^2\bar{\varepsilon}(\omega)\mathbf{E}_{\omega_1+\omega_2} = -\frac{1}{c^2}(\omega_1+\omega_2)^2\bar{\chi}^{(2)}\mathbf{E}_{\omega_1}\mathbf{E}_{\omega_2}. \tag{10.12}$$

The factor K does not appear explicitly since it is equal to unity in this case. Equation (10.12) has the form of an inhomogeneous wave equation in which the output wave is driven by the product of the input wave amplitudes appearing on the right-hand side. Because of the two tensor products in (10.12), in general rather complicated polarization selection rules apply to this and to other NLO effects. In order to concentrate on essentials, we assume the experimental geometry is such that all three waves propagate in the same direction with the $\mathbf{E}$ field transverse. An example is a tetragonal crystal (except crystal class 422) for which χ_{zyy} is non-vanishing (Yariv 1989), in which case the input waves could be y-polarized and the output wave z-polarized with all three waves propagating in the x direction.

We now assume that the wave amplitudes depend only on x and t, i.e. that the wave fronts are uniform in the transverse direction so that we can simplify ∇^2 to $\mathrm{d}^2/\mathrm{d}x^2$. In practice, of course, the input beams have a transverse profile in which typically the intensity falls off with distance from the beam centre. It will be seen in section 10.3 that when $\chi^{(3)}$ effects are strong the transverse profile can vary with distance x, but for the present we ignore this possible complication. The assumption we make is that the three wave amplitudes take the form

$$E_i(x, t) = E_i(x) \exp(ik_i x - i\omega_i t) \qquad (10.13)$$

where now the suffix i denotes the wave (1, 2 or 3 with frequency $\omega_3 = \omega_1 + \omega_2$) and the scalar amplitude is for the polarization determined by the properties of $\bar{\chi}^{(2)}$. Note that in (10.13) it is necessary to allow for the wave amplitude $E_i(x)$ to depend on distance x. This dependence corresponds to the transfer of intensity between the three beams, which is the physical effect that we are aiming to describe. The second derivative, which is required for the wave equation (10.12), is easily found from (10.13):

$$\frac{d^2 E_i}{dx^2} = \left[\frac{d^2 E_i}{dx^2} + 2ik_i \frac{dE_i}{dx} - k_i^2 E_i \right] \exp(ik_i x - i\omega_i t). \qquad (10.14)$$

10.2.2 Slowly-varying amplitude approximation and phase matching

We now make the SVA approximation. The exponential factor in (10.14) varies on the length scale of the optical wavelength $\lambda_i = 2\pi/k_i$, whereas for weak nonlinearity the amplitude $E_i(x)$ varies only on a much longer scale of distance. This means that

$$\left| \frac{d^2 E_i}{dx^2} \right| \ll \left| 2ik_i \frac{dE_i}{dx} \right| \qquad (10.15)$$

and the SVA approximation consists of neglecting the second derivative. The linear wave equation takes the form

$$k_i^2 = \varepsilon(\omega_i)\omega_i^2/c^2 \qquad (10.16)$$

where $\varepsilon(\omega_i)$ is the component of $\bar{\varepsilon}(\omega_i)$ for the polarization of wave i. Thus when (10.15) is substituted in an equation of the form of (10.12) the corresponding terms cancel. The resulting equation for the sum-frequency wave is

$$2ik_3 \frac{dE_3}{dx} \exp(ik_3 x - i\omega_3 t) = -\frac{\omega_3^2}{c^2} \chi_3 E_1 E_2 \exp[i(k_1 + k_2)x - i(\omega_1 + \omega_2)t] \qquad (10.17)$$

where χ_3 is the relevant component of the sum-frequency tensor $\bar{\chi}^{(2)}$. In (10.17) the t-dependent factors cancel since $\omega_3 = \omega_1 + \omega_2$. However, this is not true in general for the x-dependent factors since that would require $k_3 = k_1 + k_2$ or $[\varepsilon(\omega_3)]^{1/2}\omega_3 = [\varepsilon(\omega_1)]^{1/2}\omega_1 + [\varepsilon(\omega_2)]^{1/2}\omega_2$. Because of dispersion, i.e. the frequency dependence of $\varepsilon(\omega)$, this can be satisfied only in special circumstances and equation (10.17) must therefore be written as

$$2ik_3 \frac{dE_3}{dx} = -\frac{\omega_3^2}{c^2} \chi_3 E_1 E_2 \exp(i\Delta k x) \qquad (10.18)$$

where

$$\Delta k = k_1 + k_2 - k_3 \tag{10.19}$$

is a measure of the *phase mismatch* between the beams. Wave equations similar to (10.12) govern the propagation of the beams at frequencies ω_1 and ω_2 and, when the SVA approximation is applied to these, similar equations to (10.18) are derived to show that dE_1/dx and dE_2/dx are proportional to $E_2^* E_3$ and $E_1^* E_3$ respectively.

In general, therefore, the SVA approximation results in three coupled nonlinear equations for the x dependence of the three beams and the solution of these can be quite complicated. However, if the input beams ω_1 and ω_2 are intense and the nonlinearity is weak, then it is a good approximation to neglect the decrease with x of E_1 and E_2 caused by transfer of energy to the ω_3 beam. That is, *depletion* of the input beams can be neglected. In this case, (10.18) is a simple linear equation since E_1 and E_2 are constants. If the beams enter the nonlinear crystal at $x = 0$ and $E_3 = 0$ at that point, then the solution of (10.18) shows that at distance X into the crystal

$$E_3(X) = \frac{\omega_3}{2\varepsilon_3^{1/2}c} \chi_3 E_1 E_2 \frac{\exp(i\Delta k X) - 1}{\Delta k}. \tag{10.20}$$

The intensity of the sum-frequency wave is proportional to the product $E_3^* E_3$ for which (10.20) gives

$$E_3^* E_3 = \frac{\omega_3^2}{\varepsilon_3 c^2} |\chi_3|^2 |E_1|^2 |E_2|^2 \frac{\sin^2(\Delta k X/2)}{(\Delta k)^2}. \tag{10.21}$$

The prefactor here shows that the output intensity is proportional to the product of $|\chi_3|^2$ and the intensities of the input beams, as might be expected. The final term shows that the effect of phase mismatch is to limit the maximum power that can be transferred to the sum-frequency beam. This term varies between minima of zero at points $X = 2n\pi/\Delta k$ and maxima of $1/(\Delta k)^2$ at points $X = (2n + 1)\pi/\Delta k$. The distance $X_{int} = \pi/\Delta k$ between a minimum and the next maximum is called the *interaction distance*. It is worth noting that for a large sum-frequency amplitude Δk must be small, i.e. the phase mismatch should be small. This means in turn that the interaction distance should be large. However, the maximum distance that a wave can travel is limited by absorption (damping) so small damping of the linear waves is also required. Similar effects of phase mismatch are found in all NLO effects and for practical applications it is important to eliminate, or at least reduce, phase mismatch by choice of propagation directions and beam polarizations. This important question is discussed in detail in the specialized textbooks but need not be taken further here.

10.3   Third-order nonlinear effects

Following most treatments, we now discuss third-order effects; but we shall not go further into higher-order effects in the expansion (10.4). There are two reasons for paying special attention to third-order NLO effects. First, it will be recalled that second-order effects vanish in centrosymmetric materials, so that third-order effects are the leading terms in such materials. Second, it will be seen that the third-order effect known as self action or the nonlinear Kerr effect involves only one frequency, which has important consequences.

The general definition of third-order nonlinear effects follows from (10.4), and the explicit form is like (10.9) but with three input frequencies ω_1, ω_2 and ω_3. Since output frequencies $\pm\omega_1 \pm \omega_2 \pm \omega_3$ can all be generated a wide variety of possible effects might be discussed. It is sufficient for our purposes to consider only the effects that occur with a single input beam, of frequency ω say. These are *third-harmonic generation* (THG), arising from the coefficient $\chi^{(3)}_{\mu\alpha_1\alpha_2\alpha_3}(-3\omega;\omega,\omega,\omega)$ and *self action* or the *Kerr effect*, arising from $\chi^{(3)}_{\mu\alpha_1\alpha_2\alpha_3}(-\omega;\omega,\omega,-\omega)$. The first is similar to second-harmonic generation and an analysis by SVA shows that phase mismatch between the input (ω) and output (3ω) beams limits the intensity transfer in third-harmonic generation. Self action is of particular importance since it is the lowest-order NLO effect in which both input and output beams are at the same frequency. As will now be seen, self action can be interpreted in terms of an intensity-dependent contribution to the dielectric function, or equivalently to the refractive index.

It may be checked that the combinatorial factor for self action is $K = \frac{3}{4}$, so the general expression for the nonlinear polarization, (10.4), is

$$P_i^{(3)} = \varepsilon_0 K \chi^{(3)}_{ijkl} E_j E_k E_l^*. \tag{10.22}$$

This is added to the linear polarization $P_i^{(1)} = \varepsilon_0 \varepsilon_{ij} E_j$ so that the inclusion of $P_i^{(3)}$ is equivalent to defining an effective *intensity-dependent dielectric function*

$$\varepsilon_{ij}^{\text{eff}} = \varepsilon_{ij} + K \chi^{(3)}_{ijkl} E_k E_l^*. \tag{10.23}$$

It is useful to record the special case of this for a linearly polarized wave in an isotropic medium (e.g. glass). If the optical **E** field lies along x then the only non-vanishing tensor component is $\chi^{(3)} \equiv \chi^{(3)}_{xxxx}$ and (10.23) reduces to an intensity-dependent modification of the isotropic (scalar) dielectric function ε of the form

$$\varepsilon^{\text{eff}} = \varepsilon + \tfrac{3}{4}\chi^{(3)}|\mathbf{E}|^2 = \varepsilon + \eta|\mathbf{E}|^2 \tag{10.24}$$

where we introduce $\eta = \frac{3}{4}\chi^{(3)}$ to simplify the notation for later use. Equation (10.24) is sometimes written in terms of an *intensity-dependent refractive index.* If we write $n_0 = \varepsilon^{1/2}$ for the linear refractive index and define $n^{\mathrm{eff}} = (\varepsilon^{\mathrm{eff}})^{1/2}$, then we find from (10.24) by expansion up to order $|\mathbf{E}|^2$:

$$n^{\mathrm{eff}} = n_0 + n_2|\mathbf{E}|^2 \tag{10.25}$$

with

$$n_2 = 3\chi^{(3)}/8n_0 = \eta/2n_0. \tag{10.26}$$

Equations (10.24) and (10.25) are the simplest forms describing the nonlinear Kerr effect, and the phrases positive or negative Kerr medium are often used for positive or negative η. However, the increase of $\varepsilon^{\mathrm{eff}}$ cannot continue to indefinitely large values of $|\mathbf{E}|^2$. In a quantum-mechanical derivation, for example, the nonlinear term could arise from a particular atomic transition in second-order perturbation theory. At sufficiently large values of $|\mathbf{E}|^2$ the populations of the two relevant levels equalize. This is called *saturation* and is fully discussed in the texts on nonlinear optics, e.g. Butcher and Cotter (1990). In the presence of saturation $\varepsilon^{\mathrm{eff}}$ approaches a finite limit as $|\mathbf{E}|^2 \to \infty$. Two empirical forms that are often used (see e.g. Kumar *et al* 1996) are

$$\varepsilon^{\mathrm{eff}} = \varepsilon + \eta|\mathbf{E}|^2/(1 + |\mathbf{E}|^2/I_{\mathrm{S}}) \tag{10.27}$$

and

$$\varepsilon^{\mathrm{eff}} = \varepsilon + \eta I_{\mathrm{S}}[1 - \exp(-|\mathbf{E}|^2/I_{\mathrm{S}})]. \tag{10.28}$$

Both of these reduce to (10.24) for $|\mathbf{E}|^2 \ll I_{\mathrm{S}}$ whereas they have the finite limit $\varepsilon^{\mathrm{eff}} = \varepsilon + \eta I_{\mathrm{S}}$ as $|\mathbf{E}|^2 \gg I_{\mathrm{S}}$, so that I_{S} is the *saturation intensity*.

There are a number of important physical consequences of (10.24) or (10.25). The first is *self-focusing* (or *self-defocusing*) of a laser beam. A beam usually has a transverse profile in which the intensity I is a maximum at the beam centre; this profile is often approximated as a *Gaussian beam* in which I varies with radial distance r as $I = I_0 \exp(-r^2/r_0^2)$, where r_0 is a measure of the beam width. If η is positive, (10.24) shows that $\varepsilon^{\mathrm{eff}}$ is a maximum at the beam centre $r = 0$ and decreases with increasing r. This variation is the same as in the graded-index fibre, discussed briefly in section 7.4.2, and the consequence is a tendency for the beam width r_0 to decrease with propagation distance x through the medium. This effect is opposed by linear diffraction, as a result of which the beam tends to spread. If the nonlinear effect is dominant, the optical intensity becomes self-focused into a narrower beam. On the other hand, for negative η the beam is self-defocused. A full treatment of self-focusing is given in the texts previously cited, e.g. Shen (1984) or Yariv (1989). The terms self-focusing and self-defocusing medium appear in the literature as synonyms for positive and negative Kerr medium.

Two other consequences of (10.23) or (10.24) are of more importance for the discussion of superlattices in chapter 11. The first is that the Maxwell wave equation is now nonlinear and may lead to propagation of solitons, which are intense single pulses. The second is that, in the right circumstances, transmission of light through a Fabry–Pérot etalon becomes bistable or multistable. An introduction to the first of these effects will be given in section 10.4, with superlattice solitons to follow in section 11.6. Fabry–Pérot bistability is the topic of section 11.4.1.

10.4 Solitary waves and solitons

We study one important implication of the intensity dependence of the dielectric function expressed in (10.24). For a single-frequency wave propagating in the z direction, application of this form in Maxwell's equations leads to the nonlinear wave equation

$$-\frac{\mathrm{d}^2 E}{\mathrm{d}z^2} = k_2^2 E + k_0^2 \eta |E|^2 E \tag{10.29}$$

where

$$k_2^2 = \varepsilon(\omega)\omega^2/c^2 \tag{10.30}$$

in which $\varepsilon(\omega)$ is the linear dielectric function.

Like various other nonlinear wave equations, (10.29) may have solutions in the form of *solitary waves*, some of which may be *solitons*. Since some spread of frequencies is necessary to define a localized wave, the governing equation for these is derived from the time-dependent form of (10.29) by a form of the SVA approximation and is known as the *nonlinear Schrödinger equation* (NLSE). Before giving the details of the derivation we briefly summarize some of the ideas in soliton physics that will be needed in chapters 11 and 12. The subject has been studied intensively since the 1960s and for a full account the reader is referred elsewhere, for example Drazin and Johnson (1989), Rowlands (1990) or Infeld and Rowlands (1990).

The first observation of a soliton was reported by John Scott Russell in 1834 in a famous and evocative piece of scientific writing, quoted in Drazin and Johnson (1989). Scott Russell observed that when a canal barge stopped suddenly a 'hump' of water detached itself from the bow and travelled forward without change of shape; he named this hump '*the great solitary wave*' and was able to follow it on horseback along the tow path for a distance of some miles. Later study of the hydrodynamics of surface waves on water showed that to a good approximation the motion of the great solitary wave is governed by the nonlinear equation called the *Korteweg–deVries (KdV) equation*. The nonlinearity is crucial. When the wave equation is linear, it is well known that a wave packet travels at the group velocity $v_g = \mathrm{d}\omega/\mathrm{d}k$. The wave packet is a Fourier sum (or integral) of components with some range of frequencies and (unless the medium is completely dispersion free) each component travels at its own velocity $v_p = \omega/k$. The consequence is that the phase relations between the components change with the distance travelled and the wave packet broadens. As mentioned in section 10.3, however, a nonlinear term can have a focusing effect, and the soliton propagation results from a balance between broadening due to dispersion and narrowing due to nonlinearity.

The KdV equation and many of its properties were understood by the end of the 19th century. Interest revived with two major discoveries in the 1960s. First, numerical studies showed that in some cases, if two solitary waves travelling in opposite directions collided (or if a faster one overtook a slower one), the two waves could emerge from the collision without change of shape. This remarkable property is hardly to be expected in the solutions of a nonlinear equation. The usual terminology is that a solution that travels without change of shape is called a solitary wave and if it has the additional property of invariance under collisions it is called a soliton. This is not always applied precisely and sometimes 'soliton' is used for 'solitary wave'.

The second major development was the invention of the *inverse-scattering method*, which gave a systematic way of finding solutions of some nonlinear wave equations, including KdV and NLSE. This is beyond our scope, and the reader is referred to detailed accounts elsewhere, for example in Drazin and Johnson (1989) or Infeld and Rowlands (1990). In this section, we give the derivation of the NLSE from Maxwell's equations with nonlinearity included and then show a few solutions of special importance. We will also see later that the NLSE sometimes occurs when discussing magnetic solitons.

As mentioned in section 10.3, in a bulk medium the nonlinearity of (10.24) can lead to various transverse effects like self-focusing or defocusing. Here we ignore these effects and analyse a 1D wave equation. There are two possible justifications apart from simplification. The first is that even in a bulk medium diffractive broadening may balance nonlinear self-focusing, leading to a stable profile. The more important point is that our main application will be to surface or guided waves and in these self-focusing is suppressed because the transverse profile of the wave cannot change from a form of the kind shown in earlier chapters. In a fibre waveguide, for example, the transverse profile is a Bessel function solution of the type derived in chapter 7.

We start from Maxwell's equations in the form

$$\frac{\partial^2 E}{\partial z^2} - \mu_0 \frac{\partial^2 D}{\partial t^2} = \mu_0 \frac{\partial^2 P^{\mathrm{NL}}}{\partial t^2} \tag{10.31}$$

where E, D etc. are the relevant field components, e.g. x components for a plane polarized wave. The D field contains only the *linear* polarization, i.e. we have $D(z,\omega) = \varepsilon_0 \varepsilon(\omega) E(z,\omega)$ for the Fourier components at frequency

ω, and P^{NL} is the nonlinear polarization. With a view to applying the SVA we substitute

$$E(z, t) = F(z, t) \exp[\mathrm{i}(k_0 z - \omega_0 t)] \tag{10.32}$$

where ω_0 is the central frequency of the wave packet. The first term of (10.31) gives

$$\frac{\partial^2 E}{\partial z^2} = \exp[\mathrm{i}(k_0 z - \omega_0 t)]\left(-k_0^2 F + 2\mathrm{i}k_0 \frac{\partial F}{\partial z} + \frac{\partial^2 F}{\partial z^2}\right). \tag{10.33}$$

For the second term in (10.31) we express D as a Fourier integral using

$$D(z, t) = \int_{-\infty}^{\infty} D(z, \omega) \exp(-\mathrm{i}\omega t)\, \mathrm{d}\omega \tag{10.34}$$

and there is a similar definition of $F(z, \nu)$ as the Fourier transform of $F(z, t)$ at any frequency ν. Due to the factor $\exp(-\mathrm{i}\omega_0 t)$ in (10.32), $F(z, \nu)$ corresponds to the field amplitude $E(z, \omega_0 + \nu)$ in the Fourier transform corresponding to (10.34):

$$E(z, \omega_0 + \nu) = F(z, \nu) \exp(\mathrm{i}k_0 z). \tag{10.35}$$

Since $D(z, \omega_0 + \nu) = \varepsilon_0 \varepsilon(\omega_0 + \nu) E(z, \omega_0 + \nu)$, we can write

$$D(z, t) = \varepsilon_0 \exp[\mathrm{i}(k_0 z - \omega_0 t)] \int_{-\infty}^{\infty} \varepsilon(\omega_0 + \nu) F(z, \nu) \exp(-\mathrm{i}\nu t)\, \mathrm{d}\nu. \tag{10.36}$$

This expression leads to

$$-\mu_0 \frac{\partial^2 D}{\partial t^2} = \mu_0 \varepsilon_0 \exp[\mathrm{i}(k_0 z - \omega_0 t)]$$
$$\times \int_{-\infty}^{\infty} (\omega_0 + \nu)^2 \varepsilon(\omega_0 + \nu) F(z, \nu) \exp(-\mathrm{i}\nu t)\, \mathrm{d}\nu. \tag{10.37}$$

Now in linear propagation the product $\mu_0 \varepsilon_0 \varepsilon(\omega) \omega^2 = \varepsilon(\omega)\omega^2/c^2 = k^2$. We therefore introduce $k_0 + \kappa$ as the wave number corresponding to $\omega_0 + \nu$, i.e.

$$(k_0 + \kappa)^2 = \varepsilon(\omega_0 + \nu)(\omega_0 + \nu)^2/c^2 \tag{10.38}$$

so that (10.37) becomes

$$-\mu_0 \frac{\partial^2 D}{\partial t^2} = \exp[\mathrm{i}(k_0 z - \omega_0 t)] \int_{-\infty}^{\infty} (k_0 + \kappa)^2 F(z, \nu) \exp(-\mathrm{i}\nu t)\, \mathrm{d}\nu. \tag{10.39}$$

The SVA means that the spread of wave numbers in (10.39) is much less than the central wave number, $\kappa \ll k_0$, so we can expand $k_0 + \kappa = k_0 + \nu k_0' + \frac{1}{2}\nu^2 k_0''$, where

$$k_0' = \frac{\mathrm{d}k}{\mathrm{d}\omega} = 1/v_g, \qquad k_0'' = \frac{\mathrm{d}^2 k}{\mathrm{d}\omega^2} = \frac{\mathrm{d}}{\mathrm{d}\omega}(1/v_g) \tag{10.40}$$

and the derivatives are evaluated at $\omega = \omega_0$. Here v_g is the group velocity, i.e. the velocity of a wave packet in the linearized equation. The above Taylor expansion is truncated at the ν^2 term because $\nu \ll \omega_0$. It enables (10.39) to be written as

$$-\mu_0 \frac{\partial^2 D}{\partial t^2} = \exp[i(k_0 z - \omega_0 t)] \int_{-\infty}^{\infty} [k_0^2 + 2\nu k_0 k_0' + \nu^2(k_0'^2 + k_0 k_0'')]$$

$$\times F(z, \nu) \exp(-i\nu t)\, d\nu. \tag{10.41}$$

Again the factors in the integrand have been expanded up to order ν^2. Next the right-hand side of (10.41) becomes, using the facts that νF and $\nu^2 F$ are the Fourier transforms of $i\,\partial F/\partial t$ and $-\partial^2 F/\partial t^2$,

$$\exp[i(k_0 z - \omega_0 t)]\left[k_0^2 F(z, t) + 2ik_0 k_0' \frac{\partial F(z, t)}{\partial t} - (k_0'^2 + k_0 k_0'') \frac{\partial^2 F(z, t)}{\partial t^2}\right].$$

$$\tag{10.42}$$

Next we can put (10.33) and (10.42) into the nonlinear wave equation (10.31) giving

$$2ik_0 \frac{\partial F}{\partial z} + \frac{\partial^2 F}{\partial z^2} + 2ik_0 k_0' \frac{\partial F}{\partial t} - (k_0'^2 + k_0 k_0'') \frac{\partial^2 F}{\partial t^2}$$

$$= \mu_0 \exp[-i(k_0 z - \omega_0 t)] \frac{\partial^2 P^{NL}}{\partial t^2} \tag{10.43}$$

after cancellation of terms involving k_0^2. Equation (10.43) gives a familiar result in the special case of linear propagation. Thus, if we put $P^{NL} = 0$ and retain only the first-order derivatives of F, we find

$$\frac{\partial F}{\partial z} + \frac{1}{v_g} \frac{\partial F}{\partial t} = 0 \tag{10.44}$$

which has the solution $F(z, t) = \Phi(z - v_g t)$, where Φ is any function. That is, we have recovered the elementary result that in the linear approximation the modulation envelope can have any shape and travels at the group velocity.

We now return to the full equation (10.43) and make two final approximations based on the SVA. First, we use (10.44) to express the second derivatives as

$$\frac{\partial^2 F}{\partial z^2} \approx \frac{1}{v_g^2} \frac{\partial^2 F}{\partial t^2} = k_0'^2 \frac{\partial^2 F}{\partial t^2} \tag{10.45}$$

which leads to a simplification since two terms in (10.43) cancel. Second, we use the fact that the frequency spread around ω_0 is small to write

$$\frac{\partial^2 P^{NL}}{\partial t^2} \approx -\omega_0^2 P^{NL}. \tag{10.46}$$

These steps reduce (10.43) to the envelope-propagation equation:

$$i\frac{\partial F}{\partial z} + \frac{i}{v_g}\frac{\partial F}{\partial t} - \frac{k_0''}{2}\frac{\partial^2 F}{\partial t^2} = -\frac{\mu_0\omega_0^2}{2k_0}\exp[-i(k_0 z - \omega_0 t)]P^{NL}. \qquad (10.47)$$

Compared with the linear propagation equation (10.44), this contains two additional terms. The third term on the left side involves $k_0'' = d(1/v_g)/d\omega$. This is the group-velocity dispersion that was mentioned as the source of distortion of wave packets in the linear régime. The second additional term is the nonlinear polarization on the right side. As will be seen, soliton propagation results from the interplay between these two terms, *provided* the signs are such that the dispersive properties balance.

Following (10.22) and (10.24) we now write P^{NL} explicitly as

$$P^{NL} = \varepsilon_0\eta|E|^2 E = \varepsilon_0\eta\exp[i(k_0 z - \omega_0 t)]|F(z,t)|^2 F(z,t). \qquad (10.48)$$

The right side of (10.47) can now be written as $-\frac{1}{2}k_0\eta|F(z,t)|^2 F(z,t)$ since $k_0^2 = \varepsilon_0\mu_0\omega_0^2$. Hence the propagation equation (10.47) becomes a nonlinear wave equation for the envelope function:

$$i\frac{\partial F}{\partial z} + \frac{i}{v_g}\frac{\partial F}{\partial t} - \frac{k_0''}{2}\frac{\partial^2 F}{\partial t^2} = -\frac{k_0\eta}{2}|F(z,t)|^2 F(z,t) \qquad (10.49)$$

and, because of the first two terms, the envelope travels at speed v_g. This makes it convenient to transform to a time coordinate travelling at this speed. At the same time we rescale the variables to bring the equation into canonical form. We put

$$\tau = (t - z/v_g)/t_0, \qquad \zeta = z/z_0, \qquad u = F/F_0 \qquad (10.50)$$

where the scaling parameters t_0, z_0 and F_0 will be chosen to simplify the final form of the equation. With these substitutions, (10.49) becomes

$$\frac{iF_0}{z_0}\frac{\partial u}{\partial \zeta} - \frac{k_0'' F_0}{2t_0^2}\frac{\partial^2 u}{\partial \tau^2} = -\frac{k_0\eta}{2}F_0^3|u|^2 u. \qquad (10.51)$$

This leads to solitary-wave propagation provided k_0'' and η have opposite signs:

$$k_0''\eta < 0 \qquad (10.52)$$

which is known as the *Lighthill criterion*. For definiteness, we assume here that $\eta > 0$ and $k_0'' < 0$, which is the more usual case.

We now choose the scaling parameters in (10.51) in the form

$$F_0/z_0 = -k_0'' F_0/t_0^2 = \tfrac{1}{2}k_0\eta F_0^3 \equiv \alpha \qquad (10.53)$$

where α is an arbitrary multiplying constant; these relations imply that the parameters are chosen to satisfy $k_0'' z_0/t_0^2 = -1$ and $\tfrac{1}{2}k_0\eta F_0^2 z_0 = 1$. The equation for u now takes the final form

$$i\frac{\partial u}{\partial \zeta} = -\frac{1}{2}\frac{\partial^2 u}{\partial \tau^2} - |u|^2 u. \tag{10.54}$$

The derivative terms here are formally the same as those appearing in the time-dependent Schrödinger equation (with ζ as time and τ as position), and for that reason (10.54) is called the nonlinear Schrödinger equation (NLSE). However, the independent variables here have a different physical significance from those in quantum mechanics: ζ is a dimensionless position and τ is a dimensionless time measured from the moving origin $t = z/v_g$.

The derivation of the NLSE was the main objective of this section. A systematic account could continue with analysis by means of the inverse scattering method, but this would be more than is necessary for our purposes. Here we just look at an important class of soliton solutions. The simplest of these is easily found by elementary means, as we now see. We look for a separated solution of the form

$$u(\zeta, \tau) = \exp(i\kappa\zeta)w(\tau) \tag{10.55}$$

where it is assumed that w is real. Substitution in (10.54) leads to an ordinary differential equation for w:

$$-\frac{1}{2}\frac{d^2 w}{d\tau^2} + \kappa w - w^3 = 0. \tag{10.56}$$

This is a nonlinear oscillator equation that can be solved by the same method as for the linear-harmonic oscillator. Multiplication by $dw/d\tau$ and integration yields the first integral

$$\frac{1}{4}\left(\frac{dw}{d\tau}\right)^2 - \frac{1}{2}\kappa w^2 + \frac{1}{4}w^4 = C. \tag{10.57}$$

If we assume that the boundary conditions correspond to a bounded pulse, $w = 0$ and $dw/d\tau = 0$ at $\tau = \pm\infty$, then the constant of integration C is zero. In this case the second integration of (10.57) is elementary and gives

$$w(\tau) = (2\kappa)^{1/2}\text{sech}[(2\kappa)^{1/2}\tau]. \tag{10.58}$$

In terms of the original amplitude function this is

$$F(z, t) = (4qc/k_0\eta)^{1/2}\exp(iqz)\,\text{sech}[(2q/|k_0''|)^{1/2}(t - z/v_g)] \tag{10.59}$$

where $q = \kappa/z_0$. Equation (10.59) describes a sech-shaped hump travelling at the group velocity v_g with phase varying periodically in z with period $2\pi/q$. There are no restrictions on the value of the wavenumber parameter q, and it is worth noting that it plays a multiple role: (a) it determines the width $(|k_0''|/2q)^{1/2}$; (b) it determines the overall amplitude $F_{max} = (4qc/k_0\eta)^{1/2}$; (c) as stated it determines the period in z. Equation (10.59) gives the field

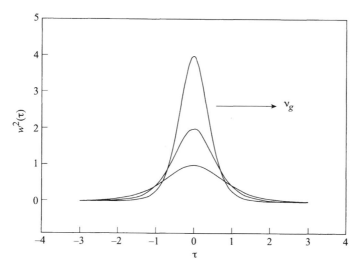

Figure 10.1. Illustration of solution (10.58) of the NLSE for $\kappa = 0.5$ (lowest curve), 1.0 and 2.0 (highest curve). The curve for 0.5 is the fundamental soliton. In terms of the solution (10.59), the pulse travels at speed v_g for any value of q and the width is proportional to $1/q$.

amplitude, whereas measurements are usually of the intensity, which is proportional to $|F|^2$. To illustrate properties (a) and (b), figure 10.1 shows graphs of w^2 versus τ for three values of κ, where w is given in (10.58).

Particular interest attaches to (10.58) and (10.59) for the special case $\kappa = \frac{1}{2}$ and $q = 1/2z_0$, namely

$$u(\zeta, \tau) = \exp(\tfrac{1}{2}i\zeta)\operatorname{sech}(\tau) \tag{10.60}$$

which corresponds to the input pulse ($z = \zeta = 0$):

$$u(0, \tau) = \operatorname{sech}(\tau). \tag{10.61}$$

This is in fact a soliton, i.e. the form is invariant through collisions, and it is called the *one-soliton solution* of the NLSE.

Systematic generalizations of the one-soliton solution by means of the inverse-scattering method were considered by Satsuma and Yajima (1974). They considered input pulses of the form

$$u(0, \tau) = N\operatorname{sech}(\tau) \tag{10.62}$$

and were able to show that for integer $N > 1$ the travelling pulse 'breathes', in the sense that its shape changes with distance of travel but the initial shape reforms after some periodic distance. In this scheme, (10.60) is the $N = 1$ soliton and is the only member of the family to travel without change of shape. The next simplest, the $N = 2$ soliton, has the explicit form

$$u(\zeta, \tau) = \frac{4\exp(-\tfrac{1}{2}i\zeta)[\cosh(3\tau) + 3\exp(-4i\zeta)\cosh(\tau)]}{\cosh(4\tau) + 4\cosh(2\tau) + 3\cos(4\zeta)}. \tag{10.63}$$

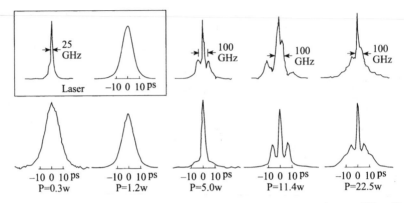

Figure 10.2. Time profiles of incident laser pulse and output pulses from a 700 m fibre for various input powers. For $P = 0.3$ W nonlinear effects are unimportant, $P = 1.2$ W corresponds to the $N = 1$ soliton in (10.62), $P = 5 \simeq 1.2 \times 2^2$ to $N = 2$, $P = 11.4 \simeq 1.2 \times 3^2$ to $N = 3$ and $P = 22.5 \simeq 1.2 \times 4^2$ to $N = 4$. The normalization of the pulse heights is arbitrary. After Mollenauer *et al* (1980).

Apart from the overall phase factor $\exp(-\tfrac{1}{2}i\zeta)$, (10.63) is periodic in ζ with period $\pi/2$. It satisfies the NLSE and the input pulse is $u(0,\tau) = 2\,\mathrm{sech}(\tau)$.

    The first experimental study of these solitons in nonmagnetic systems was carried out by Mollenauer *et al* (1980) using laser pulses of various powers propagating along a low-loss single-mode quartz fibre. The length of 700 m corresponded to $\zeta = \pi/4$ and the pulses, generated by a tunable laser, had a carrier wavelength of 1.55 µm, which is the loss minimum for quartz fibre. Since N in (10.62) is a field amplitude the pulse power P and N are related by $P \propto N^2$. Figure 10.2 shows the input pulse profile, which is a good approximation to (10.62), together with output pulses for various peak input powers. For the low-power pulse, $P = 0.3$ W, nonlinear effects are negligible and the output pulse shows dispersive broadening. $P = 1.2$ W is the $N = 1$ soliton with a significantly lower linewidth than the low-power pulse. The forms of the output pulses for $N = 2$, 3 and 4 are in good agreement with the calculations of Satsuma and Yajima (1974) for these higher-order solitons.

10.5   Fundamentals of the magnetic nonlinearities

As in previous parts of this book dealing with linear magnetic excitations (particularly chapters 4, 5 and 9), it is possible to proceed in the nonlinear case using either a macroscopic (continuum) method or a microscopic (Hamiltonian-based) method. We shall continue to employ both approaches, although the former is often more convenient and can also be developed by analogy with the nonlinear optics formalism outlined earlier in this chapter. As in those preceding sections, we shall focus here on nonlinear properties of *bulk* systems, while the applications to surface- and interface-related magnetic media are deferred to chapter 12. Some general references that cover the nonlinear properties of bulk magnetic excitations are Keffer (1966), Akhiezer *et al* (1968), L'vov (1994) and Gurevich and Melkov (1996).

10.5.1 Origin and general survey of the nonlinear effects

Up to this point our theoretical treatments of spin waves have included making a *linearization* approximation. This occurred, for example, within the microscopic approach for spin waves in bulk ferromagnets in section 2.3.2, where products of spin operators were simplified using the random-phase approximation (2.64). The S^z operators were replaced by their static thermal average assuming fluctuation effects in this component to be small. Equivalently, within the macroscopic (or continuum) approach as used in section 2.6.1 to derive the magnetic susceptibility, the torque equation of motion (2.163) for the total magnetization was linearized to give (2.164). Second-order small quantities in the fluctuating magnetization were ignored, implying for the longitudinal component $M^z = M_0$ (or $m^z = 0$).

The magnetic nonlinear effects are obtained, within either approach, when improvements are made to the above approximations. As an example within the macroscopic description we consider the following simple argument. The torque equation of motion (2.163) preserves the magnitude of the total magnetization vector **M**, so that

$$|\mathbf{M}|^2 = (m^x)^2 + (m^y)^2 + (M^z)^2 = M_0^2. \tag{10.64}$$

Therefore only two components are independent, and the third component (taken as M^z) can be expressed in terms of m^x and m^y as

$$M^z = M_0 \left[1 - \frac{(m^x)^2 + (m^y)^2}{M_0^2}\right]^{1/2} \simeq M_0 \left[1 - \frac{|\mathbf{m}|^2}{2M_0^2}\right] \tag{10.65}$$

where $\mathbf{m} = (m^x, m^y)$ is the transverse magnetization and we assume that $|\mathbf{m}| \ll M_0$. Hence a leading-order nonlinear correction is provided by the approximation

$$M^z = M_0(1 - \tfrac{1}{2}\varepsilon^2) \tag{10.66}$$

or equivalently $m^z = -\tfrac{1}{2}\varepsilon^2$, in terms of the small parameter $\varepsilon = |\mathbf{m}|/M_0$. It is clear from the above that m^z will be zero in the linear (in ε) approximation. Also it can be inferred using (10.66) that if the components of **m** vary in time sinusoidally with frequency ω, then m^z varies in time with frequency 2ω. This is an example of frequency doubling analogous to the second-order effects discussed in section 10.2 for the nonmagnetic case. It occurs here provided the spin precession is elliptical rather than circular, as is the case when there are magnetic dipole–dipole interactions (e.g. see section 2.6 in the magnetostatic regime). We generalize these arguments, making them more rigorous, in sections 10.5.2 and 10.5.3.

On a historical basis nonlinear spin–wave processes first attracted attention in ferromagnetic resonance (FMR) experiments at high power levels for the microwave-frequency exciting magnetic field (Bloembergen and Damon 1952, Bloembergen and Wang 1954). They observed a 'saturation' effect for the main FMR absorption peak and the appearance of additional (or 'subsidiary') peaks; these nonlinear effects had pronounced power thresholds. An explanation soon followed in seminal work by Suhl (1957) in terms of a *parametric instability* of the uniform-precession spin wave (the zero-wavevector bulk mode with frequency denoted by ω_B) with respect to decay into a pair of spin waves with frequencies ω_i and wavevectors $\mathbf{q}_i$ ($i = 1, 2$). The necessary conservation conditions are

$$n\omega_B = \omega_1 + \omega_2 \quad \text{and} \quad 0 = \mathbf{q}_1 + \mathbf{q}_2 \qquad (10.67)$$

where n is a positive integer. The simplest process is the decay of a single spin wave ($n = 1$) and this corresponds to a *first-order* parametric instability.

Usually in FMR the microwave field is applied transverse to the magnetization direction, i.e. the field is in the xy plane where it can couple to the components of the transverse $\mathbf{m}$. However, it was predicted and also observed by Schlömann *et al* (1960) that under *parallel pumping* by a field along the z direction the parametric excitation of spin–wave pairs may occur. A review of this early work was given by Damon (1963). The efforts towards a better understanding of this phenomenon led to the formulation of the so-called *S-theory* of nonlinear interaction of parametrically-excited spin waves (Zakharov *et al* 1975). This allowed an explanation for other nonlinear effects such as the complicated *auto-oscillations* of the magnetization (see Hartwick *et al* 1961) above the threshold of parametric excitation of spin waves. Nonlinear spin–wave effects are also manifested in the theory of magnetic solitons (see e.g. L'vov 1994).

In the following sections (and also in chapter 12) we explore these topics in more detail, first using a microscopic theory of spin–wave interactions, then in terms of a macroscopic theory in which it is convenient to include transverse and parallel pumping, and finally magnetic solitons are discussed separately.

10.5.2   Microscopic theory of spin–wave interactions

Here we shall employ a Hamiltonian approach, including both the Heisenberg exchange and magnetic dipole–dipole coupling, in order to identify the leading-order nonlinear terms that give rise to interactions and scattering processes between spin waves. At this stage attention will be directed to ferromagnets at low temperatures (where $T \ll T_C$) and the effects due to pumping fields will be omitted.

We adopt the same Hamiltonian for a bulk ferromagnet as in section 2.3.4 where the (linearized) spin–wave dispersion relation (2.94) was calculated within the random-phase approximation (RPA). In order to avoid this approximation we now follow a different method of calculation using the property that at low temperatures the spin waves (or magnons) behave like weakly interacting boson particles. This can be seen by rewriting the spin operators using the *H-P transformation* (Holstein and Primakoff 1940):

$$S_j^+ = S_j^x + iS_j^y = (2S)^{1/2}[1 - (a_j^\dagger a_j/2S)]^{1/2} a_j \tag{10.68}$$

$$S_j^- = S_j^x - iS_j^y = (2S)^{1/2} a_j^\dagger [1 - (a_j^\dagger a_j/2S)]^{1/2} \tag{10.69}$$

$$S_j^z = S - a_j^\dagger a_j. \tag{10.70}$$

Here $a_j^\dagger$ and a_j denote boson creation and annihilation operators, respectively, at any magnetic site j as defined in quantum mechanics (see e.g. Bransden and Joachain 2000). They satisfy the commutation relations

$$[a_i, a_j^\dagger] = \delta_{ij}, \qquad [a_i^\dagger, a_j^\dagger] = [a_i, a_j] = 0. \tag{10.71}$$

It is easily verified from (10.68)–(10.71) that the spin commutation relations quoted in (2.61) are reproduced, thus confirming the validity of the H-P transformation.

When $T \ll T_C$ the spins in the ferromagnet are well aligned and so $S^z \simeq S$. From (10.70) this implies that $a_j^\dagger a_j/2S$ (or, more strictly, its matrix elements between magnetic states) is small compared with unity and this property can be used to expand the square roots in (10.68) and (10.69). Keeping the leading order terms this gives

$$S_j^+ = (2S)^{1/2}[a_j - (a_j^\dagger a_j a_j/4S) + \cdots] \tag{10.72}$$

with a similar result for S_j^-. The Hamiltonian contributions from (2.58) and (2.59) can likewise be expanded in the boson operators, yielding

$$H = A_0 + H^{(2)} + H^{(3)} + H^{(4)} + \cdots. \tag{10.73}$$

Here A_0 is a constant, which, in the notation of section 2.3.4 for exchange and dipole sums, is given by

$$A_0 = -NS[g\mu_B B_0 + \tfrac{1}{2}SJ(0) + \tfrac{1}{2}SD^{zz}(0)]. \tag{10.74}$$

It is a contribution to the ground-state energy of the system, while $H^{(m)}$ denotes the term in the expansion with a product of m boson operators. The $H^{(1)}$ term is absent because it vanishes by symmetry in most crystal structures (including e.g. cubic and tetragonal).

We next examine the $H^{(2)}$ term, showing that it is equivalent to a linear spin–wave approximation. If we define Fourier transforms (in 3D) from a position to a wavevector representation by

$$a_j = N^{-1/2} \sum_{\mathbf{q}} a_{\mathbf{q}} \exp(-i\mathbf{q} \cdot \mathbf{r}_j), \qquad a_j^\dagger = N^{-1/2} \sum_{\mathbf{q}} a_{\mathbf{q}}^\dagger \exp(i\mathbf{q} \cdot \mathbf{r}_j) \qquad (10.75)$$

then it is straightforward to show that

$$H^{(2)} = \sum_{\mathbf{q}} [U(\mathbf{q}) a_{\mathbf{q}}^\dagger a_{\mathbf{q}} + \tfrac{1}{2} V(\mathbf{q}) a_{\mathbf{q}} a_{-\mathbf{q}} + \tfrac{1}{2} V^*(\mathbf{q}) a_{\mathbf{q}}^\dagger a_{-\mathbf{q}}^\dagger]. \qquad (10.76)$$

Here $U(\mathbf{q})$ and $V(\mathbf{q})$ are the same functions as defined in (2.90) and (2.91), but with $\langle S^z \rangle$ replaced by S in the present low-temperature case. Next we make a canonical transformation to a new set of boson operators $b_{\mathbf{q}}^\dagger$ and $b_{\mathbf{q}}$, chosen such that $H^{(2)}$ is in a 'diagonalized' form (apart from a constant term):

$$H^{(2)} = \sum_{\mathbf{q}} \omega_{\mathrm{B}}(\mathbf{q}) b_{\mathbf{q}}^\dagger b_{\mathbf{q}}. \qquad (10.77)$$

This has the standard form for a quantum-mechanical simple-harmonic oscillator of frequency $\omega_{\mathrm{B}}(\mathbf{q})$, which is the same as the bulk spin–wave frequency given by (2.94). In cubic crystals the transformation can be achieved by writing $a_{\mathbf{q}} = f_{\mathbf{q}} b_{\mathbf{q}} + g_{\mathbf{q}}^* b_{-\mathbf{q}}^\dagger$ and $a_{\mathbf{q}}^\dagger = f_{\mathbf{q}} b_{\mathbf{q}}^\dagger + g_{\mathbf{q}} b_{-\mathbf{q}}$, where the coefficients are

$$f_{\mathbf{q}} = \left(\frac{U(\mathbf{q}) + \omega_{\mathrm{B}}(\mathbf{q})}{2\omega_{\mathrm{B}}(\mathbf{q})} \right)^{1/2}, \qquad g_{\mathbf{q}} = \frac{V(\mathbf{q})}{|V(\mathbf{q})|} \left(\frac{U(\mathbf{q}) - \omega_{\mathrm{B}}(\mathbf{q})}{2\omega_{\mathrm{B}}(\mathbf{q})} \right)^{1/2} \qquad (10.78)$$

as may be verified as an exercise for the reader. More generally, care must be taken regarding phase factors as well as the symmetry between $\mathbf{q}$ and $-\mathbf{q}$ (see e.g. Keffer 1966, White 1983).

We may therefore conclude that by taking into account the effects of higher order terms (beyond $H^{(2)}$) in the Hamiltonian expansion in (10.73) we arrive at a description of the nonlinear effects. The leading order contributions are found to have the form

$$H^{(3)} = \sum_{\mathbf{q}_1, \mathbf{q}_2} [F_3(\mathbf{q}_1, \mathbf{q}_2) b_{\mathbf{q}_1}^\dagger b_{\mathbf{q}_2}^\dagger b_{\mathbf{q}_1 + \mathbf{q}_2} + \text{h.c.}] \qquad (10.79)$$

$$H^{(4)} = \sum_{\mathbf{q}_1, \mathbf{q}_2, \mathbf{q}_3} [F_4(\mathbf{q}_1, \mathbf{q}_2, \mathbf{q}_3) b_{\mathbf{q}_1}^\dagger b_{\mathbf{q}_2}^\dagger b_{\mathbf{q}_3} b_{\mathbf{q}_1 + \mathbf{q}_2 - \mathbf{q}_3}$$

$$+ \{F_4'(\mathbf{q}_1, \mathbf{q}_2, \mathbf{q}_3) b_{\mathbf{q}_1}^\dagger b_{\mathbf{q}_2}^\dagger b_{\mathbf{q}_3}^\dagger b_{\mathbf{q}_1 + \mathbf{q}_2 + \mathbf{q}_3} + \text{h.c.}\}] \qquad (10.80)$$

where h.c. denotes the Hermitian conjugate. Detailed expressions for the amplitude functions F_3, F_4 and F_4', which depend on the dipole–dipole and exchange terms as well as the wavevector labels, are given by Keffer (1966) and Akhiezer *et al* (1968). We now discuss their main characteristics.

Figure 10.3. Schematic representation of the three-magnon processes for (a) splitting and (b) confluence.

The $H^{(3)}$ contribution describes *three-magnon processes* in which there is either a *splitting* (when two spin waves are created and one is annihilated, with wavevector being conserved) or a *confluence* (the conjugate process in which two spin waves are annihilated and one is created). These are represented schematically in figure 10.3. They arise purely from the dipole–dipole terms in the original spin Hamiltonian.

On the other hand the $H^{(4)}$ contribution describes *four-magnon processes*. The F_4 term describes four-magnon *scattering* (in which two spin waves are annihilated and another two are created), while the F_4' term and its conjugate describe four-magnon splitting and confluence. These nonlinear processes are represented in figure 10.4. The amplitude F_4 for scattering contains contributions from both the dipole–dipole and exchange terms in the original Hamiltonian. For example, when the dipolar interactions are neglected (as in the Heisenberg model) we have

$$F_4(\mathbf{q}_1, \mathbf{q}_2, \mathbf{q}_3) = (1/4N)[2J(\mathbf{q}_1 - \mathbf{q}_3) + 2J(\mathbf{q}_2 - \mathbf{q}_3) - J(\mathbf{q}_1)$$
$$- J(\mathbf{q}_2) - J(\mathbf{q}_3) - J(\mathbf{q}_1 + \mathbf{q}_2 - \mathbf{q}_3)]. \qquad (10.81)$$

The above result, which has been obtained here by retaining the leading term in a $1/S$ expansion, is consistent with the rigorous theory by Dyson (1956) for spin–wave interactions in Heisenberg ferromagnets. By contrast the four-magnon splitting and confluence effects are specifically due to dipole–dipole interactions, and so $F_4' = 0$ for a Heisenberg ferromagnet.

We note that in general it is important to include both of the $H^{(3)}$ and $H^{(4)}$ nonlinear terms. This is because, although the $H^{(4)}$ term is of higher order in the expansion of boson operators, it can be important or even dominate if the dipolar effects are particularly small. Also, as we shall see

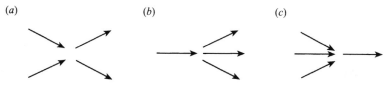

Figure 10.4. As in figure 10.3 but for four magnon processes: (a) scattering, (b) splitting and (c) confluence.

later, the two terms correspond to different parametric instabilities when a pumping field is included.

An important effect of the nonlinear terms is that they behave as small perturbations to $H^{(2)}$ that produce a frequency shift (or renormalization) and damping (or reciprocal lifetime) for the spin waves. These properties have been thoroughly studied in bulk ferromagnets with exchange or dipolar terms (or both), particularly with low-temperature theories, e.g. see the reviews by Keffer (1966), Akhiezer *et al* (1968), White (1983) and Kittel (1987). For example, in the case of a Heisenberg ferromagnet in zero applied magnetic field ($B_0 = 0$), it is known that the spin–wave energy is decreased by a term proportional to $q^2 T^{5/2}$ due to the interaction processes for small wavevectors and low temperatures. On comparison with the expansion in (2.70) for linear spin–wave energy, this is seen to be equivalent to introducing a temperature-dependent term to the spin–wave stiffness D; such effects have been seen experimentally, e.g. in light scattering (Cottam and Lockwood 1986). Extensions to higher temperatures have been made using diagrammatic perturbation methods (e.g. see Vaks *et al* 1968, Cottam 1971).

10.5.3   Macroscopic theory of spin waves including external pumping fields

We now extend the Hamiltonian description in the previous section to include the parametric instabilities and other nonlinear phenomena that arise when there is a term added for an external pumping field. Following most of the literature this will be carried out in the long-wavelength (or small-wavevector) regime where a continuum approach can be followed; see e.g. the books by Akhiezer *et al* (1968), L'vov (1994) or Gurevich and Melkov (1996).

The total external magnetic field now consists of a static part (in the z direction) and a variable pumping term with wavevector $\mathbf{q}_p$ and frequency ω_p:

$$\mathbf{B}_{total} = B_0 \hat{\mathbf{z}} + \mu_0 \mathbf{h}_p \exp(i\mathbf{q}_p \cdot \mathbf{r}) \exp(-i\omega_p t). \qquad (10.82)$$

The pumping field may, in general, have components of magnitude $h_p^{\parallel}$ or $h_p^{\perp}$ that are parallel or transverse, respectively, to the static magnetization M_0 (in the z direction). The total Hamiltonian may be expanded in terms of boson operators, which are just the semi-classical analogues of those introduced before. Then, generalizing (10.73) and (10.77)–(10.80), we have (apart from constant terms)

$$H = \sum_{\mathbf{q}} \omega_B(\mathbf{q}) b_{\mathbf{q}}^{\dagger} b_{\mathbf{q}} + H_{int} + H_p \qquad (10.83)$$

where $U(\mathbf{q})$ and $V(\mathbf{q})$ in (2.94) for the bulk spin–wave energy $\omega_B(\mathbf{q})$ reduce to the following forms in the continuum limit:

$$U(\mathbf{q}) = \omega_0 + Dq^2 + |V(\mathbf{q})|, \qquad V(\mathbf{q}) = \tfrac{1}{2}\omega_M \sin^2 \theta \exp(2i\phi) \qquad (10.84)$$

where θ and ϕ are the polar and azimuthal angles of wavevector $\mathbf{q}$ in spherical polar coordinates, and other definitions are as in sections 2.3 and 2.6. The expression for $\omega_B(\mathbf{q})$ is consistent with (2.95). The other two terms H_{int} and H_p in (10.83) are now explained.

First, H_{int} describes the leading effect of the spin–wave interactions and it consists of the three- and four-magnon processes, $H^{(3)}$ and $H^{(4)}$, respectively. These are given explicitly by (10.79) and (10.80) and are also depicted in figures 10.3 and 10.4. In the macroscopic limit there is some limited simplification to the expressions for the amplitude functions F_3, F_4 and F_4' (see e.g. Keffer 1966). However, even then, different regimes arise depending on the ratio of dipolar to exchange terms, the lattice structure, and the applied field; we shall examine the expressions in some special cases of interest later.

Next, H_p arises due to coupling between the total magnetization $\mathbf{M}$ and the pumping field term included in (10.82), i.e. it is the Zeeman energy proportional to the scalar product $\mathbf{h}_p \cdot \mathbf{M}(\mathbf{r}, t)$. Apart from a constant, it can be expressed as the sum of two terms denoted by H_{p1} and H_{p2} that involve products of one or two boson operators, respectively, as described below.

The H_{p1} term arises when there is a transverse component $h_p^\perp$ of the pumping field that couples directly to the fluctuating m^x (or m^y) in the total magnetization. This can then be expressed in terms of boson operators using the H-P transformation, and after applying the canonical transformation using (10.78) the lowest-order term is linear in the operators. Thus for the usual case of spatially homogeneous pumping, i.e. when $\mathbf{q}_p = 0$ in (10.82), this gives

$$H_{p1} \propto h_p^\perp [b_0^\dagger \exp(-i\omega_p t) + \text{h.c.}]. \tag{10.85}$$

This corresponds to the well-known phenomenon of ferromagnetic resonance (FMR). There is homogeneous precession of the magnetization, corresponding to excitation of the $\mathbf{q} = 0$ spin wave, with a decay term. This resonant process is usually treated as being analogous to a damped oscillator (see section 1.3.2). Accordingly, the amplitude is often expressed phenomenologically in Lorentzian form as (Suhl 1957)

$$|b_0| = h_p^\perp U / [(\omega_0 - \omega_p)^2 + \Gamma_0^2]^{1/2} \tag{10.86}$$

where the $\mathbf{q} = 0$ damping is written as $\Gamma_0 = \gamma \mu_0 \Delta H_0$ (with γ being the gyromagnetic ratio and $2\Delta H_0$ the FMR linewidth) and U is an oscillator strength. In most of the experiments cited in section 10.5.1 the ferrimagnet YIG (or similar ferrites) were employed because of their small ΔH_0 in combination with other properties.

The *first-order Suhl instability* mentioned in section 10.5.1 arises when the $\mathbf{q}_p = 0$ driving term in H_{p1}, namely (10.85), couples directly to the $\mathbf{q} = 0$ spin

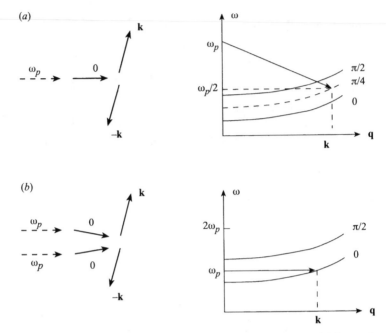

Figure 10.5. Nonlinear spin–wave processes in transverse pumping: (a) first-order Suhl process (or subsidiary resonance) and (b) second-order Suhl process (or premature saturation of the main resonance). On the left side the dashed and solid lines with arrows indicate the electromagnetic and spin–wave modes (labelled with their frequency and wavevector respectively).

wave (the uniform precession mode) that then decays via the splitting term in $H^{(3)}$ into a spin–wave pair (with wavevectors $\mathbf{k}$ and $-\mathbf{k}$ and frequencies $\frac{1}{2}\omega_\text{p}$). The appropriate part of $H^{(3)}$ for this process is proportional to

$$\sum_{\mathbf{k}\neq 0} F_3(\mathbf{k},-\mathbf{k})b_\mathbf{k}^\dagger b_{-\mathbf{k}}^\dagger b_0 \qquad (10.87)$$

arising from the dipolar interaction. In this regime we have approximately

$$F_3(\mathbf{k},-\mathbf{k}) \propto \omega_\text{M}\sin(2\theta)\exp(-i\phi) \qquad (10.88)$$

which is largest for spin waves propagating at an angle $\theta = \pi/4$. This process is represented schematically in figure 10.5(a). In the *second-order Suhl instability* a spin–wave pair with wavevectors $\mathbf{k}$ and $-\mathbf{k}$ is driven by two uniform-precession spin waves with frequency $\omega_\text{p} \approx \omega_\text{B}(\mathbf{k})$ pumped by $h_\text{p}^\perp$. This takes place via the four-magnon scattering process in figure 10.4(a), specifically from the term

$$\sum_{\mathbf{k}} F_4(\mathbf{k},-\mathbf{k},0)b_\mathbf{k}^\dagger b_{-\mathbf{k}}^\dagger b_0 b_0 \qquad (10.89)$$

contained within (10.80). The form of $F_4(\mathbf{k}, -\mathbf{k}, 0)$ in the Heisenberg (exchange only) limit can be deduced from (10.81) and it is proportional to k^2 to leading order in the continuum approximation. However, in general, the dipolar terms give other contributions to $F_4(\mathbf{k}, -\mathbf{k}, 0)$ that are proportional to ω_M multiplied by some angular factors. The role of this decay channel is to limit the growth of the FMR peak with increasing power; it is referred to as premature saturation of the main resonance.

The other pumping term H_{p2} arises when there is a parallel component $h_p^{\|}$ of the pumping field that couples to the fluctuating m^z in the longitudinal part of the total magnetization. Again the H-P transformation and the canonical transformation are used to rewrite the resulting Zeeman energy in terms of boson operators. This time, by contrast to (10.85), it is found that the leading order term involves a pair of spin–wave boson operators:

$$H_{p2} \propto h_p^{\|} \sum_{\mathbf{k}} \rho(\mathbf{k}) \, [b_{\mathbf{k}}^{\dagger} b_{-\mathbf{k}}^{\dagger} \exp(-i\omega_p t) + \text{h.c.}] \qquad (10.90)$$

where $\rho(\mathbf{k}) = \omega_M \sin^2 \theta \exp(-2i\phi)/\omega_B(\mathbf{k})$. Notice that this coupling is strongest when $\theta = \pi/2$ (i.e. when the spin–wave propagation is perpendicular to the magnetization direction). This process is a parametric instability of the first order, i.e. it corresponds to $n = 1$ in (10.67), which correctly expresses the conservation laws for homogeneous pumping. It is understood here that ω_i is shorthand for $\omega_B(\mathbf{k})$. The two excited spin waves propagate in opposite directions $\pm\mathbf{k}$ (preferentially with $\theta = \pi/2$) in accordance with wavevector conservation, and they have the same frequency $\frac{1}{2}\omega_p$ because of the property $\omega_B(\mathbf{k}) = \omega_B(-\mathbf{k})$ in most lattices. This process is represented schematically in figure 10.6.

The threshold fields for the various processes can be deduced from the above analysis by taking account of the conservation laws mentioned, together with comparisons of the different decay schemes. This would take us too far afield to cover here, but details are to be found in the references already cited.

Some of the early experiments for the spin–wave instabilities using FMR were mentioned in section 10.5.1. Most of the effort in recent years has been

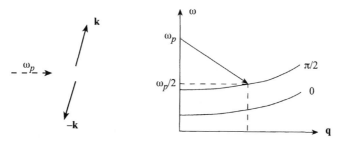

Figure 10.6. Nonlinear spin–wave process in parallel pumping, with labelling as in figure 10.5.

directed to FMR studies in thin films and in small spheres, where surface effects and spatial quantization are important. That discussion will therefore be deferred until chapter 12. However, we point out that in bulk materials the non-linear spin–wave processes have been studied by Brillouin light scattering under conditions of transverse and parallel pumping with a microwave field. The observed instabilities and their threshold effects are consistent with the FMR data. A good review of this topic is given by Borovik-Romanov and Kreines (1988). Among the materials studied by this technique are the ferrimagnet YIG and the canted antiferromagnets $CoCO_3$ and $FeBO_3$ (both of which have easy-plane anisotropy leading to a weak ferromagnetic moment).

10.5.4 Magnetic solitons

Studies of magnetic solitons fall into two rather distinct categories. On the one hand, in certain magnetic systems it has been shown that the equations of motion can be approximated to the well-known sine-Gordon (s-G) wave equation. This has soliton solutions, and applications have been made, for example, to domain walls and to some linear chain magnetic systems. On the other hand, some modulated systems and superlattices (including the magnetic ones) satisfy equations of motion approximating the nonlinear Schrödinger equation (NLSE) derived earlier, and under suitable conditions gap solitons may occur. These will be discussed first for nonmagnetic super-lattices in section 11.6; both the nonmagnetic and magnetic gap solitons are described by formalisms analogous to that given for nonlinear optics in section 10.4. It is the NLSE case that has been a topic of great interest in recent years for ferromagnetic superlattices; however, because surfaces and interfaces are involved we shall defer the treatment to chapter 12. Therefore in this section we discuss rather more briefly the applications relating to the s-G equation (for a review see e.g. White 1983, and references therein).

The s-G differential equation in 1D (with a spatial coordinate z) has the form (see e.g. Bishop and Schneider 1978 or Drazin and Johnson 1989)

$$\frac{\partial^2 \phi}{\partial z^2} - \frac{1}{c^2}\frac{\partial^2 \phi}{\partial t^2} = m^2 \sin \phi. \qquad (10.91)$$

It describes a variety of physical phenomena, one of the simplest in classical mechanics being an array of coupled restricted pendula (where ϕ is an oscillation amplitude, c is a characteristic velocity and m is a mass); see Drazin and Johnson (1989). Three distinct types of solution may be identified: spatially unbounded oscillatory solutions, spatially bounded oscillatory solutions (called breathers), and solitons. The last of these are fixed-amplitude 'kink' solutions that propagate undistorted and therefore have the same general characteristics as the solitons described in section 10.4.

It was shown by Mikeska (1978) in a study of 1D (chain) ferromagnets with easy plane single-ion anisotropy that an s-G description could be used.

He started with a spin Hamiltonian in the form

$$H = -J\sum_n \mathbf{S}_n \cdot \mathbf{S}_{n+1} + D\sum_n (S_n^z)^2 - g\mu_B B_0 \sum_n S_n^x. \qquad (10.92)$$

With anisotropy coefficient $D > 0$, the spins tend to align roughly perpendicular to the z axis while the applied field B_0 stabilizes the ordering preferentially along the x axis. Mikeska treated the spins classically, writing for the nth spin vector

$$\mathbf{S}_n = S(\sin\theta_n \cos\phi_n, \sin\theta_n \sin\phi_n, \cos\theta_n) \qquad (10.93)$$

in terms of spherical polar coordinates. Stable spin waves, i.e. small-amplitude harmonic oscillations about $\phi_n = 0$, occur in this model and can be studied by the methods in chapter 2. However, there can also be oscillations in which the angle ϕ_n ranges fully between 0 and 2π, and these are the solitons. To show this, Mikeska derived the equations of motion satisfied by θ_n and ϕ_n, and then replaced them by taking a continuum limit, i.e. putting $\phi_n(t) \to \phi(z, t)$ etc. Assuming moderate values of the applied field (such that $g\mu_B B_0 < T^2/J$), he found that an equation of the s-G form (10.91) was satisfied by angle ϕ, while θ could be obtained from

$$\theta = (1/2DS)(\partial\phi/\partial t). \qquad (10.94)$$

In this application the parameters in (10.91) correspond to $c = aS(2DJ)^{1/2}$ and $m = a(g\mu_B B_0/SJ)^{1/2}$, where a is a lattice parameter. An alternative procedure to take the classical continuum limit is discussed in the book by Mattis (1981). The soliton solutions in Mikeska's formulation are

$$\cos[\phi(z, t)] = 1 - 2\,\text{sech}^2[m\gamma_0(z - z_0 - ut)] \qquad (10.95)$$

with arbitrary centre position z_0 at $t = 0$ and arbitrary velocity $u < c$, where we have denoted $\gamma_0 = [1 - u^2/c^2]^{-1/2}$. This is sketched in figure 10.7 where the spin vectors rotate (with respect to angle ϕ) through 2π. This type of picture can be used in modelling magnetic domain walls.

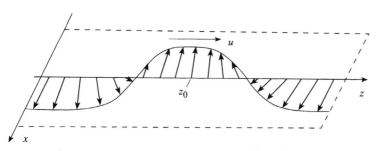

Figure 10.7. Sketch of a soliton in a 1D anisotropic ferromagnet, as represented by (10.95), showing the rotation of the spin vectors as angle ϕ changes from 0 to 2π.

The other type of solutions mentioned, called breathers, may be thought of as bound states of a soliton and an antisoliton. The latter corresponds to a decrease, rather than an increase, of ϕ by 2π.

Mikeska (1978) employed his theory to obtain estimates for the dynamic correlation functions using the ferromagnetic chain $CsNiF_3$ as an example. At around the same time Kjems and Steiner (1978) reported inelastic neutron scattering data for $CsNiF_3$ showing evidence of solitons. Further evidence with the same material was obtained from specific heat measurements (Ramirez and Wolf 1982).

References

Akhiezer A I, Bar'yakhtar V G and Peletminskii S V 1968 *Spin Waves* (Amsterdam: North-Holland)
Bishop A R and Schneider T (eds) 1978 *Solitons in Condensed Matter Physics* (Berlin: Springer)
Bloembergen N and Damon R 1952 *Phys. Rev.* **85** 699
Bloembergen N and Wang S 1954 *Phys. Rev.* **93** 72
Borovik-Romanov A S and Kreines N M 1988 in *Spin Waves and Magnetic Excitations 1* ed A S Borovik-Romanov and S K Sinha (Amsterdam: Elsevier) p 81
Bransden B H and Joachain C J 2000 *Quantum Mechanics* 2nd edition (Englewood Cliffs, NJ: Prentice-Hall)
Butcher P N and Cotter D 1990 *The Elements of Nonlinear Optics* (Cambridge: Cambridge University Press)
Cottam M G 1971 *J. Phys. C* **4** 2673
Cottam M G and Lockwood D J 1986 *Light Scattering in Magnetic Solids* (New York: Wiley)
Damon R W 1963 in *Magnetism, Vol. 1* ed G T Rado and H Suhl (New York: Academic) p 551
Drazin P G and Johnson R S 1989 *Solitons: an Introduction* (Cambridge: Cambridge University Press)
Dyson F J 1956 *Phys. Rev.* **102** 1230
Goldstone J A and Garmire E 1984 *Phys. Rev. Lett.* **53** 910
Gurevich A G and Melkov G A 1996 *Magnetization Oscillations and Waves* (Boca Raton, FL: CRC Press)
Hartwick T S, Peressini E R and Weiss M T 1961 *J. Appl. Phys.* **32** 223
Holstein T and Primakoff H 1940 *Phys. Rev.* **58** 1098
Infeld E and Rowlands G 1990 *Nonlinear Waves, Solitons and Chaos* (Cambridge: Cambridge University Press)
Keffer F 1966 *Handb. Physik* **18** 1
Kittel C 1987 *Quantum Theory of Solids* 2nd edition (New York: Wiley)
Kjems J K and Steiner M 1978 *Phys. Rev Lett.* **41** 1137
Kumar A, Kurz T and Lauterborn W 1996 *Phys. Rev. E* **53** 1166
L'vov V S 1994 *Wave Turbulence under Parametric Excitation* (Berlin: Springer)
Mattis D C 1981 *The Theory of Magnetism I* (Berlin: Springer)
Mikeska H J 1978 *J. Phys. C* **11** L29

Mills D L 1998 *Nonlinear Optics* 2nd edition (Berlin: Springer)

Mollenauer L F, Stolen R H and Gordon J P 1980 *Phys. Rev. Lett.* **45** 1095

Murgan R, Tilley D R, Ishibashi Y, Webb J F and Osman J 2002 *J. Opt. Soc. Am. B* **19** 2007

Osman J, Ishibashi Y and Tilley D R 1998 *Jap J. Appl. Phys.* **37** 4887

Popov S V, Svirko Yu. P and Zheludev N I 1995 *Susceptibility Tensors for Nonlinear Optics* (Bristol: IOPP)

Ramirez A P and Wolf W P 1982 *Phys. Rev. Lett.* **49** 227

Rowlands G 1990 *Non-linear Phenomena in Science and Engineering* (Chichester: Ellis Horwood)

Satsuma J and Yajima N 1974 *Suppl. Prog. Theor. Phys.* **55** 284

Schlömann E, Green J and Milano V 1960 *J. Appl. Phys.* **31** 3865

Shen Y R 1984 *The Principles of Nonlinear Optics* (New York: Wiley)

Suhl H 1957 *J. Phys. Chem. Solids* **1** 209

Vaks V G, Larkin A I and Pikin S A 1971 *Sov. Phys.—JETP* **26** 647

White R M 1983 *Quantum Theory of Magnetism* (Berlin: Springer)

Yariv A 1989 *Quantum Electronics* 3rd edition (New York: Wiley)

Zakharov V E, L'vov V S and Starobinets S S 1975 *Sov. Phys.—Usp* **17** 896

Chapter 11

Nonlinear properties: nonmagnetic surfaces and superlattices

In the final two chapters of this book we turn to the effects that can occur at surfaces and in superlattices when the exciting power is large enough for nonlinear properties to come into play. It is convenient to divide the material between nonmagnetic and magnetic media, as was done in the introductory account given in chapter 10. The former are the subject of this chapter and magnetic effects will then be described in chapter 12.

Nonlinear effects can occur in principle for any form of excitation but the majority of the work that has been done in the nonmagnetic case is concerned with optical excitations and we concentrate on these. The field is a very large one and we keep the length of this chapter within bounds by emphasizing basic principles. A number of reviews (Flytzanis *et al* 1991, Ostrowsky and Reinisch 1992, Dutta Gupta 1998) may be consulted for more detail on various aspects of the field. For general orientation it is helpful to recognize a number of more-or-less distinct themes.

The first issue is the way in which standard nonlinear effects are modified in the presence of one or more interfaces. The significance of this was recognized very early, starting with the calculation by Bloembergen and Pershan (1962) of second-harmonic generation from a semi-infinite medium and a film. They assumed that the medium was characterized by a $\chi^{(2)}$ coefficient that was the same as in the bulk material. Later it was realized, as mentioned already in section 10.1, that the presence of interfaces breaks inversion symmetry, so that surface-specific $\chi^{(2)}$ effects can arise even in samples for which $\chi^{(2)}$ is zero for the bulk material. This issue is discussed in section 11.1.

One of the main qualitative points to recognize is that nonlinear effects can be greatly enhanced by the electric field concentrations that can occur in films and superlattices. For example, as will be seen in section 11.2.1, the transmission spectrum of a thin film with partially reflecting mirrors, i.e. a Fabry–Pérot optical resonator, consists of a series of equally spaced peaks. Each peak corresponds to a 'leaky' standing wave with a large field amplitude. Since $\chi^{(2)}$ effects, say, are proportional to the square of the field amplitude they can be larger at the peaks than in the corresponding bulk medium. Similar effects can occur in superlattices. These effects are discussed in section 11.2.

Surface and guided modes may be used to transmit signals and for optical signal processing nonlinear effects in which surface modes interact with one another may be used. An example would be sum-frequency generation in which both input waves and the output wave are surface or guided modes. An introduction to this topic is given in section 11.3.

Much of the literature is concerned with the Kerr-effect nonlinearity which, as seen in section 10.3, can be interpreted as an intensity-dependent refractive index. One of the first topics to be explored was transmission through a Fabry–Pérot resonator containing a Kerr nonlinear medium. The key result is that for a sufficiently intense input beam the transmission becomes bistable or multistable. Because this result is central to much subsequent development the first part of section 11.4 is devoted to a detailed review of the analysis of the bistability by means of the SVA approximation used earlier. Other approaches and the extension to superlattices are discussed later in section 11.4 and it will be shown that within the SVA the propagation through a superlattice can be expressed in terms of a *nonlinear transfer matrix*, generalizing the treatment introduced in section 8.1.1.

A second Kerr effect that has received much attention is the self-guiding of surface polaritons at the surface of a nonlinear medium. It may be recalled from (6.67) in section 6.2 that within linear optics the guiding condition for a wave to be localized on a film is that its refractive index should be larger than the refractive index of either bounding medium. Now consider the surface between vacuum and a nonlinear medium with a positive Kerr effect (n increasing with intensity). If a surface polariton can be established at the surface then the refractive index $n = n_0 + n_2|E|^2$ in the surface region exceeds n_0 so that the guiding condition is satisfied; hence the surface polariton becomes self-guided. A more detailed discussion of this is given in section 11.5.

Not surprisingly, considerable attention has been given to soliton propagation, particularly in superlattices. One of the key ideas is the *gap soliton*. The gap in question is one of the stop bands for optical propagation through a superlattice, as shown in figure 8.9. Just below and just above the gap the derivative k_0'' varies rapidly so that the Lighthill criterion (10.52) can be satisfied in some frequency interval in the superlattice, even if it is not in the bulk medium. The resulting mode is the gap soliton and is discussed in section 11.6.

11.1   Single surfaces: second harmonic generation

As mentioned already, the first calculation of a surface nonlinear optic effect appeared in the early 1960s (Bloembergen and Pershan 1962). They calculated the second harmonic generation from a semi-infinite medium with a non-vanishing $\chi^{(2)}$ coefficient, assuming no depletion of the incident beam. In some ways the calculation is similar to that of normal-incidence Brillouin scattering from a semi-infinite medium that was presented in section 3.3.2, but Bloembergen and Pershan deal with the general case of oblique incidence. Here we briefly outline the calculation, denoting the medium (which occupies the half space $-\infty < z < 0$) by suffix 2 on relevant quantities. In the first stage the incident light is transmitted into the medium; this is described by a Fresnel factor like that in (3.41). The propagating wave in the medium may be written as $\mathbf{E}_2 \exp(-i\mathbf{k}_2 \cdot \mathbf{r} - i\omega t)$ and in line with (10.9) this generates a second-harmonic polarization

$$\mathbf{P}_2 = \varepsilon_0 K \ddot{\chi}^{(2)} \mathbf{E}_2 \mathbf{E}_2 \exp(-2i\mathbf{k}_2 \cdot \mathbf{r} - 2i\omega t). \qquad (11.1)$$

Here the combinatorial factor for second-harmonic generation is $K = \frac{1}{2}$ and the tensor product of $\ddot{\chi}^{(2)}$ with the two $\mathbf{E}$ fields is implied. The polarization $\mathbf{P}_2$ acts as a driving force for the wave equation at frequency 2ω, which has a similar form to (3.46) in the Brillouin-scattering problem. The solution of this equation to yield the radiated second-harmonic field in $z > 0$ and the transmitted field in $z < 0$, both at frequency 2ω, involves application of the electromagnetic boundary conditions at $z = 0$ in a similar way to the derivation of (3.49). The question of phase matching, introduced in section 10.2.2, appears in this calculation as a difference between twice the incident wave vector, $2\mathbf{k}_2(\omega)$, and the second-harmonic wave vector $\mathbf{k}_2(2\omega)$. It is worth pointing out a key feature of this calculation, namely that the propagation of both the incident and the SHG radiation is described within *linear* optics. The only place where the nonlinearity appears is in (11.1), which serves as the bridge between the two linear parts of the problem. Many of the calculations that will be described in this chapter have this character, and it will be seen that they make heavy use of the apparatus of linear optics that has appeared in earlier chapters.

Bloembergen and Pershan go on to develop the formalism for second-harmonic generation by a nonlinear film. This central problem has been addressed in many subsequent papers and will be discussed in section 11.2.

We pointed out in section 10.1 that because of the reduced symmetry at a surface, surface-specific $\chi^{(2)}$ effects can occur even in centrosymmetric materials. Most attention has been given to second-harmonic generation (SHG). Important experimental studies (Driscoll and Guidotti 1983, Guidotti *et al* 1983) on Ge and Si single crystals showed that the intensity of SHG depends on the orientation of the crystal face. A careful discussion of the origin of surface SHG was given by Guyot-Sionnest *et al* (1986), who distinguished two effects. First, since the inversion symmetry is broken at the surface, a non-vanishing $\chi^{(2)}$ in the sense of (10.4) or (10.9) can arise in the surface layer. This is called a structural-discontinuity effect. Second, the electromagnetic boundary condition that D_z is continuous, where z is the direction normal to the surface, means that E_z is discontinuous. Here **D** and **E** are the usual fields of macroscopic electrodynamics (see e.g. Landau and Lifshitz 1960) and a discontinuity in the macroscopic E_z means that the corresponding microscopic field has a rapid variation with z. Consequently the spatial dispersion terms, which in the interest of simplicity were omitted from the formulation of section 10.1, become important near the surface. This is because spatial dispersion means a **q** dependence in $\chi^{(2)}$ which when Fourier transformed means a dependence on the spatial derivatives of **E**. Guyot-Sionnest *et al* (1986) argue that the field and structural discontinuity contributions are distinct, pointing out that in a hypothetical experiment where the two media have identical refractive indices the field-discontinuity term would vanish but the structural-discontinuity effect would still be there.

Some progress has been made with microscopic calculations of the surface SHG coefficients (see e.g. Reining *et al* 1994, Wijers *et al* 1995), although the difficulties are considerable. Here we review only the phenomenological formulations of SHG (for example Guyot-Sionnest *et al* 1986, Sipe *et al* 1987, Mizrahi and Sipe 1988, Lüpke *et al* 1994); these may be seen as the extension of the formalism of section 10.1 to surfaces.

The field-gradient term arises from spatial dispersion, which in a coordinate space representation gives an SHG polarization (Sipe *et al* 1987, Lüpke *et al* 1994)

$$(P_{2\omega}^{(2)})_i = \varepsilon_0 \Gamma_{ijkl} E_j \nabla_k E_l. \tag{11.2}$$

This can be non-vanishing in the bulk of a centrosymmetric crystal, and in fact for a cubic crystal (Sipe *et al* 1987)

$$\Gamma_{ijkl} = a_1 \delta_{ijkl} + a_2 \delta_{ij}\delta_{kl} + a_3 \delta_{ik}\delta_{jl} + a_4 \delta_{il}\delta_{jk}. \tag{11.3}$$

Here the first Kronecker delta means that all four suffices are equal, and in the other products $(ij) \neq (kl)$ and so on. As implied by (11.3), the SHG polarization (11.2) can be non-vanishing in the bulk of the sample, although it is expected to be much larger in the surface region because of the field discontinuity. One question in experiments, obviously, is how to distinguish between the bulk and surface contributions.

The structural-discontinuity contribution arises just at the surface. A key point of the phenomenological theory is that both the field and structural terms can be combined in a single *nonlinear* polarization:

$$(P_{2\omega}^{2S})_i = \varepsilon_0 \Delta_{ijk} E_j E_k \delta(z+h) \tag{11.4}$$

where the E components are evaluated inside the nonlinear medium. This is represented by the final delta function, which also allows for the possibility that the active region is displaced by a distance h of order a monolayer thickness into the medium. The form of the tensor $\bar{\Delta}$ is restricted by the 2D point group of the surface and the non-vanishing elements for some important surfaces are listed by Sipe *et al* (1987) and by Lüpke *et al* (1994).

Given the nonlinear polarization (11.4), the radiated second-harmonic field can be calculated within classical electrodynamics, and again the outline is similar to the Brillouin-scattering calculation of section 3.3.2. First the Fresnel factors for the incident field are used to determine the field components E_j and E_k at frequency ω appearing on the right of (11.4). The polarization in (11.4) is then used as a driving term in the inhomogeneous wave equation for the fields at frequency 2ω in both media, and the solution of this equation gives the radiated SHG directions and intensities. The details of this substantial calculation are given by Sipe *et al* (1987) and Lüpke *et al* (1994).

Because SHG on centrosymmetric materials, notably Ge and Si, is surface-specific in the way described, it has developed into an important tool of surface analysis. However, further discussion is outside the scope of this book.

11.2   Field enhancement effects

Nonlinear effects are proportional to a power E^2 or E^3 of the electric field amplitude. In transmission through a layered system E varies with position and, where E is large, E^2 and E^3 are disproportionately larger than in transmission through a bulk medium. In consequence NLO effects can be enhanced above the magnitude observed in bulk. The simplest example is transmission through a Fabry–Pérot resonator and a more general one is transmission through a superlattice. We take these two examples in turn.

11.2.1   Single films

A schematic diagram of a Fabry–Pérot oscillator is shown in figure 11.1. It is simply a slab of a transparent dielectric, often glass, coated with partially transmitting mirrors. If the mirrors were 100% reflecting then the slab would support standing waves at frequencies f for which the thickness L is an integral number of half wavelengths:

$$L = n\lambda/2 = nc/2\varepsilon^{1/2}f. \tag{11.5}$$

In typical applications, λ is an optical wavelength, say $\lambda \leq 1\,\mu\text{m}$, and L is at least some mm, so the mode number n is large. Thus the electric field pattern in the slab contains a large number of nodes and antinodes. When the mirrors are partially transmitting, the resonances described by (11.5) become broadened into leaky standing waves with a finite linewidth. Even so, the electric-field amplitudes at the field maxima in the slab are large

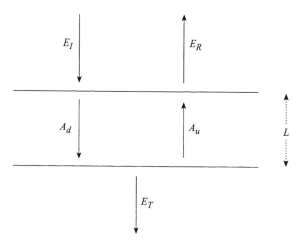

Figure 11.1. Fabry–Pérot oscillator. A slab of thickness L is coated with partially reflecting mirrors. The figure shows the notation for the calculation of reflection and transmission of normally incident light. The amplitudes within the medium are denoted A_d and A_u to be consistent with the notation in section 11.4.1.

when (11.5) is satisfied, or nearly satisfied. An important consequence is that the intensity transmitted through the slab becomes large around frequencies satisfying (11.5). As may be seen from section 1.4.1, the linear Fabry–Pérot is a basic spectroscopic instrument, and an account is given in optics texts, e.g. Born and Wolf (1986). Within linear optics calculation of the transmitted and reflected intensities is a standard problem. For a symmetric resonator of thickness L, with the upper and lower media identical, the transmission coefficient $T = |E_T/E_I|^2$ for normal incidence is (see, for example, Born and Wolf 1980)

$$T = 1/[1 + F \sin^2(k_2 L)] \qquad (11.6)$$

where k_2 is the wave number in the film and the factor F is the *finesse* of the resonator, which is given by

$$F = 4R/(1 - R)^2. \qquad (11.7)$$

Here R is the power reflectivity of one of the partial mirrors, namely

$$R = |r_{21}|^2 \qquad (11.8)$$

and r_{21} is the complex amplitude reflection coefficient of either mirror. Graphs to illustrate (11.6) are shown in figure 11.2. As can be seen, T is periodic in $k_2 L$ with period π, and varies between maxima of 1 at $k_2 L = n\pi$ and minima of $(1 + F)^{-1}$ at $k_2 L = (n + \frac{1}{2})\pi$. The main qualitative point about these graphs is that the transmission peaks correspond to 'leaky' standing

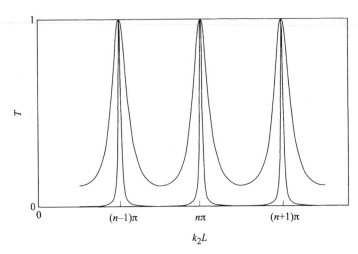

Figure 11.2. Transmission spectra T versus $k_2 L$ calculated from (11.6) for the linear Fabry–Pérot interferometer. Upper curve, $R = 0.5$ and $F = 8$; lower curve, $R = 0.9$ and $F = 360$.

waves within the film and the maximum field in these can be large. This means in turn that the driving polarization P^{NL} for nonlinear effects is large so that NLO effects are enhanced, particularly for large finesse F.

In outline, provided that depletion of the incident beam can be ignored, the calculation of a given nonlinear effect can be carried out by a generalization of the technique used by Bloembergen and Pershan (1962) for a single interface, as described for SHG in section 11.1. Indeed, as mentioned there, Bloembergen and Pershan did present the basic formalism for SHG from a film, though without much elaboration. For SHG, for example, the driving polarization P^{NL} at frequency 2ω is given by a form like (10.9) with $\omega_1 = \omega_2$ and the E field given by a superposition of the up and down waves shown in figure 11.1. For normal incidence this would be of the form $E = A_u \exp(ik_z z) + A_d \exp(-ik_z z)$. As pointed out in section 11.1, once the driving polarization is known calculation of the radiated fields involves solution of a *linear* (though inhomogeneous) wave equation, and this can be done exactly. Various forms of the calculation have appeared in the literature, perhaps the most comprehensive treatment of SHG and THG being that of Hashizume *et al* (1995), who develop a combination of Green-function and transfer-matrix techniques to analyse transmission through a general multilayer sample. Their formalism is complete for the case when surface- and interface-specific harmonic generation of the kind discussed in section 11.1 are unimportant, i.e. when all the layers are characterized by the bulk NLO coefficients.

As implied above, other things being equal, the SHG intensity will reach a maximum at one of the transmission peaks in figure 11.2. As seen from

(11.6), these are at thickness L or wave number k_2 satisfying the standing wave condition

$$k_2(\omega)L = m\pi \qquad (11.9)$$

for which the incident field intensity within the film attains maximum amplitude. We may call (11.9) the condition for *spatial resonance*; it is of course equivalent to (11.5). Now other things are not equal: the SHG will also be enhanced if the condition

$$k_2(2\omega)L = n\pi \qquad (11.10)$$

for spatial resonance in the output beam is satisfied. However, because of spatial dispersion (11.9) and (11.10) are not simultaneously satisfied in general. Therefore the SHG output is a combination of ω fringes arising from (11.9) and 2ω fringes arising from (11.10). By way of illustration, figure 11.3(b) shows the results of SHG measurements from a range of III–V MBE-grown epilayer structures of the form shown in figure 11.3(a). The source was a pulsed (Q-switched) Nd:YAG laser at wavelength 1.064 μm. The optical polarizations indicated in figure 11.3(a) are determined by the symmetry-allowed SHG tensor elements for these cubic materials. It is seen that as expected the measurements have the form of a complicated fringe pattern. A simple interpretation is not possible, but as shown the calculated curves are in very good agreement with the experiment.

We draw attention to the application of these ideas and the formalism of Hashizume *et al* (1995) by Shoji *et al* (1997). They are concerned with the determination of the absolute values of NLO coefficients and point out that most earlier work ignored the complication of multiple reflection that has just been discussed. They report measurements on a range of widely used materials and an analysis by means of the formalism outlined above leads to a downward revision of accepted values.

11.2.2 Multilayers and composites

As happens for films, field concentrations in transmission through multilayer or superlattice samples can lead to enhancement of nonlinear optical effects. An analysis could be made within the formalism of Hashizume *et al* (1995) and some of the other formalisms that have been developed will be described in section 11.4.2 for the special case of Kerr nonlinearities. Here we make use of the generalization to nonlinear effects of the effective-medium approximation that was used for linear effects in section 8.3.2. The first discussion was given in the original paper of Agranovich and Kravtsov (1985); a later derivation is due to Boyd and Sipe (1994). The general formalism has to be worked through separately for each tensor component of each NLO tensor so the number of possible formulae is quite large. From the latter paper we quote two of the simpler results, for the SHG component χ_{zzz}^{eff}

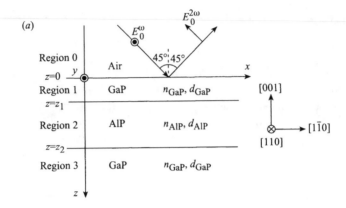

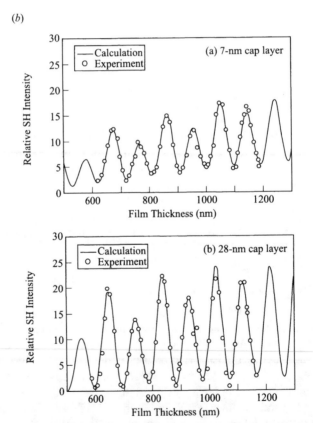

Figure 11.3. (a) Depiction of the samples and optical geometry used for SHG by Hashizume *et al* (1995). The samples are AlP grown by MBE on GaP substrates with a GaP capping layer. (b) Dependence of SHG intensity on AlP layer thickness for samples with (above) 7 nm and (below) 28 nm capping layers. The open circles are experimental results and the curves are theoretical. After Hashizume *et al* (1995).

and the Kerr-effect component χ_{zzzz}^{eff}. Here, as in section 8.3.2, the z axis is taken normal to the layers. The expressions are

$$\chi_{zzz}^{\text{eff}} = \frac{\dfrac{f_1\chi_1}{\varepsilon_1(2\omega)\varepsilon_1^2(\omega)} + \dfrac{f_2\chi_2}{\varepsilon_2(2\omega)\varepsilon_2^2(\omega)}}{\left[\dfrac{f_1}{\varepsilon_1(\omega)} + \dfrac{f_2}{\varepsilon_2(\omega)}\right]^2 \left[\dfrac{f_1}{\varepsilon_1(2\omega)} + \dfrac{f_2}{\varepsilon_2(2\omega)}\right]} \tag{11.11}$$

and

$$\chi_{zzzz}^{\text{eff}} = \frac{\dfrac{f_1\chi_1}{|\varepsilon_1(\omega)|^2\varepsilon_1^2(\omega)} + \dfrac{f_2\chi_2}{|\varepsilon_2(\omega)|^2\varepsilon_2^2(\omega)}}{\left|\dfrac{f_1}{\varepsilon_1(\omega)} + \dfrac{f_2}{\varepsilon_2(\omega)}\right|^2 \left[\dfrac{f_1}{\varepsilon_1(\omega)} + \dfrac{f_2}{\varepsilon_2(\omega)}\right]^2}. \tag{11.12}$$

Here f_1 and f_2 are the volume fractions of the two constituents, i.e. $f_i = d_i/D$ in the notation of section 8.3.2, so $f_1 + f_2 = 1$. Also ε_1 and ε_2 are the linear dielectric functions at the frequencies indicated, and χ_1 and χ_2 stand for the tensor components corresponding to the left-hand side, i.e. χ_{zzz} in (11.11) and χ_{zzzz} in (11.12). The main qualitative point that Boyd and Sipe make about these formulae is for the special case when only one of the media, say 1, has a substantial nonlinear coefficient. Then if frequency dispersion is neglected, i.e. $\varepsilon_i(2\omega) = \varepsilon_i(\omega)$, the nonlinear coefficient of the multilayer can exceed that of material 1, $\chi^{\text{eff}} > \chi_1$, when $\varepsilon_1 < \varepsilon_2$. The reason is that the field enhancement in medium 1 more than compensates for the decrease in the volume fraction of 1.

A demonstration experiment to confirm these ideas was reported by Fischer *et al* (1995). The nonlinear constituent 1 was a polymeric solid PBZT and constituent 2 was TiO$_2$. The experiment was concerned with the $\chi^{(3)}$ tensor, so that (11.12) applies, and the effective $\chi^{(3)}$ of a multilayer sample was determined from the nonlinear phase shift of a 1.9 µm laser beam transmitted through the sample. Figure 11.4 shows the measured and calculated nonlinear response (phase shift versus angle of incidence) for a multilayer sample with d_1 (PBZT, nonlinear) $= 40$ nm and d_2 (TiO$_2$, linear) $= 50$ nm. The enhancement predicted by (11.12) occurs when the z components of the field **E** are involved and it is seen in figure 11.4 that enhancement is seen in p-polarization, where **E** includes a z component, but not in s-polarization, where it does not. The theoretical curves are drawn for $\varepsilon_2/\varepsilon_1 = 1.77$, which is in fair agreement with bulk values.

It is worth pointing out the simplifying assumptions made in this analysis. First, as in the formalism of Hashizume *et al* (1995), possible interface-specific contributions are neglected, as is seen from the appearance of only the bulk NLO coefficients in (11.11) and (11.12). Second, the use of an effective-medium description introduces further approximations. However,

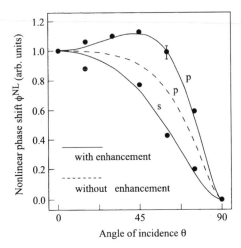

Figure 11.4. Measured (points) and calculated (curves) nonlinear phase shift in transmission of a 1.9 μm beam through a TiO_2/PBZT superlattice. For p-polarization the theoretical curves are shown both with and without the field enhancement effect resulting from (11.12); for s-polarization there is no field enhancement. After Fischer *et al* (1995).

the main qualitative point of the dominance of field-enhancement over volume effects is quite clear.

11.3 Nonlinear optics via surface modes

In the introduction to nonlinear effects given in sections 10.2 and 10.3 it was assumed that the various waves travel in an unbounded medium. Optical communications, however, rely on guided waves, either in fibres as discussed in section 7.4.2, or in some cases in films as discussed in section 6.2.1. For this reason, a great amount of work has been done on nonlinear optics in such bounded media. For detailed discussions of this work the reader may refer to the book by Agrawal (1989) for fibres and to the review by Stegeman and Seaton (1985) for films and plane interfaces. Here we concentrate on surface-polariton and surface-plasmon type modes. In a single-surface mode the exponential decay of field amplitude with distance from the surface corresponds to a substantial field enhancement near the surface, so that enhanced NLO effects are possible.

In section 10.2.2 we made the point, in connection with (10.20), that in order for significant nonlinear effects to be observed a long interaction length X_{int} is necessary, so that attenuation of the modes involved must be small. For many surface modes, damping is large enough to lead to a fairly short attenuation distance of the mode and this reduces the advantage of field enhancement. A notable exception is the long-range surface polariton

(LRSP) predicted by Sarid (1981) for a thin metallic film sandwiched between identical non-lossy media. As discussed in sections 6.2.1 and 6.3.3, in this mode the electric-field distribution is antisymmetric about the mid-plane of the film so that the field amplitude in the film is small and the damping in the metallic film has relatively little effect. Calculations of various NLO effects involving the LRSP followed rapidly on the original prediction (Deck and Sarid 1982, Stegeman *et al* 1982, for example).

The first experiment to demonstrate NLO via the LRSP was the observation of SHG by Quail *et al* (1983). The experiment used an ATR geometry as depicted in figure 11.5(a). A thin (13.5 nm) silver film was deposited on a quartz substrate and the combination was separated from the ATR prism by a liquid layer. This was chosen to have a refractive index close to that of quartz so that the assumption of identical media used in the prediction of the LRSP is satisfied. As indicated in figure 11.5(a), the direction of the expected second-harmonic (2ω) beam is offset (by $5°$) from the specular direction because of the optical dispersion of the prism material. Since quartz has a much larger $\chi^{(2)}$ coefficient than silver, the experiment corresponds to a linear film on a nonlinear substrate. The incident beam, from a pulsed Nd-glass laser, is p-polarized to correspond with the LRSP, and the quartz-crystal axes (x and y in standard notation in figure 11.5(a)) are chosen so that the second-harmonic polarization is also p-polarized and can excite the 2ω LRSP.

The variation with angle of incidence of the detected SHG intensity for a liquid-layer thickness of 3 μm is shown in figure 11.5(b). This thickness is calculated to correspond to optimal coupling to the ω LRSP in the sense described in section 6.3.1. It is seen that the 2ω intensity peaks to a value some six orders of magnitude above the lowest detectable level at the angle for exact coupling. The fact that the interaction is indeed via the LRSP is supported by the theoretical curve shown dashed, which is calculated for a single-interface surface plasmon.

Figure 11.5(c) shows similar results for a silver thickness of 18.5 nm and a liquid-layer thickness of 1 μm. This latter value corresponds to optimum coupling to the 2ω LRSP and overcoupling to the ω LRSP. There are two differences from figure 11.5(b). First, because of the overcoupling the peak due to the ω mode broadens and has a lower maximum value; this is the same as the overcoupling effect shown for linear ATR reflectivity on curve c of figure 6.13. Second, the optimum coupling to the 2ω mode means that a second peak is seen around $\theta = 56°$ due to resonance with the output-frequency mode.

The theoretical curves in figure 11.5 were calculated by Deck and Sarid (1982) by a similar method to that described in section 11.1 for SHG at a single interface and in section 11.2 for a film. In the first part of the calculation, linear optics is applied to calculate the field distribution at frequency ω. The polarization $\mathbf{P}_{2\omega}(x)$ at frequency 2ω is calculated from this distribution

(a)

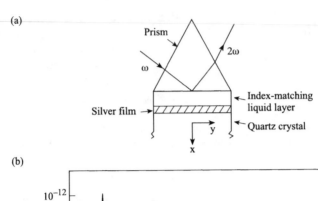

(b)

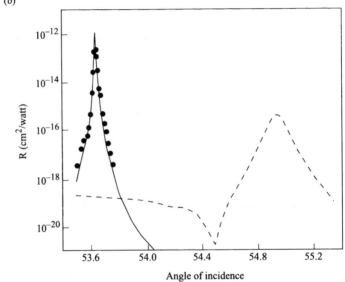

(c)

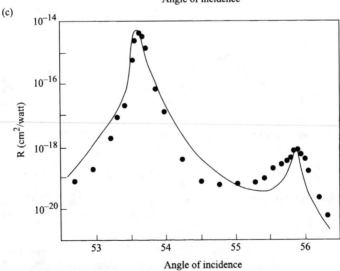

and the SHG coefficients of bulk quartz and applied as a driving term in a second linear calculation of the 2ω field distribution. This calculation yields the SHG beam amplitude and hence the intensity.

11.4 Kerr effect bistability and multistability

The important Kerr effect, arising from a $\chi^{(3)}$ nonlinearity, was introduced in section 10.3. It is interpreted as an intensity-dependent dielectric function or refractive index, (10.24) or (10.25). The way in which the effect can lead to soliton propagation in a bulk material was discussed in section 10.4. We now turn, in the following sections 11.4.1 and 11.4.2, to the way in which transmission through a Fabry–Pérot and through a multilayer are affected by the intensity dependence.

11.4.1 Single film and Fabry–Pérot resonator

In this section we discuss the properties of a film and of a Fabry–Pérot (FP) resonator when the dielectric has a third-order nonlinearity leading to an intensity-dependent dielectric function of the form of (10.23). To be a useful instrument, the FP is coated with partial mirrors of nearly 100% reflectivity, and it is helpful to distinguish between a FP in this sense and a film defined to have no mirrors or only weakly-reflecting mirrors. It is suffi-cient to give the detailed calculations for an isotropic medium, e.g. glass, and normally incident plane-polarized light, so that the nonlinearity is given by (10.24). The notation has been defined in figure 11.1 and we assume that regions 1 and 3 are vacuum, as is usually the case in applications. A brief review of the linear properties of the resonator was given in section 11.2.1. The basic physical point is that for the nonlinear problem the dielectric func-tion ε in (10.24) is a function of the field strength in the slab and this means that the standing-wave condition, (11.5) or (11.9), depends on the intensity I_0 of the incident wave. In consequence, the transmission and reflection coeffi-cients depend on I_0; it is because of this that bistability can arise. An analysis of the nonlinear FP by the SVA was first given by Marburger and Felber (1978) and we give a version of their work. The approach is the standard one for such optical problems: for a given frequency we find the solutions of Maxwell's equations in each of the three regions in figure 11.1, then

---

Figure 11.5. (a) Experimental arrangement for second-harmonic generation via the long-range surface plasmon in an Ag film. (b) Second-harmonic reflection coefficient (points) for Ag thickness 13.5 nm and liquid-layer thickness 3 µm. Calculated curves are for coupling via the LRSP (solid) and via a single-surface plasmon (dashed). (c) As in (b) but for Ag thickness 18.5 nm and liquid-layer thickness 1 µm. After Quail *et al* (1983).

apply the electromagnetic boundary conditions to find the required reflection and transmission coefficients. The boundary conditions used by Marburger and Felber are approximate, as will be discussed later.

At frequency ω, the forms of the fields in regions 1 and 3 are

$$E = E_1 \exp(-ik_0 z) + E_R \exp(ik_0 z), \qquad z > 0 \qquad (11.13)$$

$$E = E_T \exp[-ik_0(z + L)], \qquad z < -L \qquad (11.14)$$

where we denote $k_0^2 = \omega^2/c^2$. Here the incident-wave amplitude E_1 is determined by the incident-wave intensity and the aim of the calculation is to find the reflected (E_R) and transmitted (E_T) amplitudes. Inside the slab, $-L < z < 0$, the field is a solution of the nonlinear wave equation

$$-\frac{d^2 E}{dz^2} = k_2^2 E + k_0^2 \eta |E|^2 E \qquad (11.15)$$

where $k_2^2 = \varepsilon(\omega)\omega^2/c^2$. We write the solution of (11.15) as a sum of a 'down' and an 'up' wave:

$$E = E_d(z) + E_u(z) = A_d(z) \exp(-ik_2 z) + A_u(z) \exp(ik_2 z). \qquad (11.16)$$

In order to apply the slowly-varying amplitude approximation, we substitute (11.16) into (11.15) and retain only the first derivatives of $A_d(z)$ and $A_u(z)$. This leads to

$$2ik_2 \frac{dA_d}{dz} = k_0^2 \eta [|A_d|^2 + 2|A_u|^2] A_d \qquad (11.17)$$

$$-2ik_2 \frac{dA_u}{dz} = k_0^2 \eta [2|A_d|^2 + |A_u|^2] A_u. \qquad (11.18)$$

In order to make progress, we must now assume that η is real; this is so as long as the frequency is well away from any optical resonance region. It then follows simply from (11.17) and (11.18) that $|A_d|^2$ and $|A_u|^2$ are constants independent of z. The terms multiplying A_d in (11.17) and A_u in (11.18) are therefore constants so that the solutions for $A_d(z)$ and $A_u(z)$ become simple exponentials. The resulting field expressions, together with (11.13) and (11.14) for the fields in the external regions, can be substituted in the boundary conditions. Following one of the clearest accounts of the linear FP (Landau and Lifshitz 1960), it is convenient to express these as relations between the outgoing and incoming waves at each interface since this can be done in terms of *single-interface* transmission and reflection coefficients. The resulting equations are

$$E_d(0) = t_{12} E_1 + r_{21} E_u(0) \qquad (11.19)$$

$$E_R = r_{12} E_1 + t_{21} E_u(0) \qquad (11.20)$$

$$E_u(-L) = r_{23} E_d(-L) \qquad (11.21)$$

$$E_T = t_{23} E_d(-L) \qquad (11.22)$$

where the notation is that t_{12} is the complex amplitude transmission amplitude for a wave incident in medium 1 on the interface with medium 2, r_{21} is the corresponding reflection amplitude for a wave in 2 incident on the same interface, and so on. For example, in the absence of partially reflecting mirrors

$$r_{12} = \frac{\varepsilon_1^{1/2} - \varepsilon_2^{1/2}}{\varepsilon_1^{1/2} + \varepsilon_2^{1/2}} = \frac{1 - [\varepsilon(\omega)]^{1/2}}{1 + [\varepsilon(\omega)]^{1/2}}. \tag{11.23}$$

Mirrors are easily modelled by a simple modification of (11.23) and similar formulae (Lim *et al* 1997). As can be seen from figure 11.1, in (11.19) and (11.20) the outgoing amplitudes $E_d(0)$ and E_R at the upper interface are expressed in terms of the incoming amplitudes E_1 and $E_u(0)$ at that interface. Equations (11.21) and (11.22) have a similar character. In the original derivation, and in many subsequent accounts, it was assumed that like (11.23) the reflection and transmission coefficients in (11.19) to (11.22) are the same as in the linear transmission calculation.

The unknowns in this calculation are E_R, E_T and the amplitudes A_d and A_u; (11.19) to (11.20) are linear equations giving these in terms of the incident amplitude E_1. The solution is straightforward, but it is sufficient here to quote the expression for the transmission coefficient $T = |E_T/E_1|^2$, namely

$$T = \frac{t_{12}^2 t_{23}^2}{(1 - r_{21}r_{23})^2 + 4r_{21}r_{23}\sin^2(\theta/2)} \tag{11.24}$$

where

$$\theta = \theta_0 + 3B(|A_u|^2 + |A_d|^2)L \tag{11.25}$$

with $B = k_0^2\eta/2k_2$ and θ_0 is a constant which physically represents the detuning of the frequency from the nearest leaky standing wave.

Equations (11.24) and (11.25) are implicit equations for T since θ and T both depend on the incident power $|E_1|^2$ through the appearance of the term $|A_u|^2 + |A_d|^2$ in θ. For the case when the reflectivities at $z = 0$ and $z = -L$ are the same they can be rewritten in the form

$$T = 1/[1 + F\sin^2(\phi/2)] \tag{11.26}$$

where

$$\phi = 2k_2 L + P|E_1|^2 T \tag{11.27}$$

with $P = 3k_0^2\eta L(1 + R)/[2k_2(1 - R)]$. The factor F that appears in (11.26) is the *finesse* of the FP cavity, as defined previously in (11.7). The physical meaning of ϕ in (11.27) is that it is the phase shift for a double transit of the slab and the second term shows the change in this phase shift due to the nonlinearity.

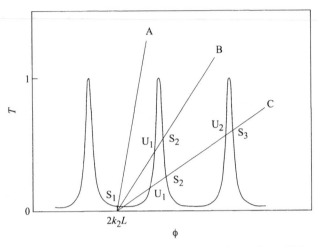

Figure 11.6. Graphical solution of (11.26) and (11.27) for the nonlinear Fabry–Pérot. The construction is shown for a positive value of the NLO coefficient η in (11.15). Lines A, B and C correspond to (11.27) with increasing values of $P|E_1|^2$.

Equations (11.26) and (11.27) are the forms that are usually quoted for this problem (e.g. Shen 1984). Although as has been mentioned the derivation contains some approximations, notably the use of linearized boundary conditions, the implications of these equations are qualitatively similar to those of more exact theories and are therefore worth discussing.

Obviously the transmission of the linear FP is included as a special case; putting $\eta = 0$ implies $P = 0$ in (11.27), so that (11.26) goes over into the explicit form (11.6) that was quoted previously and illustrated in figure 11.2.

In the nonlinear problem, (11.26) and (11.27) are simultaneous equations for T and ϕ. The general form of the solution is easily appreciated from the graphical construction shown in figure 11.6. The graph of (11.27) is shown for positive η, i.e. positive $\chi^{(3)}$; the analysis for negative $\chi^{(3)}$ is similar. In the (T, ϕ) plane the graph of (11.26) reproduces the spectrum of figure 11.2. The graph of (11.27) is a straight line with slope $1/P|E_1|^2$ and provided this slope is small enough, i.e. provided $P|E_1|^2$ is large enough, the two graphs have multiple intersections. As can be seen, the exact form of the solution depends on the value of $2k_2L$, i.e. on the closeness of the low-intensity solution to a transmission peak. For small incident intensity, i.e. for small $P|E_1|^2$, curve A, the operating point S_1 is close to that of the linear régime. For larger intensity, e.g. curve B, the intersections may be shown to correspond to two stable operating points, S_1 and S_2, and one unstable point, U_1. For still larger intensities, e.g. curve C, further stable and unstable points occur. The reason for the multiple roots in curves B and C can be seen from the fact that ϕ is the phase shift for a double transit of the slab. S_2 is closer than S_1 to a transmission maximum and

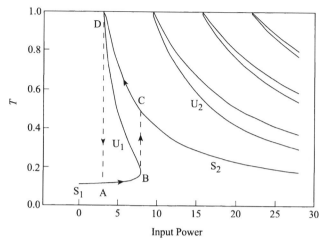

Figure 11.7. Transmissivity T versus input power for the nonlinear Fabry–Pérot, taking $R = 0.5$ ($F = 8$) and $2k_2L$ equal to an odd-integer multiple of π. The horizontal axis is the value of $P|E_1|^2$ as defined in (11.27).

therefore the average value of $|E|^2$ is higher than at S_1. It follows from (10.24) that the effective dielectric function ε^{eff} is larger than at S_1 and therefore, self-consistently, the phase shift ϕ is larger.

Further insight into the solutions is gained from graphs of the transmissivity T versus input power at fixed frequency of the form shown in figure 11.7. This is derived from figure 11.6 in the following way. For a given frequency, the phase detuning $2k_2L$ is determined. As seen from (11.27), the input power $|E_1|^2$ is inversely proportional to the slopes of the lines A, B, C etc. in the (ϕ, T) plane. As indicated, the curves in figure 11.7 are the loci of the T values of the corresponding points in figure 11.6 as this slope varies. As seen, above some threshold value of input power the transmission becomes bistable, or in general multistable. An important consequence is the appearance of hysteresis in T as a function of input power. For example, as the input power is increased and then decreased within the appropriate range, the system can trace out the hysteresis loop ABCD marked. The bistability and associated hysteresis are the basis of one proposed technique for data storage in optical computing.

Since the slope of the line corresponding to (11.27) is $1/P|E_1|^2$ it is clear that the intensity $|E_1|^2$ for the onset of bistability is proportional to $1/P$. It follows from the expression for P that bistability is favoured by large nonlinearity η and by the mirror reflectivity R being close to unity.

The above discussion was for $\eta > 0$, but as mentioned the behaviour for $\eta < 0$ is very similar. The lines corresponding to (11.27) now have negative slope, but with increasing $|E_1|^2$ multiple intersections occur in the same way as for positive η.

The calculation that has been reviewed here contains two basic approximations. The first is the use of the SVA in (11.17) and (11.18). The second concerns the boundary conditions (11.19) to (11.22). The derivation of these uses the expression for the magnetic field of the wave, which in principle should be derived from the full nonlinear field equations. The use of formulae like (11.23), which are the standard forms for the linear calculation, means that the exact boundary conditions have been linearized. Both of these approximations are adequate for a high-finesse FP as used in the optical or near-infrared spectral region. Typically, the thickness L is large compared with the optical wavelength, so that the SVA should be adequate. With high-reflectivity mirrors, $R \approx 1$ in the final formulae (11.26) and (11.27), the refinement of including nonlinearity in the boundary conditions would make little difference in practice.

The two approximations are distinct and can be removed separately. In fact, Marburger and Felber (1978) also give a derivation that does not use the SVA but still applies linearized boundary conditions. The starting equation (11.15) can be solved exactly in terms of Jacobian elliptic functions. With the resulting expressions for E the derivation can be repeated and the result is that formulae like (11.26) and (11.27) apply with the difference that the sine function is replaced by an elliptic function.

As implied, for a film without mirrors the use of linearized boundary conditions is expected to be a poor approximation. In the next section, we give a version of the treatment of films and multilayers first introduced by Dutta Gupta and Agarwal (1987). This retains the SVA but uses the full non-linear boundary conditions; the formalism leads to a generalization of the transfer-matrix method first introduced in section 8.1.1. Since a single layer is a special case of this formalism we defer the account.

The first exact treatments of the nonlinear FP, without SVA and using the full boundary conditions, were given by Chen and Mills (1987a) for normal incidence and by Leung (1989) for oblique incidence. These treatments differ in details, but both are quasi-analytic in the sense that the solution is reduced to the numerical evaluation of certain integrals. These integrals contain four unknowns, which ultimately are calculated from the electromagnetic boundary conditions. For a full account, the reader is referred to the original papers or to the review by Dutta Gupta (1998).

The nonlinear FP has attracted a great deal of attention over the years both because of the obvious implications for data storage of the optical bistability and because of its significance in the study of nonlinear dynamics. A full and detailed account of work up to the date of their review is given by Reinisch and Vitrant (1994).

The general framework of the analyses reviewed above was called into question by Goldstone and Garmire (1984). As described in section 10.1, the conventional approach to nonlinear optics is to assume that the polarization **P** can be expanded as a Taylor series in the optical field **E**, as in

(10.3) and (10.4). However, as remarked at the end of section 10.1, the origin of the nonlinear optics lies in the nonlinear dynamics of the material, so that like (10.8) the underlying equation of motion is nonlinear in **P** (or in the ionic displacements) but linear in **E**. As mentioned in section 10.1, for weak nonlinearity the inversion of the equation of motion to give the conventional formalism is valid. However, for strong nonlinearity, in particular near a resonance in the equation of motion for **P**, inversion is questionable. What Goldstone and Garmire discussed was optical transmission across the interface from a linear to a nonlinear medium, the latter being described by a nonlinear equation for **P**. They applied the SVA to this equation, and were able to show that an intrinsic bistability of the transmission coefficient can occur.

An extension of this approach to a film was given by Chew *et al* (2001). Normal incidence and plane polarization is assumed, so that **P** and **E** are given by scalar amplitudes P and E. As in Goldstone and Garmire the starting point is the equation of motion for P:

$$\frac{\partial^2 P}{\partial t^2} + \Gamma \frac{\partial P}{\partial t} + \omega_0^2 P + 3b|P|^2 P = \gamma E. \qquad (11.28)$$

This nonlinear equation, in which the linear part is resonant at frequency ω_0, is often known as the Duffing-oscillator equation. Chew *et al* treat it in the single-frequency approximation $\partial/\partial t \rightarrow -i\omega$. P and E are also related by the wave equation

$$-\frac{\partial^2 E}{\partial z^2} = \omega^2 \mu_0 (\varepsilon_0 \varepsilon_\infty E + P). \qquad (11.29)$$

If (11.29) is used to replace E in (11.28), the result is two coupled equations for P and the complex conjugate P^*. Chew *et al* devise a numerical scheme for the integration of these equations across the thickness of the film and the end point is a series of graphs for transmission coefficient versus input power at various frequencies. Qualitatively these show the same multistability as graphs like figure 11.7, but unlike the conventional approach the method is valid right through the resonance region.

11.4.2 Multilayers and superlattices

The exact treatment of the nonlinear film by Chen and Mills (1987a) was later extended by the same authors (Chen and Mills 1987b) to multilayers. As might be expected, the calculations are technically quite difficult. In this section, therefore, we choose to give an introduction to the 'nonlinear transfer matrix' method for multilayers introduced by Dutta Gupta and Agarwal (1987), which as mentioned retains the SVA but applies the exact boundary conditions.

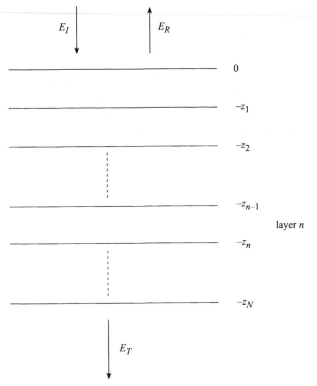

Figure 11.8. Notation for calculation of reflection and transmission by a nonlinear multilayer.

We consider a general multilayer with notation as defined in figure 11.8. Layer n, of thickness d_n, extends from $z = -z_{n-1}$ to $z = -z_n$, and the bounding media in $z > 0$ and $z < -z_N$ are taken as vacuum. The aim of the calculation is to find the complex reflection and transmission amplitudes E_R/E_I and E_T/E_I. For simplicity we deal with normal incidence and polarization $\mathbf{E} = (E, 0, 0)$ and $\mathbf{H} = (0, H, 0)$. Within layer n we apply the SVA in the form of (11.16) with the amplitudes satisfying (11.17) and (11.18). As before, it can be shown that $|A_d|^2$ and $|A_u|^2$ are constants so that (11.17) and (11.18) have exact solutions in terms of exponentials. The E field in layer n can therefore be written in the form

$$E_n = \alpha_n \exp[-\mathrm{i}p_n(z + z_n)] + \beta_n \exp[\mathrm{i}q_n(z + z_n)] \qquad (11.30)$$

and it follows from Maxwell's equations that

$$\mu_0 \omega H_n = p_n \alpha_n \exp[-\mathrm{i}p_n(z + z_n)] - q_n \beta_n \exp[\mathrm{i}q_n(z + z_n)]. \qquad (11.31)$$

Here we introduce

$$p_n = k_2 + k_0^2\eta(|\alpha_n|^2 + 2|\beta_n|^2)/2k_2 \tag{11.32}$$

$$q_n = k_2 + k_0^2\eta(2|\alpha_n|^2 + |\beta_n|^2)/2k_2 \tag{11.33}$$

where k_0, η and k_2 are defined in the previous section. As in (8.3), the position coordinate z in the phase factors in (11.30) and (11.31) is measured from the local origin $-z_n$. As seen from (11.30) and (11.31), the fields depend on the two complex amplitudes α_n and β_n; because of (11.32) and (11.33) this dependence is nonlinear.

We need also the expressions for the fields in the upper and lower media, which are for $z > 0$:

$$E = E_I \exp(-ik_0z) + E_R \exp(ik_0z) \tag{11.34}$$

$$\mu_0\omega H = k_0E_I \exp(-ik_0z) - k_0E_R \exp(ik_0z) \tag{11.35}$$

and for $z < -z_N$

$$E = E_T \exp[-ik_0(z + z_N)] \tag{11.36}$$

$$\mu_0\omega H = k_0 E_T \exp[-ik_0(z + z_N)]. \tag{11.37}$$

Again it simplifies subsequent expressions to measure z from the local origin $-z_N$.

The key to solving the optical problem is that there is no incoming wave in the lower medium, and for this reason we apply the boundary conditions starting at the lowest interface. Equating the values of E and H at $z = -z_N$ we find

$$\alpha_N + \beta_N = E_T \tag{11.38}$$

$$p_N\alpha_N - q_N\beta_N = k_0E_T. \tag{11.39}$$

Using the definitions of p_N and q_N the above are two nonlinear equations in α_N and β_N as unknowns. They can be solved (we suppose) to give these as functions of E_T:

$$\begin{pmatrix} \alpha_N \\ \beta_N \end{pmatrix} = \mathbf{F}_N \begin{pmatrix} E_T \\ 0 \end{pmatrix} \tag{11.40}$$

where $\mathbf{F}_N$ is an operator between two column vectors as shown; it may be called a nonlinear matrix. For the general interface at $z = -z_{n-1}$ the boundary conditions are

$$\alpha_{n-1} + \beta_{n-1} = \alpha_n \exp(-ip_nd_n) + \beta_n \exp(iq_nd_n) \tag{11.41}$$

$$p_{n-1}\alpha_{n-1} - q_{n-1}\beta_{n-1} = p_n\alpha_n \exp(-ip_nd_n) - q_n\beta_n \exp(iq_nd_n). \tag{11.42}$$

These determine α_{n-1} and β_{n-1} in terms of α_n and β_n, say

$$\begin{pmatrix} \alpha_{n-1} \\ \beta_{n-1} \end{pmatrix} = \mathbf{F}_{n-1} \begin{pmatrix} \alpha_n \\ \beta_n \end{pmatrix} \tag{11.43}$$

where $\mathbf{F}_{n-1}$ is another nonlinear matrix. Finally the boundary conditions at $z = 0$ are

$$E_I + E_R = \alpha_1 \exp(-ip_1 d_1) + \beta_1 \exp(iq_1 d_1) \tag{11.44}$$

$$k_0(E_I - E_R) = p_1 \alpha_1 \exp(-ip_1 d_1) - q_1 \beta_1 \exp(iq_1 d_1) \tag{11.45}$$

with solution

$$\begin{pmatrix} E_I \\ E_R \end{pmatrix} = \mathbf{F}_0 \begin{pmatrix} \alpha_1 \\ \beta_1 \end{pmatrix}. \tag{11.46}$$

Combining (11.40), (11.43) and (11.46), we obtain

$$\begin{pmatrix} E_I \\ E_R \end{pmatrix} = \mathbf{F}_0 \mathbf{F}_1 \dots \mathbf{F}_n \dots \mathbf{F}_N \begin{pmatrix} E_T \\ 0 \end{pmatrix}. \tag{11.47}$$

Formally, this involves a product of nonlinear matrices but really it represents a calculational scheme in which two simultaneous nonlinear equations are to be solved at each of the $N - 1$ stages indicated. Assuming that this numerical problem is tractable, the result is that E_I and E_R are determined given a value of E_T:

$$E_I = G(E_T), \qquad E_R = H(E_T) \tag{11.48}$$

where G and H are numerically defined functions. In practice, of course, it is E_I, not E_T, that is determined by the experimental conditions, so the final stage of the calculation involves the inversion of (11.46) to give the required complex reflection and transmission coefficients:

$$r = E_R/E_I = H(E_T)/G(E_T) \tag{11.49}$$

$$t = E_T/E_I = E_T/G(E_T). \tag{11.50}$$

We have outlined the calculational scheme for the simplest case, normal incidence as shown in figure 11.8. However, it can be seen that the calculational structure remains the same for more complicated cases such as oblique incidence because in any geometry the fields within each layer are given in terms of two complex amplitudes like α_n and β_n. Since all the steps represented by (11.47) are nonlinear none of the results of linear algebra can be applied to the product of operators. Similarly, there is no simplification for a periodic superlattice compared with a general, aperiodic multilayer since there is no rigorous equivalent of Bloch's theorem for a nonlinear system. Evaluation of the reflection and transmission coefficients therefore has to be numerical and a key point is the need to devise an efficient calculational scheme for the process of (11.47). This issue is addressed in the review by Dutta Gupta (1998), which also contains a discussion of various applications of the formalism.

In section 11.4.1 we discussed in detail only the results obtained for a single film with the approximation of linearized boundary conditions. The transmission and reflection properties of a single film with application of the full, nonlinear boundary conditions, but within the SVA, can be found as the $N = 1$ special case of the formalism presented here.

11.5   Self guiding of surface modes

It was mentioned at the beginning of this chapter that, with a positive Kerr effect ($\eta > 0$), self guiding can occur at the interface between two media with positive refractive indices. Loosely speaking, this is because the presence of an intense wave at the surface of a nonlinear medium leads to an increase in refractive index near the surface, so that the guiding condition (6.67) is satisfied in a self-consistent manner. Interest in this problem was stimulated by the work of Kaplan (1977), who investigated a number of effects arising at the interface between a linear and a nonlinear medium. Subsequently a number of authors studied the properties of the self-guided s-polarized mode at such an interface. It will be recalled from section 6.1 that the linear surface polariton is p-polarized and that a necessary condition for its existence is that the dielectric function of one of the media should be negative. This is not the case for the self-guided mode, so it is quite distinctive. Self guiding can also occur in p-polarization and the first exact calculation is due to Leung (1985b).

The earlier literature on self guided modes is reviewed by Stegeman *et al* (1986) and by Wendler (1986) and later references are given by Dutta Gupta (1998). Here we give the details of the calculation for the s-polarized mode, since this contains the essential ideas, then briefly discuss other work. Our treatment is based on that given by Wendler (1986) and is also similar to that of Stegeman *et al* (1985). The aim is to find the dispersion relation $\omega(q_x)$. However, since the mode is nonlinear, the intensity I enters the dispersion relation as a parameter, so that more precisely $\omega = \omega(q_x; I)$.

We use the axes and notation defined in figure 11.9 for the interface between the two media. We assume that the lower medium 2 is nonlinear with dielectric constant $\varepsilon_{20} + \eta|E|^2$ and that the upper medium is linear

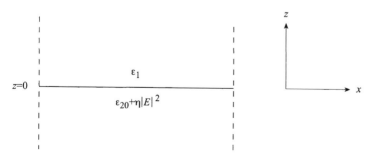

Figure 11.9. Notation for calculation of the self-guided s-polarized surface polariton.

with dielectric constant ε_1. Although the formal results to be derived are general, we simplify the later discussion by assuming that ε_1 and ε_{20} are positive constants. As might be supposed in this case, self guiding requires also that η be positive, so we assume this too. The formalism for negative η is no harder; details may be found in the references cited.

We look for an s-polarized mode, $\mathbf{E} = (0, E, 0)$, propagating as a plane wave along the x direction, so that all field quantities are proportional to $\exp(iq_x x - i\omega t)$. With these assumptions and with the Kerr nonlinearity of (10.24), the wave equation in medium 2 is

$$\frac{\mathrm{d}^2 E}{\mathrm{d}z^2} - \left(\kappa_2^2 - \frac{\omega^2}{c^2}\eta|E|^2\right)E = 0 \qquad (11.51)$$

where

$$\kappa_2^2 = q_x^2 - \varepsilon_{20}(\omega^2/c^2). \qquad (11.52)$$

It emerges that (11.51) has solutions with E real, so we assume this to be so. If (11.51) is multiplied by $\mathrm{d}E/\mathrm{d}z$, it can be integrated to give the energy integral

$$\left(\frac{\mathrm{d}E}{\mathrm{d}z}\right)^2 - \kappa_2^2 E^2 + \tfrac{1}{2}\eta\frac{\omega^2}{c^2}E^4 = 0. \qquad (11.53)$$

The constant of integration has been set equal to zero here on the assumption that the solution is localized at the interface so that $E \to 0$ and $\mathrm{d}E/\mathrm{d}z \to 0$ as $z \to -\infty$. Equation (11.53) can be integrated in terms of elementary functions, and the result (for positive η) is

$$E = \frac{2^{1/2}c}{\eta^{1/2}\omega}\kappa_2 \operatorname{sech}[\kappa_2(z + z_2)] \qquad (11.54)$$

where z_2 is a second constant of integration.

In the upper medium, the solution of the wave equation that satisfies the condition $E \to 0$ as $z \to \infty$ is

$$E = E_1 \exp(-\kappa_1 z) \qquad (11.55)$$

where

$$\kappa_1^2 = q_x^2 - \varepsilon_1(\omega^2/c^2) \qquad (11.56)$$

and E_1 is arbitrary since medium 1 is linear. Here again, in later discussion we make the simplifying assumption that ε_1 is a positive constant. Since the parallel component of magnetic field H_x is proportional to $\mathrm{d}E/\mathrm{d}z$ the boundary conditions at $z = 0$ are that E and $\mathrm{d}E/\mathrm{d}z$ are continuous. Application of these leads to

$$\frac{2^{1/2}c}{\eta^{1/2}\omega}\kappa_2 \operatorname{sech}(\kappa_2 z_2) = E_1 \qquad (11.57)$$

and

$$\frac{2^{1/2}c}{\eta^{1/2}\omega}\kappa_2^2 \operatorname{sech}(\kappa_2 z_2)\tanh(\kappa_2 z_2) = \kappa_1 E_1. \qquad (11.58)$$

Division of (11.58) by (11.57) eliminates E_1:

$$\kappa_2 \tanh(\kappa_2 z_2) = \kappa_1. \qquad (11.59)$$

We seek the dependence of q_x on ω. For a given ω, q_x enters (11.59) because of (11.52) and (11.56) for κ_1 and κ_2. However, (11.59) also contains the second unknown z_2 so that (11.59) does not give the complete solution. The reason is that the intensity is a nontrivial parameter in this problem. We therefore evaluate the power flow, namely the time average of the Poynting vector, in order to find a second equation between q_x and z_2. Before we do this, however, it is helpful to note some constraints on the form of the solution. First, since the tanh function never exceeds unity, (11.59) implies $\kappa_1 < \kappa_2$ or equivalently

$$\varepsilon_1 > \varepsilon_{20} \qquad (11.60)$$

as a necessary condition for a solution. Second, κ_1 and κ_2 are required to be real and positive. It follows from (11.52), (11.56) and (11.60) that when the upper medium is not optically active, $\varepsilon_1 > 0$, the condition for this is

$$q_x > \varepsilon_1^{1/2}\omega/c. \qquad (11.61)$$

In other words, the dispersion curve lies to the right of the medium-1 light line.

The Poynting vector $\mathbf{S} = \mathbf{E} \times \mathbf{H}$ has x and z components, but the latter contains a factor $\cos(q_x x - \omega t)\sin(q_x x - \omega t)$ so that its time average is zero. The results for the x component are found to be

$$S_x = \frac{2c^2 q_x \kappa_2^2}{\mu_0 \eta \omega^3}\operatorname{sech}^2[\kappa_2(z + z_2)]\cos^2(q_x x - \omega t) \qquad (11.62)$$

for $z < 0$ and

$$S_x = \frac{q_x}{\mu_0 \omega}E_1^2 \exp(-2\kappa_1 z)\cos^2(q_x x - \omega t) \qquad (11.63)$$

for $z > 0$. The average of the time-dependent terms gives $\frac{1}{2}$, and the time-averaged power flow per unit length in the y direction is given by the integral of S_x:

$$P = \frac{1}{2}\int_{-\infty}^{\infty} S_x^0 \, dz. \qquad (11.64)$$

Here the $\frac{1}{2}$ is the time average and S_x^0 denotes the time-independent factors in (11.62) and (11.63). The integrals are elementary and with use of (11.57) for E_1 one finds

$$P = \frac{c^2 q_x \kappa_2}{\mu_0 \eta \omega^3} \left[\frac{\kappa_2}{2\kappa_1} \operatorname{sech}^2(\kappa_2 z_2) + 1 + \tanh(\kappa_2 z_2) \right]. \qquad (11.65)$$

Use of (11.59) together with the identity $\operatorname{sech}^2 u = 1 - \tanh^2 u$ reduces this to

$$P = \frac{c^2 q_x \kappa_2}{\mu_0 \eta \omega^3} \left[1 + \frac{1}{2} \left(\frac{\kappa_1}{\kappa_2} + \frac{\kappa_2}{\kappa_1} \right) \right]. \qquad (11.66)$$

We have now completed the formulation of the problem. For given values of ω and P, (11.59) and (11.66) are simultaneous equations for the unknowns z_2 and q_x. The former enters the field profile via (11.54), but here we concentrate on the latter since this gives the dispersion relation in the form $q_x = q_x(\omega; P)$. Similar expressions to (11.59) and (11.66) hold for $\eta < 0$.

Equations (11.59) and (11.66) hold for any signs and any frequency dependence of ε_1 and ε_{20}. However, we now specialize in the case when ε_1 and ε_{20} are positive and frequency independent. Graphical analysis of (11.66) is then straightforward. We define the dimensionless variables

$$\Pi = \mu_0 \eta P \omega \qquad (11.67)$$

$$\xi = c q_x / \omega \qquad (11.68)$$

in terms of which (11.66) becomes

$$\Pi = \xi (\xi^2 - \varepsilon_{20})^{1/2} \left[1 + \frac{(\xi^2 - \varepsilon_1)^{1/2}}{2(\xi^2 - \varepsilon_{20})^{1/2}} + \frac{(\xi^2 - \varepsilon_{20})^{1/2}}{2(\xi^2 - \varepsilon_1)^{1/2}} \right]. \qquad (11.69)$$

The dispersion curves can be represented as a graph of Π versus ξ; figure 11.10 shows the equivalent graph of $\Pi\xi$ versus ξ. As can be seen from (11.69), this has the properties $\Pi\xi \to \infty$ as $\xi \to \varepsilon_1^{1/2}$ and $\Pi\xi \propto \xi^2$ for large ξ. Because of the definition (11.67) this graph is universal in the sense that for given material parameters ε_1 and ε_{20} it holds for all values of the power P with the nonlinear coefficient η appearing only as a scaling factor on P. The usual kind of dispersion curve is obtained as the graph of Π versus $\Pi\xi$ which in terms of the physical variables is $\mu_0 \eta P \omega$ versus $\mu_0 \eta P c q_x$, i.e. scaled frequency versus scaled wavenumber. This curve exists only for ω above a threshold value ω_m, and for $\omega > \omega_m$ propagation is bistable in the sense that two modes with different q_x values can propagate at the same frequency.

The above is a simplified version of the analysis given by Wendler (1986). His discussion goes further in a number of ways. First, he allows for the possible presence of a thin polarizable dipole-active layer at the interface by modifying the H-field boundary condition in the way that was done in (6.24) for a charge sheet. Second, he gives the formulae for $\eta < 0$

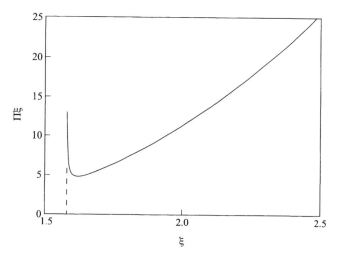

Figure 11.10. Dispersion curve for propagation of a s-polarized self-guided wave at the interface between a nonlinear medium 1 and a linear medium 2. Linear dielectric constants are $\varepsilon_1 = 2.5$, $\varepsilon_{20} = 2.0$.

as well as for $\eta > 0$. Third, in discussing and illustrating the results Wendler deals with dipole-active media in which the linear dielectric functions have the 'reststrahl' form of (2.146). As has been done here, however, Wendler excludes the effects of damping from the numerical illustrations.

Stegeman *et al* (1985) and Ariyasu *et al* (1985) use a similar approach to Wendler (1986) but go further in some respects. They include interfaces with metals as well as dielectrics and develop the formalism for the case when both media are nonlinear. Furthermore, they include the effects of damping in their numerical illustrations.

An analysis of the s-polarized self-guided modes of a nonlinear film bounded by linear media was given by Boardman and Egan (1985). The method is an extension of the formalism that has been presented here. The main complication is that since there is no boundary condition at infinity for the film the constant of integration can no longer be set equal to zero, as was done in (11.53). Consequently the fields in the film are found in terms of elliptic functions, of which the hyperbolic functions appearing in (11.54) and related equations are special cases. Another special geometry for which analytic solutions were found consists of a finite linear superlattice bounded by nonlinear media (Mihalache and Wang 1991); this is solved by a combination of the transfer-matrix technique for the superlattice, together with solutions of the form of (11.54) for the nonlinear media. Finally a comprehensive account of s-polarized nonlinear modes in arbitrary multilayer structures was given by Vassiliev and Cottam (2000). Their formalism, which allows all media in an N-layer structure to be nonlinear in general, makes use of a phase-trajectory method to simplify the analysis and to classify the modes.

The assumption made in the discussion so far is that the mode propagates parallel to the surface as a plane wave $\exp(iq_x x - i\omega t)$. However, there are other possibilities and in particular it is of interest to consider soliton-like propagation along x; this has been done by a number of groups. Boardman *et al* (1986) applied the slowly-varying amplitude approximation (SVA) as reviewed in section 10.4 to show that such propagation is indeed possible. Likewise, Mihalache *et al* (1988) applied the SVA to study the evolution of an s-polarized soliton along the three-layer structure comprising a linear substrate, a linear film and a thick nonlinear overlayer. They include a fifth-order nonlinearity so that (10.25) is replaced by the more general

$$n^{\mathrm{eff}} = n_0 + n_2|E|^2 + n_4|E|^4. \tag{11.70}$$

The additional term is expected to come into play as the pulse duration in pulsed lasers decreases.

One obvious weakness of the method that has been reviewed and the various developments of it is that they rely on the fact that the nonlinear wave equation (11.51) can be integrated analytically so that the treatment is restricted to the non-saturating nonlinearity of (10.24). One exception is the particular saturable form postulated by Wood *et al* (1988) for which an analytic form can be found for the fields. An alternative approach of much greater generality was put forward by Leung (1985a,b). One of the main advantages of this method is that it can be used for p- as well as s-polarized modes and we therefore give a brief introduction to the formalism in the context of the p-polarized self-guided wave.

Even for the non-saturating nonlinearity of (10.24) a simple attempt to extend the previous formalism to p-polarization runs into the difficulty that the electric field now contains both components E_x and E_z, so that the non-linear term in (10.24) takes the form $\eta(|E_x|^2 + |E_z|^2)$. The wave equation (11.51) is replaced by coupled equations in E_x and E_z, but because of the form of the nonlinearity these are not tractable in any simple way. A number of early treatments sought to overcome this difficulty by postulating an anisotropic nonlinearity so that, for example, ε_{xx} would be taken as linear while ε_{zz} would be taken in the form of (10.24) with E replaced by E_z. This has the advantage that the coupled equations become tractable, but there is no obvious physical justification for this form of $\tilde{\varepsilon}$.

Leung's analysis for p-polarization uses the **B** field since this has only the one component B_y, written as B for simplicity. For an isotropic nonlinear medium, the dielectric constant is written as a general function:

$$\varepsilon = \varepsilon_1 + \varepsilon_2(|\mathbf{E}|^2) \tag{11.71}$$

so that (10.24) and the saturating forms (10.27) and (10.28), for example, are included as special cases. For the case of a plane wave $\exp(iq_x x - i\omega t)$ it follows from Maxwell's equations that

$$(B'/\varepsilon)' = [(\xi^2/\varepsilon) - 1]B \tag{11.72}$$

where the primes denote differentiation with respect to z. Here ε is given by (11.71) and ξ is the variable introduced in (11.68). In terms of B, the argument in (11.71) is

$$E^2 = \xi^2(B/\varepsilon)^2 + (B'/\varepsilon)^2. \tag{11.73}$$

The crucial step in Leung's analysis is to replace (11.71) by the inverse function, defined in the sense that

$$|\mathbf{E}|^2 = I(\varepsilon - \varepsilon_1) \tag{11.74}$$

is the solution of (11.71) for $|\mathbf{E}|^2$. After further analysis, which is very clearly given in Leung (1985b) and therefore not repeated here, the boundary conditions can be expressed in terms of this function. For the interface between a linear and a nonlinear medium, the result is

$$\xi^2 = \frac{\varepsilon_b\varepsilon^2(0)\{[\varepsilon(0) - \varepsilon_b]I(\varepsilon(0) - \varepsilon_1) - J(\varepsilon(0) - \varepsilon_1)\}}{\varepsilon(0)[\varepsilon^2(0) - \varepsilon_b^2]I(\varepsilon(0) - \varepsilon_1) - [\varepsilon^2(0) + \varepsilon_b^2]J(\varepsilon(0) - \varepsilon_1)} \tag{11.75}$$

where ε_b is the dielectric constant of the bounding medium, J is defined as

$$J(\varepsilon) = \int_0^\varepsilon I(\eta)\,d\eta \tag{11.76}$$

and $\varepsilon(0)$ is the value of the dielectric function (11.71) at the surface of the nonlinear medium.

Equation (11.75) gives ξ as a function of the surface value $\varepsilon(0)$, which as seen from (11.71) depends on the electric-field magnitude at the surface. As in the simpler approach reviewed earlier, this ultimately depends on the wave intensity. As before, the next step in the analysis is therefore to apply Poynting's theorem in order to relate $\varepsilon(0)$ to the intensity. The details are given in the original paper and again not repeated here. Leung's solution is exact in the sense that the problem is reduced to quadratures. As is clear, it applies for any form of functional dependence in (11.71) and in particular it can be used for saturating nonlinearities. The method may also be used for s-polarization (Leung 1985a).

For the simple form (10.24) and weak nonlinearity, $\eta|E|^2 \ll \varepsilon$, the general result can be expanded to first order in $|E|^2$. It is found (Leung 1985b) that

$$\xi = \left(\frac{\varepsilon_b\varepsilon_1}{\varepsilon_b + \varepsilon_1}\right)^{1/2}\left(1 + \frac{\varepsilon_b\eta|\mathbf{E}(0)|^2}{4\varepsilon_1(\varepsilon_b + \varepsilon_1)}\right) \tag{11.77}$$

where $\mathbf{E}(0)$ is the field at the surface. Equation (11.77) will be recognized as a generalization of the linear result (6.7). A similar empirical expression was applied by Chen and Carter (1982, 1984) to analyse experimental results on this mode.

As mentioned, Leung's analysis applies to a general form of the functional dependence in (11.71). However, the detailed discussion in the original papers is for the non-saturating form (10.24). An explicit application for the case of the saturating nonlinearity (10.27) is given by Groza and Strizhevskii (1991).

11.6   Gap solitons and related modes

As was seen in section 10.4, the Lighthill criterion is a necessary condition for the existence of solitons. Figure 8.9(a) shows that for frequencies near to a stop band for propagation through a periodic superlattice, the second derivative k_0'' varies rapidly with frequency and has opposite signs just below and just above the stop band, positive and negative respectively. It might therefore be expected that soliton propagation would occur in a periodic nonlinear superlattice at such frequencies. A clear indication that this should be the case came from a numerical study by Chen and Mills (1987c) of optical propagation at normal incidence through a finite nonlinear superlattice. The unit cell was taken as two films, one linear with dielectric constant $\varepsilon_1^{(0)}$ and the other a nonlinear Kerr medium with dielectric constant $\varepsilon_2 = \varepsilon_2^{(0)}[1 + \lambda|E|^2]$. For the numerical studies the authors took $\varepsilon_1^{(0)} = 2.25$ and $\varepsilon_2^{(0)} = 4.5$ with corresponding thicknesses $d_1 = d_2 = 0.125$ in units of the vacuum wavelength. The formalism was an extension of the method applied by Chen and Mills (1987a) for transmission through a single nonlinear film, and numerical results for transmission through a superlattice with 20 unit cells are shown in figure 11.11.

The inset to figure 11.11(a) shows the linear dispersion curve ($\lambda = 0$) of an infinite superlattice in terms of the dimensionless frequency $\Omega = \omega[\varepsilon_1^{(0)}]^{1/2}(d_1 + d_2)/c$. In the transmission study, the incident frequency was taken to correspond to point a, $\Omega = 0.75\pi$, just above the gap edge at $\Omega = 0.74\pi$. In the linear régime the transmission would be zero, or very small, because a is in the stop band. However, loosely speaking, for negative λ the nonlinearity lowers ε_2 so that the gap edge rises above a and transmission becomes possible. The results of the transmission calculation are shown as the main part of figure 11.11(a). The horizontal axis is $\tilde{\lambda} = \lambda E_0^2$ where E_0 is the incident field strength, so that $\tilde{\lambda}$ is a dimensionless measure of the incident intensity. The results have the very striking property that 100% transmission occurs at a series of points P_1, P_2 etc. These points are at negative values of $\tilde{\lambda}$, where the Lighthill criterion is satisfied, and Chen and Mills propose that they are associated with excitation of a stationary soliton within the superlattice. They support this contention, to begin with, by the numerical result for the field intensity $|E|^2$ as a function of position in the superlattice, as shown in figure 11.11(b). The envelope of this graph is close to a sech^2 function, so comparison with (10.60) suggests that the intensity is indeed a soliton.

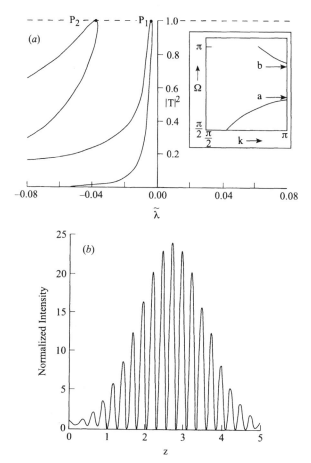

Figure 11.11. (a) Calculated transmission in a finite nonlinear superlattice, with insert showing a portion of the linear dispersion curve. The notation and numerical values are given in the text. (b) Field intensity at transmission point P_1 as a function of position in the superlattice. After Chen and Mills (1987c).

Chen and Mills mention further that similar effects occur for incident frequencies near the lower edge of the next band, point b, but now for positive λ in keeping with the Lighthill criterion. Furthermore, they show numerical results rather like figure 11.11(b) for their calculation of the field amplitude in a localized excitation in an infinite superlattice; here again the result has a sech-function amplitude.

The first analytic discussion of these results was given by Mills and Trullinger (1987) and later treatments were given by de Sterke and Sipe (1988, 1989a). The latter start from the observation that the intensity curve in figure 11.11(b) and the field-amplitude curve for the infinite superlattice that is also calculated by Chen and Mills (1987c) are reminiscent of a modulated Bloch function in electron band theory. The analysis of de Sterke and Sipe is for a nonlinear dielectric function $\varepsilon(x) + \eta(x)|E(x)|^2$ in which both the linear and nonlinear coefficients are periodic along the axis x normal to the superlattice layers. The Bloch functions $\phi_m(x)\exp(-i\omega_m t)$ of the linear lattice with dielectric function $\varepsilon(x)$ are used as basis functions for expansion. The modulated Bloch wave in the nonlinear material is written as

$$E(x,t) = a(x,t)\phi_m(x)\exp(-i\omega_m t) + \text{c.c.} \tag{11.78}$$

and the key result is that within the slowly-varying-amplitude approximation the envelope function $a(x,t)$ satisfies the nonlinear Schrödinger equation (10.54). As in the detailed discussion of section 10.4, the envelope propagates at the group velocity $v_g = d\omega_m/dk$. This is consistent with the numerical discovery of a stationary soliton by Chen and Mills (1987c) since v_g is zero at the extremities of the stop band.

The question of the necessary conditions for the experimental observation of gap solitons was addressed in a number of subsequent papers, for example de Sterke and Sipe (1989b) and de Sterke (1992). The first experimental results were reported by Mohideen et al (1995). The observations were on a grating produced by ultraviolet laser interference on a GeO_2-doped silica glass fibre. The grating period was close to the central wavelength 1053 nm of the pulsed laser used to observe the soliton. The insert to figure 11.12(a) shows the calculated transmission spectrum of the corresponding infinite superlattice which is seen to have a stop band near the laser wavelength. The main part shows a calculated transmission spectrum for the actual finite superlattice together with the Fourier transform of the 60 ps pulses that were used. The exact wavelength range covered by the stop band could be varied by longitudinal strain on the fibre which altered the grating period. The transmitted power is shown in figure 11.12(b) for three different tunings of the stop band. The main result concerns the circles in figure 11.12(b); these show that when the central wavelength of the pulse coincides with the 50% transmission point at the stop-band edge, as in figure 11.12(a), the transmitted power increases in a more than linear way with the input intensity. The interpretation of this as evidence of gap–soliton propagation is supported in Mohideen et al (1995) by a comparison between calculated and observed transmitted pulse shapes. For completeness, it should be said that in figure 11.12(b) the squares correspond to strain tuning so that the pulse falls within the pass band; the increase in attenuation with increasing input intensity is attributed to nonlinear loss mechanisms. The lowest set of points (diamonds) are for pulse wavelength in the centre of the stop band where transmission is less than 20%.

The discovery of gap solitons led to a range of related studies. As has been seen, the original work was concerned with propagation just within the stop band of the linear structure. Feng and Kneubühl (1993) studied normal-incidence propagation through a medium with periodic linear

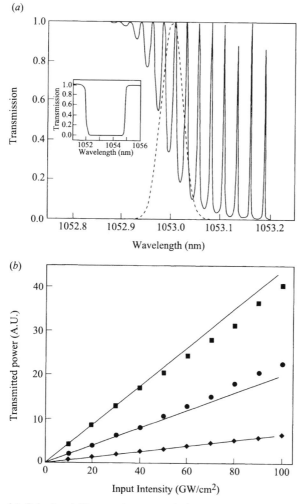

Figure 11.12. (a) Calculated fibre grating transmission spectrum used in observation of gap solitons, together with (dotted) Fourier transform of incident pulse. Inset shows a calculated transmission curve for the corresponding infinite superlattice. (b) Transmitted power as a function of incident intensity for three different strain tunings of stop band relative to central laser wavelength λ_0. ■, λ_0 in pass band; ●, λ_0 on stop-band edge, as in (a); ◆, λ_0 in centre of stop band. After Mohideen *et al* (1995).

response described by the refractive index variation

$$n(z) = n_0 + n_1 \cos(2k_B z) \qquad (11.79)$$

and a constant Kerr nonlinearity. They apply an analysis similar to that applied in section 11.4.1 for the nonlinear FP oscillator, so they assume that the solution to the nonlinear wave equation takes the form of a sum of forward and backward waves:

$$E(z, t) = E_f(z, t) \exp(-i\omega t + ik_B z) + E_b(z, t) \exp(-i\omega t - ik_B z). \quad (11.80)$$

This is similar to (11.16) with the inclusion of time dependence. Application of the SVA leads to the coupled equations

$$\frac{\partial E_f}{\partial z} + \frac{n_0}{c}\frac{\partial E_f}{\partial t} = i\Delta k E_f + i\kappa E_b + i\alpha(|E_f|^2 + 2|E_b|^2)E_f \quad (11.81)$$

$$\frac{\partial E_b}{\partial z} - \frac{n_0}{c}\frac{\partial E_b}{\partial t} = -i\Delta k E_b - i\kappa E_f - i\alpha(2|E_f|^2 + |E_b|^2)E_b \quad (11.82)$$

for the amplitudes. Here $\kappa = k_B n_1/2n_0$ accounts for the linear coupling due to the periodicity, $\omega_B = k_B c/n_0$ is the Bragg frequency, $\Delta k = (\omega - \omega_B)n_0/c$ is a measure of the frequency detuning from the Bragg frequency and α is proportional to the Kerr coefficient η. Equations (11.81) and (11.82) are a generalization of (11.17) and (11.18). The analysis of these coupled nonlinear equations shows that 'in-gap' solitons with ω in the stop band can propagate; these are analogous to the original Chen–Mills solitons. In addition, the authors find new solutions in the form of 'out-gap' solitons propagating within a pass band.

The work discussed so far uses the standard continuum formulation of nonlinear optics as reviewed in section 10.1. A connection with the origin of nonlinear effects in the underlying atomic transitions was made by Kozhekin and Kurizki (1995). They considered a medium with periodic linear properties, similar to (11.79), in which the propagating light interacts with a two-level atomic system. It is well known (see e.g. Shen 1984) that in the equivalent uniform medium *self-induced transparency* (SIT) can occur, leading to soliton propagation near to the atomic resonance frequency. Kozhekin and Kurizki show that a similar SIT involving soliton propagation can occur within the band gap of the periodic medium.

As mentioned at the beginning of this chapter, a large proportion of the work on nonlinear superlattices has been concerned with optical properties, partly because of their obvious practical importance, and these have been the focus of the chapter. However, we may mention here the investigation of gap solitons in discrete periodic media that was initiated by Kivshar and collaborators (Kivshar 1992, 1993, Kivshar *et al* 1994). The fundamental point is that discrete nonlinear systems can display a richer variety of dynamical behaviour than their continuum counterparts. These authors consider a variety of models, a typical one being described by the equation of motion (Kivshar *et al* 1994):

$$m\frac{d^2 u_n}{dt^2} = K(u_{n+1} + u_{n-1} - 2u_n) - m\omega_0^2 u_n - \alpha u_n^2 - \beta u_n^3. \quad (11.83)$$

In comparison with the equation of motion (2.1) for the linear monatomic lattice, this contains the additional linear term $-m\omega_0^2 u_n$ so that the long-wavelength ($q = 0$) mode has frequency ω_0 rather than zero, as in (2.4). More importantly, the final two terms describe single-site nonlinearity. Arguably, a more physical model would include nonlinear forces between neighbouring sites rather than single-site nonlinearity, so that the nonlinear terms might be replaced by

$$F_{NL} = K_4[(u_{n-1} - u_n)^3 + (u_{n+1} - u_n)^3]. \tag{11.84}$$

Kivshar (1993) does raise this possibility, but mentions that the resulting dynamics is the same as that found from (11.83).

Kivshar *et al* (1994) apply the SVA to (11.83) and find that the envelope function $\psi_n(t)$ satisfies the *discrete nonlinear Schrödinger equation*

$$i\frac{d\psi_n}{dt} + Q(\psi_{n+1} + \psi_{n-1} - 2\psi_n) + \lambda|\psi_n|^2\psi_n = 0 \tag{11.85}$$

where $Q = K/2m\omega_0$ and

$$\lambda = \frac{1}{2m\omega_0}\left(\frac{10\alpha^2}{3\omega_0^2} - 3\beta\right). \tag{11.86}$$

Equation (11.85) leads to a considerable variety of dynamical behaviour for which reference may be made to the original papers. Here we focus on two results of particular relevance to the present section. First, a *nonlinearity-induced gap* can appear in the spectrum. This is at wavenumber $q = 2\pi/a$ (where a is the lattice constant) for which all even-n displacements are equal, as are all odd-n displacements. The frequency gap is

$$\Delta\omega = |\lambda(w_0^2 - v_0^2)| \tag{11.87}$$

where w_0 and v_0 are defined as follows. On one side of the gap, all the odd displacements are zero and all the even displacements are equal to v_0. On the other side, the situation is reversed, with the odd displacements equal to w_0. In effect, the nonlinearity leads to a distinction between the odd and even sites so that the system resembles the linear diatomic lattice, as discussed in section 2.1.2. The second result is that within the gap a localized gap soliton is found as a solution of (11.83); the pattern of displacements in this mode is illustrated in figure 11.13.

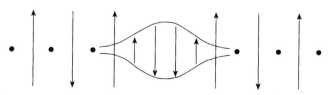

Figure 11.13. Pattern of site displacements in the self-induced gap soliton. After Kivshar (1993).

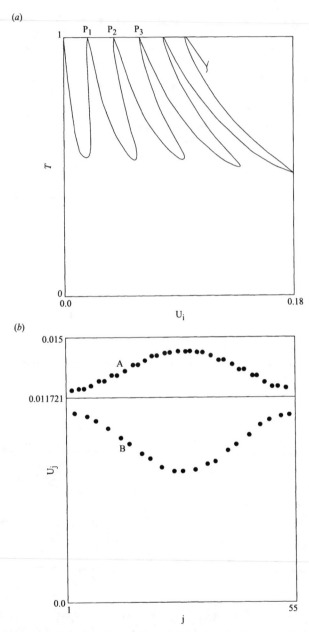

Figure 11.14. Calculated results for a 55-layer Fibonacci superlattice with linear dielectric functions $\varepsilon_A = 4$, $\varepsilon_B = 9$ and normal incidence in vacuum. Frequency ω and layer thicknesses d_A, d_B chosen so that $\delta_A = (\omega/c)\varepsilon_A^{1/2}d_A = \delta_B = (\omega/c)\varepsilon_B^{1/2}d_B = 2\pi$. (a) Transmitted intensity versus incident intensity. (b) Total intensity versus layer number at point P_1 of (a). A layers upper points, B layers lower points. After Dutta Gupta and Ray (1989).

The examples discussed so far have been based on periodic superlattices and the question arises whether the periodicity is essential for the appearance of the gap soliton. This has been addressed by a number of groups. Kahn *et al* (1989) carried out numerical simulations for both periodic and quasiperiodic (Fibonacci) nonmagnetic/antiferromagnetic Kerr type superlattices, but were not able to find stationary solitons resembling gap solitons in the quasiperiodic structures. (This work is part of a series of studies on nonlinear properties of antiferromagnets which will be discussed further in chapter 12.) However, Dutta Gupta and Ray (1989, 1990) were able to show the existence of gap solitons in Fibonacci superlattices. Their study was of normal-incidence transmission through a superlattice defined as in section 8.5, so that successive members of the family are given by (8.63). Both A and B building materials were taken as Kerr-type nonlinear media and the numerical studies used the nonlinear transfer-matrix formalism reviewed in section 11.4.2. The calculated transmission coefficient for a 55-layer sequence, i.e. S_9, is shown in figure 11.14(a) and the resemblence to the original Chen–Mills result for a periodic superlattice, figure 11.11(a), is obvious. The 100% transmission points, P_1, P_2, P_3 etc. are candidates for stationary solitons. Indeed this is what they are, as can be seen from figure 11.14(b). This shows the total intensity, i.e. the sum of the forward and backward intensities, on layers A and B at point P_1. It can be seen that these do have a $sech^2$ shape. Dutta Gupta and Ray (1989) mention in addition that points P_2, P_3 correspond to intensity distributions like 2 and 3 soliton structures. These results are not specific to the 55-layer structure, as Dutta Gupta and Ray demonstrate by showing similar results for the 233-layer structure S_{12}.

References

Agranovich V M and Kravtsov V E 1985 *Solid State Commun.* **55** 85
Agrawal G P 1989 *Nonlinear Fiber Optics* (San Diego: Academic)
Ariyasu J, Seaton C T, Stegeman G I, Maradudin A A and Wallis R F 1985 *J. Appl. Phys.* **58** 2460
Bloembergen N and Pershan P S 1962 *Phys. Rev.* **128** 606
Boardman A D and Egan D 1985 *IEEE J. Quant. Electron.* QE **21** 1701
Boardman A D, Cooper G S, Maradudin A A and Shen T P 1986 *Phys. Rev.* B **34** 8273
Born M and Wolf E 1986 *Principles of Optics* (Oxford: Pergamon)
Boyd R W and Sipe J E 1994 *J. Opt. Soc. Am.* B **11** 297
Chen W and Mills D L 1987a *Phys. Rev.* B **35** 524
Chen W and Mills D L 1987b *Phys. Rev.* B **36** 6269
Chen W and Mills D L 1987c *Phys. Rev. Lett.* **58** 160
Chen Y J and Carter G M 1982 *Appl. Phys. Lett.* **41** 307
Chen Y J and Carter G M 1984 *J. Phys. Paris Colloq.* **45** C5-261
Chew K-H, Osman J and Tilley D R 2001 *Opt. Commun.* **191** 393
Deck R T and Sarid D 1982 *J. Opt. Soc. Am.* **72** 1613

De Sterke C M 1992 *Phys. Rev. A* **45** 2012

De Sterke C M and Sipe J E 1988 *Phys. Rev. A* **38** 5149

De Sterke C M and Sipe J E 1989a *Phys. Rev. A* **39** 5163

De Sterke C M and Sipe J E 1989b *Optics Lett.* **14** 871

Driscoll T A and Guidotti D 1983 *Phys. Rev. B* **28** 1171

Dutta Gupta S 1998 *Progress in Optics* ed E Wolf **38** 1

Dutta Gupta S and Agarwal G S 1987 *J. Opt. Soc. Am. B* **4** 691

Dutta Gupta S and Ray D S 1989 *Phys. Rev. B* **40** 10604

Dutta Gupta S and Ray D S 1990 *Phys. Rev. B* **41** 8047

Feng J and Kneubühl F K 1993 *IEEE J. Quant. Electron.* **29** 590

Fischer G L, Boyd R W, Gehr R J, Jenekhe S A, Osaheni J A, Sipe J E and Weller-Brophy
L A 1995 *Phys. Rev. Lett.* **74** 1871

Flytzanis C, Hache F, Klein M C, Ricard D and Roussignol P 1991 *Progress in Optics*
ed E Wolf **29** 321

Goldstone J A and Garmire E 1984 *Phys. Rev. Lett.* **53** 910

Groza A D and Strizhevskii V L 1991 *Phys. Stat. Sol. (b)* **163** 381

Guidotti D, Driscoll T A and Gerritson H J 1983 *Solid State Commun.* **46** 337

Guyot-Sionnest P, Chen W and Shen Y R 1986 *Phys. Rev. B* **33** 8254

Hashizume N, Ohashi M, Kondo T and Ito R 1995 *J. Opt. Soc. Am B* **12** 1894

Kahn L M, Huang K and Mills D L 1989 *Phys. Rev. B* **39** 12449

Kaplan A E 1977 *Sov. Phys. JETP* **45** 896

Kivshar Yu. S 1992 in *Future Directions of Nonlinear Dynamics in Physical and Biological
Systems* ed P L Christiansen, J C Eilbeck and R D Parmentier (London: Plenum)

Kivshar Yu. S 1993 *Phys. Rev. Lett.* **70** 3055

Kivshar Yu. S, Haelterman M and Sheppard A P 1994 *Phys. Rev. E* **50** 3161

Kozhekin A and Kurizki G 1995 *Phys. Rev. Lett.* **74** 5020

Landau L D and Lifshitz E M 1960 *Electrodynamics of Continuous Media* (Oxford:
Pergamon)

Leung K M 1985a *Phys. Rev. A* **31** 1189

Leung K M 1985b *Phys. Rev. B* **32** 5093

Leung K M 1989 *Phys. Rev. B* **39** 3590

Lim S C, Osman J and Tilley D R 1997 *J. Phys.: Condens. Matter* **9** 8297

Lüpke G, Bottomley D J and van Driel H M 1994 *J. Opt. Soc. Am. B* **11** 33

Marburger J H and Felber F S 1978 *Phys. Rev. A* **17** 335

Mihalache D, Mazilu D, Bertolotti M, Sibilia C and Fedyanin V K 1988 *Solid State
Commun.* **66** 517

Mihalache D and Wang R P 1991 *J. Appl. Phys.* **69** 1892

Mills D L and Trullinger S E 1987 *Phys. Rev. B* **36** 947

Mizrahi V and Sipe J E 1988 *J. Opt. Soc. Am. B* **5** 660

Mohideen U, Slusher R E, Mizrahi V, Erdogan T, Kuwata-Gonokami M, Lemaire P J,
Sipe J E, de Sterke C M and Broderick N G R 1995 *Opt. Lett.* **20** 1674

Ostrowsky D B and Reinisch R 1992 *Guided Wave Nonlinear Optics* Dordrecht: Kluwer

Quail J C, Rako J G, Simon H J and Deck R 1983 *Phys. Rev. Lett.* **50** 1987

Reining L, Del Sole R, Cini M and Jiang G P 1994 *Phys. Rev. B* **50** 8411

Reinisch R and Vitrant G 1994 *Prog. Quant. Electr.* **18** 1

Sarid D 1981 *Phys. Rev. Lett.* **47** 1927

Shen Y R 1984 *The Principles of Nonlinear Optics* (New York: Wiley)

Shoji I, Kondo T, Kitamoto A, Shirane M and Ito R 1997 *J. Opt. Soc. Am. B* **14** 2268

Sipe J E, Moss D J and van Driel H M 1987 *Phys. Rev. B* **35** 1129

Stegeman G I, Burke J J and Hall D G 1982 *Appl. Phys. Lett.* **41** 906

Stegeman G I and Seaton C T 1985 *J. Appl. Phys.* **58** R57

Stegeman G I, Seaton C T, Ariyasu J, Wallis R F and Maradudin A A 1985 *J. Appl. Phys.* **58** 2453

Stegeman G I, Seaton C T, Hetherington W M, Boardman A D and Egan P 1986 *Electromagnetic Surface Excitations* ed R F Wallis and G I Stegeman (Berlin: Springer) p 261

Vassiliev O N and Cottam M G 2000 *Surface Rev. and Lett.* **7** 89

Wendler L 1986 *Phys. Stat. Solidi (b)* **135** 759

Wijers C M J, de Boeij P L, van Hasselt C W and Rasing Th. 1995 *Solid State Commun.* **93** 17

Wood V E, Evans E D and Kenan R P 1988 *Optics Commun.* **69** 156

Chapter 12

Nonlinear properties: magnetic surfaces and superlattices

By analogy with what was done in chapter 11 for the nonmagnetic case, we now extend the results for magnetic materials covered in chapter 10, specifically section 10.5, to include the nonlinear spin–wave and soliton effects associated with surfaces, films and (to a very limited extent) super-lattices. At the time of writing, this is a rapidly developing and changing area of research, both experimentally and theoretically, and so we will not be able to provide a complete treatment. Instead we shall provide an introduction and highlight certain developments, so that readers have the basis for further studies. Some books and review articles covering particular aspects of this topic are by Cottam (1994), Srinivasan and Slavin (1995), Marcelli and Nikitov (1996) and Hillebrands (2000).

In general, the inclusion of effects due to surfaces and interfaces means that there are boundary conditions (usually themselves in a nonlinear form) to take into account, and also (for a film or a superlattice) spatial quantization will occur as an additional feature to those described for the bulk nonlinear magnetic media in section 10.5. Apart from thin films and superlattices, other geometries that have attracted attention (and will also be included briefly in our discussions) are spheres and circular discs.

12.1 Interactions between spin waves

We begin by discussing within a dipole-exchange model the nonlinear three-magnon and four-magnon interaction processes and associated parametric instabilities in the case of limited geometries. The development in theory, whether carried out within a macroscopic or microscopic formalism, is broadly similar to that described in section 10.5 for bulk materials. The main difference comes from the expansion of the fluctuating magnetization for each spatial dimension in which there is no longer any translational symmetry. The following discussion will be largely qualitative because of the mathematical complexity involved, and most of the specific examples of the theory will be the simplest case, namely a ferromagnetic thin film.

The macroscopic (continuum) nonlinear theory for a thin film was developed in a series of papers, mainly by Kalinikos (1983), Slavin and Kalinikos (1987), Boardman and Nikitov (1988), Melkov (1988) and Kalinikos and Slavin (1991). Reviews are given by Cottam and Slavin (1994) and Kalinikos (1994). The microscopic theory of interactions between the dipole-exchange spin–wave in ultrathin ferromagnetic films was developed by Costa Filho *et al* (1998, 2000) and later extended to anti-ferromagnetic materials by Pereira and Cottam (2003).

12.1.1 Macroscopic nonlinear theory

We therefore begin with an outline of the macroscopic nonlinear theory for ferromagnetic thin films before moving on to other cases. In contrast to the linear dipole-exchange theory, based on the approach introduced by Wolfram and Dewames (1972) and described in section 5.4, it has proved to be more useful for nonlinear applications to use the alternative formalism (so far mentioned briefly in section 5.4), based on the tensorial Green-function approach of Vendik and Chartorizhskii (1970). Details are to be found in the papers referenced above by Kalinikos, Slavin and co-workers.

Many of the nonlinear processes in films turn out to be strongly direc-tional. Thus we adopt here a more general film geometry than considered in chapter 5 by allowing the applied magnetic field to be in an arbitrary direction, causing the static magnetization vector $\mathbf{M}_0$ to be inclined at a variable angle θ to the film normal. This is depicted in figure 12.1, where the two sets of coordinate axes (ξ, ρ, ζ) and (x, y, z) are defined. We start by Fourier transforming the variable magnetization $\mathbf{m}(\xi, \mathbf{r}_{\|}, t)$ of the film with respect to the in-plane coordinates, where $\mathbf{r}_{\|} = (\rho, \zeta)$ is a 2D vector:

$$\mathbf{m}(\xi, \mathbf{r}_{\|}, t) = \int \mathbf{m}_{\mathbf{q}_{\|}}(\xi) \exp[i(\mathbf{q}_{\|} \cdot \mathbf{r}_{\|} - \omega t)] \, \mathrm{d}^2 \mathbf{q}_{\|}. \tag{12.1}$$

In the linear approximation, with both the electromagnetic and exchange boundary conditions at the film surfaces taken into account, the magnetiza-tion distribution $\mathbf{m}_{\mathbf{q}_{\|}}(\xi)$ across the thickness $(-\frac{1}{2}L < \xi < \frac{1}{2}L)$ can be expanded in the form

$$\mathbf{m}_{\mathbf{q}_{\|}}(\xi) = \sum_n \mathbf{m}_{n\mathbf{q}_{\|}} \Phi_n(\xi) \tag{12.2}$$

where integer $n \, (= 0, 1, 2, \ldots)$ labels the discrete spin–wave branches and the $\Phi_n(\xi)$ are spin–wave eigenfunctions (Kalinikos 1983, Slavin and Kalinikos 1987). For example, in the exchange-dominated case for a perpendicularly magnetized film $(\theta = 0)$, the z and ξ axes coincide (see figure 12.1), and it is

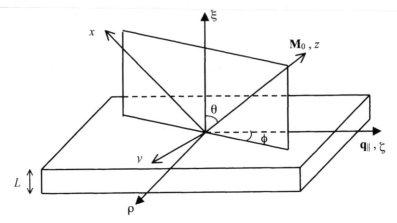

Figure 12.1. The assumed film geometry showing the coordinate axes (ξ, ρ, ζ) and (x, y, z). These are defined such that ξ and ζ are along the film normal and the in-plane wavevector $\mathbf{q}_{\parallel}$, respectively, while z is along the magnetization direction.

found from a straightforward generalization of results in section 4.2.2 that

$$\Phi_n(z) \propto \{\cos[q_{zn}(z + \tfrac{1}{2}L)] + (\eta'/q_{zn}) \sin[q_{zn}(z + \tfrac{1}{2}L)]\}. \tag{12.3}$$

We are using η and η' to denote the pinning parameters at the upper and lower film surfaces in figure 12.1. The discrete wavevector components q_{zn}, which depend on both η and η', are found from an equation of the same form as (4.40).

In the dipole–exchange case of interest here (and for $\theta \neq 0$) the results are much more complicated. Nevertheless, we can illustrate the main ideas by taking the case of zero pinning ($\eta = \eta' = 0$) at the film surfaces. It then follows (see e.g. Slavin and Kalinikos 1987) that the approximate dispersion equation for the discrete spin–wave branches is

$$\omega_n(\mathbf{q}_{\parallel}) = \{[\omega_0 + D(q_{\parallel}^2 + q_{\xi n}^2)][\omega_0 + D(q_{\parallel}^2 + q_{\xi n}^2) + \omega_{\mathrm{M}} F_{nn}]\}^{1/2} \tag{12.4}$$

where the quantized wavevector components $q_{\xi n}$ are $n\pi/L$ and the notations for the exchange factor D and the characteristic frequencies ω_0 and ω_{M} are the same as in chapter 5. The above quantity F_{nn} has the form

$$F_{nn} = (1 - P_{nn} \cos^2 \phi) \sin^2 \theta$$
$$+ P_{nn}\{\cos^2 \theta + [\omega_{\mathrm{M}}/\Omega_n(\mathbf{q}_{\parallel})](1 - P_{nn}) \sin^2 \theta \sin^2 \phi\} \tag{12.5}$$

where P_{nn} is a matrix element of the dipole–dipole interaction, $\Omega_n(\mathbf{q}_{\parallel})$ is the spin–wave frequency in the exchange limit, i.e. it is (12.4) with ω_{M} replaced by zero, and the angles are defined in figure 12.1. The above results are derived by treating the dipole–dipole effects perturbatively; they follow from using standard first-order perturbation theory and the exchange eigenfunctions $\Phi_n(\xi)$. It is also found that

$$P_{00}(q_{\parallel}L) = 1 - [\{1 - \exp(-q_{\parallel}L)\}/q_{\parallel}L] \tag{12.6}$$

while for $n \neq 0$

$$P_{nn}(q_{\parallel}L) = \beta_n(1 - 2\beta_n[\{1 - (-1)^n \exp(-q_{\parallel}L)\}/q_{\parallel}L]) \tag{12.7}$$

where $\beta_n = q_{\parallel}^2 L^2/(q_{\parallel}^2 L^2 + n^2\pi^2)$.

Before proceeding further, it is helpful to pause to interpret the above results and relate them to the dipole-exchange theory used in chapter 5. We recall that in section 5.4 the exchange terms were introduced into the previous equation of motion for the magnetostatic (dipolar) theory, yielding a higher-order differential equation for the field variables. By contrast, in the present case the dipole–dipole terms are introduced perturbatively, so the approach is quite different. Nevertheless, some similarities are immediately obvious between, for example, (5.37) for the bulk spin–wave frequencies of an unpinned film with in-plane magnetization and (12.4) for an unpinned film with arbitrary magnetization direction. The only formal difference is in the factor multiplying ω_M. However, for $q_{\parallel}L \gg 1$ (which is the case for large in-plane wavevector or a thick film) it can be seen from (12.7) that $P_{nn} \to 1$ ($n \neq 0$), and so F_{nn} reduces to

$$F_{nn} = 1 - \sin^2\theta\cos^2\phi = \sin^2\theta_{q_{\parallel}} \tag{12.8}$$

where θ_q denotes the angle between the direction of $\mathbf{q}_{\parallel}$ and the internal magnetic field. Since the Voigt geometry corresponds to $\theta_q = \pi/2$, the consistency between (5.37) and (12.4) is demonstrated. More generally, for arbitrary $q_{\parallel}L$ and with the case of $n = 0$ included, it can be shown that F_{nn} lies outside the range from 0 to 1 for certain values of θ and ϕ. This corresponds to spin waves outside the bulk region, i.e. to quasi-surface spin waves that can be related to the Damon–Eshbach mode discussed in chapter 5. A good discussion of this behaviour is given by Kalinikos (1994). It was mentioned in chapter 5 (see e.g. figure 5.10 and related discussion) that 'crossover' effects occur between bulk and surface dipole-exchange spin waves. In these specific regions of strong hybridization (or mixing) the above results do not directly apply because non-degenerate perturbation theory has been assumed. A generalization of this method to include hybridized spin waves is described by Cottam *et al* (1995).

We now return to *nonlinear* spin–wave processes in films for which the above formalism is particularly suited. The next stages are analogous to those described in section 10.5.3 for the case of bulk magnetic media. Starting from the expansion for the variable magnetization in (12.2), we rewrite the dipole-exchange Hamiltonian as an expansion in terms of boson operators that either create or annihilate a linear spin wave of frequency $\omega_n(\mathbf{q}_{\parallel})$ in branch n and with in-plane wavevector $\mathbf{q}_\parallel$. Thus (10.83) for the bulk case generalizes to

$$H = \sum_{n,\mathbf{q}_\parallel} \omega_n(\mathbf{q}_\parallel) b_{n\mathbf{q}_\parallel}^+ b_{n\mathbf{q}} + H_{\text{int}} + H_{\text{p}} \qquad (12.9)$$

where H_{int} and H_{p} represent the effects of spin–wave interactions (three-magnon and four-magnon processes) and pumping (by a transverse or parallel field), as we shall describe shortly. The quantities $U_n(\mathbf{q}_\parallel)$ and $V_n(\mathbf{q}_\parallel)$, which are the analogues of $U(\mathbf{q})$ and $V(\mathbf{q})$ defined in (2.90) and (2.91) for the bulk case, are given now by

$$U_n(\mathbf{q}_\parallel) = \Omega_n(\mathbf{q}_\parallel) + \tfrac{1}{2}\omega_M \sin^2\theta + \tfrac{1}{2}\omega_M P_{nn}[\cos^2\theta - \sin^2\theta\cos^2\phi] \qquad (12.10)$$

$$V_n(\mathbf{q}_\parallel) = U_n(\mathbf{q}_\parallel) - \Omega_n(\mathbf{q}_\parallel) - \omega_M P_{nn}\sin^2\phi + \tfrac{1}{2}i\omega_M P_{nn}\cos\theta\sin 2\phi. \qquad (12.11)$$

The property that $\omega_n^2(\mathbf{q}_\parallel) = U_n^2(\mathbf{q}_\parallel) - |V_n(\mathbf{q}_\parallel)|^2$ follows, by analogy with (2.94).

The forms taken by H_{int} and H_{p} in (12.9) can be developed in terms of the boson operators, just as in section 10.5 for the bulk case. An essential difference is that the single 3D-wavevector label $\mathbf{q}$ for the operators previously is now replaced by the 2D-wavevector label $\mathbf{q}_\parallel$ and a branch index n. Since a film does not possess translational symmetry in the direction perpendicular to the surfaces (the ξ axis), it follows that the spin–wave interaction vertices, which schematically are still of the same form as depicted in figures 10.3 and 10.4, now allow for scattering between *different* spin–wave branches (different n) as well as between the same branches. This dependence will be reflected in the form taken by the amplitude coefficients, i.e. the analogues of F_3, F_4 and F_4' appearing in (10.79) and (10.80). Expressions are quoted in the references cited earlier, and we shall comment on particular cases later.

It is relevant at this point to emphasize that $|V_n(\mathbf{q}_\parallel)|$, as defined in (12.11), provides a measure of the ellipticity of the precessional spin–wave modes. Its value is very important for the theory of parametric instability, which we discussed in section 10.5 for the case of a bulk medium. We note, in particular, that the spin–wave modes having circular polarization cannot be excited parametrically. For example, the instability for the case of parallel pumping (the process represented schematically by figure 10.6) will only occur if there is a fluctuating part to M^z, the longitudinal magnetization. The simplified argument presented in section 10.3.1, leading to (10.66), indicates that the fluctuating part, i.e. the term proportional to the square of $|\mathbf{m}|/M_0$, will reduce to a constant if the precession is circular. This can be confirmed by calculating the threshold pumping-field amplitude $(h_{\text{p}}^\parallel)_{\text{th}}$ for the onset of the instability, as was done by Schlömann *et al* (1960) for the bulk case. For a film geometry their result becomes

$$(h_{\text{p}}^\parallel)_{\text{th}} = \omega_{\text{p}}\Gamma_n(\mathbf{q}_\parallel)/\gamma|V_n(\mathbf{q}_\parallel)| \qquad (12.12)$$

for the spin wave specified by $\mathbf{q}_{\parallel}$ and n. Here $\Gamma_n(\mathbf{q}_{\parallel})$ is a damping term as in section 10.5.3. It can now be seen from (12.12) that if $|V_n(\mathbf{q}_{\parallel})|$ vanishes in any particular geometry then the threshold $(h_p^{\parallel})_{th}$ becomes infinite, i.e. the parametric instability does not occur. In the case of a film magnetized perpendicularly to its surface ($\theta = 0$) it follows from (12.10) and (12.11) that

$$|V_n(\mathbf{q}_{\parallel})| = \tfrac{1}{2}\omega_M P_{nn}. \tag{12.13}$$

Since P_{nn} takes its maximum value of unity when $q_{\parallel}L \to \infty$, it follows from (2.12) and (12.13) that this case corresponds to the minimum value of the threshold field. The $q_{\parallel} = 0$ standing spin wave modes, however, have $|V_n(\mathbf{q}_{\parallel})| = 0$ and therefore cannot be excited parametrically.

By contrast, for a film magnetized parallel to the surfaces ($\theta = \pi/2$), it follows that (see also Vendik and Chartorizhskii 1970)

$$|V_n(\mathbf{q}_{\parallel})| = \tfrac{1}{2}\omega_M[1 - P_{nn}(1 + \sin^2 \phi)]. \tag{12.14}$$

Hence in this case it is the standing spin wave modes (with $q_{\parallel} = 0$ and $P_{nn} = 0$) that have the largest ellipticity and are excited at the threshold. The resulting threshold curve, i.e. for $(h_p^{\parallel})_{th}$ versus static applied field B_0, consists of a series of sharp minima corresponding to excitation of the discrete standing modes with different n.

The above case of an in-plane magnetized film is qualitatively quite different from the behaviour occurring more generally in finite magnetic samples whenever there is a component of the magnetization perpendicular to the surface, e.g. as for a perpendicularly magnetized film, a small sphere, or a finite disc. In these latter cases the threshold curve is relatively smooth (often with a characteristic 'butterfly' shape). We now discuss examples of some measurements made for a 1 mm-diameter sphere of the ferrimagnet YIG magnetized along the [111] direction using a pumping field with $\omega_p/2\pi = 9.5\,\text{GHz}$ (see Rezende *et al* 1994). The data obtained in the case of parallel pumping are shown in figure 12.2(a), while figure 12.2(b) corresponds to the subsidiary resonance (first-order Suhl instability). The latter is the signature of the spin–wave interactions in the case of the pumping field being transverse to the magnetization, as explained in section 10.5 for the bulk case. The experiments were performed at room temperature with the single-crystal spherical YIG sample located at the centre of a high-Q (~2000) cavity, which was placed between the poles of an electromagnet to provide the static field $B_0 = \mu_0 H_0$. The cavity could be rotated so that the microwave pumping field was arranged to be either parallel or perpendicular to H_0. At low power levels the pulse reflected from the cavity has essentially the same shape as the incoming microwave (or pumping) pulse, but the nonlinear effects become evident as the power is increased. We first describe the case of parallel pumping (figure 12.2(a)). As the power is increased, an abrupt change in the reflected

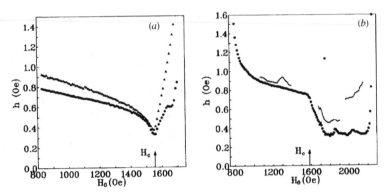

Figure 12.2. Threshold microwave fields $(h_p)_{th}$ versus d.c. applied field for the onset of spin–wave instability (lower curves) and auto-oscillations (upper curves or fragments) in a 1 mm diameter YIG sphere for (a) parallel pumping at 9.5 GHz with field H_0 $(= B_0/\mu_0)$ along [111]; (b) subsidiary resonance at 8.87 GHz with H_0 along [110]. After Rezende *et al* (1994).

pulse shape is observed when the value of $h_p^{\parallel}$ reaches a minimum threshold $(h_p^{\parallel})_{th}$. This corresponds to the onset of the parametric process shown schematically in figure 10.6, in which a pair of spin waves with wavevectors $\pm\mathbf{q}$ and frequency $\frac{1}{2}\omega_p$ are excited. The dependence of $(h_p^{\parallel})_{th}$ on the static field is relatively smooth for spheres of this diameter; it is represented by the lower curve in figure 12.2(a). Further increases in $h_p^{\parallel}$ leads to the spontaneous appearance of lower-frequency (~ 0.01–1 MHz) amplitude modulations in the microwave reflection. These are indicated by the upper curve in figure 12.2(a); they are the so-called auto-oscillations and will be discussed later (in section 12.2).

The general shape of the 'butterfly' curve for $(h_p^{\parallel})_{th}$ versus H_0 (or B_0) in figure 12.2(a) can be understood as follows (see Rezende *et al* 1994). To the left of the minimum (at H_c) the excitation of parametric spin waves with $\theta_q = \pi/2$ is energetically possible (we recall that θ_q denotes the angle between $\mathbf{q}$ and the static applied field). Since this configuration can be shown to have the lowest threshold, the monotonic and smooth variations of $(h_p^{\parallel})_{th}$ with H_0 come mainly from the $\mathbf{q}$-dependence of the spin–wave damping in the analogous result to (12.12) for a sphere. When H_0 exceeds H_c, energy conservation allows only spin–wave pairs with $\theta_q < \pi/2$ to be excited. This gives a reduction in the denominator term (the V coefficient) in the analogue of (12.12), causing $(h_p^{\parallel})_{th}$ to increase rather rapidly with H_0.

Similar arguments can be developed to explain the data for the subsidiary resonance in figure 12.2(b), which were obtained with the same YIG sample but with the pumping field in a transverse orientation. As mentioned in section 10.5.3, this process takes place via a three-magnon scattering term and is represented in figure 10.5(a). It was mentioned before that, in bulk, it takes place preferentially for spin waves with $\theta_q = \pi/4$. This corresponds to a spin wave frequency in the middle of the band, rather than at the top of the band ($\theta_q = \pi/2$) as in parallel pumping. A consequence is that the threshold curve has a different shape and is less structured. This holds even in the case of a 1 mm sphere; it is seen from figure 12.2(b) that there is now a shoulder at the H_c field (rather than a sharp minimum as in figure 12.2(a)). There are some fine-structure effects for $H_0 > H_c$ that are characteristic of the spherical geometry. The upper set of curves in figure 12.2(b) again refers to the onset of auto-oscillations; they will be discussed in the next section.

Apart from the case of spherical samples, there have also been quite extensive studies of the spin–wave parametric processes in disk-shaped samples (e.g. with diameters of the order of 1 mm and thicknesses typically 1 μm or less). As might be expected (e.g. by analogy with discussion in section 7.4), the modes depend on the radial coordinate through Bessel functions. Studies that include both microwave pumping experiments and theory were reported by McMichael and Wigen (1990) and Wigen *et al* (1992). An excellent review has been given by Wigen (1994).

All of the forgoing discussion has been for ferromagnets (or ferri-magnets), since the small spin–wave energy gap in these materials make them ideal for excitation by a microwave signal. Antiferromagnets have been of less interest for these studies because their energy gap is typically much larger (e.g. in the infrared region), as noted in chapter 2. However, some antiferromagnets with an easy-plane type of anisotropy have a low-frequency spin–wave branch with a gap energy that corresponds to the microwave region. The parametric processes under microwave pumping in such materials have been studied both theoretically and experimentally (for CsMnF$_3$) by Smirnov (1996). Pumping threshold effects analogous to those in ferromagnets were observed.

Finally, we mention that there have been experiments to investigate the effect of spatial inhomogeneities in the amplitude of the microwave pumping field across the plane of ferromagnetic films. In many experiments the spin waves were excited by a strip-line transducer of width W, so the pumping field of this transducer was also localized to a width of order W. The experi-ments (see e.g. Vugal'ter 1990, Kalinikos and Slavin 1991) and associated theory (see e.g. Adam and Stitzer 1980, Srinivasan *et al* 1987, Zil'berman *et al* 1990) were concerned, in particular, with the influence of the localized pumping field on the threshold of spin–wave parametric instability. One consequence of the localization of h_p to a width W is that the in-plane trans-lational symmetry is removed, and so previous conservation laws for in-plane wavevectors, such as that quoted in (10.67), no longer hold exactly. Instead, they are fulfilled to an accuracy of order

$$\Delta \mathbf{q}_{\|,p} \approx (2\pi\delta/W)\,\mathbf{i}_W \qquad (12.15)$$

where the dimensionless constant $\delta \leq 1$ depends on the distribution profile of h_p across the film plane, and $\mathbf{i}_W$ is a unit vector along the width of the strip line.

It follows from the above that a localized pumping field can excite spin–wave pairs having a *nonzero* wavevector in the range from zero to order $\Delta \mathbf{q}_{\|,p}$. Thus for parallel pumping, instead of (10.67), we would have $|\mathbf{q}_{\|,1} + \mathbf{q}_{\|,2}| \leq |\Delta \mathbf{q}_{\|,p}|$, while the frequency conservation law (for order $n = 1$) is replaced by

$$\omega_p \approx \omega_1 + \omega_2 + (\mathbf{v}_{g1} + \mathbf{v}_{g2}) \cdot \Delta \mathbf{q}_{\|,p}. \tag{12.16}$$

Here $\mathbf{v}_{gi} = \nabla_{\mathbf{q}_{\|,i}} \omega_i$ $(i = 1, 2)$ is the spin-wave group velocity calculated in the absence of inhomeneities (i.e. with $\Delta \mathbf{q}_{\|,p}$ set equal to zero). From results such as (12.16) the effects of the localization in modifying the thresholds $(h_p)_{th}$ in the cases of either parallel or transverse pumping can be studied. It can be shown that the parametrically excited spin waves may carry energy away from the zone of pumping localization, leading to considerable increases in $(h_p)_{th}$ in some situations. In general, the lowest thresholds correspond to those spin waves having the smallest in-plane group velocity. The good overall agreement between theory and experiment was confirmed in the references cited earlier.

In summary, we have discussed how parametric spin–wave processes in limited ferromagnetic media differ from those in bulk ferromagnets in the following main respects:

1. The discreteness of the spin–wave spectrum in limited geometries leads to a fine structure in absorption spectra under microwave pumping, e.g. as observed in the threshold curves for parametric processes. In most cases the additional selection rules result in increased stability under the external pumping.
2. Spatial inhomogeneities in the pumping field across a film surface allow the possibility of direct excitation of spin–wave pairs with nonzero in-plane wavevector and often result in higher threshold values $(h_p)_{th}$.

12.1.2 Microscopic nonlinear theory

While most of the nonlinear theory of spin waves in magnetic films has been carried within the macroscopic approach, we briefly discuss some dipole-exchange calculations within a microscopic formulation in the absence of a pumping field.

We consider the case of ultrathin ferromagnetic films (e.g. where the number N of atomic layers is less than about 100). Previously, in section 5.4.2, we described a *linearized* version of a spin–wave theory due to Costa Filho *et al* (1998, 2000) for this case. These authors actually used an expansion in terms of boson creation and annihilation operators in the same spirit as the analysis presented in section 10.5.2 for bulk ferromagnets, but with the obvious difference that wavevector Fourier transforms could be made only in the 2D parallel to the layers. However, an expansion having the same form as (10.73) was obtained; e.g. it was found for the 'linear' part (after 'diagonalization') that

$$H^{(2)} = \sum_{n, \mathbf{q}_{\parallel}} \omega_n(\mathbf{q}_{\parallel}) b_{n\mathbf{q}_{\parallel}}^{\dagger} b_{n\mathbf{q}_{\parallel}} \tag{12.17}$$

which is a generalization of (10.77). Here the linear spin wave frequencies $\omega_n(\mathbf{q}_{\parallel})$ (for $n = 1, 2, \ldots, N$) in the Voigt geometry correspond precisely to the N positive frequencies given by the determinantal condition in (5.46). Costa Filho *et al* went on to obtain the explicit expressions for the terms $H^{(3)}$ and $H^{(4)}$, describing the three- and four-magnon interaction processes, respectively. These still have an analogous form to (10.79) and (10.80), along with their diagrammatic representations (see figures 10.3 and 10.4). However, the amplitude terms are much more complicated because scattering can take place (subject to certain selection rules) between the same or different spin wave branches; this is again a consequence of the absence of translational symmetry in the direction perpendicular to the atomic layers.

Costa Filho *et al* (1998, 2000) used their expressions for the nonlinear terms to calculate the damping and frequency shift resulting from a perturbation of the discrete spin–wave branches, just as was previously done in the bulk case (see section 10.5.2). This was carried out using a Green-function method with a representation in terms of Feynman diagrams. These diagrams have 'interaction vertices' that are schematically of the same form as in figures 10.3 and 10.4.

We now describe some numerical results deduced for films of ferromagnetic EuO in the Voigt configuration with applied field $B_0 = 0.3\,\text{T}$ and two values of the thickness corresponding to $N = 10$ and 20. For $N = 10$ the lowest spin wave (branch 1) is the analogue of the Damon–Eshbach (DE) surface mode, while the next spin wave (branch 2) is a bulk mode; their respective frequencies are 28.1 and 51.1 GHz when $q_x = 0$. However, for $N = 20$ the spin waves are closer together, so the lowest branches 1 and 2 (occurring at 28.2 and 34.4 GHz, respectively, when $q_x = 0$) are hybridizations between the DE surface mode and the lowest bulk mode. Similar effects were seen in figure 5.13. Curves for the relative damping calculated for some of these modes in a low-temperature limit are shown in figure 12.3 as a function of wavevector across the Brillouin zone. In this case the dominant damping for a spin wave from any branch n comes from three-magnon splitting and confluence processes whereby

$$\omega_n(\mathbf{q}_{\parallel}) = \omega_{n'}(\mathbf{q}'_{\parallel}) \pm \omega_{n''}(\mathbf{q}_{\parallel} - \mathbf{q}'_{\parallel}) \tag{12.18}$$

with conservation of energy and in-plane wavevector. The branch labels n' and n'' may be different from n, and it is necessary to sum over all

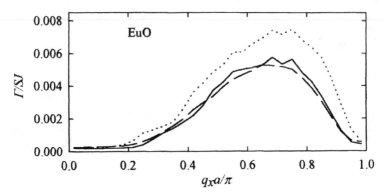

Figure 12.3. The relative spin–wave damping Γ/SJ for an EuO film in the Voigt geometry. The full line refers to $N = 10$, branch 1; the dotted line to $N = 10$, branch 2; and the dashed line to $N = 20$, branch 1. After Cottam and Costa Filho (1999).

combinations. The damping is slightly larger for branch 2 in figure 12.3, mainly because there are adjacent branches both above and below it, giving more ways to satisfy the conservation conditions. Some corresponding results calculated for the energy shift of the spin waves in EuO are shown in figure 12.4. In this case the inter-branch scattering effects are much more important; it is found that the energy shift has a weaker dependence on q_x and is larger for the thicker film.

Some extensions of the above microscopic nonlinear theory to the case of antiferromagnetic films (with applications to MnF_2) have been made by Pereira and Cottam (2003).

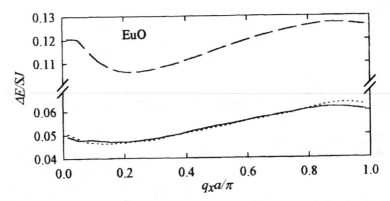

Figure 12.4. The relative spin–wave energy shift $\Delta E/SJ$ for a EuO film in the Voigt geometry. The full line refers to $N = 10$, branch 1; the dotted line to $N = 10$, branch 2; and the dashed line to $N = 20$, branch 1. After Cottam and Costa Filho (1999).

12.2   Nonlinear dynamics of spin waves: auto-oscillations

We now consider other aspects of the nonlinear dynamics of spin waves in limited samples. Our focus will be on two areas, both of which are conventionally analysed within the macroscopic description. We start here by considering what happens when the microwave pumping field is employed to drive a magnetic system further into instability. This is characterized by the auto-oscillations mentioned earlier and the onset of various features associated with chaotic behaviour. Then in section 12.3 we consider another aspect of the nonlinearity in which magnetic solitons (gap and envelope solitons) may arise in certain systems.

Often it is found to be the case experimentally that when the threshold $(h_p)_{th}$ of parametric excitation of spin waves is exceeded, either in parallel or transverse pumping, a new stationary state is *not* established. Instead, it has been observed since about the 1960s that the magnetization in a ferromagnet undergoes complicated auto-oscillations (see e.g. Hartwick *et al* 1961, Jantz and Schneider 1971). An excellent and thorough review of auto-oscillations, and their role as a precursor to chaotic behaviour, has been given by Rezende *et al* (1994). A broadly analogous behaviour in $CsMnF_3$, an antiferromagnet with an easy-plane type of anisotropy, was reported by Smirnov (1996).

Some of the main experimental features of auto-oscillations (e.g. as observed in samples of YIG) are:

1. Auto-oscillations appear typically with a well-defined threshold at pumping levels that are 0.1 to 1 dB higher than the threshold $(h_p)_{th}$ associated with parametric excitation of spin waves. The threshold depends on the wavevectors of the excited spin waves.
2. Auto-oscillations typically have frequencies in the range 10 to 100 kHz depending on the static applied magnetic field. For a small excess over threshold the spectrum of auto-oscillations consists of a single line, but new lines appear as the pumping level is increased, until eventually a noise-like chaotic spectrum of auto-oscillations is observed.
3. The auto-oscillations exhibit very pronounced directional and crystallographic anisotropies.

Some observations due to Rezende *et al* (1994) for small YIG spheres are included in figures 12.2(a) and (b). The features appear to have a more complex form in the case of transverse pumping (subsidiary resonance), which we will comment on later.

The theoretical interpretation of these auto-oscillations has proved to be especially challenging because of their sensitivity to many different factors in the experiments. Several models were proposed, but it was not until the work of Zakharov *et al* (1975) that the fundamental role of the nonlinear dynamics gained recognition. From their pioneering work, and subsequent developments, it was established that the auto-oscillations arise from a back-and-forth dynamical interchange of power between parametric modes due to the three- and four-magnon interaction terms. In fact, the auto-oscillations can be shown to arise naturally in the solution of the nonlinear differential equations (in the macroscopic model) that describe the interacting spin waves in the regime above the $(h_p)_{th}$ threshold. Their onset can be linked to a *bifurcation* in the spin–wave equations. It would take us too far afield to explore this connection, but we mention that a bifurcation in this context is an analogue of a phase transition in equilibrium themodynamics; it corresponds to a qualitative change in the dynamic state of the system (see e.g. Drazin 1992, Strogatz 1994).

The experiments on auto-oscillations and their related effects have mainly been performed for small-diameter spheres and for thin films (including disk-shaped samples); reviews for these geometries are given by Rezende *et al* (1994) and Wigen (1994), respectively. Some further details are to be found in the book edited by Marcelli and Nikitov (1996).

For either of the above sample geometries, the nonlinear parameters (the analogues of the coefficients F_3, F_4 and F_4' introduced in section 10.5.2 in the bulk case) depend sensitively on both the magnitude and direction of the excited spin waves, and thus one expects the nature of the bifurcations to change with the conditions of the experiment. This is the reason for the different profiles for the auto-oscillation thresholds in cases of parallel and transverse pumping fields (see figures 12.2(a) and (b), respectively, for YIG spheres in these configurations). The same point is illustrated by reference to figure 12.5, which shows oscilloscope traces of the auto-oscillations in the two pumping configurations (Rezende *et al* 1991). In figure 12.5(a) (parallel pumping) for $H_0 = 1.57\,\text{kOe}$ the amplitude of oscillations increases smoothly as h_p is increased, while the frequency remains constant ($\simeq 220\,\text{kHz}$). Rezende *et al* were able to deduce that the auto-oscillation amplitude A for h_p greater than the critical field h_p^{Auto} scaled as

$$A \propto [(h_p/h_p^{\text{Auto}}) - 1]^\beta \tag{12.19}$$

with index $\beta \simeq 0.5$. This is typical of a so-called Hopf bifurcation (which is a kind of analogue of a second-order phase transition). The behaviour in figure 12.5(b) (transverse pumping) for $H_0 = 1.95\,\text{kOe}$ is quite different. In this case the auto-oscillation pattern sets in with a finite amplitude and vanishing frequency (close to the critical field). This is typical of a so-called homoclinic bifurcation (which is a kind of analogue of a first-order phase transition). Also, in figure 12.5(b), the shape of the auto-oscillations is evolving and becoming more complex as h_p increases.

Auto-oscillations are also observed in ferromagnetic thin films and disks. In the latter case, where the spin waves have a spatial quantization ('standing' mode) characteristic with respect to both the film thickness and

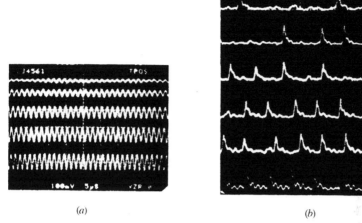

(a)   (b)

Figure 12.5. Oscilloscope traces showing the behaviour of the auto-oscillations in a YIG sphere with increasing microwave pumping field h_p above the threshold $(h_p)_{th}$ for parametric excitation: (a) parallel pumping with $h_p/(h_p)_{th} = 1.224$, 1.225, 1.293, 1.321, 1.342 from top to bottom; (b) transverse pumping (subsidiary resonance) with $h_p/(h_p)_{th} = 1.862$, 1.872, 1.883, 1.894, 1.905, 1.916 from top to bottom. After Rezende *et al* (1991).

the radial distance, they exhibit exotic patterns of behaviour. An example, taken from Wigen (1994), is shown in figure 12.6 for a YIG disk of radius $300\,\mu$m and thickness $0.56\,\mu$m. The lower part of the figure simply shows the microwave absorption versus applied field at low power (well below

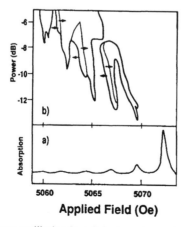

Figure 12.6. Fingers of auto-oscillation in a disk-shaped sample of YIG (radius $= 300\,\mu$m, thickness $= 0.56\,\mu$m). Hysteresis is indicated by the arrows and thinner lines within the fingers. After Wigen (1994).

any nonlinear threshold); this was used to characterize the sample. The upper part represents the behaviour at much higher pumping power, showing domains of auto-oscillations and a tendency towards chaos, again as a function of applied field. The domains of auto-oscillations are divided into regions called 'fingers', and they can be classified in terms of the standing spin waves in the disk. In some cases parts of the fingers have a pronounced hysteresis effect, depending on whether the applied field is being increased or decreased, and is indicated by the arrows.

While it appears to be relatively straightforward to observe auto-oscillations in magnetic materials with almost any shape and orientation relative to the applied magnetic field provided the pumping field is strong enough, some of the other manifestations of the nonlinear dynamics are not easy to study reliably. These include effects such as period doubling, intermittency and chaotic sequences. However, in particular, the experimental discoveries of period-doubling bifurcations in the pattern of spin–wave auto-oscillations, first under parallel pumping (Yamazaki 1984, Rezende *et al* 1986) and then transverse pumping (Wiese and Benner 1990), created an intense interest in this topic and spurred on further experimental and theoretical work. Further details are to be found in several excellent reviews (see Cherepanov and Slavin 1995, Wiese and Benner 1995, Peterman and Wigen 1996).

12.3 Magnetic solitons

We previously presented a brief discussion of magnetic solitons in bulk magnetic materials in section 10.5.3, where an analysis based on the sine-Gordon nonlinear equation was appropriate. By contrast, in magnetic thin films and periodic structures, it is shown that a description in terms of the nonlinear Schrödinger equation (NLSE) emerges. Basically, this difference arises because in the bulk case the nonlinearity depended on the competition between the single-ion anisotropy term in (10.92) and the applied field, causing nonlinear terms in the longitudinal magnetization to be important. In the case of a film we shall explicitly take account of the analytic form of the amplitudes describing four-magnon interaction processes in this geometry; in other words, the spatial quantization due to the finite film thickness is important.

The outcome is that the soliton description for magnetic films will have some similarities with the case of nonlinear optics (see sections 10.4 and 11.6). We begin by discussing gap solitons associated with individual homogeneous magnetic films; then in the following section we generalize this to periodically-modulated films (or lateral superlattices). Some useful review articles are by Boardman *et al* (1996) and Slavin *et al* (1994, 2002).

12.3.1 Envelope solitons in films

The terminology of envelope solitons has already been introduced in the non-magnetic context of nonlinear optics in section 11.6. In the case of magnetic materials the first observation of dipole-exchange spin–wave envelope solitons was made by Kalinikos *et al* (1983) using thin films of YIG. These same authors went on to pioneer other experimental developments concerning solitons, including showing that a layered structure consisting of two magnetic films with different thicknesses was advantageous for further developments (Kalinikos *et al* 1991). Other observations of solitons in magnetic films were reported by De Gasperis *et al* (1987) and Chen *et al* (1993a, 1994).

Before describing these experiments further, we shall first outline the arguments to show that the solitons in a film can be described in terms of the nonlinear Schrödinger equation (NLSE) in the same form (10.54) as for the nonlinear optics application. To simplify the arguments, we restrict attention to the situation in which the nonlinearities arise due to four-magnon scattering processes, i.e. the three-magnon decay processes are prohibited by the conservation laws. Using the Hamiltonian (12.9) with the pumping term H_{p}, the operator equation of motion for the amplitudes $b_{n\mathbf{q}_{\parallel}}$ can then be written in the form

$$\left(\frac{\partial}{\partial t} + \mathrm{i}\omega_n(\mathbf{q}_{\parallel})\right)b_{n\mathbf{q}_{\parallel}} + \mathrm{i}\sum_{n_1,n_2,n_3}\sum_{\mathbf{q}_{\parallel 1},\mathbf{q}_{\parallel 2},\mathbf{q}_{\parallel 3}} T(n,\mathbf{q}_{\parallel},n_1,\mathbf{q}_{\parallel 1},n_2,\mathbf{q}_{\parallel 2},n_3\mathbf{q}_{\parallel 3})$$

$$\times b^{\dagger}_{n_1\mathbf{q}_{\parallel 1}}b_{n_2\mathbf{q}_{\parallel 2}}b_{n_3\mathbf{q}_{\parallel 3}}\delta_{\mathbf{q}_{\parallel}+\mathbf{q}_{\parallel 1}-\mathbf{q}_{\parallel 2}-\mathbf{q}_{\parallel 3},0} = 0 \tag{12.20}$$

where the linear-spin wave frequency has the form (12.4). In the above the term T denotes the coefficient of the four-magnon scattering, i.e. the analogue for the film of the coefficient F_4 in (10.80). The Kronecker delta ensures conservation of the in-plane wavevector in the scattering. In the case of pulsed excitation of a spin wave, (12.20) leads to the formation of envelope solitons as we now show.

We assume that a narrow packet of spin waves for mode n, having a central in-plane wavevector $\mathbf{q}_0$ and central frequency ω_{n0}, is linearly excited by an external pulse. The conservation laws (for frequency and in-plane component of wavevector) governing the four-wave interaction processes within this packet are expressible as

$$2\omega_{n0} = \omega_{n1} + \omega_{n2}, \qquad 2\mathbf{q}_0 = \mathbf{q}_1 + \mathbf{q}_2 \tag{12.21}$$

where we introduce the shorthand notations $\omega_{ni} \equiv \omega_{n\mathbf{q}_i}$ and $\mathbf{q}_i \equiv \mathbf{q}_{\parallel i}$ with $i = 1, 2, 3$. For simplicity we assume also that all the waves taking part in this process belong to the same branch (i.e. have the same n) and are described by same distribution, $\Phi_n(\xi)$ in (12.2), across the film thickness. Approximate forms of the equations of motion for b_{n0}, b_{n1}, b_{n2} can then be deduced from (12.20), e.g.

$$\left(\frac{\partial}{\partial t} + i\tilde{\omega}_{n0}\right)b_{n0} = 0 \tag{12.22}$$

where

$$\tilde{\omega}_{n0} = \omega_{n0} + T_{00}|b_{n0}|^2 \tag{12.23}$$

represents a renormalized (and nonlinear) dispersion relation. In (12.23) the four-wave coefficient is $T_{00} \equiv T(n, \mathbf{q}_0, n, \mathbf{q}_0, n, \mathbf{q}_0, n, \mathbf{q}_0)$ and the final term represents an expectation value in quantum mechanics (becoming a scalar magnitude in the classical limit). There are similar results to (12.23) for $\tilde{\omega}_{ni}$ $(i = 1, 2)$:

$$\tilde{\omega}_{ni} = \omega_{ni} + 2T_{i0}|b_{n0}|^2, \qquad i = 1, 2 \tag{12.24}$$

where now $T_{i0} \equiv T(n, \mathbf{q}_i, n, \mathbf{q}_0, n, \mathbf{q}_i, n, \mathbf{q}_0)$.

The T-coefficients in the film geometry are analytic functions of the wavevector, so when $\mathbf{q}_1$ and $\mathbf{q}_2$ are both close to $\mathbf{q}_0$ in the wave packet (denoting $\mathbf{q}_{1,2} \equiv \mathbf{q}_0 \pm \mathbf{q}$ where $q \ll q_0$) we can take both T_{10} and T_{20} to be approximately equal to T_{00}. Next we expand the nonlinear dispersion relations $\tilde{\omega}_{ni}$ $(i = 1, 2)$ in a Taylor series near the point corresponding to the central wavevector of the wave packet $\mathbf{q}_\parallel = \mathbf{q}_0$ (or $\mathbf{q} = 0$):

$$\tilde{\omega}_{n1,2} = \omega_{n0} \pm v_g q + \tfrac{1}{2}D_0 q^2 + \cdots + 2T_{00}|b_{n0}|^2 \tag{12.25}$$

where

$$v_g = \left(\frac{\partial \omega_{nq_\parallel}}{\partial q_\parallel}\right)_{q_\parallel = q_0}, \qquad D_0 = \left(\frac{\partial^2 \omega_{nq}}{\partial q_\parallel^2}\right)_{q_\parallel = q_0} \tag{12.26}$$

are the group velocity and the dispersion (both calculated at the centre of the packet).

By going back to (12.22) and the analogous equations for the other waves, it can be shown by a similar Taylor expansion that the condition for the four-wave parametric process to occur (i.e. for there to be an exponential growth representing an instability) is

$$D_0 q^2 (D_0 q^2 + 4T_{00}|b_{n0}|^2) < 0. \tag{12.27}$$

Clearly this can only be the case provided $T_{00}D_0 < 0$, which is just the *Lighthill criterion*. We shall see that it is an analogue of (10.52) in nonlinear optics, with the four-wave interaction coefficient T_{00} and the dispersion term D_0 playing the same roles as η and k_0'' respectively.

Since the spin–wave packet being considered is spectrally narrow (with $q \ll q_0$) we can analyse the evolution of this packet by considering slow variations of its envelope. This is in the same spirit as the so-called slowly varying amplitude approximation (SVA) of nonlinear optics, as introduced in section 10.2.2 and then employed to derive the NLSE in section 10.4. Here, to obtain the equation for the envelope, we expand the spin–wave dispersion relation in the film in another Taylor expansion about the point $\mathbf{q}_{\parallel} = \mathbf{q}_0$ (or $\mathbf{q} = 0$), substitute this back into (12.20), and introduce 'slow' variables

$$\phi_n(q, t) = \left(\frac{\gamma}{2M_0}\right)^{1/2} b_{n1}(q_0 + q, t) \exp[-i\omega_{n0}(q_0)t] \tag{12.28}$$

where b_{n1} represents spin–wave amplitude as before. We are assuming here, for simplicity, that $\mathbf{q}$ is collinear to $\mathbf{q}_0$. Now, by using the 'slow' Fourier transform

$$\varphi_n(\zeta, t) = \int dq\, \phi_n(q, t)\, e^{iq\zeta} \tag{12.29}$$

where ζ is the in-plane spatial coordinate defined in figure 12.1, together with (12.25) and the analytic properties of the T-coefficients mentioned earlier, we can transform (12.20) from the wavevector (q) representation to the spatial (ζ) representation. The final expression is (see e.g. Slavin *et al* 1994 for more details)

$$\left[i\left(\frac{\partial}{\partial t} + v_g \frac{\partial}{\partial \zeta}\right) + \frac{1}{2}D_0 \frac{\partial^2}{\partial \zeta^2} - T|\varphi|^2\right]\varphi_n(\zeta, t) = 0 \tag{12.30}$$

where we define $T = 2M_0 T_{00}/\gamma$ to simplify the notation. The above equation is what we have set out to prove. It is now recognizable as a form of the NLSE, as can be seen most directly by comparison with (10.49). With the definition of T the Lighthill criterion becomes simply

$$TD_0 < 0. \tag{12.31}$$

The analytic solutions in the various special cases discussed in section 10.4 can be written down for the magnetic envelope solitons by direct analogy. Thus, for example, the one-soliton solution analogous to the sech function in (10.61) is

$$\varphi(\zeta, t) = \varphi_0 \operatorname{sech}[\varphi_0(T/D_0)^{1/2}(\zeta - v_0 t)] \tag{12.32}$$

where φ_0 and v_0 are arbitrary independent constants used to denote the amplitude and velocity of a soliton.

Although we have demonstrated here the formal similarities between solitons in the optical and film-thin magnetic applications, there are considerable quantitative differences due to their distinctive physical characteristics. Some data to emphasize this are shown in table 12.1. In most of the experiments cited at the beginning of this section the solitons were observed using a magnetostatic wave (MSW) delay line of the type depicted schematically in figure 12.7. It consists of a thin film of the magnetic material (YIG), magnetized by an externally applied static field and deposited on a gadolinium gallium garnet (GGG) substrate. Excitation of a microwave

Table 12.1. Quantitative comparisons between solitons in optics and thin-film magnetics (typical of YIG). After Boardman *et al* (1996).

Physical property	Optics	Thin-film magnetics
Pulse width dispersion	1 ps	100 ns
Observable within	50 m	1 cm
Group velocity	$2 \times 10^8 \, \text{ms}^{-1}$	$1.8 \times 10^4 \, \text{ms}^{-1}$
Damping coefficient	~ 0	$6 \, \mu\text{s}^{-1}$

signal takes place through one of the metal striplines embedded in an alumina (Al_2O_3) layer as its substrate. Under favourable conditions, the pulse produced in the YIG film will undergo a rapid evolution to a soliton (or multi-soliton) state, which is detected using the other stripline. The experimental evidence for soliton formation is very convincing, and includes the following: (i) sech-shaped pulses have been observed; (ii) collision experiments show preservation of pulse shapes; and (iii) the measured threshold of nonlinearity is inversely proportional to the square of the pulse width (as predicted).

In figure 12.8 we show some measurements due to Kalinikos *et al* (1990) on a YIG film of thickness 7 μm. The input and output are represented by the upper and lower traces, respectively, and the effect of increasing the input pulse duration ΔT from 4 to 20 ns is studied. In a favourable range of ΔT, as exemplified by figure 12.8(b), a single spin wave envelope soliton is formed. When ΔT is increased further (figure 12.8(c)), a regime of nonlinear pulse compression is observed in which a high and narrow central peak is

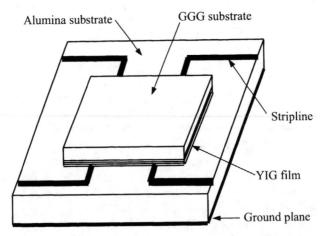

Figure 12.7. Schematic form of a magnetostatic wave (MSW) delay line as used to excite solitons in a YIG film.

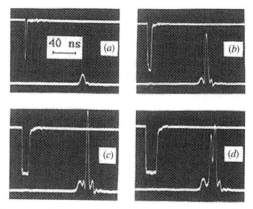

Figure 12.8. Oscillograms of the envelopes of the input (upper traces) and output (lower traces) spin–wave pulses demonstrating soliton formation in a YIG film. The input power was fixed at 280 mW and the different input pulse durations are (a) 4 ns, (b) 6 ns, (c) 12 ns and (d) 20 ns. After Kalinikos *et al* (1990).

formed on the output pulse profile. With a further increase of ΔT the output pulse splits and there is a regime of multi-soliton formation (figure 12.8(d)).

Before moving on to consider solitons in periodic structures, we briefly mention some other aspects associated with magnetic solitons in films. First, there have been considerable efforts on the theoretical side to model the evolution of the soliton output profiles corresponding to different values of input power and pulse duration; we mention as examples the work of Slavin and Dudko (1990) and Chen *et al* (1993b), in which a fair degree of success was achieved. Second, there has been work relating to so-called 'bright' and 'dark' solitons (see e.g. Chen *et al* 1993a, Slavin and Rojdestvenski 1994). In essence, bright solitons are the case that we have discussed so far, while dark solitons correspond to $TD_0 > 0$, i.e. the Lighthill criterion is *not* satisfied, but provided $\varphi^{-1}(\partial^2\varphi/\partial\zeta^2) < 0$ and the envelope shape is concave upwards the NLSE still has a soliton-like solution. This latter case (the dark soliton) appears as a 'dip' rather than a 'spike' on the background of a stable constant-amplitude spin wave. Third, and finally, there has been some discussion of possibilities for soliton formation in antiferromagnetic films (see e.g. Boardman *et al* 1993 for applications to MnF_2).

12.3.2 Envelope solitons in periodic structures

By comparison with the considerable amount of work on solitons and nonlinear spin waves in single films and double films, the work on magnetic solitons in superlattices and other periodic structures is relatively sparse. However, this may well change in the near future.

We note that formation of the dipole-exchange spin–wave envelope solitons in films (see section 12.3.1) was observed in the spectral regions of high dispersion near the dipole 'gaps' in the spin–wave spectrum. This is a different situation from the cases of nonlinear wave propagation in the periodic structures that were described in section 11.6. There we covered nonmagnetic examples of 'gap' solitons that formed when the frequency of the input signal lies within the gap (or stop band) in the spectrum originating from the structural periodicity. We might expect that an analogous type of 'gap' soliton should occur in periodic magnetic structures; in practice, however, this realization is far from straightforward because specific materials and structural requirements need to be satisfied for the effects to be observable.

In the next few paragraphs we outline a theory due to Chen *et al* (1992) for gap solitons in a periodic magnetic structure. Specifically, these authors considered a magnetic film (e.g. of YIG) magnetized in the direction perpendicular to its surfaces and with a periodic modulation in its magnetic properties in a direction parallel to the surfaces. They analysed the conditions for soliton formation to occur when the input microwave signal lies in the spectral regions of strong dispersion *near* the periodic gaps (but not actually within them). The periodic modulation in the in-plane direction is most conveniently achieved by introducing a modulation in the applied (static) magnetic field, e.g. this can be done simply by attaching to the surface of a YIG film a piece of standard magnetic recording tape with a periodic signal recorded on it (Voronenko *et al* 1988).

We start by considering a magnetic film of thickness L with surfaces in the xy-plane and magnetized perpendicularly in the z direction. The static applied field B_0 in the z direction is modulated by another z-directed field that has a periodicity d in the x direction:

$$b(x) = b_0 \cos(2\pi x/d) \qquad (12.33)$$

and we later assume that $b_0 \ll B_0$.

If the spins at the surfaces of the film are assumed to be unpinned (for simplicity), the linear spin–wave spectrum of the unperturbed film (i.e. without the modulating field) is well known from the theory described in section 12.1.1. For the assumed geometry the frequencies of the discrete branches are

$$\omega_n(q_{||}) = \{\Omega_n(q_{||})[\Omega_n(q_{||}) + \omega_M P_{nn}(q_{||}L)]\}^{1/2} \qquad (12.34)$$

where

$$\Omega_n(q_{||}) = \omega_0 + Dq_{||}^2 + D(n\pi/L)^2, \qquad n = 0, 1, 2, \ldots. \qquad (12.35)$$

The same notation is employed as before, and the relevant expressions for the dipolar matrix elements P_{nn} are quoted in (12.6) and (12.7).

The next step is to take account of the periodic modulation defined in (12.33). Since it is assumed that $b_0 \ll B_0$, we may employ the so-called two-mode mixing approximation (Voronenko *et al* 1988), which is formally analogous to the nearly-free-electron approximation described in solid-state physics textbooks (see e.g. Kittel 1995), to calculate the modified spin–wave dispersion relations near the energy gaps at in-plane wave numbers $\pm\pi/d$, $\pm 3\pi/d$ etc. Thus, for $q_\parallel = (\pi/d) \pm \delta$ where $\delta > 0$ is small (compared with π/d), we estimate that the two branches are

$$\omega_n^{(\pm)}(q_\parallel) = \tfrac{1}{2}[\omega_n^{(\pm)}(\pi/d + \delta) + \omega_n^{(\pm)}(\pi/d - \delta)]$$
$$\pm \{\tfrac{1}{4}[\omega_n^{(\pm)}(\pi/d + \delta) - \omega_n^{(\pm)}(\pi/d - \delta)]^2 + [G_n \pm J_n]^2\}^{1/2}. \quad (12.36)$$

Here we have denoted

$$G_n = \tfrac{1}{2}\gamma b_0[u_n^{(+)}(\pi/d + \delta)u_n^{(+)}(\pi/d - \delta)$$
$$+ u_n^{(-)}(\pi/d + \delta)u_n^{(-)}(\pi/d - \delta)] \quad (12.37)$$
$$J_n = \tfrac{1}{2}\gamma b_0[u_n^{(+)}(\pi/d + \delta)u_n^{(-)}(\pi/d - \delta)$$
$$+ u_n^{(-)}(\pi/d + \delta)u_n^{(+)}(\pi/d - \delta)] \quad (12.38)$$

with

$$u_n^{(\pm)}(q_\parallel) = \pm\{[\Omega_n(q_\parallel) + \tfrac{1}{2}\omega_M P_{nn}(q_\parallel)] \pm \omega_n(q_\parallel)]/2\omega_n(q_\parallel)\}^{1/2}. \quad (12.39)$$

There is a further simplification if the length scales are chosen such that $d \gg L$; this inequality implies that $q_\parallel L \ll 1$ for the wavevectors of interest, which means that typically $P_{nn} \ll 1$ and the J_n terms in (12.36) can be neglected.

Using (12.36)–(12.39) Chen *et al* made numerical calculations assuming film thickness $L = 10\,\mu m$, periodicity length $d = 200\,\mu m$, applied field $B_0 = 0.295\,T$, modulating field $b_0 = 0.0015\,T$, and other parameters as for YIG. In figure 12.9 plots are shown for the frequency, group velocity and dispersion parameter for a range of in-plane wavevectors near the gap at π/d ($= 157\,cm^{-1}$). Chen *et al* used these data, together with their estimates of the nonlinear coefficient T entering into the NLSE (12.30), to deduce favourable conditions for envelope soliton formation. They concluded that the point marked A (at $143\,cm^{-1}$) in figure 12.9 should satisfy the main criteria. For example, the dispersion coefficient there is negative, allowing the Lighthill criterion to be satisfied. Also the group velocity is reasonably large, which means that the soliton formation time can be kept small (compared with the decay time from spin–wave damping). Other criteria deduced by Chen *et al* relate to the duration ΔT of the input pulse, so overall the conditions involve a careful choice of the materials parameters and experimental conditions. This field is likely to receive more attention

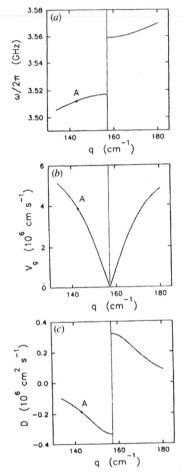

Figure 12.9. Plots of (a) spin–wave frequency $\omega/2\pi$, (b) group velocity v_g and (c) dispersion D versus in-plane wavevector $q \equiv q_{\parallel}$ for a YIG film with a periodic modulation (see the text for parameter values). The vertical line at $q = 157\,\mathrm{cm}^{-1}$ corresponds to the periodic gap at π/d. Point A (at $143\,\mathrm{cm}^{-1}$) is used in the discussion. After Chen *et al* (1992).

in the future, with possibly other means being explored to introduce a periodic modulation (e.g. modulating the film thickness).

Finally, we mention that theoretical studies of nonlinear superlattices made up of alternating layers of an antiferromagnet (such as FeF_2) and a nonmagnetic dielectric (such as ZnF_2) have been reported (see e.g. Almeida and Mills 1987, Chen and Mills 1987, Kahn *et al* 1989). In such studies the antiferromagnetic layer is characterized by a magnetic susceptibility function that includes a nonlinear part proportional to the square of the magnetic field; in other words it is the direct analogue of the Kerr-type nonlinearity in the dielectric function where there is a term involving the square of the electric field (as discussed in chapters 10 and 11). Therefore these studies provide information on the nonlinear optics of structures with magnetic materials, rather than the nonlinear dynamics of the spin waves (as we have been considering elsewhere in this chapter). By calculating the transmissivity through an FeF_2/ZnF_2 periodic superlattice (assuming the latter material to be characterized by a linear dielectric function), Kahn *et al* (1989) identified gap-soliton-mediated resonances that were attributed to Fabry–Pérot oscillations. These authors also made similar studies of Fibonacci quasiperiodic superlattices involving the same materials, but they found no evidence of solitons in the stop gap for this case.

12.4   Experimental studies by light scattering

So far in this chapter we have focused on microwave (including pumping) techniques as providing the experimental evidence for nonlinear spin–wave processes and soliton-like behaviour. This was mainly done because of the historical developments in this field. However, another important experimental method has been based on inelastic light scattering (mainly Brillouin scattering), and we now describe some of the developments and results relevant to nonlinear magnetic systems in surface-related geometries. Useful accounts of this topic are included in the review articles by Hillebrands (2000), Demokritov *et al* (2001) and Slavin *et al* (2002).

We introduced the Brillouin light-scattering method in a general way in section 1.4 and numerous applications have been given throughout this book. In the present context there are two additional features to mention. First, it is sometimes of interest to carry out the light scattering measurements under conditions of microwave pumping, and for this purpose the sample would be mounted inside a microwave cavity; see e.g. Borovik-Romanov and Kreines (1988) for some early examples. Second, some types of measurements (e.g. relating to solitons) require very specific spatial and/ or temporal resolution characteristics. This has become achievable only relatively recently through pioneering work by Hillebrands (1999). A pulse generator produces pulses typically of 10–30 ns duration with a repetition rate of 1 MHZ; these are sent to a microwave switching device to create a pulsed microwave field and to generate a spin–wave pulse. On the detection side, the device can handle more than 10^6 events per second on a continuous basis and a lower limit on time resolution of about 2 ns is set by the Fabry–Pérot spectrometer (which is a multi-pass tandem instrument of the same type as used in other Brillouin scattering experiments).

The first Brillouin scattering observations for propagating magneto-static spin waves with $q_\parallel \neq 0$ were reported by Srinivasan *et al* (1987). Using an MSW stripline (of the generic type shown in figure 12.7) to excite

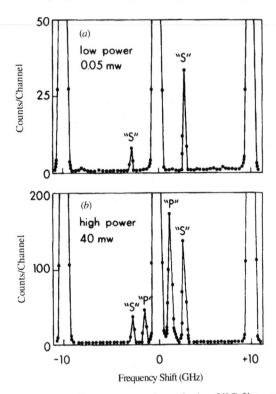

Figure 12.10. Brillouin spectra for magnetostatic modes in a YIG film under conditions of microwave excitation at frequency 3 GHz. The film was magnetized parallel to the surfaces with a static field $B_0 = 0.04$ T. The two different power levels are indicated; the peaks labeled 'S' and 'P' correspond to excitation of the surface mode and parametric bulk modes respectively. After Srinivasan *et al* (1987).

the spin waves, they detected the parametric decay of the magnetostatic (Damon–Eshbach) surface mode with $q_\parallel \approx 10^2 \, \text{cm}^{-1}$ in a YIG film into two magnetostatic 'backward' bulk modes with much larger wavevectors ($q_\parallel \approx 10^4 - 10^5 \, \text{cm}^{-1}$). As explained in section 5.1.1, these bulk modes (so-called because their group velocity is negative) correspond to a film magnetized parallel to the surface and have propagation along the magnetization direction. We show in figure 12.10 some Brillouin spectra obtained using 3 GHz microwave excitation at two different power levels. Figure 12.10(a) corresponds to the linear regime (below threshold) and the peaks labelled 'S' correspond to the Damon–Eshbach surface mode on the Stokes and anti-Stokes sides of the spectra. At the higher power level (figure 12.10(b)) the parametric excitation of a pair of bulk magnetostatic modes becomes possible; these are degenerate in frequency (each having half of the frequency of the surface mode from which they decayed). These additional parametric peaks are labelled 'P' in figure 12.10(b).

Examples of further Brillouin light-scattering experiments to study spin–wave instabilities and parametric processes are by Wilber *et al* (1988) and Kabos *et al* (1994); both papers related to the subsidiary microwave absorption in YIG films. Among other nonlinear processes that have been investigated by Brillouin scattering are spin–wave beam shaping (which includes the effects of self-channelling and self-focusing), the formation of spin–wave solitons and so-called spin 'bullets', and collisions between non-linear spin–wave pulses. A brief discussion of these is given below with some examples included, while a detailed account is to be found in the review by Slavin *et al* (2002).

Experimental evidence in garnet films showing stationary (i.e. without time dependence) self-channelling and initial stages of self-focusing was reported by Boyle *et al* (1996). They observed, both for the magnetostatic surface mode and the backward bulk modes, nonlinear beam shaping and some evidence of self-channelling. Soon afterwards, Bauer *et al* (1997) presented the first evidence of nonlinear self-focusing, using the configuration for the backward bulk modes. Some data for stationary self-focusing are shown in figure 12.11 where a relatively wide (18 mm) film of a Bi-substituted

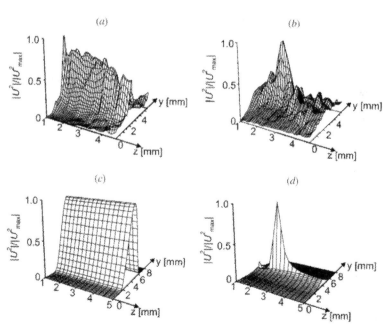

Figure 12.11. Stationary self-focusing of beams of backward bulk magnetostatic waves in a Bi-substituted iron garnet film. The normalized intensity, denoted here as $|U|^2/U_{max}^2$, is plotted in the film plane (the yz-plane): (a), experimental data for the low-power linear regime ($P_{in} = 10\,\text{mW}$); (b), experimental data for the nonlinear regime ($P_{in} = 600\,\text{mW}$); (c) and (d), results of numerical calculations. After Büttner *et al* (2000).

iron garnet ($Lu_{0.96}Bi_{2.04}Fe_5O_{12}$) with thickness $1.5\,\mu m$ was used (Büttner *et al* 2000). The advantage of this material is its relatively high Brillouin light scattering signal. The experimental measurements are given in figures 12.11(a) and (b) for the linear and nonlinear regimes (in terms of the input power). These plots show the (normalized) scattered intensity, which is proportional to the square of the spin–wave amplitude, for a spin–wave wave packet propagating along the length of the film (in the z direction). By choosing to use a wide film Büttner *et al* also measured the spatial dependence of the intensity in the transverse (y) direction. The corresponding numerical simulations are shown in figures 12.11(c) and (d), respectively. In the linear case (figure 12.11(a)) the spin–wave amplitude changes slightly due to the effect of diffraction, which causes a beam divergence during its propagation (along the z direction). By contrast, in the nonlinear case (figure 12.11(b)) the beam no longer diverges; it converges to small lateral dimensions while the spin–wave amplitude increases. There is a clear focus near the point $z = y = 2\,mm$. This is the result of competition between the diffraction and nonlinearity, and so it can be thought of as a spatial soliton. The spin–wave beam loses energy due to dissipation during propagation, so eventually (for $z > 3\,mm$) it diverges when its amplitude becomes so small that diffraction effects dominate.

The extension of the above Brillouin-scattering work to include time dependence as well as spatial dependence, i.e. the study of spatiotemporal self-focusing, has produced further evidence for spin–wave solitons and for related entities called *spin–wave bullets* (Bauer *et al* 1998, Büttner *et al* 2000). An example is shown in figure 12.12 for a large-area YIG film (width $18\,mm$, length $26\,mm$ and thickness $7\,\mu m$). In this experiment there were five pulses, each of duration $29\,ns$, created in rapid succession at the antenna, and spatiotemporal Brillouin measurements were made as they propagated away from the antenna (i.e. in the z direction). The times corresponding to these pulses are indicated in figures 12.12(a) and (b), which represent the experimental data obtained in the linear and nonlinear regimes (in terms of the input power), respectively. The corresponding numerical simulations appear in figures 12.12(c) and (d). The upper part of each panel shows a 3D plot for the relative intensities, while the lower part shows the cross sections (at half-maximum intensity). In the linear case (figure 12.12(a)) diffraction and dispersion cause a broadening of each initial wave packet, and the amplitude decreases with propagation due to dissipation. In this sample diffraction has a considerably larger effect than dispersion, so the broadening is more pronounced in the transverse (y) direction than in the propagation (z) direction, as can be seen from the cross-section profiles. The observed behaviour of the wave packets in the nonlinear regime (figure 12.12(b)) is very different. Here the initial high-amplitude wave packet starts to converge and its amplitude increases. One might expect that this would continue (i.e. in the absence of a stable

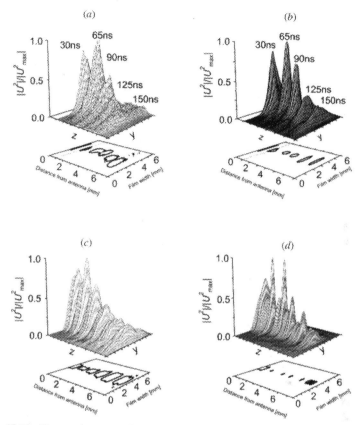

Figure 12.12. Non-stationary self-focusing of beams of backward bulk magnetostatic waves in a YIG film for an experiment in which five pulses (each of duration 29 ns) were created at the five times (as indicated) after pulse launch from the antenna. The upper part of each panel shows the normalized intensity plotted in the film plane (the yz-plane), while the lower part shows the cross sections of the propagating pulses at half-maximum intensity: (a), experimental data for the low-power linear regime ($P_{in} = 10$ mW); (b), experimental data for the nonlinear regime ($P_{in} = 460$ mW); (c) and (d), results of numerical calculations. After Büttner *et al* (2000).

equilibrium in 2D among dispersion, diffraction and nonlinearity), leading to a wave collapse with all the energy concentrated at one spatial point. Of course, in a real material with dissipation this is avoided. Instead, in a certain range along the propagation distance, nonlinear collapse is stabilized by the dissipation, and a quasi-stable, strongly-localized 2D wave packet (known as a spin–wave bullet) is formed. Interestingly, an analogous effect had previously been predicted for optical wave packets (Silberberg 1990), but it has apparently not been observed in materials with a Kerr-type nonlinearity at the present time. Eventually, in figure 12.12(b), the energy in the wave packet is so small due to dissipation that the influence of the non-linearity becomes negligible, and the wave packet then (for times greater than about 110 ns) diverges as in the linear case. Later developments regarding spin–wave solitons and bullets are described by Slavin *et al* (2002).

In both figures 12.11 and 12.12 we have included panels for numerical simulations, and overall it is seen that the agreement with the experiments is encouraging. The simulations involve solving the nonlinear Schrödinger equation (NLSE) in the two spatial dimensions (y and z) and time t. Appropriate boundary conditions were assumed at the lateral edges of the film and for the form of the input pulse launched at the antenna. In other contexts, the NLSE is often solved by the method of fast Fourier transforms (FFT), but this turns out not to be convenient here because of the complicated boundary

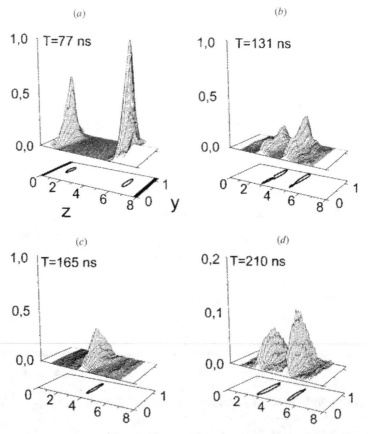

Figure 12.13. Formation and collision of quasi-1D spin–wave envelope solitons in a narrow YIG film waveguide. The upper and lower frames show intensity and cross sections as in figure 12.12. The events are shown at four different values T of the propagation time: (a) 77 ns; (b) 131 ns; (c) 165 ns; (d) 210 ns. After Büttner *et al* (1999).

conditions at the antenna; instead finite-difference methods were employed. Details are to be found in the original paper (Büttner *et al* 2000).

Finally, it is worthwhile emphasizing the difference between the two forms of nonlinear spin–wave packets that we have discussed in this chapter: the *quasi-1D envelope solitons* and the *2D self-focused wave packets* (or *bullets*). The former are created in spin–wave waveguides and in *narrow* YIG films; the latter are created in *wide* YIG films. They exhibit quite different properties when collisions are studied, either between solitons or between bullets (see Büttner *et al* 1999). The solitons have the property of retaining their shape after collision, as mentioned earlier. This has been very nicely illustrated by Brillouin scattering in a narrow YIG waveguide (see figure 12.13 for a sequence of four time intervals). In similar experiments on wide YIG films (and therefore involving bullets) there is no observed shape preservation after collision; the bullets are essentially destroyed by the collision (Büttner *et al* 1999).

References

Adam J D and Stitzer S N 1980 *Appl. Phys. Lett.* **36** 485

Almeida N S and Mills D L 1987 *Phys. Rev. B* **36** 2015

Bauer M, Mathieu C, Demokritov S O, Hillebrands B, Kolodin P A, Sure S, Dötsch H and Slavin A N 1997 *Phys. Rev. B* **56** 8483

Bauer M, Büttner O, Demokritov S O, Hillebrands B, Grimalsky Y, Rapoport Y and Slavin A N 1998 *Phys. Rev. Lett.* **81** 3769

Boardman A D and Nikitov S A 1988 *Phys. Rev. B* **38** 11444

Boardman A D, Nikitov S A and Waby N A 1993 *Phys. Rev. B* **48** 13602

Boardman A D, Putman R C J, Xie K, Metha H M and Nikitov S A 1996 in *Nonlinear Microwave Signal Processing: Towards a New Range of Devices* ed R Marcelli and S A Nikitov (Dordrecht: Plenum) p 277

Borovik-Romanov A S and Kreines N M 1988 in *Spin Waves and Magnetic Excitations I* ed A S Borovik-Romanov and S K Sinha (Amsterdam: Elsevier) p 81

Boyle J W, Nikitov S A, Boardman A D, Booth J G and Booth K 1996 *Phys. Rev. B* **53** 12175

Büttner O, Bauer M, Demokritov S O, Hillebrands B, Kostylev M P, Kalinikos B A and Slavin A N 1999 *Phys. Rev. Lett.* **82** 4320

Büttner O, Bauer M, Demokritov S O, Hillebrands B, Kivshar Y S, Grimalsky Y, Rapoport Y and Slavin A N 2000 *Phys. Rev. B* **61** 11576

Chen M, Tsankov M A, Nash J M and Patton C E 1993a *Phys. Rev. Lett.* **70** 1707

Chen M, Nash J M and Patton C E 1993b *J. Appl. Phys.* **73** 3906

Chen M, Tsankov M A, Nash J M and Patton C E 1994 *Phys. Rev. B* **49** 12773

Chen N N, Slavin A N and Cottam M G 1992 *IEEE Trans. Magnetics* **28** 3306

Chen W and Mills D L 1987 *Phys. Rev. Lett.* **58** 160

Cherepanov V B and Slavin A N 1995 in *High Frequency Processes in Magnetic Materials*, ed G Srinivasan and A N Slavin (Singapore: World Scientific) p 250

Costa Filho R N, Cottam M G and Farias G A 1998 *Solid State Comm.* **108** 439

Costa Filho R N, Cottam M G and Farias G A 2000 *Phys. Rev. B* **62** 6545

Cottam M G, ed 1994 *Linear and Nonlinear Spin Waves in Magnetic Films and Superlattices* (Singapore: World Scientific)

Cottam M G and Slavin A N 1994 in *Linear and Nonlinear Spin Waves in Magnetic Films and Superlattices* ed M G Cottam (Singapore: World Scientific) p 1

Cottam M G, Rojdestvenski I V and Slavin A N 1995 in *High Frequency Processes in Magnetic Materials* ed G Srinivasan and A N Slavin (Singapore: World Scientific) p 394

Cottam M G and Costa Filho R N 1999 *NanoStructured Mat.* **12** 395

De Gasperis P, Marcelli R and Miccoli G 1987 *Phys. Rev. Lett.* **59** 481

Demokritov S O, Hillebrands B and Slavin A N 2001 *Phys. Reports* **348** 441

Drazin P G 1992 *Nonlinear Systems* (Cambridge: Cambridge University Press)

Hartwick T S, Peressini E R and Weiss M T 1961 *J. Appl. Phys.* **32** 223

Hillebrands B 1999 *Rev. Sci. Instrum.* **70** 1589

Hillebrands B 2000 in *Light Scattering in Solids VII* ed M Cardona and G Güntherodt (Berlin: Springer) p 174

Jantz W and Schneider J 1971 *Solid State Comm.* **9** 69

Kabos P, Patton C E, Dima M O and Church D B 1994 *J. Appl. Phys.* **75** 3553

Kahn L M, Huang K and Mills D L 1989 *Phys. Rev. B* **39** 12449

Kalinikos B A 1983 *Sov. Tech. Phys. Lett.* **9** 325

Kalinikos B A 1994 in *Linear and Nonlinear Spin Waves in Magnetic Films and Superlattices* ed M G Cottam (Singapore: World Scientific) p 89

Kalinikos B A, Kovshikov N G and Slavin A N 1983 *Sov. Phys.—JETP Lett.* **38** 413

Kalinikos B A, Kovshikov N G and Slavin A N 1990 *Phys. Rev. B* **42** 8658

Kalinikos B A, Kovshikov N G and Slavin A N 1991 *J. Appl. Phys.* **69** 5712

Kalinikos B A and Slavin A N 1991 *IEEE Trans. Magnetics* **27** 5444

Kittel C 1995 *Introduction to Solid State Physics* 7th edition (New York: Wiley)

Marcelli R and Nikitov S A (eds) 1996 *Nonlinear Microwave Signal Processing: Towards a New Range of Devices* (Dordrecht: Plenum)

McMichael R D and Wigen P E 1990 *Phys. Rev. Lett.* **64** 64

Melkov G A 1988 *Sov. Phys.—Solid State* **30** 1458

Mills D L and Trullinger S E 1987 *Phys. Rev. B* **36** 947

Pereira J M and Cottam M G 2003 *Phys. Rev. B* **68** 104429

Peterman D W and Wigen P E 1996 in *Nonlinear Microwave Signal Processing: Towards a New Range of Devices* ed R Marcelli and S A Nikitov (Dordrecht: Plenum) p 355

Rezende S M, de Aguiar F M and de Alcantara Bonfim O F 1986 *J. Mag. Magn. Mat.* **54-57** 1127

Rezende S M, Azevedo A, Koiller J and Cascon A 1991 *J. Appl. Phys.* **69** 5430

Rezende S M, Azevedo A and de Aguiar F M 1994 in *Linear and Nonlinear Spin Waves in Magnetic Films and Superlattices* ed M G Cottam (Singapore: World Scientific) p 335

Schlömann E, Green J and Milano V 1960 *J. Appl. Phys.* **31** 3865

Silberberg Y 1990 *Opt. Lett.* **15** 1282

Slavin A N and Kalinikos B A 1987 *Sov. Phys.—Tech Phys.* **32** 1446

Slavin A N and Dudko G M 1990 *J. Mag. Magn. Mat.* **86** 115

Slavin A N and Rojdestvenski I V 1994 *IEEE Trans. Magnetics* **30** 37

Slavin A N, Kalinikos B A and Kovshikov N G 1994 in *Linear and Nonlinear Spin Waves in Magnetic Films and Superlattices* ed M G Cottam (Singapore: World Scientific) p 421

Slavin A N, Demokritov S O and Hillebrands B 2002 in *Spin Dynamics in Confined Magnetic Structures I* ed B Hillebrands and K Ounadjela (Berlin: Springer) p 35

Smirnov A I 1996 in *Nonlinear Microwave Signal Processing: Towards a New Range of Devices* ed R Marcelli and S A Nikitov (Dordrecht: Plenum) p 253

Srinivasan G, Patton C E and Emtage P R 1987 *J. Appl. Phys.* **61** 2318

Srinivasan G and Slavin A N (eds) 1995 *High Frequency Processes in Magnetic Materials* (Singapore: World Scientific)

Strogatz S H 1994 *Nonlinear Dynamics and Chaos* (Reading: Addison-Wesley)

Vendik O G and Chartorizhskii D N 1970 *Sov. Phys.—Solid State* **12** 2357

Voronenko A V, Gerus S V and Kharitonov V D 1988 *Sov. Phys.—JETP* **31** 245

Vugal'ter G A 1990 *Sov. Phys.—JETP* **70** 1072

Wiese G and Benner H 1990 *Z. Physik B* **79** 119

Wiese G and Benner H 1995 in *High Frequency Processes in Magnetic Materials* ed G Srinivasan and A N Slavin (Singapore: World Scientific) p 284

Wigen P E, Ball K D and Shields P J 1992 in *Proc. 1st Experimental Chaos Conf.* (Singapore: World Scientific)

Wigen P E 1994 in *Linear and Nonlinear Spin Waves in Magnetic Films and Superlattices* ed M G Cottam (Singapore: World Scientific) p 375

Wilber W D, Booth J G, Patton C E, Srinivasan G and Cross R W 1988 *J. Appl. Phys.* **64** 5477

Wolfram T and Dewames R E 1972 *Prog. Surf. Sci.* **2** 233

Yamazaki H 1984 *J Phys. Soc. Japan* **53** 1155

Zakharov V E, L'vov V S and Starobinets S S 1975 *Sov. Phys.—Usp.* **17** 896

Zil'berman P E, Golubev N S and Temiryazev A G 1990 *Sov. Phys.—JETP* **70** 353

Appendix

Green functions and linear-response theory

In section 1.3.2 we introduced the basic ideas of linear-response theory by using a simple mechanical example. A linear-response function was defined there in terms of the effect produced on a system by a small applied stimulus. Then it was shown to have certain useful properties in terms of the excitations and their power spectrum. The purpose of this appendix is to formalize that treatment in a way that will be useful for dealing with surface problems. We summarize some basic results concerning Green functions and linear-response functions, and the connection between these quantities.

Comprehensive treatments of Green functions are to be found in the numerous text books on many-body theory; for example, Fetter and Walecka (1971), Rickayzen (1980) and Negele and Orland (1988). General accounts of the linear-response method have been given, for example, by Forster (1975), Landau and Lifshitz (1980) and Lovesey (1986). Its specific application to surface problems has been reviewed by Cottam and Maradudin (1984).

A.1 Basic properties of Green functions

For any two quantum-mechanical operators A and B we shall define the Green functions $\langle\langle A(t); B(t')\rangle\rangle$, with time labels t and t', by

$$\langle\langle A(t); B(t')\rangle\rangle = i\theta(t - t')\langle[A(t), B(t')]_\varepsilon\rangle. \tag{A.1}$$

Here $\theta(t - t')$ is the unit step function corresponding to

$$\theta(t - t') = \begin{cases} 1 & (t > t') \\ 0 & (t < t') \end{cases} \tag{A.2}$$

and the operators are in the Heisenberg representation, i.e.

$$A(t) = \exp(iH_0 t) A \exp(-iH_0 t) \tag{A.3}$$

where H_0 is the Hamiltonian describing the system. The angular brackets in (A.1) denote a thermal average, evaluated according to equilibrium statistical mechanics. Finally

$$[A(t), B(t')]_\varepsilon = A(t)B(t') - \varepsilon B(t')A(t) \tag{A.4}$$

where ε is a parameter which can be *chosen* as either 1 or -1. The conventional choice in boson and fermion problems is to take $\varepsilon = 1$ and $\varepsilon = -1$, respectively, but this is not essential.

The definition in (A.1) is the same, apart from the overall sign, as that first used by Zubarev (1960) for retarded Green functions. Other Green functions, known as advanced and causal Green functions, can also be introduced but we shall not require them here. Basically the definition of the Green function involves thermal averages of products of operators, i.e. it can provide information about correlations in the physical system.

One of the simplest properties of $\langle\langle A(t); B(t')\rangle\rangle$ is that it depends on t and t' only through the time difference $(t - t')$. This follows from (A.1), (A.3) and the definition of thermal average; the proof is given in most standard texts on many-body theory. It allows Fourier components $\langle\langle A; B\rangle\rangle_\omega$ of $\langle\langle A(t); B(t')\rangle\rangle$ to be defined by

$$\langle\langle A(t); B(t')\rangle\rangle = \int_{-\infty}^{\infty} \langle\langle A; B\rangle\rangle_\omega \exp[-i\omega(t - t')] \, d\omega. \tag{A.5}$$

For most cases of practical interest the Green function $\langle\langle A; B\rangle\rangle_\omega$ can only be evaluated by making approximations, and several different approaches are available. One method is based on writing down the equation of motion of the Green function (e.g. see Zubarev 1960):

$$\omega\langle\langle A; B\rangle\rangle_\omega = (1/2\pi)\langle[A, B]_\varepsilon\rangle + \langle\langle[A, H]; B\rangle\rangle_\omega. \tag{A.6}$$

This generates another Green function $\langle\langle[A, H]; B\rangle\rangle_\omega$, whose equation of motion can itself be written down. After a certain number of stages of this process a *decoupling approximation* (e.g. like the random-phase approximation, or RPA, applied to operator products in section 2.3.2) is used to simplify the Green function in the final equation and to yield a closed set of equations that can be solved for the required Green function. Another method, which is particularly appropriate for some applications to nonlinear systems (e.g. see discussion in section 12.1), involves using perturbation theory in a diagrammatic expansion (Feynman diagrams). Details are to be found in the references given earlier. For most of the applications considered in this book we employ yet another approach, namely linear-response theory, which has a more direct physical appeal and enables appropriate Green functions to be deduced using macroscopic arguments.

A.2   Linear-response theory

A simple illustrative example of the method is given in section 1.3.2, and we now follow this up by presenting the formal theory. We make use of the concept of the density matrix ρ, which may be defined by (see e.g. Landau and Lifshitz 1980)

$$\rho = \sum_m |m\rangle p_m \langle m| \qquad (A.7)$$

where p_m is the probability of the system being in the state denoted by $|m\rangle$ (assumed to be orthogonalized and normalized). The summation is over all states. The density matrix is useful for applications to nonequilibrium situations where the time-dependence needs to be studied.

We consider a system with total Hamiltonian H given by

$$H = H_0 + H_1(t) \qquad (A.8)$$

where H_0 is independent of time and $H_1(t)$ is an external perturbation taking the form

$$H_1(t) = -Bf(t). \qquad (A.9)$$

Here $f(t)$ denotes an external field that couples linearly with a system variable represented by the operator B, giving the interaction term (A.9) by analogy with (1.37). For example, if B represents an atomic displacement then $f(t)$ would be a mechanical force (as in section 2.2.3). We suppose that at time $t = -\infty$ the system is in equilibrium, described by the Hamiltonian H_0 and the corresponding density matrix

$$\rho_0 = \exp(-H_0/k_{\mathrm{B}}T)/\mathrm{tr}[\exp(-H_0/k_{\mathrm{B}}T)] \qquad (A.10)$$

where tr denotes the trace of the operators. The perturbation H_1 is then increased adiabatically from zero, so that at a later time t the density matrix is equal to

$$\rho = \rho_0 + \rho_1. \qquad (A.11)$$

The density matrix has the equation of motion (see Landau and Lifshitz 1980)

$$i\frac{d\rho}{dt} = [H, \rho] \qquad (\hbar = 1) \qquad (A.12)$$

which can be proved from the definition (A.7). If (A.8) and (A.11) are substituted into (A.12) and only the terms which are linear in the perturbing field $f(t)$ are retained we find

$$i\frac{d\rho_1}{dt} = [H_0, \rho_1] - [B, \rho_0]f(t) \qquad (A.13)$$

where we have used the property that H_0 and ρ_0 commute. Equation (A.13) can be employed to show that

$$\frac{d}{dt}\{\exp(iH_0t)\rho_1\exp(iH_0t)\} = i\exp(iH_0t)[B,\rho_0]\exp(-iH_0t)f(t). \quad (A.14)$$

On integrating both sides of (A.14) with respect to t and rearranging, the formal solution for ρ_1 is obtained as

$$\rho_1 = i\int_{-\infty}^{t} \exp\{iH_0(t'-t)\}[B,\rho_0]\exp\{-iH_0(t'-t)\}f(t')\,dt'. \quad (A.15)$$

From the above expression, which is linear in f, we may deduce the response of the system to the perturbation (A.9). This response can be expressed in terms of the change $\bar{A}(t)$ that it produces in the mean value of any system variable denoted by the operator A. From the definition (A.7) it can be shown that the mean value of A is simply given by $\mathrm{tr}(\rho A)$. Hence we have in the present case

$$\bar{A}(t) = \mathrm{tr}(\rho_1 A). \quad (A.16)$$

On substituting (A.15) into (A.16) and using the cyclic invariance property for products of operators within the trace, we may express $\bar{A}(t)$ as

$$\bar{A}(t) = i\int_{-\infty}^{t} \langle[A(t), B(t')]\rangle f(t')\,dt'. \quad (A.17)$$

Here $\langle\ldots\rangle = \mathrm{tr}(\rho_0\ldots)$ denotes a thermal average with respect to the unperturbed system. When use is made of (A.1) for the case of a *commutator* Green function ($\varepsilon = 1$), (A.17) gives

$$\bar{A}(t) = \int_{-\infty}^{\infty} \langle\langle A(t); B(t')\rangle\rangle f(t')\,dt'. \quad (A.18)$$

In terms of frequency Fourier components the above result is simply

$$\bar{A}_\omega = \langle\langle A; B\rangle\rangle_\omega F(\omega) \quad (A.19)$$

where $\langle\langle A; B\rangle\rangle_\omega$ is given by (A.5), and

$$F(\omega) = \frac{1}{2\pi}\int_{-\infty}^{\infty} f(t)\exp(i\omega t)\,dt \quad (A.20)$$

with a similar definition for $\bar{A}_\omega$ as the Fourier transform of $\bar{A}(t)$.

Hence by explicitly calculating the response $\bar{A}_\omega$ for a given externally applied field $F(\omega)$, we have a convenient method of deducing from the relationship (A.19) the required Green function, or response function, $\langle\langle A; B\rangle\rangle_\omega$. This is essentially what was done in section 1.3.2 for the mechanical analogue presented there. Apart from the frequency the Green function will typically depend on other quantities such as position or wave vector labels, depending on the nature of the problem. In surface problems it is often convenient to assume that the external field takes the form corresponding to a delta-function stimulus $f(t)\delta(\mathbf{r} - \mathbf{r}')$ at position $\mathbf{r}'$ within the system. The interaction energy with a position-dependent operator $B(\mathbf{r})$ then becomes

$$H_1 = -\int B(\mathbf{r})\delta(\mathbf{r} - \mathbf{r}')\,\mathrm{d}^3 r f(t) = -B(\mathbf{r}')f(t). \qquad (A.21)$$

From the linear response in another operator $A(\mathbf{r})$ at position $\mathbf{r}$ we are able to deduce, by analogy with (A.19), the Green function $\langle\langle A(\mathbf{r}); B(\mathbf{r}')\rangle\rangle_\omega$. Examples for specific systems are given in the text (e.g. in section 2.2.3).

A.3 The fluctuation–dissipation theorem

An important property of Green functions (response functions) is that they give the power spectrum of the excitations of the system. This is achieved by application of the fluctuation–dissipation theorem, which relates the mean-square fluctuations in the excitation amplitude to the imaginary part of an appropriately defined Green function. We now outline the quantum-mechanical derivation of this result: for an alternative classical formulation (valid at high temperatures) we refer to Landau and Lifshitz (1980).

The Green-function definition in (A.1) involves the correlation functions

$$C_{AB}(t - t') = \langle A(t)B(t')\rangle, \qquad C_{BA}(t - t') = \langle B(t')A(t)\rangle. \qquad (A.22)$$

Their dependence on $(t - t')$ can be proved in the same way as for the Green function. It is easily shown, using the expression for the thermal average and the cyclic invariance property of operators under the trace, that the two correlation functions are related to one another by

$$C_{AB}(t - t') = C_{BA}(t - t' + \mathrm{i}\beta) \qquad (A.23)$$

where $\beta = 1/k_B T$. If frequency Fourier transforms $\langle AB\rangle_\omega$ and $\langle BA\rangle_\omega$ are defined for $C_{AB}(t - t')$ and $C_{BA}(t - t')$ respectively, as in (A.20), it is easy to show that (A.23) implies

$$\langle BA\rangle_\omega = \exp(\beta\omega)\langle AB\rangle_\omega. \qquad (A.24)$$

When both sides of (A.1) are Fourier-transformed to the frequency representation, making use of (A.5) and (A.24), the result is

$$\langle\langle A; B\rangle\rangle_\omega = \int_{-\infty}^{\infty} \frac{\langle BA\rangle_{\omega'}[1 - \exp(-\beta\omega')]\,\mathrm{d}\omega'}{\omega - \omega' + \mathrm{i}\eta} \qquad (A.25)$$

where η is a positive infinitesimal quantity $(\eta \to 0)$. In obtaining the above-result it is helpful to use the integral representation (see Zubarev 1960)

$$\theta(t - t') = \int_{-\infty}^{\infty} \frac{\exp[-ix(t - t')]\, dx}{x + i\eta} \tag{A.26}$$

which can be proved by contour integration.

Equation (A.25) can be separated into its real and imaginary parts by applying the symbolic mathematical relationship

$$\frac{1}{\omega - \omega' + i\eta} = P\left(\frac{1}{\omega - \omega'}\right) - i\pi\delta(\omega - \omega') \tag{A.27}$$

where P denotes that the principal value is taken in any integration over ω'. Recall that the principal value means that we avoid the singularity at $\omega' = \omega$ in (A.27) by splitting the integration into two parts, namely from $-\infty$ to $\omega - \rho$ and $\omega + \rho$ to ∞, and then letting the positive $\rho \to 0$. Provided $\langle BA\rangle_\omega$ is a real quantity (e.g. as in the case when $B = A^+$) we may take the imaginary part of each side of (A.25) to obtain

$$\mathrm{Im}\langle\langle A; B\rangle\rangle_\omega = \pi[1 - \exp(-\beta\omega)]\langle BA\rangle_\omega. \tag{A.28}$$

On rearrangement of the terms, this becomes

$$\langle BA\rangle_\omega = (1/\pi)\{n(\omega) + 1\}\mathrm{Im}\langle\langle A; B\rangle\rangle_\omega \tag{A.29}$$

where $n(\omega)$ is the Bose–Einstein factor defined in (1.51). Using (A.24) we also have

$$\langle AB\rangle_\omega = (1/\pi)n(\omega)\,\mathrm{Im}\langle\langle A; B\rangle\rangle_\omega. \tag{A.30}$$

Equations (A.29) and (A.30) are the standard forms of the fluctuation–dissipation theorem. With appropriate choices for operators A and B, the theorem can be used to deduce the power spectrum of the excitations (e.g. as in section 2.2.3) or a scattering cross section for comparison with experiment (e.g. see section 3.3.2). At high temperatures (such that $\beta\omega \ll 1$) both (A.29) and (A.30) reduce to the classical limit

$$\langle AB\rangle_\omega = \langle BA\rangle_\omega = (k_B T/\pi\omega)\,\mathrm{Im}\langle\langle A; B\rangle\rangle_\omega. \tag{A.31}$$

By taking the real part of each side of (A.25) we obtain the additional relationship

$$\mathrm{Re}\langle\langle A; B\rangle\rangle_\omega = P\int_{-\infty}^{\infty} \frac{\langle BA\rangle_{\omega'}[1 - \exp(-\beta\omega')]\, d\omega'}{\omega - \omega'}. \tag{A.32}$$

A.4 The Kramers–Kronig relations

It was pointed out when discussing (2.107) for the dielectric function that, because $\gamma_{ij}(\mathbf{r} - \mathbf{r}', t - t')$ is zero for $t' < t$, the response function

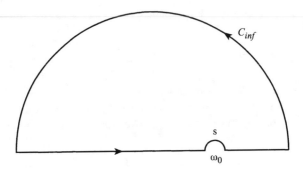

Figure A.1. Contour for the integral in (A.33). The real-axis path is indented by a small semicircle S around the point ω_0.

$\chi(\omega) = \varepsilon(\omega) - 1$ is analytic in the upper half frequency plane. From our definition, the same property of causality holds for the Green function $\langle\langle A(t); B(t')\rangle\rangle$ and it follows that the Fourier transform $\langle\langle A; B\rangle\rangle_\omega$ is analytic in the upper half plane. We now apply this analyticity property to derive relations between the real part $\chi'(\omega)$ and the imaginary part $\chi''(\omega)$ of the response function; for simplicity we give the proof only for scalar functions.

We start with the integral in the complex ω plane

$$I \equiv \int_C \frac{\chi(\omega)}{\omega - \omega_0}\, d\omega = 0 \qquad (A.33)$$

where C is the contour drawn in figure A.1. The value of the integral is zero because the integrand is analytic within the contour. By analogy with (2.116) we have $\chi(\omega) \to 0$ as $\omega \to \infty$, and provided the convergence is sufficiently rapid (faster than $1/\omega$), which we assume to be the case, the integral around the large semicircle C_{inf} vanishes. The integral along the real axis can be evaluated by means of the identity (A.27), so that (A.33) leads to

$$P\int_{-\infty}^{\infty} \frac{\chi(\omega)}{\omega - \omega_0}\, d\omega - i\pi\chi(\omega_0) = 0. \qquad (A.34)$$

Separating the real and imaginary parts of this equation, we obtain the Kramers–Kronig (KK) relations:

$$\varepsilon'(\omega_0) - 1 = \frac{1}{\pi} P\int_{-\infty}^{\infty} \frac{\varepsilon''(\omega)}{\omega - \omega_0}\, d\omega \qquad (A.35)$$

$$\varepsilon''(\omega_0) = -\frac{1}{\pi} P\int_{-\infty}^{\infty} \frac{\varepsilon'(\omega) - 1}{\omega - \omega_0}\, d\omega. \qquad (A.36)$$

We have expressed these in terms of the dielectric function since this is the usual situation in optics. Since the proof uses only analyticity in the upper half plane, the relations hold for any response function like $\chi(\omega)$ or $\langle\langle A; B\rangle\rangle_\omega$.

The KK relations are discussed in detail in advanced optics texts (see e.g. Lipson *et al* 1995) but some brief comments are in order here. Equation (A.35) shows that the frequency dependence of ε' (dispersion) results from a nonzero value of ε'' (absorption). Since the relations involve integrals over frequency, however, the absorption can be in a different frequency range from the dispersion. An everyday example is the fact that glass is dispersive but transparent in the visible. The derivation of the KK relations uses only the property of causality, so any experiment or calculation that does not satisfy the KK relations must be incorrect. Very often in an experiment only one part, say ε', of the dielectric function is measured. The relation (A.36) may then be used to calculate ε''. The difficulty, obviously, is that (A.36) involves an integral over all frequencies, so it can only be applied if the measurements of ε' cover a sufficiently wide frequency range.

References

Cottam M G and Maradudin A A 1984 in *Surface Excitations* ed V M Agranovich and R Loudon (Amsterdam: North-Holland)

Fetter A L and Walecka J D 1971 *Quantum Theory of Many-Particle Systems* (New York: McGraw-Hill)

Forster D 1975 *Hydrodynamic Fluctuations, Broken Symmetry and Correlation Functions* (New York: Benjamin)

Landau L D and Lifshitz E M 1980 *Statistical Physics* (Oxford: Pergamon)

Lipson SG, Lipson H and Tannhauser DS 1995 *Optical Physics* 3rd edition (Cambridge: Cambridge University Press)

Lovesey S W 1986 *Condensed Matter Physics: Dynamic Correlations* (New York: Benjamin)

Negele J W and Orland H 1988 *Quantum Many-Particle Systems* (Redwood City: Addison-Wesley)

Rickayzen G 1980 *Green's Functions and Condensed Matter* (London: Academic)

Zubarev D N 1960 *Usp. Fiz. Nauk.* **71** 71 [*Sov. Phys.—Usp.* **3** 320 (1960)]

Subject index

Materials index

Abbreviations and acronyms

ABC	additional boundary condition
AES	Auger electron spectroscopy
AFMR	antiferromagnetic resonance
ATR	attenuated total reflection
BLS	Brillouin light scattering
CVD	chemical vapour deposition
DE	Damon–Eshbach
EELS	electron energy loss spectroscopy
FFT	fast Fourier transform
FIR	far infrared
GGG	gadolinium gallium garnet
GMR	giant magnetoresistance
H-P	Holstein–Primakoff
HREELS	high-resolution electron energy loss spectroscopy
IPE	inverse photoemission
KDP	potassium di-hydrogen phosphate
KD*P	deuterated potassium di-hydrogen phosphate
KdV	Korteweg–deVries
KK	Kramers–Kronig
LB	Langmuir–Blodgett
LEED	low-energy electron diffraction
LO	longitudinal optic
LPE	liquid-phase epitaxy
LR	longitudinal resonance
LRSP	long-range surface plasmon
LST	Lyddane–Sachs–Teller

MBE	molecular-beam epitaxy
MOCVD	metallo-organic chemical vapour deposition
MOKE	magneto-optical Kerr effect
MQW	multiple quantum well
MSW	magnetostatic surface wave
NLO	nonlinear optics
NLSE	nonlinear Schrödinger equation
PBC	periodic boundary condition
PBG	photonic band gap
PE	photoemission
RHEED	reflection high-energy electron diffraction
RKKY	Ruderman–Kittel–Kasuya–Yosida
RPA	random-phase approximation
RW	Rayleigh wave
SEMPA	scanning electron microscopy with polarization analysis
SERS	surface-enhanced Raman scattering
SEXAFS	surface extended x-ray absorption fine structure
SHG	second harmonic generation
s-G	sine-Gordon
SGFM	surface Green function matching
SIMS	secondary ion mass spectroscopy
SIT	self-induced transparency
SMOKE	surface magneto-optical Kerr effect
SNOM	scanning near-field optical microscopy
SPEELS	spin-polarized electron energy loss spectroscopy
SPIPE	spin-polarized inverse photoemission
SPP	surface plasmon–polariton
SVA	slowly-varying amplitude (approximation)
SWR	spin wave resonance
TEM	transmission electron microscopy
THG	third-harmonic generation
TO	transverse optic
TOF	time of flight
UHV	ultrahigh vacuum
UPS	ultraviolet photoelectron spectroscopy
XPS	x-ray photoelectron spectroscopy
YIG	yttrium iron garnet, $Y_3Fe_5O_{12}$
1D, 2D, 3D	one-dimensional, two-dimensional, three-dimensional
2DEG	two-dimensional electron gas

Printed in the United States
by Baker & Taylor Publisher Services